Wind Power Plants

Robert Gasch • Jochen Twele

Editors

Wind Power Plants

Fundamentals, Design, Construction and Operation

Second Edition

 Springer

Editors

Prof. Dr.-Ing. Robert Gasch
TU Berlin
Fak. V Verkehrs- und
Maschinensysteme
Institut für Luft- und Raumfahrt
Marchstr. 12-14
10587 Berlin
Germany
robert.gasch@gmx.de

Prof. Dr.-Ing. Jochen Twele
HTW Berlin
Wilhelminenhofstr. 75A
12459 Berlin
Germany

ISBN 978-3-642-22937-4   e-ISBN 978-3-642-22938-1
DOI 10.1007/978-3-642-22938-1
Springer Heidelberg Dordrecht London New York

Library of Congress Control Number: 2011937488

Printed on acid-free paper

Springer is part of Springer Science+Business Media (www.springer.com)

Preface

In 1991, when we sent our manuscript of the first German edition „Wind Power Plants" to the publishing house, our lecturer Dr. Schlemmbach was asked by his colleagues "Do you think, anyone will buy and read this book ?" It lasted only one year, until we had to prepare the second edition.
The reason for this unexpected success was the first "feed in act" that had passed the German parliament in 1991. Everybody was now allowed to produce electricity from renewable energies and to feed it into the grid for a fixed price guaranteed over twenty years. This political decision initiated the boom of the German wind energy industry - similar to the Danish political decision ten years before.

In 1991 most of the authors were members of a research group at the Technical University of Berlin, students, research fellows, postgraduate students and post-docs. Now many of them hold prominent positions in the wind energy industry. This has let to a tightly knit professional network, that helps to keep the book up-to-date.

The first English edition (Solar-Praxis, Berlin and James &James, London 2002), based on the 3rd German edition from 1996, was translated by Dörte Müller and Thomas Ackermann, living in Stockholm. Richard Holmes, Berlin, translated Max Frisch`s Questionnaire.

This new English edition is based on the completely revised and extended 5th German edition, Teubner, 2006. It was translated by Christoph Heilmann and reviewed by Wilson Rickerson and Karl E. Stoffers, both from the United States, Jeremy Dunn (Great Britain), Moran Seamus (Ireland) and Simon Cowper (Great Britain).

Heike Müller organized with her skilful hands the graphical work and the final layout.
Robert Gasch, Jochen Twele and Christoph Heilmann, Berlin, kept in touch with the co-authors to coordinate the work.

We sincerely would like to thank all the contributors for their efforts. We also would like to say thank you to the sponsors and to Dr. Merkle and Dr. Baumann from the Springer Publishing House for their patience.

The editors   Berlin, September 2011

Chapters and Authors

Chapter 0	Questionnaire 87	Max Frisch
Chapter 1	Introduction	Prof. Dr.-Ing. R. Gasch, Prof. Dr.-Ing. J. Twele, Dipl.-Ing. K. Ohde
Chapter 2	Historical development of windmills	Prof. Dr.-Ing. R. Gasch, Dipl.-Ing. M. Schubert
Chapter 3	Design and components	Prof. Dr.-Ing. J. Twele, Dr.-Ing. C. Heilmann, Dipl.-Ing. M. Schubert
Chapter 4	The wind	Dipl.-Ing. W. Langreder, Dr.-Ing. P. Bade
Chapter 5	Blade geometry	Prof. Dr.-Ing. R. Gasch, Dr.-Ing. J. Maurer, Dr.-Ing. C. Heilmann
Chapter 6	Calculation of performance characteristics	Dr.-Ing. J. Maurer, Dr.-Ing. K. Kaiser, Dr.-Ing. C. Heilmann
Chapter 7	Scaling wind turbines and rules of similarity	Prof. Dr.-Ing. R. Gasch
Chapter 8	Structural dynamics	Prof. Dr. Dipl.-Ing. M. Kühn, Prof. Dr.-Ing. R. Gasch, Dipl.-Ing. B. Sundermann
Chapter 9	Guidelines and analysis procedures	Prof. Dr.-Ing. A. Reuter
Chapter 10	Wind pump systems	Dr.-Ing. P. Bade, Prof. Dr.-Ing. J. Twele, Dr.-Ing. R. Kortenkamp
Chapter 11	Electricity generation	Dipl.-Ing. W. Conrad, Prof. Dr.-Ing. R. Gasch

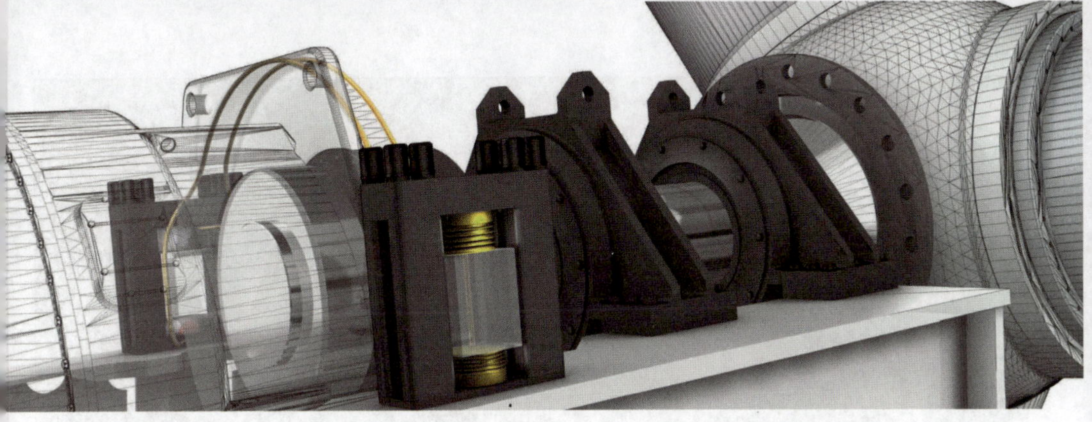

Vibration-related components in wind turbines

ESM is the world's leading specialist for vibration technology in modern wind turbines and supplies the elastic connections in the drive train.

One example is the hydraulic gearbox support.

The detached power train is often employed in large wind turbines nowadays. ESM offers an advanced hydraulic torque support for this type of the drive train. It reduces systemic constrained forces in the drive train by more than 90%, preventing damage from occurring to the gearbox and other vital components.

The elastomeric parts are equipped with hydraulic chambers that also give them the function of a hydraulic piston cylinder. The hydraulic chambers of the four elastomeric parts are connected crosswise with hoses, forming two enclosed hydraulic systems (Fig. 1).

Thus, the hard springs II act when torque is applied, while the vertical forced movement of the gearbox only has to be countered by the comparably lower forces of the soft spring I (Fig. 1).

The high degree of rigidity in the direction of rotation for the safe transmission of rotor torque and the great resilience in all lateral directions is shown as an important system feature in the force deflection diagram (Fig. 2).

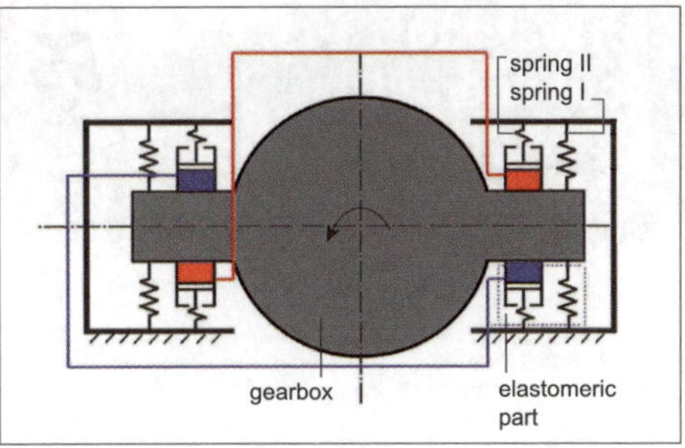

Fig. 1 Functional schematic

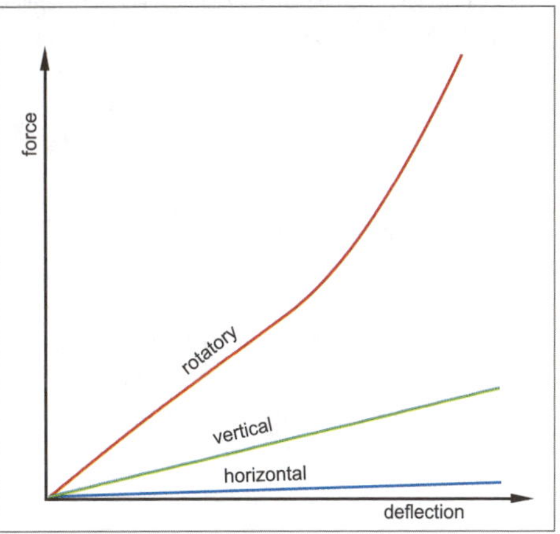

Fig. 2 Force deflection diagram

Phone: +49 (0) 6253/9885-0
Fax: +49 (0) 6253/9885-50
E-Mail: info@esm-gmbh.de

Auf der Rut 5
D-64668 Rimbach

ESM Energie- und Schwingungstechnik Mitsch GmbH
www.esm-gmbh.de

Content

REpower
Systems

Mutual respect, trust
and openness characterize
the way we work.

REpower Systems SE · Überseering 10 · 22297 Hamburg
Tel.: +49-40-5 55 50 90-0 · E-mail: info@repower.de · www.repower.de

10 years

Questionaire 87

Instead of a foreword, we reprint here, by kind permission, 25 questions posed by late swiss writer and architect Max Frisch on receiving an honorary doctorate at the Technische Universität Berlin on 29 June 1987.

1. Are you sure that the preservation of Mankind really matters to you, once you and all the people you know no longer exist?
2. And if it does, why don't you act differently?
3. What has changed human society more: the French Revolution or a technological invention, electronics for example?
4. Bearing everything in mind that we have to thank the technological arms race for, even just considering, for example, the sector of kitchen appliances, do you think that we should be grateful to the technologists too, and thus also to the defence ministers who provided them with our tax money for their research?
5. As a non-expert, what would you like to have discovered in the near future (be brief)?
6. Can you imagine a human existence (i.e. in the first world) without computers?
7. And if you can, does the thought make you shudder, does it make you feel nostalgic, or doesn't it make you do anything the computer can't do for you?
8. In the brief period you have been alive, which appliances have come onto the market without anybody ever having had a need for them previously (avoid using brand names when referring to the appliances), and why do you purchase such appliances:
   a) To generate economic growth?
   b) Because you believe advertisements?
9. The dinosaurs existed for 250 million years. What do you think 250 million years of economic growth will be like (be brief)?
10. If an engineer is apolitical, in the sense of not caring which rulers make use of the technological inventions: What do you think of this person?
11. Assuming that you approve of our existing society, because a better one does not exist anywhere: do you think that, in an age of objective constraints, to which those in power are constantly referring, governments are still necessary at all?
12. If a contemporary has heard of lasers but has no idea at all what a laser actually is, be frank: Can you as a scientist take the views of such lay people and their political demonstrations seriously?
13. Do you believe in a Scholar Republic?
14. When did technology begin no longer to ease our human existence (Which was the original purpose of tools) but to establish over us an

extra human domination, and to take away from us the nature that it sub-jugates?

15. Do you think technomania is irreversible? Assuming that the catastrophe is avoidable?
16. Can you imagine a society in which scientists would have to accept liability for the crimes that had only become possible as a result of their inventions, a theocracy for example?
17. Assuming that you not only approve of the existing society, but that you respond with tear gas if someone else calls it into question: aren't you afraid that without some greater utopia People will become more and more stupid, or is that precisely the reason why you feel so post modernly comfortable?
18. In view of the technological feasibility of the Apocalypse, what is your attitude today to the Biblical metaphor of the forbidden fruit from the tree of Knowledge?
    a) Do you believe in the freedom of research?
    b) Do you side with the Pope who forbade Galileo from claiming that the earth rotates round the sun?
19. If you were working on an invention which would make it impossible to lie in public: Can you imagine who might want to fund your daring research?
20. What would you like not to have invented?
21. Does it ever happen that once a technological invention has been developed it is not suitable for applications which do not correspond to the intentions of its inventor?
22. Do you think it is possible that the human mind that we educate is basically aimed at the self destruction of the species?
23. What, apart from wishful thinking, could argue against this?
24. Do you know what makes you do research?
25. Do you believe as a scientist in a responsible technological field, that is in technological research with a UNIVERSITAS HUMANITAS, or put more plainly: do you believe in a Technological University in Berlin?

EICOGEAR®
Gearboxes for Wind Turbines

- Robust design
  for low life cycle costs

- Customer-specific design
  considering the individual need

- Excellent efficiency
  for high yield of current

- Production in highest quality

- Tests under extreme conditions

Eickhoff Antriebstechnik GmbH · Hunscheidtstrasse 176 · 44789 Bochum · Germany
Telephone +49 234 975-0 · Telefax +49 234 975-2579 · www.eickhoff-bochum.de

Who should you partner
with to build a reliable future?

The answer is Suzlon, the world's fifth largest* wind turbine manufacturer. With over 17,000 MW of wind energy capacity installed in 28 countries, achieving outstanding availability, Suzlon is committed to providing superior, efficient and customer focused wind solutions. Through its latest S9X turbine platform, Suzlon brings - cost-effective, robust and proven wind turbine technology to the world. Think of it as partnering with a company whose services are just as reliable as the energy it provides.

SUZLON
POWERING A GREENER TOMORROW

Commissioned by Suzlon, Iberdrola's windfarm in North Dakota, USA

www.suzlon.com

World's fifth largest* wind turbine manufacturer with installed capacity of over 17 GW | Over 13,000 employees in 32 countries across six continents | Manufacturing units on three continents | R&D facilities in Denmark, Germany, India and The Netherlands

Source: *BTM Consult ApS – A part of Navigant Consulting – World Market Update 2010

1 Introduction to Wind Energy

1.1 Wind Energy in the year 2010

In the history of industrial development, the golden age of heavy machinery is long since past. We now live in the era of information technologies, where the rate of technological advance is extremely rapid. Though computer industry growth rates make the industries of the past seem obsolete, there is one modern machine industry whose growth rate over the past two decades have been comparable to that of the IT sector: wind power plants.

The rapid increase in size and capacity of commercially manufactured wind turbines between the years 1982 and 2006 is illustrated in Fig. 1-1. Fig. 1-2 charts the expansion of the installed capacity from wind turbines during the same period.

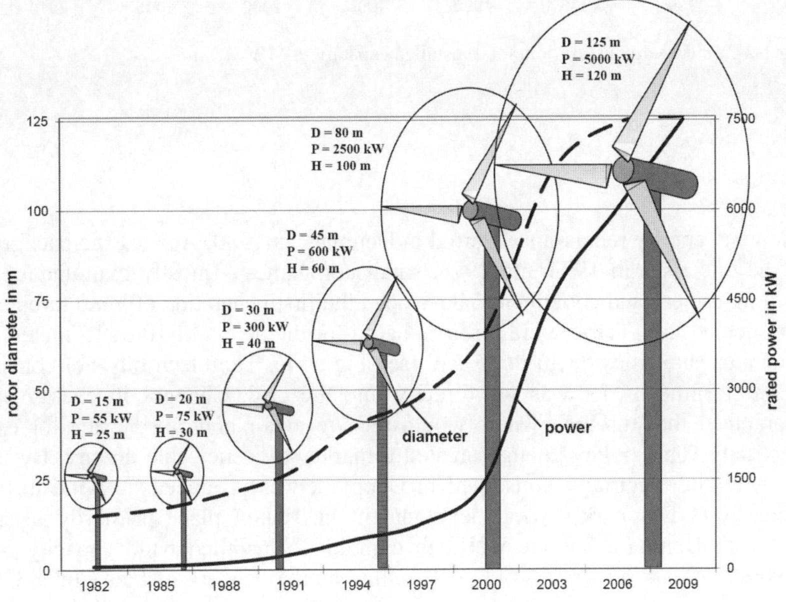

Fig 1-1 Size and power increases of commercially produced wind turbines over time

Within a very short time, a mature and reliable energy technology has been developed. Both the growth rate of the installed capacity and the increase in turbine size have been remarkable. In 2010, for example, the largest commercial machines had a capacity of 7.5 MW and a diameter of 126 meters.

The examples below summarise the most important events in the recent history of wind energy development in key markets, (see also Fig 1-1 and Fig. 1-2), section 1.2 contains a more detailed discussion of wind market development.

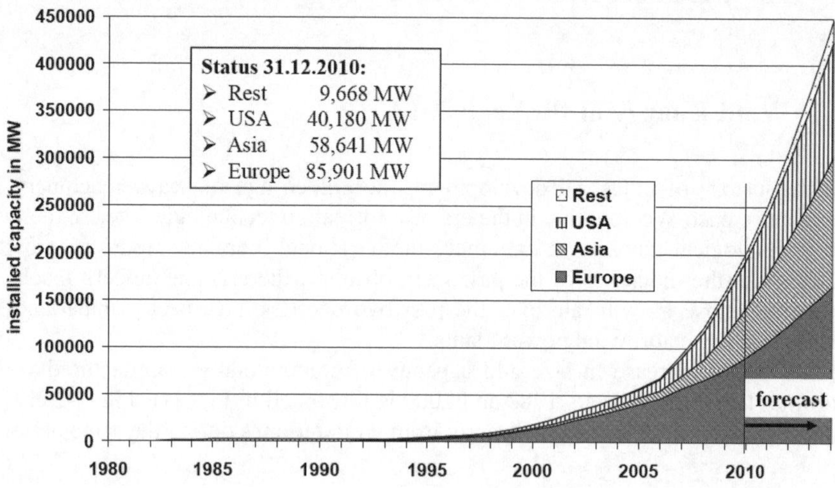

Fig 1-2 Wind energy utilisation, total installed capacity in MW [1, 2]

Denmark

The wind energy renaissance started in Denmark in 1980. Against the background of the oil crises in 1973 and 1978, small companies – mostly manufacturers of rural machinery and equipment – developed the first generation of wind turbines for commercial use. These wind turbines had rotor diameters of 10 to 15 meters, and generator capacities of 30 to 55 kW (see Fig. 1-1). The electricity not consumed by the turbine owner was fed directly into the grid. Changes in energy policy guaranteed that turbine owners would receive a fair and fixed price for excess electricity. This policy change created a market for renewable energy. By 2007, 25 % of the electricity consumed in Denmark was produced by wind turbines. Since 2003 Denmark's wind development has taken place primarily offshore. Although Denmark led the world in cumulative installed wind capacity in the 1990s, Denmark now ranks only 10th behind market leaders such as China, United States, Germany, Spain and others.

The United States of America

The US wind energy boom in California began in 1980/81. The boom resulted in a total installed capacity of 1,600 MW by 1987. Californian wind farms employed large numbers of small turbines (35 to 75 kW) that were either manufactured in the US or imported from Denmark. As in Denmark, the extreme rise of oil prices in the 70s led the government to favour renewable energies, including solar, wind and geothermal technologies. As a result, commercial solar-thermal power plants, with a total capacity of 350 MW were constructed in California. These positive trends were stopped, however, when the Democratic Governor of California, Jerry Brown, lost his majority to the Republicans in 1987. With the oil crisis subsiding, the new governor changed the energy laws to favour the cheapest offer. Fossil-fuelled power plants, with their large CO_2 and greenhouse gas emissions, once again became the predominant power generation technology. By 2001, however, wind energy had begun to make a comeback in the US, and impressive 9,922 MW of new capacity was installed in 2009.

Germany

Germany did not experience rapid wind market growth until 1991. In that year, a federal law called the Electricity Feed Law (EFL) guaranteed both grid access and a fair, fixed-price to wind energy generators. During the next eight years, 3,000 MW of new capacity came online. By 1998, the coastal provinces of Niedersachsen and Schleswig-Holstein supplied about 7 % of their electricity demand through wind power. Ten years later, this share had increased to 40 %. The Renewable Energy Law (REL), which came into force in April 2000, and replaced the EFL, encouraged the development of inland sites and laid the regulatory groundwork for offshore installations. Following this legislation, a record for new installations was set in 2002 with 3,247 MW. The market has cooled somewhat, with between 1,500 MW and 2,000 MW installed each year since 2002. There are high expectations, however, for market growth offshore in the future. .

Spain

There has been a rapid increase in the number of wind turbines in Spain during the past few years. In 2010, yearly installations amounted to 1,516 MW. The total installed capacity had increased to 20,676 MW. Spain is considered one of Europe's fastest growing markets.

India

India has experienced a wind energy boom since 1993. Although few turbines had been installed by 1990, 200 MW had been installed by 1994. By 1998, this number had increased to 1,000 MW, and by 2010, India's installed capacity reached 13,065 MW. This development has been fuelled by India's enormous electricity demand. Government and industry want to use wind farms to end the frequent industrial production stoppages caused by electricity shortages.

China

China´s wind market recorded (new installations 16,500 MW) the highest growth rates in the world in 2009 of 113 % (see Fig. 1-3), leaving the country total installed capacity 42,287 MW at the end of 2010. Beside this, in the northern grasslands of the country (Inner Mongolia), about 150.000 wind powered battery-chargers are used by nomadic groups. These wind turbines have an average capacity of approximately 100 Watts.

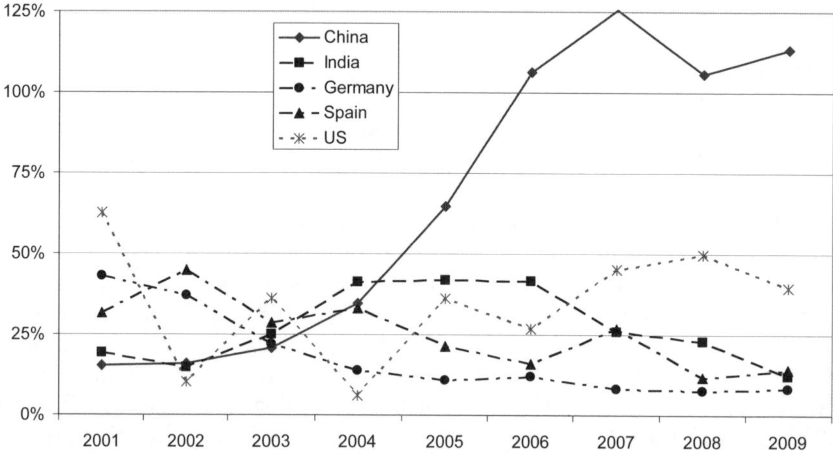

Fig 1-3 Growth rates in top 5 most important markets

1.2 The Demand for Electricity

Since the demand for energy, and specifically electricity, has increased so dramatically over the last 100 years (Fig 1-4), it has now become important to consider the environmental impacts of energy production. In the past, high standards of living and "modern lifestyles" were based on increased energy consumption. Today, statistics from highly developed countries show that standards of living can increase independently of energy consumption if energy efficiency measures are introduced.

In 2008, the global demand for electricity was about 20.2 10^{12} kWh. This demand was met mainly through fossil fuels and nuclear power (see Fig. 1-5). Renewable energies, other than hydropower, only had a 2.8 % share. This marks a slight increase compared to 1999, despite to fast increasing total demand.

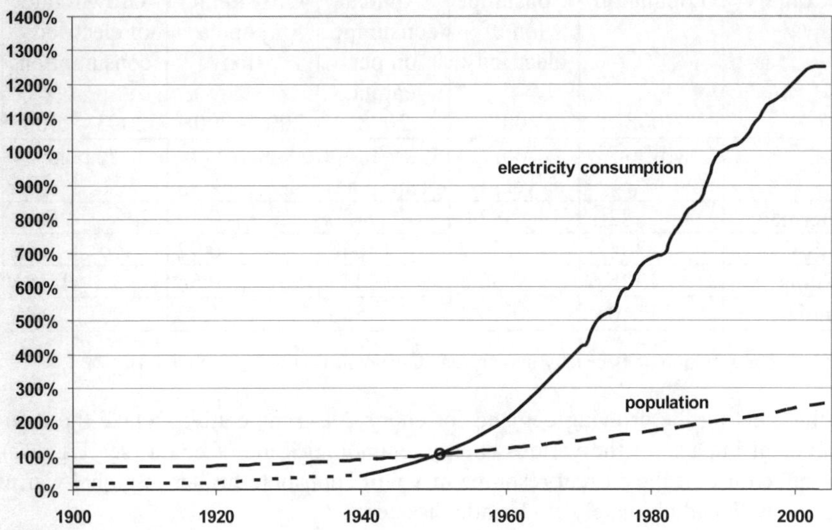

Fig. 1-4 Development of world-wide population and consumption of electricity   [3]

$1950 = 100\%$; population $= 2.55 \ 10^{9}$; annual electricity consumption $= 1.2 \ 10^{12}$ kWh

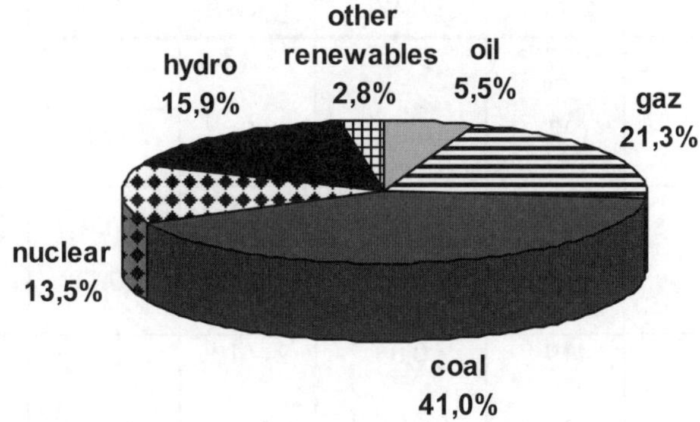

Fig 1-5 Share of the global electricity supply in 2008   [3]

The large increase in electrical energy demand has strongly influenced global energy market projections. Using statistics from 2003 to 2008, Fig. 1-6 shows that countries with expanding populations and growing economies have experienced enormous growth in electricity consumption. Highly developed industrial countries, however, have started to limit their energy consumption, without decreasing their living standards, by encouraging energy conservation and energy-efficient technologies.

Country	Population	Consumption of electricity	Annual consumption per capita	Growth rate of population	Growth rate of electricity consumption
	2008	2008	2008	2003 - 2008	2003 - 2008
	Million	TWh / year	kWh / cap. p.a.	% p.a.	% p.a.
Germany	82.1	587.0	7150	- 0.10	0.62
USA	304.5	4155.9	13648	0.92	1.85
China	1325.6	3252.3	2453	0.58	15.84
India	1140.0	645.3	566	1.42	7.85

Fig 1-6 Consumption of electricity in representative countries (2003-2008) [3, 4]

While meeting the growing demand for energy, it is imperative to take the environmental impacts of the various energy technologies into account (see Fig 1-7). Continued use of the current resource mix will contribute further to global warming and will lead ultimately to climatic disaster.

	energy resource	CO_2	SO_2	NO_x	ash	nuclear waste
		g/kWh	g/kWh	g/kWh	g/kWh	mg/kWh
fossil fuels	Carbon (with pollution control)	977 [1] (977)	5 – 9 (0.8) [1]	3 – 6 (0.8) [1]	25 [2] (0.1) [1]	- -
	Oil (with pollution control)	730 (730)	1 - 4.2 [3]- 12 (0.8) [1]	2 – 5 [4] (0.8)	(0.1)	- -
	gas (with pollution control)	419 [1] (419)	0.05 (0.01) [1]	2 – 4 [4] (0.7) [1]	(0.01) [1]	- -
	nuclear					4

[1] DEWI [5]   [2] Strauß [6]   [3] refers to 1% sulphur in the oil   [4] Heitmann [7]

Fig 1-7 Comparison of the environmental impact of different fossil and non-fossil fuel energy resources. The values in brackets ( ) are obtained using modern pollution control technologies.

An increased reliance on the energy from the wind, water and sun will decrease the chance of environmental disaster since renewables do not emit greenhouse gasses, and obviously produce no nuclear waste.

Furthermore, wind energy is not a land-intensive energy technology, which has become an important point of discussion in increasingly crowded northern

European countries. Fig. 1-8 shows land used by wind energy in terms of power produced per square meter when compared to other types of power stations.

	Site	Data	
Hydropower	Itaipu, 1985 (Brazil)	12.600 MW $H = 200$ m	6 W / m^2
	Spiez, 1986 (Switzerland)	23 MW $H = 65$ m	70 W / m^2 (per m^2 flooded area)
Brown coal (lignite) fired plants	Schkopau, 1996 (Germany)	1.000 MW	8 W / m^2
	Schwarze Pumpe, 1998 (Germany)	1.600 MW	16 W / m^2
	Buschhaus, 1985 (Germany)	380 MW	31 W / m^2 (per m^2 mining area)
Wind power plants	Germany	$v_{Wind} = 4.5\text{-}6.0$ m/s	50 - 120 W / m^2 (per m^2 rotor area) foundation area is 10 times less

Fig. 1-8 Electrical power produced per square meter land use

There are other advantages to renewable energies besides the fact that they emit no greenhouse gases, and are less land intensive than conventional fuels. Within a few months, renewable energy plants are able to produce enough energy to pay back the amount of energy used to manufacture them. This so-called energy amortisation is shown in Fig 1-9.

	Wind			Solar			Water		
	4.5 m/s	5.5 m/s	6.5 m/s	Mono	Multi	Amorph	Large	Small	Micro
Energy amortis ation (in months)	6 - 20	4 - 13	2 - 8	28 - 55	19 - 38	14 - 28	5 - 6	8 - 9	9 - 11

Fig 1-9 Energy amortisation of different renewable energy sources [8]

The economic benefits associated with a decentralised energy supply, e.g. job creation, create an opportunity for sustainable development that strengthens local economies. Compared with an energy supply based on large and centralised conventional power stations, a decentralised energy supply creates steady local employment, because of:
- jobs created by the planning and construction of the plants
- the craftsmanship involved (i.e. the installation of solar thermal or PV plants)

-   the higher rate of employment per kWh associated with the operation
-   and maintenance (O&M) of renewable energy plants.

Fig.1-10 shows the average cost of electricity generation from newly built power plants. The total cost varies from 0.036 € / kWh, for electricity generated by gas-driven power plants, to 0.057 € / kWh for nuclear and wind power plants.

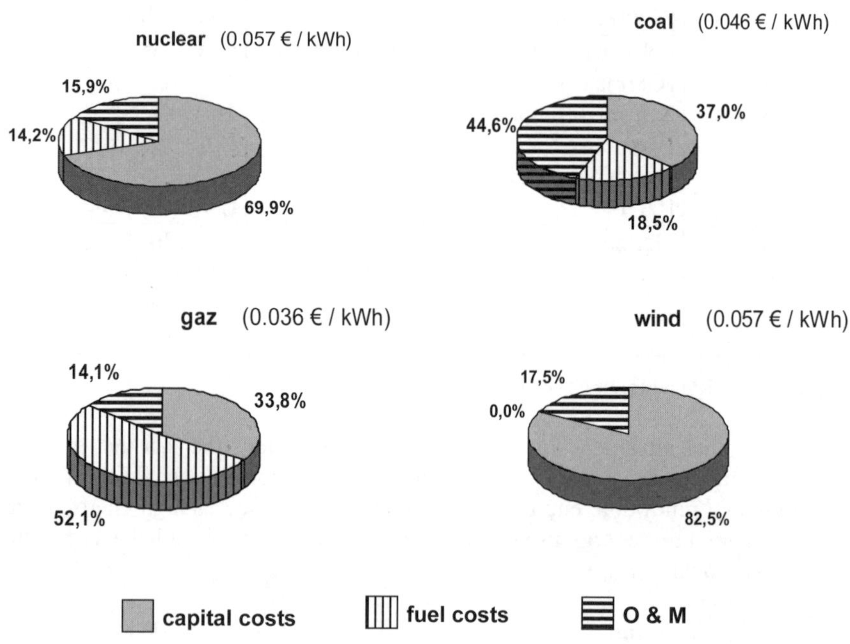

Fig. 1-10 Cost distribution for different power sources [9]

For the different types of electricity generation, these costs are distributed among three categories:

-   capital costs - the total upfront investment
-   fuel costs
-   costs of operation and maintenance (O&M costs)

Though capital costs are lower for fossil fuel plants than they are for nuclear or wind plants, fuel costs are higher.

The Kyoto Protocol is the first attempt to limit carbon dioxide emissions through an international convention. Wind energy, and other renewable energy sources, provide a way to meet increasing energy demand without compromising the environment and contributing to climatic disaster. Fig.1-11 is a scenario, designed by Shell, for achieving a more sustainable global energy supply.

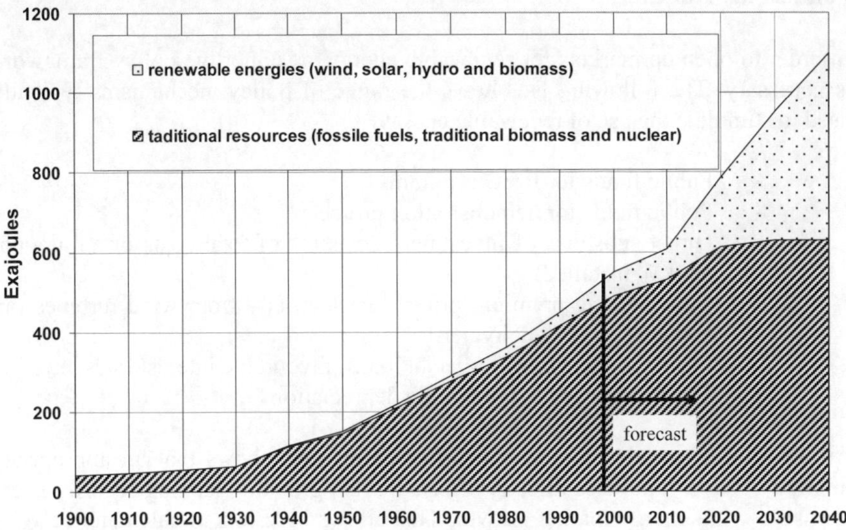

Fig. 1-11 A scenario for meeting future global energy demand [10]

The next section will illustrate how policy can be used to encourage renewable energy use.

1.3 Energy Policy and Governmental Instruments

When analysing wind energy markets, two principal market types can be identified: those where governmental support is motivated by environmental concerns (e.g. Europe) and those where governmental support is based on energy need (e.g. Asia) (see Fig. 1-12).

Environment-driven markets	**Energy-driven markets**
- No need for additional capacity	- Immediate need for additional
- Financing available	energy - capacity shortfall
- Wind energy only contributes a	- Shortage of foreign currency
small part of total energy supply	- Dependant on importing fossil fuels
- Desire and obligation to reduce CO_2	- Moderate to high economic growth
- Wind energy development is not	- Need for technology transfer and
very sensitive to variations in	local production
international fuel prices	- Very sensitive to variations in inter
	national fuel prices

Fig. 1-12 Typical markets for wind energy [9]

Political instruments

In order to open up markets for renewable energy, a supportive policy framework is necessary. The following list shows the range of policy mechanisms typically used to stimulate the use of renewable energy:

- Public funds for R&D programs
- Public funds for demonstration projects
- Direct subsidies of investment costs (% of total costs or an amount per kW installed)
- Guaranteeing premium prices for electricity from wind turbines (an amount per kWh delivered)
- Financial incentives – special loans, favourable interest rates, etc.
- Tax incentives i.e. favourable depreciation

An analysis of the development in different markets shows that an appropriate mixture of several policies directly affects market growth. Planning security is essential to wind energy market growth. This allows investors to calculate turnover and profit over a wind power plant's life time -- i.e. for 20 years.

Fig. 1-13 compares the wind energy policies of European governments. Some countries guarantee a fixed price to wind power, like Germany through Renewable Energy Feed-In Tariffs (FITs), while other countries employ a bidding system that sets a fixed yearly quota for new capacity that is to be installed. Obviously, countries with a guaranteed price per kWh fed into the grid (Germany, Spain) have stimulated their markets more than those operating with capacity quotas.

		country	installed capacity 31.12.2010	installation in 2010	share of continental installation	installed capacity per area	installed capacity per capita
			MW	MW	%	kW / km^2	W / p.c.
environment-driven	fixed price regulation	Germany	27,214	1,343	31.7	75.8	331.5
		Spain	20,676	1,516	24.1	41,0	453.5
		total	**47,890**	**2,859**	**55.8**	**55.5**	**375.0**
	quota regulation	UK	5,204	868	6.1	21.3	84.8
		Ireland	1,428	624	1.7	20.3	321.6
		total	**6,632**	**1,492**	**7.8**	**21.2**	**100.8**
energy-driven		US	40,180	5,115	90.9	4.2	132.0
		India	13,065	2,196	22.3	4.0	11.5
		China	42,287	16,500	72.1	4.4	31.9
		total	**55,352**	**18,696**	**94.4**	**4.3**	**22.4**

Fig. 1-13 Overview of governmental instruments and their effects in some European markets

The liberalisation of the European energy market will affect the development of renewable energies in the future. But liberalisation will not benefit the environment unless it is combined with penalties for CO_2 emissions and/or the production of nuclear waste. This penalty could be enacted through a tax on conventional energies. Such a tax would encourage renewable energies to become competitive without the use of subsidies.

Fig 1-14. gives a summary of the wind power capacity installed in Europe at the end of 2010.

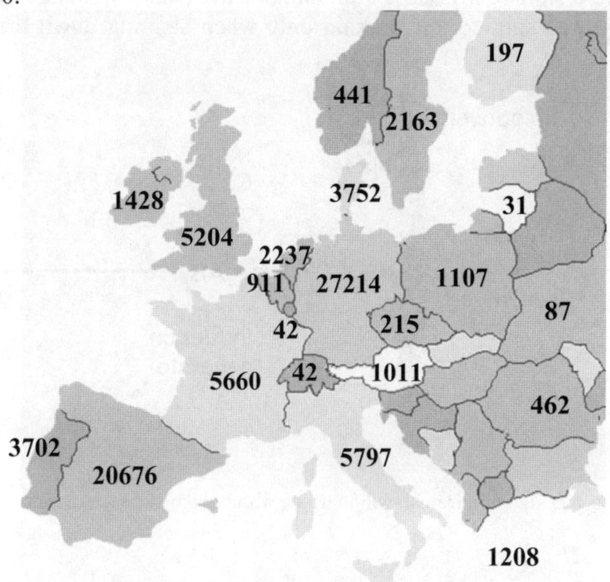

Fig.1-14 Installed wind power capacity in Europe at the end of 2010 [2]

1.4 Technological development

The following columns of words summarise the debate between competing wind power plant design options during the last two decades:

- 3-bladed rotor	- 2-bladed rotor
- constant rotational speed	- variable rotational speed
- stall control	- pitch control
- induction generator	- synchronous generator
- electrical excited synchronous generator	- permanent magnetic synchronous generator
- gearbox	- no gearbox, direct drive
- glass fibre blades	- metal, wooden blades
- direct grid connection	- ac-dc-ac conversion
- hydraulic actuators	- electrical actuators

In the 80s, stall-controlled wind turbines that were connected to the grid through induction generators were the predominant commercial technology. The rotational speed of these "classic Danish design" turbines was kept nearly constant since the turbine was directly coupled to the grid via the generators. Similarly, since these turbines were stall-controlled, no pitching of the blades was necessary. The blade tips are used as spoilers for braking only when shutting down the machine,(see Fig1-15).

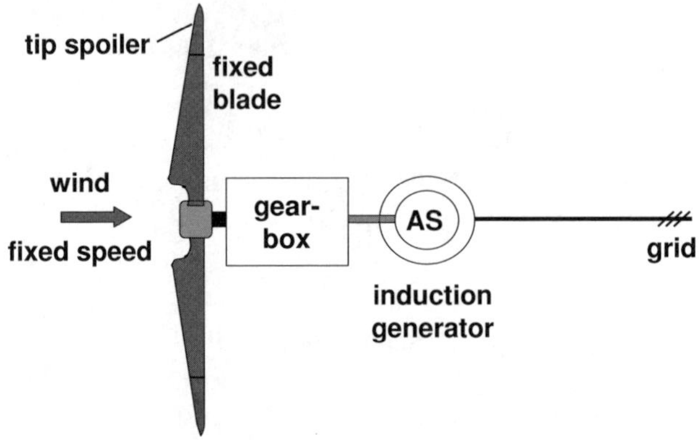

Fig. 1-15 Basic design of a Danish wind turbine with induction generator and constant rotational speed

In the 90s, and mainly in Germany, variable speed wind turbines were developed and produced on a large scale. Usually, the rotor is pitch-controlled to avoid over-speeding. Either a large synchronous generator – often directly coupled to the rotor (no gearbox) - produces variable frequency electricity (50 or 60 Hz) via an ac-dc-ac conversion (see Fig 1-16), or a double-fed, gearbox-driven induction generator is used.

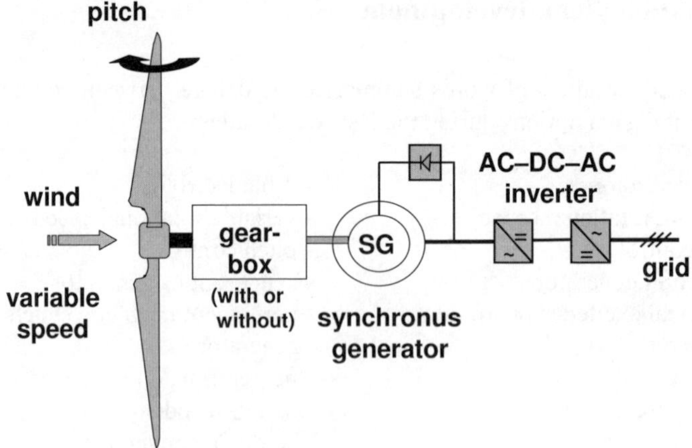

Fig. 1-16 Pitch-controlled variable speed wind turbine with synchronous generator and ac-dc-ac power conversion

The classic Danish design is very simple and robust. Variable speed machines are more flexible in their adaptation to the grid, but they require a more sophisticated control technology. Both designs are used even in huge megawatt machines with rotor diameters of 70 meters and more. For bigger machines, many manufacturers use double fed induction generators with a small ac-dc-ac inverter running with variable speed to avoid high loads coming from gusts.

With these designs, small but innovative wind manufacturers have - step by step -succeeded in building wind turbines of a size that the big aerospace industries in the United States (Mod 1, Mod 2 turbines) and Germany (Growian) failed to accomplish 20 years ago.

Looking into the future: offshore wind farms will be the next important step in wind energy since offshore development eliminates land-use issues and takes advantage of the more favourable wind regime at sea.

Fig. 1-17 shows the early near-shore windfarm of Vindeby (1991). Currently, much effort is put into the development of suitable and cheap foundations for offshore windfarms.

Fig 1-17 Offshore wind turbines, Vindeby, Denmark (1991)

References

[1] Global Wind Energy Council, *Global Wind Statistics 2010*, February 2011

[2] European Wind Energy Association, *Wind in power – 2010 European Statistics*, February 2011

[3] IEA, *key world energy statistics 2010*

[4] IEA, *key world energy statistics 2005*

[5] Hinsch, C.; Rehfeld, K.; *Die Windenergie in verschiedenen Energiemärkten*; DEWI Magazin Nr. 11, Aug. 1997

[6] Strauß, K.; *Kraftwerkstechnik*, Springer Verlag Berlin, 3. Auflage, 1997

[7] H.G. Heitmann, *Praxis der Kraftwerks-Chemie*, Vulkan Verlag Essen, 1997

[8] Quaschning, V.: *Regenerative Energiesysteme*, 5. edition, Hanser Verlag, 2007

[9] EWEA, *Wind Energy – The Facts*, European Commission, 2004

[10] Shell, 1998

2 Historical development of windmills

2.1 Windmills with a vertical axis

According to historians, the first machines utilising wind energy were operated in the orient. As early as 1,700 B.C., it is mentioned that Hammurabi used windmills for irrigation in the plains of Mesopotamia [1]. There is written evidence of the quite early utilisation of wind power in Afghanistan: Documents from 700 AD confirm that the profession of a millwright was one of high social esteem there [1]. Even today, ruins of these windmills that were running for centuries can be found in Iran and Afghanistan (cf. Fig. 2-1).

Fig. 2-1 Ruins of a vertical axis windmill in Afghanistan, 1977 [3]

The world's oldest windmills had a vertical axis of rotation. Braided mats were attached to the axis. The mats caused drag forces and, therefore, were "taken along with the wind". In Persian windmills, an asymmetry was created by screening half the rotor with a wall. This way the drag forces could be utilised for driving the rotor (Fig. 2-2a).

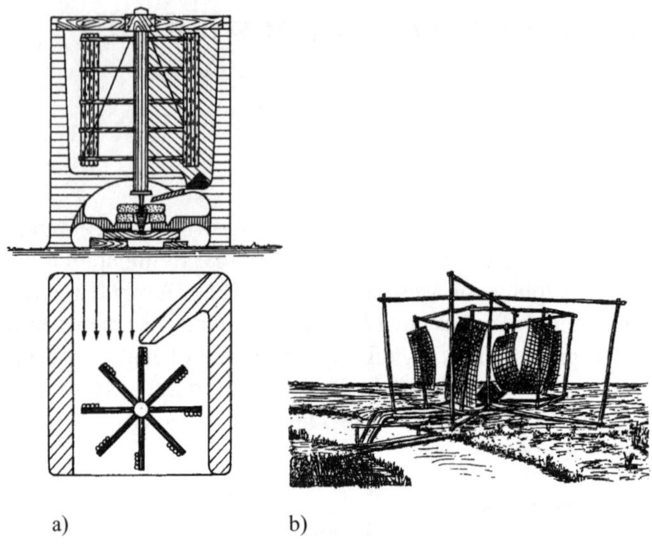

a)   b)

Fig. 2-2 a) Persian windmill [3]   b) Chinese windmill with flapping sails [4]

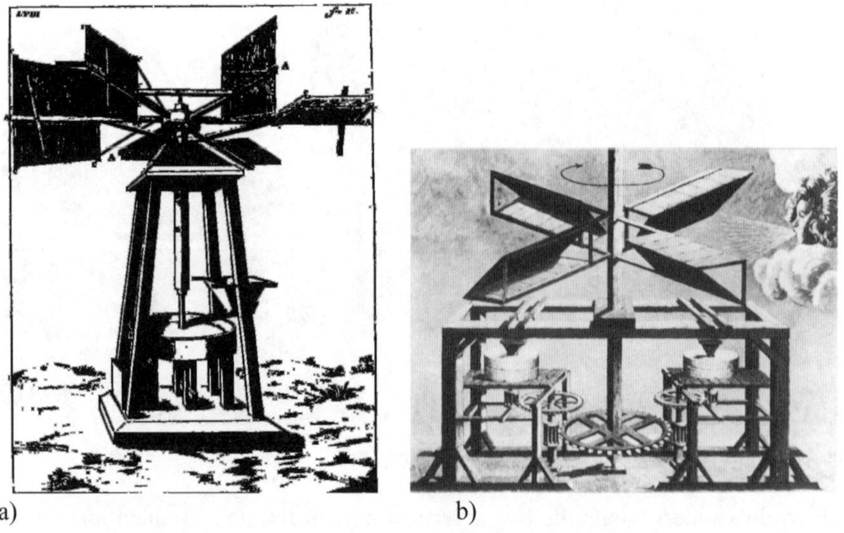

a)   b)

Fig. 2-3 Later design of vertical axis windmills: a) with flapping sails, France 1719 [2]; b) with bodies driven by drag forces, Italy, approx. 1600 [4]

In Chinese windmills - which also date back a long time -, a similar asymmetry is created by sails which rotate out of the wind on their way „back", i.e. when they advance into the wind (Fig. 2-2b). These *Chinese drag wheels* date back to approx. 1000 AD. Similar to the Persian mills, they had a vertical axis and used braided mats as „sails". However, in contrast to the Persian mills, they had the typical advantage of vertical axis windmills to utilise the wind independent of its direction.

The simplicity of this construction can be appreciated in Fig. 2-3a which shows a later version of a vertical axis mill with flapping sails: The millstone is attached directly to the vertical drive shaft without redirecting the rotational movement and without an intermediate gear. The more recent horizontal axis windmills, such as the Dutch smock mills (Fig. 2-8), designed for a higher tip speed ratio, require a far more sophisticated construction not only for redirecting and back-gearing the rotational movement from the horizontal to the vertical axis, but also for the much more complicated bearing of the faster and heavier horizontal shaft. The windmill of Veranzio (Fig. 2.3b) like the cup anemometer (Fig. 2.19a), belongs to drag driven rotors with a relatively low tip speed ratio. For more details on the operation of these devices see section 2.3.2.

The simplicity of the vertical axis design is also central to the Savonious rotor (1924, Fig. 2.4a) and the Darrieus rotor (1929, Fig. 2.4b). But as late „occidental" versions of the vertical axis principle they utilise - partially or exclusively - the lift force as their driving power, see 2.3.3 for a more detailed discussion.

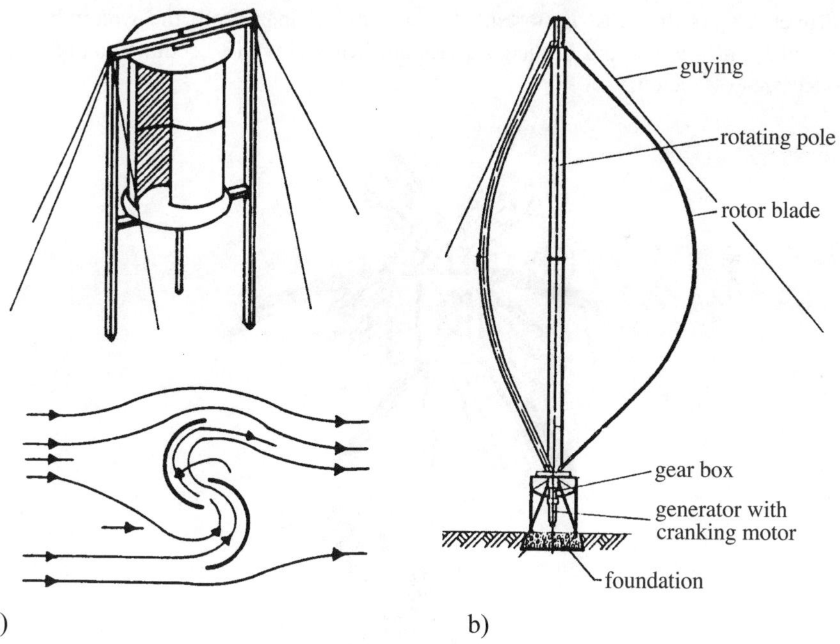

Fig. 2-4 a) Savonius rotor [5]; b) Darrieus rotor [6]

2.2 Horizontal axis windmills

2.2.1 From the post windmill to the Western mill

In the Occident, a windmill type different from the oriental vertical axis mills was developed, albeit very much later. The most prominent distinctive feature is the rotor with a horizontal axis whose sails rotate in a plane vertical to the wind, just like an aircraft propeller. In that case a driving principle different from the blade area obstruction of the drag driven rotors has to operate.

The first theoretical descriptions regarding lift forces of blades intercepting the wind, i.e. the driving power of horizontal axis devices, date back only to the beginning of the 20th century. Millwrights in earlier centuries may have used the idea that a wheel intercepts the airflow just like a screw („airscrew").

The oldest construction of a lift-driven horizontal axis device is the *post wind-mill*. A picture was found in an English prayer-book from the 12th century (Fig. 2-5a). It is also mentioned at this time in the statutes of the French city of Arles (Provence). As the most important driving engine apart from the waterwheel, it spread from England and France via Holland, Germany (13th century) and Poland to Russia (14th century).

Fig. 2-5a Drawing of a post mill in an English prayer-book of 1270 [2]

It is disputed among historians who invented it and where it came from. However, there now seems to be a general agreement, „unlike previously believed, that the Crusaders did not come across windmills in Syria, but took them there themselves." [13] The post mill consists of a timber support holding the vertical central post around which the boxlike buck (i.e. the mill house) turns on a pivot, Fig. 2-5b. Using a tail pole, the buck together with the rotor was oriented into the wind.

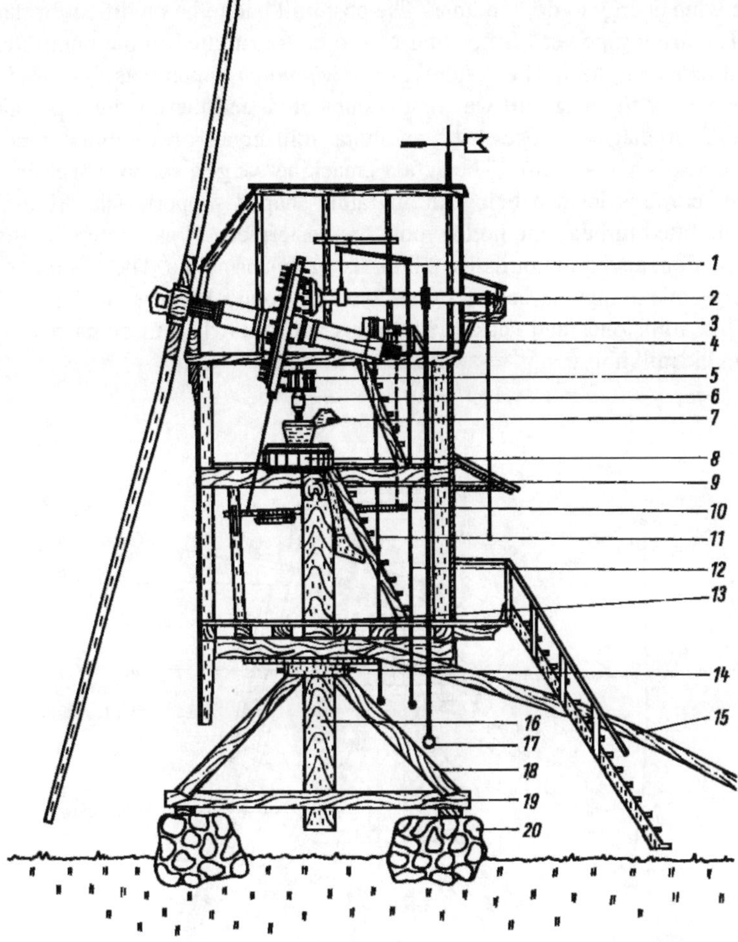

Fig. 2-5b Section of a post mill [3]

1 gear wheel with brake, 2 shaft for sack hoist, 3 hand-driven hoist, 4 rotor shaft, 5 lantern gear, 6 quant, 7 hopper, 8 millstones, 9 traverse beam, 10 brake lever, 11 brake rope, 12 hoist operating rope, 13 floor for flour, 14 saddle, 15 tailpole, 16 central post, 17 sack hoist, 18 quarter bars, 19 cross trees, 20 foundation

The main shaft with the rotor is almost horizontal. The brake wheel drives via the lantern gear the vertical shaft with the millstone. Only from the 19th Century onwards, post windmills were equipped with two lantern gears for parallel milling operations of two sets of millstones. The post windmill was exclusively applied for grinding grain.

In Holland there was an economic interest in the reclamation of land by draining the polders already in the 15th century. Therefore, first attempts were made to use the wind energy to drive pumps. The post mill had to be modified for that purpose. The driving power of the wind had to be transmitted to the pump that was situated under the mill. The result was the *wipmolen* which was first used about 300 years after the post mill was first documented, and these were especially designed for drainage purposes. The revolving mill house of a wipmolen contains only the gearbox (Fig. 2-6). The actual „machine", e.g. a scoop wheel or Archimedean screw, is located below the pyramid-shaped support. The driving shaft had to be fitted through the hollow post - a masterpiece of carpenters' craftsmanship! Later on, also grain mills were built using this principle. There is the obvious advantage of having the set of stones on the ground because no longer heavy loads, like millstones and sacks with grains and flour, had to be carried up and down in the mill house.

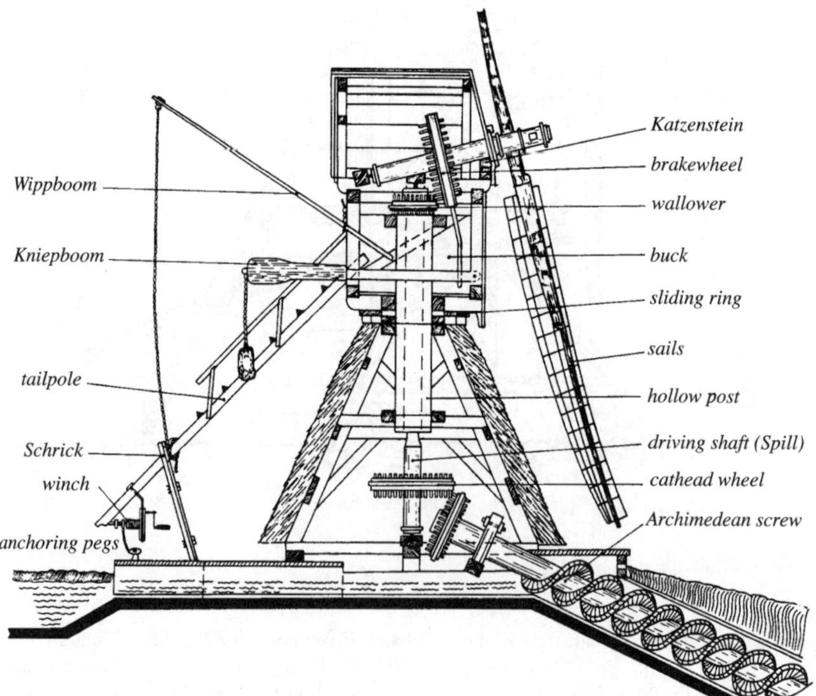

Fig. 2-6 Section of a wipmolen [7]

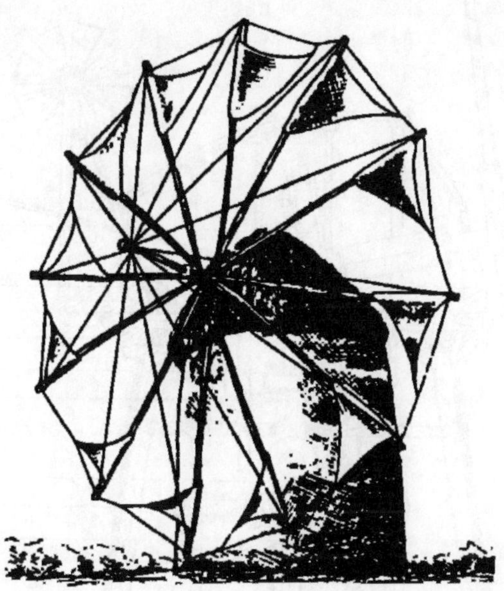

Fig. 2-7 Mediterranean tower mill with sails - an early version of the tower mill [8]

In Southern Europe, the post mill did not gain popularity. Another mill type was wide-spread there: the *tower mill*. Already very early on, the first wind mills of this kind were used for irrigation. The first documentation of these mills dates back to the 13th Century [1]. Main features of the older Mediterranean type are the cylindrical stone built mill house, a fixed thatched roof, and a guyed rotor with eight or more sails (Fig. 2-7). Later versions, mainly in Southern France, had a turnable wooden cap and a four-bladed wooden rotor like the post mills.

The turnable cap is the main characteristics of the *Dutch smock mill* which came into use in the 16th Century (Fig. 2-8). It is a further development of the tower mill as the lighter wooden construction of the octagonal tower could be easier erected on the wet Dutch marshland than the heavy stone construction of the tower mill. In Holland, the Dutch smock windmills were mainly used for the drainage of the polders, often arranged in a series to lift the water mill by mill over the embankments. In the rest of Europe, they were applied preferably for grinding grain.

With tens of thousands of Dutch smock mills being built, the use of wind power experienced its heyday in the Netherlands in the 18th and 19th Century. The large number of mills lead to a standardisation of its construction which was unusual for that time. Even in special versions such as the *gallery windmill* with its multi-storey socle (Fig. 2-9), the basic type of the Dutch smock mill can be easily recognised.

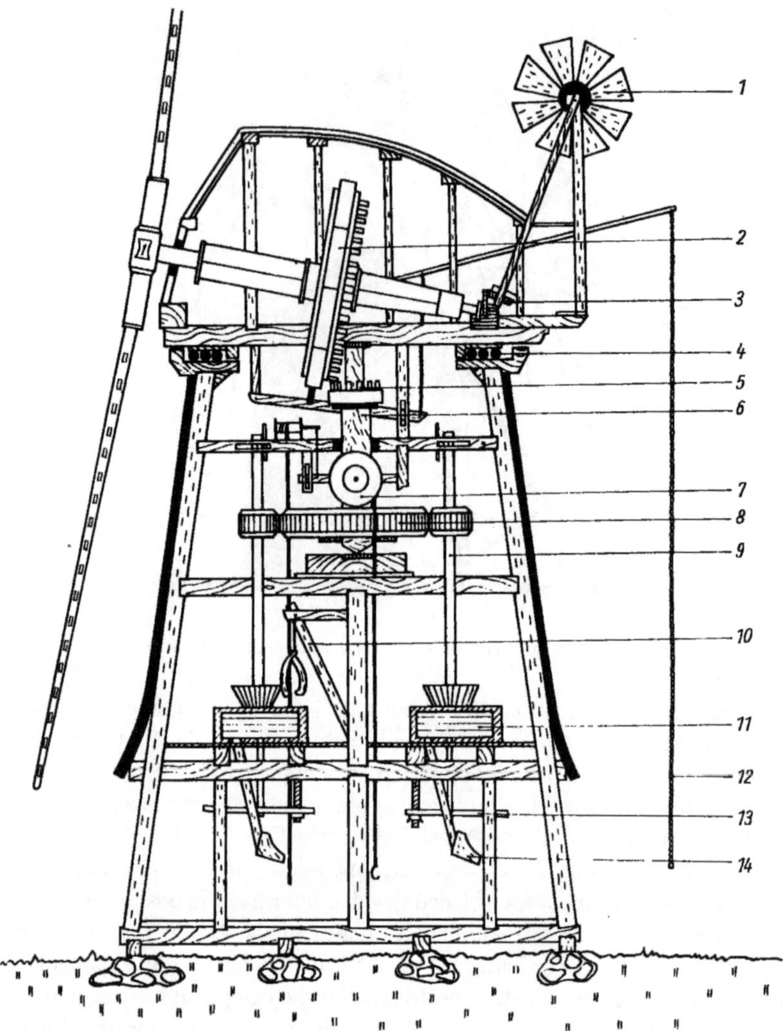

Fig. 2-8 Section of a Dutch smock mill [3]

1 fantail; 2 gear wheel with brake; 3 gear for cap rotation; 4 rollers; 5 wallower; 6 main shaft; 7 sack hoist; 8 great spur wheel; 9 spindle drive; 10 millstone crane; 11 millworks with chute; 12 brake chain; 13 stone adjustment; 14 flour chute

A somewhat exotic development is the 17th century *Paltrock mill* which shows that wind energy can be utilised universally as a driving force. The whole mill (as the cap of the Dutch smock mill) rested on a live ring. This way an entire sawmill can be driven by a windwheel (Fig. 2-11).

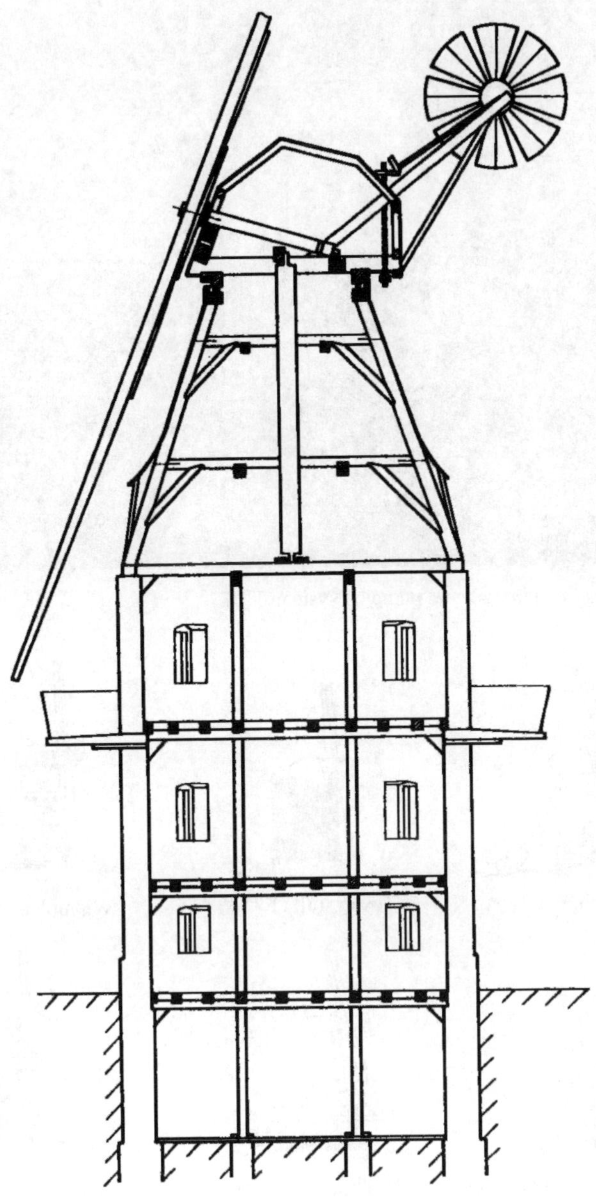

Fig. 2-9 Sketch of a gallery windmill [[9]

Fig. 2-10 Western mills as wind pumping systems [10]

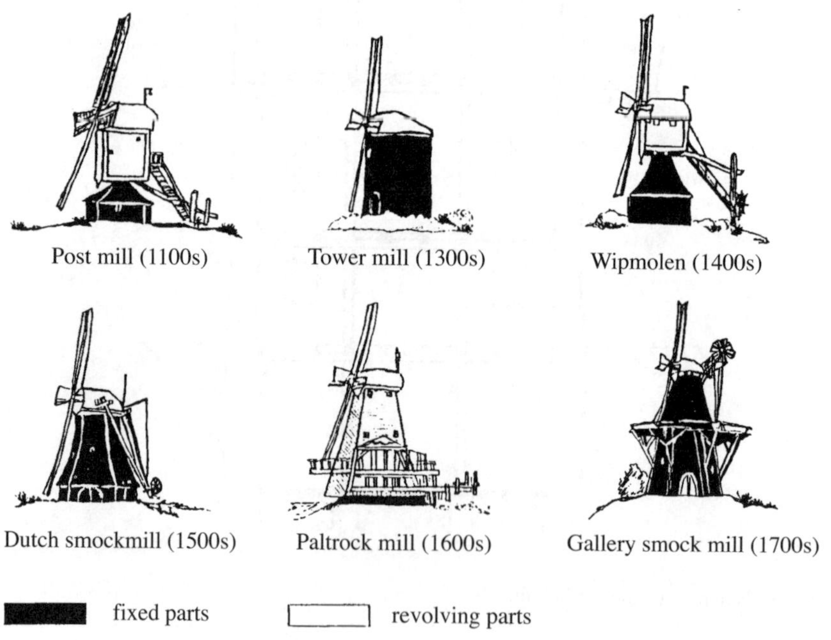

Fig. 2-11 Overview of historical horizontal axis windmill types [11]

The last type of historical windmills is the American farm windmill ('Western mill'), which was developed in the mid-19th century. The Western mill was mainly used for providing drinking water for both people and cattle in North America. Moreover, it assured the water supply for the steam locomotives of the new railways expanding into the West. The main characteristics of this windmill is the is the "rotor rosette" of a diameter between 3 and 5 m, with more than 20 metal sheet blades, situated on top of a metal lattice tower. It uses a crank shaft to drive a piston pump (Fig. 2-10).

The Western mill was the first windmill type with a fully automatically controlled yaw system including a storm control (see chapter 12). So the Western mill is still nowadays a "modern" machine of which tens of thousands are installed with a nearly unchanged design in Australia, Argentina and the USA.

2.2.2 Technical innovations

In contrast to the modern wind turbines, the old windmills needed the permanent attendance of a miller who was not only responsible for milling but also for the safe operation of the windmill. The main tasks in windmill operation were: adjusting the wind mill rotor to the wind direction, hoisting or reefing the sails according to the wind speed for power control and braking the rotor early enough when a storm was brewing. Only the Western mill was the first type without need for an operator.

The miller or his donkey pulled at the so-called tailpole and turned the mill into the wind. Later, winches were attached to the tailpole. This way, the tailpole could be pulled towards pegs which were situated in a circle around the mill (Fig. 2-6). Even later still, a small fantail, that was situated at right angles to the large rotor, drove the winch. Therefore, it always intercepted the wind, when there were skew winds at the rotor. This mechanism was much easier to use with Dutch smock mills as the fantail could be mounted directly to the cap (from approx. 1750, Fig. 2-12).

The adaption of the power consumption by the rotor to the prevailing wind conditions was far more crucial. For this purpose, more or less sail was on the sail frames of the windmill blade (variation of the area a, in eq. 2-13). This was a precarious issue especially when the miller under-estimated the wind, and suddenly a stronger breeze or even a storm came up and there was the danger of an over speeding of the rotor. Then, to reef the sails completely the miller had to brake the rotor as fast as possible using a wooden brake acting on the wooden gear wheel. Due to excessive frictional heat many windmills burnt down because the miller had started braking the rotor too late.

Therefore, the invention of the spring sails (in the 17th century, Fig. 2-12), which were easily controlled by a lever, brought a significant relief for the millers. They allowed braking the rotor even during a storm because even under full operation, the shutters of the spring sails could be opened completely from inside the mill house so that the wind could blow through the blade surface.

The discovery of the blade twist was most relevant for an increase in the rotor efficiency. This innovation was studied by John Smeaton who presented the results of his wind rotor experiments to the Royal Society of England in 1759 [1]. With a sophisticated test rig (Fig. 2-13), which nowadays is replaced by the wind tunnels, he verified and improved the existing rules of wind mill construction. He recommended a blade pitch angle between the blade and the rotor plane of 18° at the rotor hub and 7° at the rotor tip. He also found out that for a given rotor diameter an increase of sail surface beyond a certain surface size no longer increases the rotor power output.

Moreover, Smeaton determined the power and the tip speed ratio - the ratio of circumferential speed at the blade tip to the wind speed - whose values were in the range between 2.2 and 4.3.

Fig. 2-12 Drive train of a large corn mill with fantail yawing, spring sails, and three sets of millstones [10]

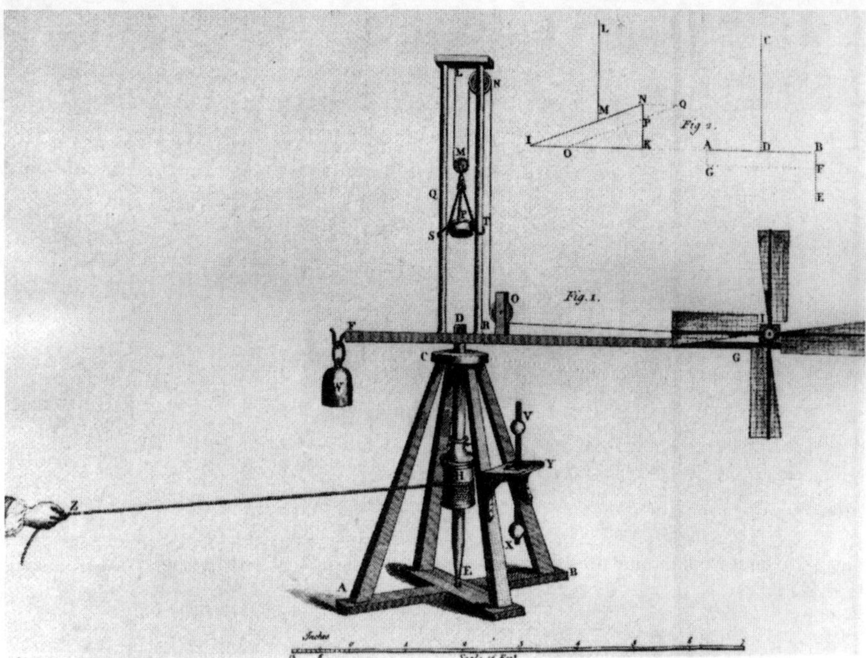

Fig. 2-13 Test rig by Smeaton for measuring the power characteristics of windmill rotors [12]

With the development of the Western mill in the 19th century, a completely new era began for the application of wind energy: it reflects the industrialisation of the wind energy application.

The Western mill was not only the first wind turbine in industrial series production and made from metal, it was the first wind turbine with a completely automatic control requiring no supervision by a human operator: A sophisticated system of wind vanes served for yawing and protecting against overspeed during stormy weather, see chapter 12. Therefore, the turbines operated self-governed on vast pasture land. Naturally, it became obvious to apply this "stand-alone" Western mill also for generating electricity, first experiments started in 1890 in the USA [20].

Paul LaCour, professor at the Askov School of engineering in Denmark, investigated since 1891 systematically the application of wind energy for electricity generation – and noticed at once that due to its low tip speed ratio the Western mill was not very suitable for this purpose. He developed a very perfect self-governing four-bladed wind turbine with a DC generator for remote dwellings. During the First World War (1914-18) more than 250 turbines of this type operated in Denmark. For more than 50 years, these "LaCour machines" were produced [21].

Fig. 2-14 First wind turbine for the production of direct current by Paul LaCour, Askov, Jütland, 1891 [21]

2.2.3 Begin and end of the wind power era in the Occident

From the 12th century to the early 20th century, water and wind power were the only relevant sources of mechanical energy. Braudel remarks the following [13]:

„In the 11th, 12th and 13th centurys, the Occident experienced its first mechanical revolution. We refer to 'revolution' as the total of all changes that were caused by the increased number of water and windmills. Though the output of these 'primary drives' was very limited (between 2 to 5 HP for a watermill, 5 to max. 10 HP for a windmill), in a world with a very poor power supply, this constituted a considerable increase in power and was decisive for the first growth phase of Europe.“

In the 19th century, the steam engines and combustion engines began to replace the wind and watermills. However, the low speed of the second mechanical revolution in the area of driving engines is shown by the 1895 industrial census of the German Reich [7]:

18,362 wind turbines
54,529 water turbines
58,530 steam engines
21,350 combustion engines and others.

130 years after the invention of the steam engine, half of the drive units were still of traditional origin!

2.2.4 *The period after the First World War until the end of the 1960s*

After the First World War (1914-18) the scientific understanding of wind turbine design made a great leap forward, partially based on the experience of propeller design for military and civil aeroplanes.

In 1920 Betz applied the actuator disc theory to the wind turbine and found that a maximum of 59% of the kinetic wind energy can be converted into mechanical energy by a free-stream turbine [18]. This was previously discovered by Lanchester in England [19]. But Betz continued, linking these considerations, rooted in the theory of linear momentum and the energy conservation law, with the airfoil theory (Blade element momentum theory). This resulted in simple design rules for the blade geometry of optimised wind turbine rotors. With small modifications these basics developed by Betz are still used in wind turbine design.

With this new theoretical background for wind turbines, many promising approaches for the modern wind turbine design emerged, e.g. in France, Germany (Bilau, Kleinhenz-MAN, Honeff and others) and Russia (Sabinin, Yurieff and others). In Crimea close to Yalta, the wind turbine WIME D-30 with a diameter of 30 m and a power of 300 kW was operated from 1931 to 1942 [23], feeding into a small 20 MW grid.

But the start of the Second World War by the German National Socialists quickly disrupted these beginnings. Only in the USA the development of wind turbines was continued during the war. The engineer Palmer C. Putnam designed together with the water turbine firm "Smith" the first grid-connected Megawatt wind turbine ($D = 53$ m; 1,250 kW). Well-known scientists participated in the development of its concept (Fig. 2-15 d). It was commissioned in 1941 and operated until 1945. Unfortunately the economic balance showed that the power production costs were 50% higher than that for the conventional power generation. Therefore, the improvements of the technical concept proposed by Putnam were not put into practice.

With the reconstruction of Europe after the Second World War and the growing realisation that coal reserves were continuously decreasing, interest in wind energy application arose again in the 1950s. Through the „Organisation for European Economic Cooperation (OEEC), Working Group 2", experts from England (Golding), Denmark (Juul), Germany (Hütter), France (Vadot) and others met to discuss their experiences in the wind turbine design.

In Germany, Hütter followed with the "Research Association Wind Power" a very innovative turbine concept, leading in 1958 to the prototype W34 ($D = 34$ m, 100 kW). It had fibre glass blades, an electro-hydraulic pitch control and produced 50 Hz AC power with a synchronous generator, Fig. 2-15c. A teetering hub dampened the dynamics of this two-bladed rotor. With many interruptions this wind turbine operated until 1968.

a) Gedser wind turbine
(200 kW, D = 24 m, DK 1957)

b) TVIND wind turbine
(2,000 kW, D = 54 m, DK 1977)

c) Hütter wind turbine
(100 kW, D = 34 m, D 1958)

d) Smith-Putnam wind turbine
(1,250 kW, D = 53 m, USA 1941)

Fig. 2-15 Historical wind turbine prototypes

Johannes Juul in Denmark followed a completely different course. His aim was a simple and robust turbine concept for the grid connection. He was a "wind electrician" trained by Paul LaCour in Askov. Later, he became the leader (Linjemester) of the grid expansion division of the Sjaelandic electricity supplier (SEAS). With SEAS he erected the famous Gedser wind turbine (Fig. 2-15a, $D = 24$ m, 200 kW). This wind turbine provided electricity to the grid for thousands of working hours from 1957 to 1962.

His electrical concept was ingenious: an asynchronous motor was pushed by the rotor into the super synchronous rotational speed range and thus became a generator without any effort of synchronisation. The rotor, although of simple design (plywood profiles on a guyed steel spar), had a skilful aerodynamic design which caused flow separation in the strong wind range acting as a completely passive power limitation. The turnable blade tips served as braking flaps and were activated by centrifugal forces at a grid failure.

However, at the beginning of the 1960s, cheap oil from the Near East came to Europe. The calculations by Juul himself showed that the wind generated electricity was too expensive to compete with the conventional fossil generation. This caused the breakdown of the second break-up.

2.2.5 *The Renaissance of the wind energy after 1980*

The oil price shocks in 1973 and 1978 initiated again a reflection on the future energy supply. In 1977, even the Gedser wind turbine of J. Juul was re-activated for research purposes and again coupled to the grid.

But the Renaissance of wind energy began with a tremendous crash. In the USA, Germany, Sweden and some other countries, supported by the governments, giant wind turbines were designed by the aerospace industry, Fig. 2-16. Nearly all of them failed after some hundred hours of operation due to technical problems: too early, too big and too expensive.

An exception was the Maglarp wind turbine WTS-3 ($D = 78$ m, 3.000 kW) which operated grid-connected more than 20,000 hours, and also the Tvind wind turbine, developed by "amateurs", Fig. 2-15b, which is still operating today (but with two thirds of its original rated power).

In contrast to this, the small Danish manufacturers of agricultural machines (Vestas, Bonus, Nordtank, Windworld, etc.) were very successful in the beginning of the 1980s with wind turbines consisting of a rotor diameter between 12 and 15 m produced in series and equipped with an asynchronous machine, according to the concept of Juul. However, the blades of these turbines were manufactured with fibre glass, following Hütter's blade design. With a rated power of 30, 55 or 75 kW they were technically and economically successful because an appropriate feed-in tariff was set and granted by the Danish government. This first small market grew continuously.

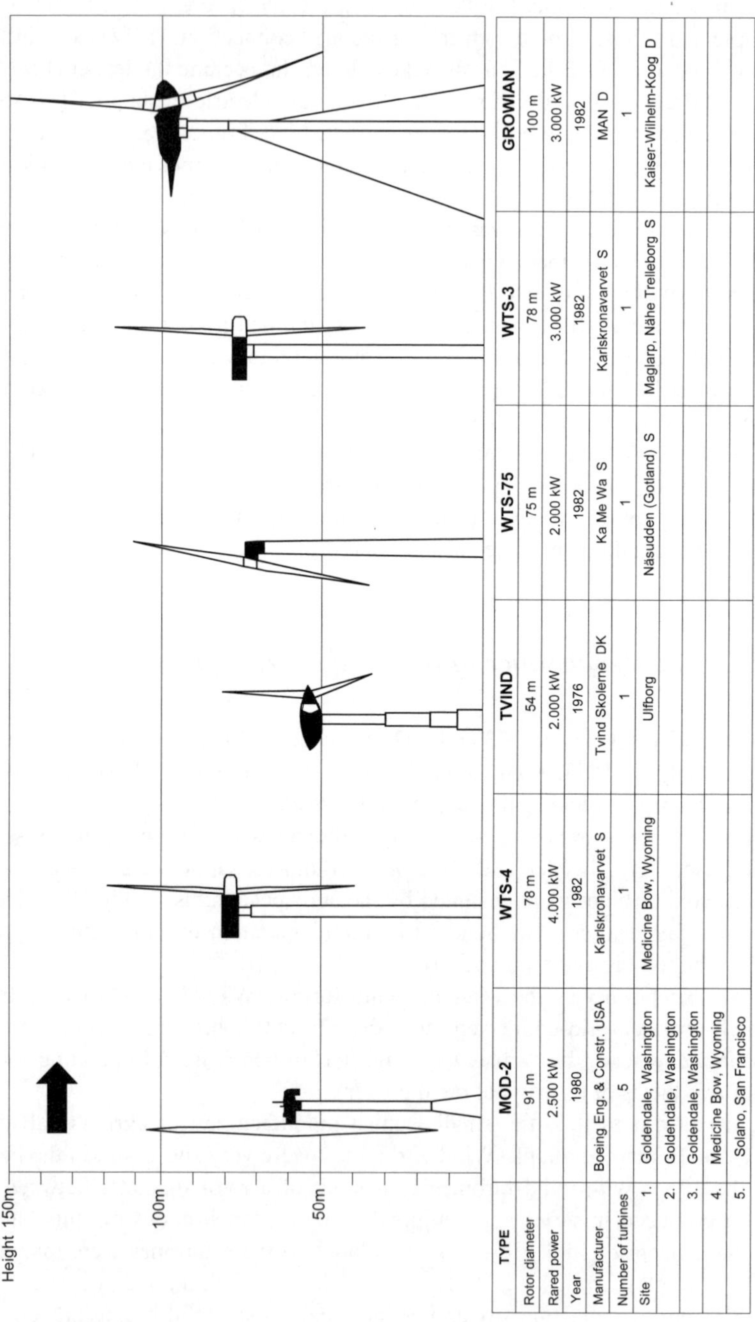

TYPE	MOD-2	WTS-4	TVIND	WTS-75	WTS-3	GROWIAN
Rotor diameter	91 m	78 m	54 m	75 m	78 m	100 m
Rared power	2.500 kW	4.000 kW	2.000 kW	2.000 kW	3.000 kW	3.000 kW
Year	1980	1982	1976	1982	1982	1982
Manufacturer	Boeing Eng. & Constr. USA	Karlskronavarvet S	Tvind Skolerne DK	Ka Me Wa S	Karlskronavarvet S	MAN D
Number of turbines	5	1	1	1	1	1
Site 1.	Goldendale, Washington	Medicine Bow, Wyoming	Ulfborg	Näsudden (Gotland) S	Maglarp, Nähe Trelleborg S	Kaiser-Wilhelm-Koog D
2.	Goldendale, Washington					
3.	Goldendale, Washington					
4.	Medicine Bow, Wyoming					
5.	Solano, San Francisco					

Fig. 2-16 The Multi-Megawatt class in the beginning of the 1980s [24]

Today, after more than 30 years of continuous research and development on the wind turbines these formerly "small-scale manufacturers" produce successfully in the rotor diameter range from 80 to 126 m, where formerly the aerospace industries failed.

2.3 The physics of the use of wind energy

2.3.1 Wind power

The power of the wind that flows at a velocity v through an area A is

$$P_{\text{wind}} = \frac{1}{2}\, \rho A v^3 .$$
(2.1)

It is proportional to the air density ρ, the cross sectional area A (perpendicular to v) and the third power of the wind velocity v. The third power of the wind velocity can be explained as follows: the power P_{wind} in the wind is the kinetic energy

$$E = \frac{1}{2}\, m v^2$$
(2.2)

of the air mass m, passing through the area A in a given time. Since the resulting mass flow

$$\dot{m} = A\rho \frac{dx}{dt} = \rho A v$$
(2.3)

itself is proportional to the wind velocity (Fig. 2-17), the power (energy per unit of time) is expressed by

$$P_{\text{wind}} = \dot{E} = \frac{1}{2}\, \dot{m} v^2 = \frac{1}{2}\rho\, A v^3 .$$
(2.4)

The power of the wind is converted into mechanical power of the rotor by deceleration of the flowing air mass. On one hand, it cannot be converted completely, since this would decelerate the mass flow to zero, it would block the cross sectional (rotor) area A for the following air masses.

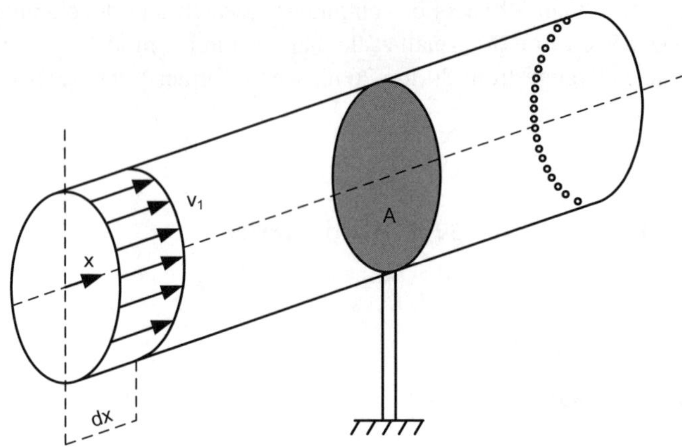

Fig. 2-17 Scheme of a stream tube with mass flow through the cross sectional area A

On the other hand, if the air flows through the area without any deceleration of the wind velocity, there would be no power conversion as well. So there must be an optimum of wind energy conversion through flow deceleration between these two extremes.

Betz [14] and Lanchester [19] discovered that the maximum power is extracted by a free (i.e. unshrouded) wind turbine if the original upstream wind velocity v_1 is reduced to a velocity $v_3 = v_1/3$ far downstream the rotor.

Then, the resulting velocity in the rotor plane is $v_2 = 2v_1/3$ (Fig. 2-18). In that case of a theoretically maximum power extraction, the result is

$$P_{\text{Betz}} = \frac{1}{2}\,\rho\,A\,v^3\,c_{\text{P.Betz}} \tag{2.5}$$

with the maximum power coefficient $c_{\text{P.Betz}} = {}^{16}/_{27} = 0.59$. Even in this best case of power extraction without any losses, only 59 % of the wind power is extractable.

Real power coefficients c_P are lower. For the drag driven rotors they are below $c_\text{P} = 0.2$, and for lift driven rotors with good airfoil profiles they reach up to $c_\text{P} = 0.5$. Betz' theory is discussed in detail in chapter 5.

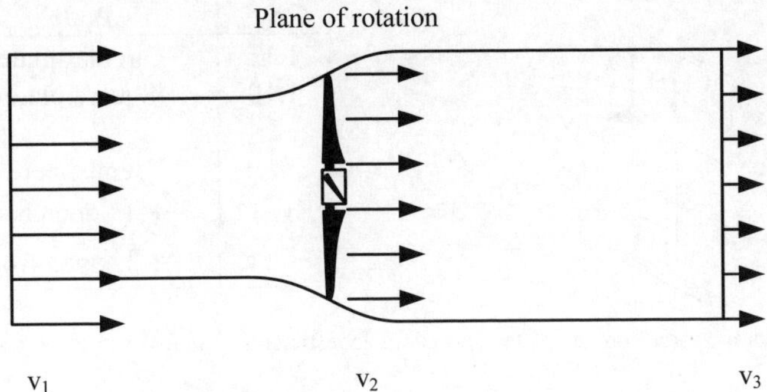

Fig. 2-18 Air flow through the rotor of a wind turbine, divergence of the stream tube resulting from the flow deceleration

2.3.2 Drag driven rotors

The drag devices utilise the force that acts on an area a perpendicular to the wind direction, Fig. 2-19, This force, referred to as

$$D = c_\mathrm{D}\ \frac{\rho}{2}\ a\,v^2 \qquad\qquad (2.6)$$

is proportional to the area a, to the air density ρ and to the square of the wind velocity v.

This formula is also used to describe the drag of other bodies placed in a flow, where a is their projected area perpendicular to the flow velocity.

The drag coefficient c_D is the proportional constant and describes the „aerodynamic quality" of the body: the higher the aerodynamic quality of a body, the lower is c_D and thus the corresponding drag force (Fig. 2-19).

The torque, rotational speed and power of the early Persian (or Chinese) vertical axis windmill, using the drag principle, can be easily estimated, based on the assumption that the torque of the simplified model shown in Fig. 2-20b is equivalent to that of the real windmill in Fig. 2-20a. The simplified model neglects the coming and going of the blades and the effect of the preceding and the following blades, respectively.

In that case, the actual relative velocity $w = v - u$ at the rotating plate is the difference between wind velocity v and the circumferential velocity $u = R_\mathrm{M}\,\Omega$ at the mean radius R_M of the area a . The angular velocity due to the rotational speed n is $\Omega = 2\,\pi\,n$.

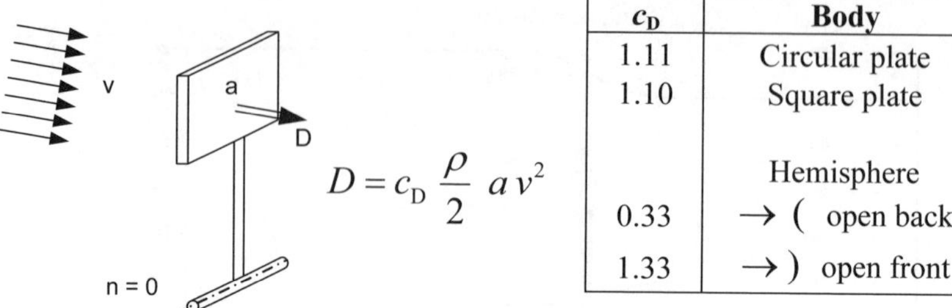

$$D = c_D \, \frac{\rho}{2} \, a \, v^2$$

c_D	Body
1.11	Circular plate
1.10	Square plate
	Hemisphere
0.33	$\rightarrow$ (open back
1.33	$\rightarrow$) open front

Fig. 2-19   Drag force on a plate and drag coefficients c_D for some typical bodies

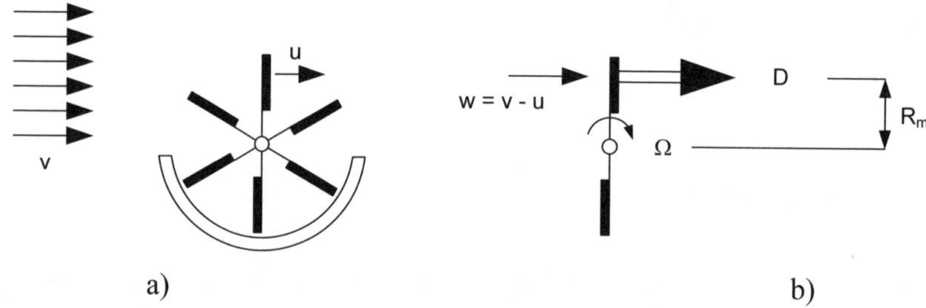

Fig. 2-20 a) Principle of a Persian windmill, b) simplified model

Thus, the resulting drag force on the rotating plate is

$$D = c_D \, \frac{\rho}{2} \, a \, w^2 = c_D \, \frac{\rho}{2} \, a \, (v-u)^2 . \tag{2.7}$$

Hence, the mean driving mechanical power – which in reality is slightly pulsating – amounts to

$$P = D \, u = \frac{\rho}{2} \, a \, v^3 \left\{ c_D \cdot \left(1 - \frac{u}{v}\right)^2 \cdot \frac{u}{v} \right\} = \frac{\rho}{2} \, a \, v^3 \, c_P . \tag{2.8}$$

This driving power in eq. (2.8) is - like the power in the wind - proportional to the projected area a and to v^3, the cube of the wind velocity.[1] The expression in the braces is equal to the power coefficient c_P, the aerodynamic efficiency of the rotor. It gives the portion of the wind power which is converted into mechanical power.

---

[1] Usually, the reference area is the swept rotor area A, not the projected „drag" area a of this simplified consideration. The area A would be in this case the rotor height multiplied by half the rotor diameter (unshaded side). The resulting $c_P(\lambda)$ would even be lower by the factor a/A.

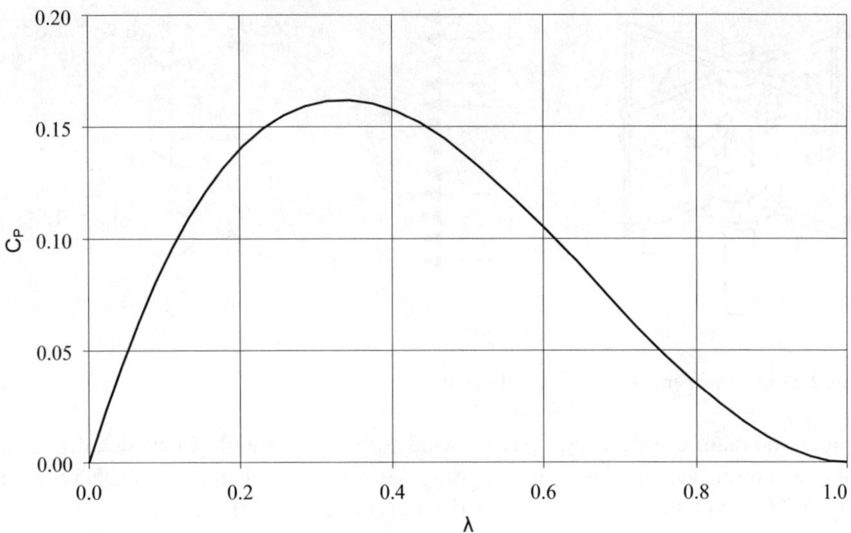

Fig. 2-21 Power coefficient versus tip speed ratio $\lambda = \Omega\, R_M/v$ of the Persian windmill (approximation for the simplified model)

This coefficient must be lower than the theoretical maximum value $c_{P.Betz} = 0.59$ determined by Betz. It depends on the ratio tip speed ratio $\lambda = u\,/\,v$, which was introduced in chapter 1, of the circumferential velocity $u = \Omega \cdot R_M$ to the wind velocity v.[2]

For a given wind velocity v, the diagram of $c_P(\lambda) = c_P(\Omega \cdot R_M\,/\,v)$ shows which portion of the wind power $(\rho/2)\,a\,v^3$ can be extracted. It depends on the circumferential speed u, respectively the angular velocity Ω (i.e. the rotational speed n).

Fig. 2-21 shows such a diagram for the simplified model of the *Persian windmill* (Fig. 2-20) using the drag coefficient $c_D = 1.1$ of the square plate. At complete standstill ($\lambda = 0$) no mechanical power is extracted from the wind. Neither it is at idling with maximum rotational speed ($\lambda = \lambda_{idle} = 1$), where the circumferential velocity is equal to the wind velocity. In between these extreme cases, the maximum power coefficient $c_{P.max} \approx 0.16$ is reached at a tip speed ratio of about $\lambda_{opt} \approx 0.33$. Merely 16% of the wind energy can be converted to mechanical energy.

Even worse is the power output of the *cup anemometer* (Fig. 2-22): On the "way back into the wind", the cup has to be pushed against the drag resulting from the relative velocity $w = v + u$, causing additional losses.

---

[2] For the horizontal axis machines which are the main topic of this book, the tip speed ratio is defined as ratio of the circumferential velocity at the blade tip to the undisturbed upstream wind velocity.

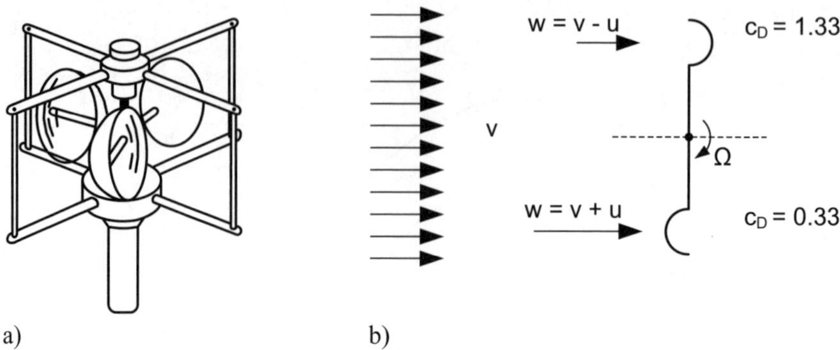

Fig. 2-22 a) Cup anemometer, b) simplified model

The aerodynamic efficiency of this "wind turbine" is roughly calculated with a similar simplified model (for the drag coefficients of the hemispheres see Fig. 2-22). From the difference of the driving (dr) drag force

$$D_{dr} = c_{D.dr}\,\frac{\rho}{2}\,a\,w^2 = 1.33\,\frac{\rho}{2}\,a\,(v-u)^2 \qquad (2.9)$$

and the slowing (sl) drag force

$$D_{sl} = 0.33\,\frac{\rho}{2}\,a\,(v+u)^2 \qquad (2.10)$$

results the mechanical power

$$P = (D_{dr} - D_{sl})\,u = \frac{\rho}{2}\,av^3\left\{\lambda\left(1 - 3.32\,\lambda + \lambda^2\right)\right\}. \qquad (2.11)$$

Again, the expression in the braces corresponds to the power coefficient $c_P\,(\lambda)$. Its maximum of $c_{P.max} \approx 0.08$ (at $\lambda_{opt} = 0.16$, Fig. 2-23) is even lower than the one of the Persian windmill (Fig. 2-21). Therefore, this type of "wind turbine" is not used for extracting power from the wind. But running idle, it is suitable for measuring the wind velocity (see chapter 4). The tip speed ratio $\lambda_{idle} \approx 0.34$ gives the "calibration factor" between rotational speed n and wind velocity v, since from $\lambda = \Omega\,R_M/\,v\,= 2\,\pi\,R_M\,n/\,v$ results

$$v = \Omega\left(\frac{R_M}{\lambda_{idle}}\right) = 2\pi\left(\frac{R_M}{\lambda_{idle}}\right)n\,. \qquad (2.12)$$

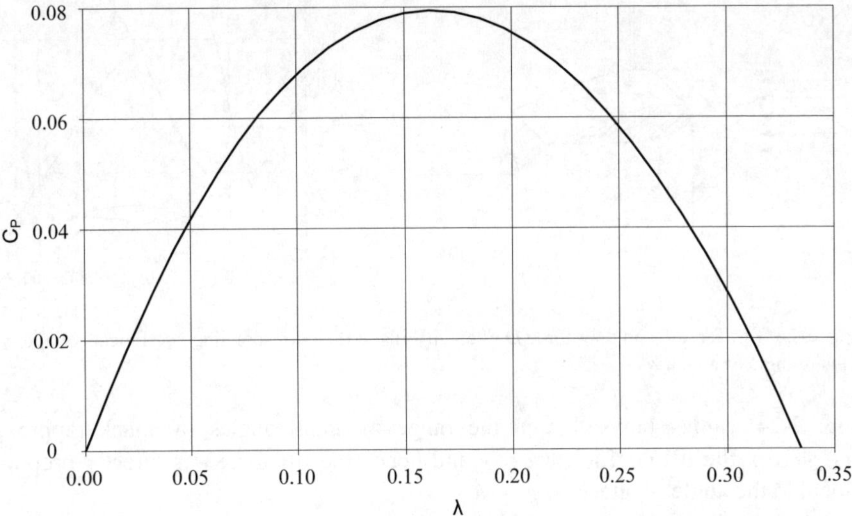

Fig. 2-23 Power coefficient versus tip speed ratio of a cup anemometer (approximation for the simplified model)

This roughly calculated value of $\lambda_{\text{idle}} \approx 0.34$ corresponds quite well to the values found in measurements [16]. But for wind measurements, each anemometer should be calibrated correctly in a wind tunnel.

2.3.3 Lift driven rotors

For many bodies, like airfoils or also the flat inclined plate, the force resulting from the attacking flow has not only a drag component D parallel to the direction of the flow velocity w, but also a component L perpendicular to it (Fig. 2-24), the lift force

$$L = c_{\text{L}} \, \frac{\rho}{2} \, a \, w^2 .$$

(2.13)

Similar to the drag force, it is proportional to the projected area $a = c\,b$ and the dynamic pressure $(\rho / 2)\, w^2$. For small angles of attack α_{A} the lift force L acts at approx. a quarter of the cord length c behind the leading edge.

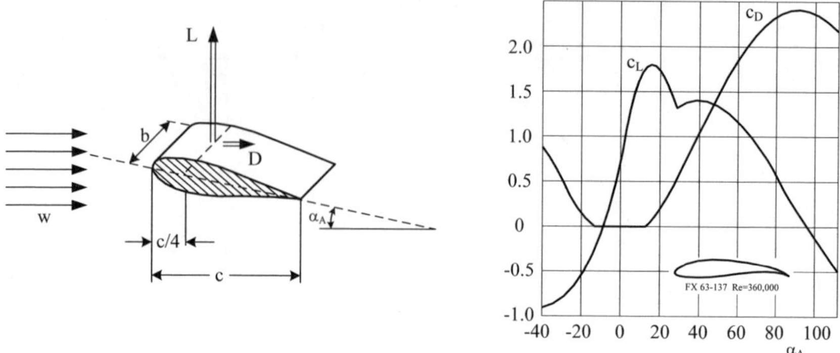

Fig. 2-24 Lift force L and drag force D of an airfoil and the corresponding coefficients c_L and c_D versus angle of attack α_A

Fig. 2-24, right, shows that in the range of small angles of attack (approx. $\alpha_A \leq 10°$), the lift coefficient c_A - and hence the lift force - is directly proportional to the angle of attack:

$$L = c_L \left(\alpha_A\right) \frac{\rho}{2} \, a \, w^2 \, , \qquad (2.14)$$

with $c_A \left(\alpha_A\right) \approx c_A{}' \, \alpha_A$ for $\alpha_A \leq 0.1745 \, (\approx 10°)$.

In the case of an ideal thin rectangular plate of infinite width b, $c_A{}'$ is 2π. Real values of actual profiles are slightly lower, approx. $c_A{}' \approx 5.5$.

Of course, there is also a drag force D existing, but it is very small for good quality aerodynamic profiles in the range of a small angle of attack:

$c_D = (0.01 \dots 0.20) \, c_A$. Only beyond approx. $\alpha_A = 15°$ the drag coefficient begins to grow drastically, Fig. 2-24, right.

The lift L_V force is the driving force of the wind turbines based on the lift principle. In order to distinguish them clearly from drag driven rotors discussed in the previous section 2.3.2, the basic principle is explained using the example of the Darrieus rotor (Fig. 2-25; cf. Fig 2-4b). Though being a turbine working with the lift principle, the Darrieus rotor has a vertical axis, which is typical for the drag driven rotors; and it has a tip speed ratio (i.e. the ratio of the circumferential blade velocity to the wind velocity) significantly higher than the discussed drag driven rotors reaching at maximum $\lambda_{max} = 1$.

At the Darrieus rotor, due to its high tip speed ratio of 4 or 5, the two blades considered in Fig. 2-25 are attacked by the relative velocity nearly tangentially at their leading edge. The lift force is very much higher than the drag force, and therefore also the relevant force (L and L') driving the rotor. By definition, the lift force is perpendicular to the relative velocity and causes with the levers (h and h') the necessary driving torque.

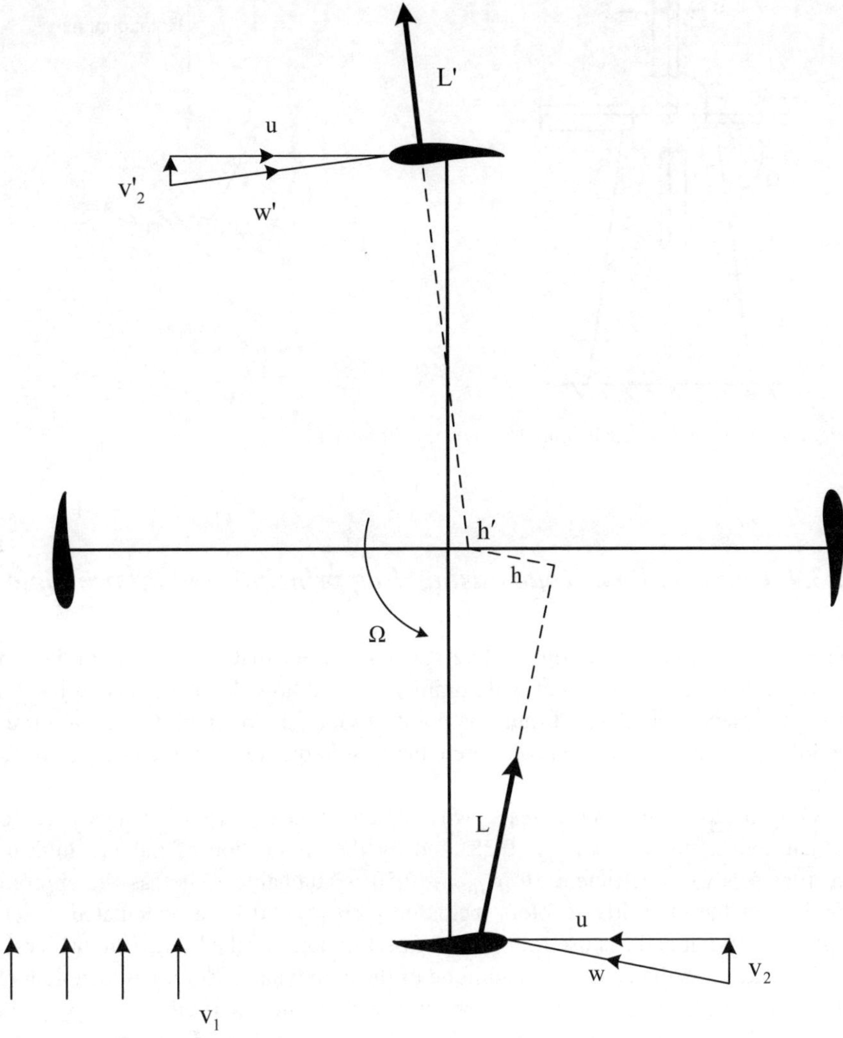

Fig. 2-25 Lift forces L and L' at the Darrieus rotor producing the driving torque

All horizontal axis wind turbines, like the post windmill, the Dutch smock mill or the Mediterranean sail windmill as well, are driven by the lift principle (Fig. 2-26). Their power coefficients are in the range of $c_{P.max} \approx 0.25$ and therefore significantly higher than the maximum values of the drag driven rotors. Modern horizontal axis wind turbines with good aerodynamic profiles (which show small drag coefficients) reach power coefficients up to $c_{P.max} = 0.5$. So they are already very close to the limit value of $c_{P.Betz} = 16 / 27 = 0.59$ found by Betz and Lanchester.

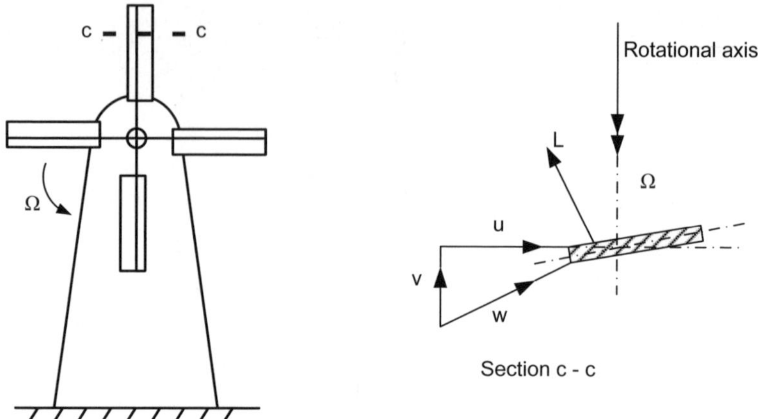

Fig. 2-26 Horizontal axis windmill driven by the lift force L

2.3.4 *Comparison of rotors using drag principle and lift principle*

The theoretical consideration of Betz and Lanchester that the maximum 59% of wind power is extractable leaves it completely open how the power is extracted in the rotor plane (Fig. 2-18). From this point of view, the oriental drag driven rotor and the occidental lift driven rotor are equally suitable for the wind energy utilisation.

Only a closer analysis reveals, why already Smeaton in 1759 measured for Dutch smock mills a $c_{P.max} = 0.28$, and by the application of modern high-lift profiles a power coefficient of $c_{P.max} = 0.50$ is attainable. Whereas the approximation for the drag driven rotors yields only $c_{P.max} = 0.16$, as calculated in section 2.3.2. What is the cause for the better performance of the lift driven turbines?

The reason is the different magnitude of the aerodynamic forces attainable with the same blade area a. The maximum aerodynamic coefficients $c_{W.max}$ and $c_{A.max}$ are in the same order (Fig. 2-24 and 2-27), but there is a fundamental difference in the magnitudes of the attacking relative velocity w. For the drag driven rotor, the relative velocity $c = v - u = v \cdot (1-\lambda)$ is always lower than the wind velocity since it is reduced by the circumferential velocity u.

The lift driven rotor has a relative velocity w that results from the geometrical addition of wind velocity v and circumferential velocity u which means it is always higher than the wind velocity: $w = \left(v^2 + u^2\right)^{1/2} = v^2(1+\lambda)^{1/2}$. Depending on the tip speed ratio it achieves the multiple of the wind velocity.

Drag devices	Lift devices
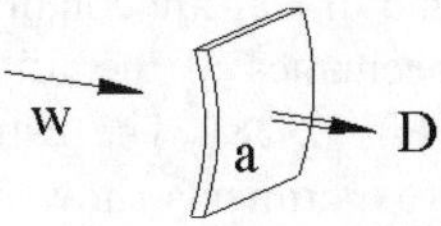	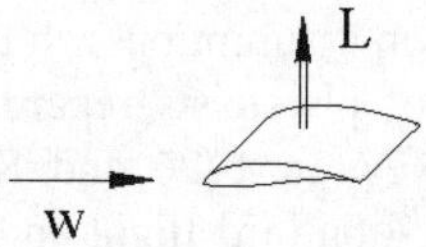
$D = \frac{\rho}{2} w^2 a c_D$	$L = \frac{\rho}{2} w^2 a c_L$
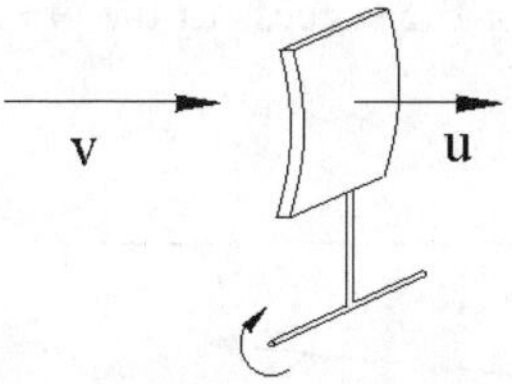	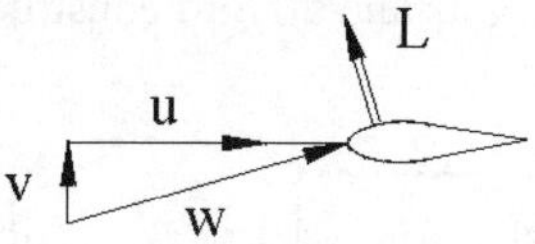
$w = v - u = v(1 - \lambda)$	$w = \sqrt{v^2 + u^2} = v\sqrt{1 + \lambda^2}$
$\lambda < 1$	λ = 1 to 15

arched plate (10%)	NACA 4415
$c_{D\,max} \approx 1.2$ $c_{L\,max} \approx 1.2$	$c_{D\,max} \approx 1.2$ $c_{L\,max} \approx 1.4$
tip speed ratio λ =	blade tip speed u / windspeed v

Fig. 2-27 Comparison of rotors driven by drag and lift

Therefore, with the same area *a*, the magnitude of the aerodynamic force - being proportional to the square of the attacking relative velocity - of the lift force driven rotor is a multiple of the force attainable by the drag force driven rotors. The aerodynamic forces obtained by the drag principle in the "active rotor plane" (see Fig. 2-18) are too small to come even a little close to the optimum extractable power of 59%. The fact that also the lift driven rotors do not completely attain the ideal power coefficient is due to some losses under real flow conditions, neglected in the considerations of Betz and Lanchester (see chapter 5).

It is worthwhile noting that the lift principle which is the basic principle of all wind turbines with a horizontal axis of rotation – from the post windmill to the Western mill - was used intelligently and efficiently for more than 700 years without being explained by a technical and physical theory. Still in 1889, Otto Lilienthal correctly notes: "Technical handbooks give for this kind of aerodynamic drag force (the technical term Lilienthal uses for lift and aerodynamic forces in general - the author) such formulae which are mostly the result of theoretical considerations and are based on assumptions which cannot be found in real-life conditions."

The ideas of physicists regarding the fluid mechanics of the lift force were wrong (e.g. Newton 1726 and Rayleigh 1876, Fig. 2-28). The estimations of Lilienthal based on bird flight and his subsequent experiments showed how powerful the lift forces of plane and cambered plates with a small angle of attack really were.

Only in 1907, long after the glides of Lilienthal and four years after the first successful motorized flights of the Wright brothers, Joukowski using the potential theory, found a sufficient theoretical explanation for the success of the practicians, e.g. the millwrights and aircraft constructors.

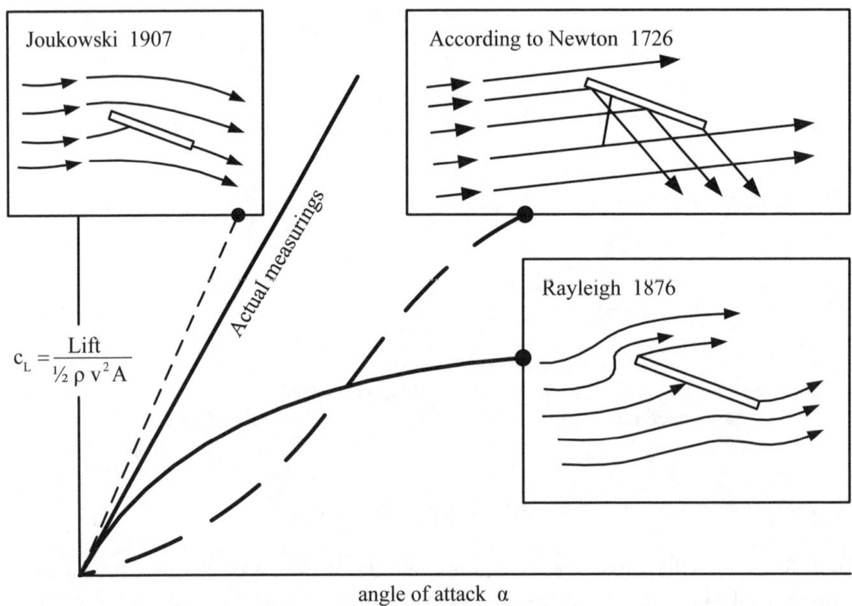

Fig. 2-28 Historical theories on the aerodynamic lift force, [17]

References

[1]   Golding, E.W.: *The Generation of Electricity by Windpower*, Auflage 1955; Reprint with additional material, E.&F.N. Spon Ltd., London 1976

[2]   Rieseberg, H.J.: *Mühlen in Berlin (Mills in Berlin)*, Medusa Verlagsges., Berlin-Wien 1983

[3]   Bennert, W. und Werner, U.-J.: *Windenergie (Wind energy)*, VEB Verlag Technik, Berlin 1989

[4]   König, F.v.: *Windenergie in praktischer Nutzung (Practical application of wind energy)*, Udo Pfriemer Verlag, München 1976

[5]   Le Gouriérès, D.: *Wind Power Plants, Theory and Design*, Pergamon Press GmbH, Frankfurt 1982

[6]   Dornier: Firmenprospekt (*company brochure*)

[7]   Mager, J.: *Mühlenflügel und Wasserrad (Mill wings and water wheels)*, VEB-Fachbuchverlag; Leipzig 1986

[8]   natur: *Im Windschatten der Anderen (In the wind shade of the others)*, Heft 1/85

[9]   Herzberg, H. und Rieseberg, H.J.: *Mühlen und Müller in Berlin (Mills and millers of Berlin)*, Werner-Verlag, Düsseldorf 1987

[10]   Reynolds, J.: *Windmills and Watermills*, Hugh Evelyn, London 1974

[11]   Prospekt des Internationalen Wind- und Wassermühlenmuseums Gifhorn (*Brochure of the international windmill and watermill museum in Gifhorn, Germany*)

[12]   Varchmin, J. und Radkau, J.: *Kraft, Energie und Arbeit (Power, energy and Work)*, Rowohlt Taschenbuch Verlag, Reinbek 1981

[13]   Braudel, F.: *Sozialgeschichte des 15. bis 18. Jahrhunderts; Band 1: Der Alltag (Social history from the 15th to the 18th century, vol. 1: everyday life)*, Deutsche Ausgabe, Kindler-Verlag, München 1985 und Büchergilde Gutenberg, Ffm

[14]   Betz, A.: *Windenergie und ihre Ausnutzung durch Windmühen (Wind energy and its application by windmills)*, Vandenhoekk and Rupprecht, Göttingen 1926

[15]   Glauert, H.: *Windmills and Fans*. In: Durand,W.F. "Aerodynamic Therory 4" (1935)

[16]   Schrenck: *Über die Trägheitsfehler des Schalenkreuzanemometers bei schwankender Windstärke (On the errors of a cup anemometer due to inertia at fluctuating wind speed)*, Zeitschrift technische Physik Nr.10 (1929), Seite 57-66

[17]   Allen, J.E.: *Aerodynamik - eine allgemeine moderne Darstellung (Aerodynamics – a general, modern description)*, H. Reich Verlag, München 1970

[18]   Betz, A.: *Das Maximum der theoretisch möglichen Ausnutzung des Windes durch Windmotoren (The maximum of the theoretically possible exploitation of the wind by wind motors)*, Zeitschrift f. d. gesamte Turbinenwesen, V.17, Sept. 1920

[19]   Lanchester, F.W.: *A Contribution to the theory of propulsion and the screw propeller*, Trans. Inst. Naval Arch., Vol. LVII, 1915

[20]   Hills, R.L.: *Power from the Wind – A history of windmill technology*, Cambridge University Press, 1996

[21]   Petersen, F., Thorndahl, J. et al.: *Som vinden blaeser*, ISBN 87-89292-14-6, Elmuseet, 1993

[22]   Thorndahl, J.: *Danske elproducerende vindmoeller 1892 – 1962*, ISBN 87-89292-36-7 Elmuseet, 1996

[23]   Hau, E.: *Windkraftanlagen (Wind power plants)*, 2. Auflage, Kap. 1 und 2, Springer Verlag Berlin, 1996

[24]   Zelck, G.: *Windenergienutzung (Application of wind energy)*, 1985

3 Wind turbines - design and components

Wind turbines are energy converters. Independent of their application, type or detailed design all wind turbines have in common that they convert the kinetic energy of the flowing air mass into mechanical energy of rotation. As already discussed in chapter 2, two *aerodynamic principles* are suitable for this purpose, the lift and the drag, see Fig. 3-1. Drag driven rotors reach only, as mentioned, moderate power coefficients and are of no major importance to the technical applications, apart from the anemometers.

The main distinguishing feature of the group of the lift driven rotors is the *orientation of the rotor shaft.* Wind turbines with a vertical axis of rotation have the advantage that they operate independently of the wind direction, so they do not need a yaw system for orientation into the wind. But larger wind turbines of this type did not establish due to important disadvantages like its nervous dynamics and the weak wind close to the ground. The vertical axis turbines are presented in chapter 2 for historical reasons, in the following they will not be discussed further. Readers interested in Darrieus rotors can find out more about this topic in chapter 13 of the 3rd German edition of this book. The following sections concentrate on wind turbines with a horizontal axis of rotation. Fig. 3-1 shows a typology of the main features of wind turbines. Their *type and design* are strongly influenced by the specific application:

- Direct mechanical operation: Driving millstones, saws, hammers or presses
- Conversion into hydraulic energy: Water pumping
- Conversion into thermal energy: Heating and cooling
- Conversion into electrical energy: feeding into an electrical grid, operating independent of a grid in combination with a battery storage system or forming an independent hybrid system grid, e.g. in combination with a back-up diesel engine or photovoltaics.

Since the most important application of modern wind turbines is the *generation of electrical energy*, this chapter is devoted to this particular topic.

One of the first commercial wind turbines to feed electricity into the grid was the Vestas V-15 with a rated power of 55 kW, Fig. 3-2. Already in the beginning of the 1980s, it was manufactured and installed in large numbers. It already had all essential *components of grid-connected wind turbines*:

- Rotor: rotor blades, aerodynamic brake and hub,
- Drive train: rotor shaft, bearings, brake, gearbox and generator,
- Yaw system between nacelle and tower: yaw bearing and yaw drive,
- Supporting structure: tower and foundation and
- Electrical components for control and grid connection.

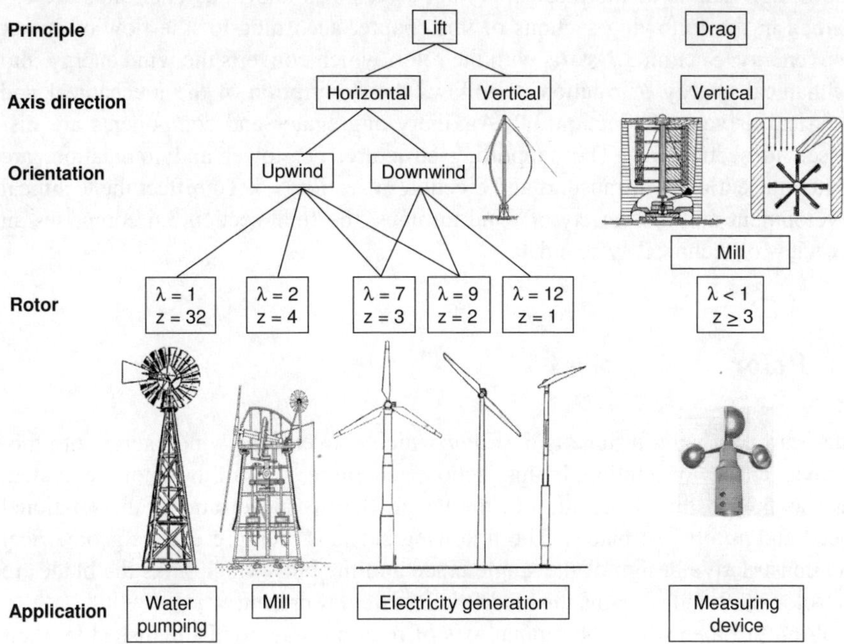

Fig. 3-1 Typology of wind turbines and typical applications

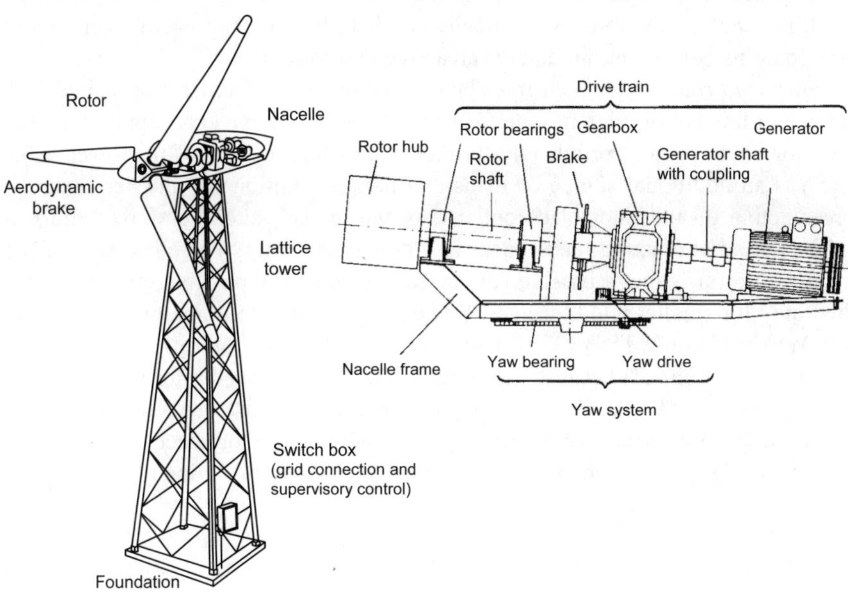

Fig. 3-2 VESTAS V15, general view and nacelle section [1]

The design details of the most important assemblies and functional units are explained in the following sections of this chapter according to the flow of power resp. energy. Section 3.1 starts with the rotor, which converts the wind energy into mechanical energy of rotation. It follows the description of the mechanical and electrical drive train, section 3.2. Auxiliary aggregates and components are discussed in section 3.3. The supporting structure, i.e. tower and foundation, are treated in section 3.4, transport and erection in section 3.5. To reflect the technical development and the variety of wind turbines, the final section 3.6 comprises an overview of technical system data.

3.1 Rotor

The heart of a wind turbine is the *rotor* which converts the wind energy into mechanical energy of rotation. In this section, general features of the rotor are treated, such as the position in relation to the tower, the tip speed ratio or the rotational speed and number of blades. The following sections then present the geometrical and constructive design of the single blade and the hub, which links the blades to the rotor shaft and contains the blade pitch system if present.

Wind turbines with a horizontal axis of rotation may be distinguished by their rotor position in relation to the tower. Today's market is dominated by *upwind rotors:* relative to the wind their rotor is located in front of the tower (windward). The flow reaching the rotor is practically not disturbed by the tower (except tower dam), but this configuration requires an active yaw system.

Downwind rotors have their rotor behind the tower (leeward). The main disadvantage of this configuration, in terms of loads, permission and acceptance, is that the rotor blade passes periodically through the disturbed flow of the tower wake which is an additional source of loads and noise. Additionally, the aerodynamic forces acting on the rotor blade collapse in the tower wake, therefore additional and increased alternating loads occur at a downwind rotor. The advantage of the downwind rotor is that a passive yaw system is generally possible, but this is applied only for smaller wind turbines. Some early big wind turbine prototypes, e.g. GROWIAN [2] or WTS-3 [3, 4], had a downwind rotor (but with an active yaw system, cf. section 3.3) because they had elastic and/or hinged rotor blades, which that way could not hit the tower or the guy wires when a gust was passing.

The *rotational speed n of the rotor* (in the following rotor speed) is one of the principal design parameters of a wind turbine. The power of a wind turbine

$$P = M\,\Omega = M\,2\,\pi\,n \qquad (3.1)$$

is the product of rotor torque M and the angular rotational speed $\Omega = 2\,\pi\,n$. The rotor speed and the wind speed are linked by the tip speed ratio

$$\lambda = 2\,\pi\,n\,R/\,v_1 = \Omega\,R/\,v_1 \quad , \tag{3.2}$$

which is the ratio of circumferential speed at the blade tip to the wind speed v_1 far upwind the rotor. The tip speed ratio is the most significant parameter in the aerodynamic design of the rotor blades (cf. chapter 5). The wind turbine which has a low design tip speed ratio (Design tip speed ratio $\lambda_D \approx 1$, e.g. Western mill with piston pump) provides a high torque while running at a low rotor speed. By contrast, a grid-connected wind turbine, designed with a tip speed ratio in the range of $\lambda_D = 5$ to 8, provides at the same power P a small torque and a high rotor speed required for the generators (Fig. 3-1 and 3-3).

Wind turbines which operate at a *constant rotor speed* have a decreasing tip speed ratio with increasing wind speed, due to $\lambda \sim n/\,v_1$. Therefore, they reach the optimum value λ_D of the aerodynamic design only at one specific wind speed. *Variable-speed* wind turbines, on the contrary, operate within a large wind speed range at their design tip speed ratio λ_D if the driven load is varied appropriately. Therefore, the speed-variable operation is advantageous for the rotor efficiency, but a significant effort is required on the part of the electrical converters (AC-DC-AC) to keep the frequency constant (50 resp. 60 Hz) for feeding into a grid.

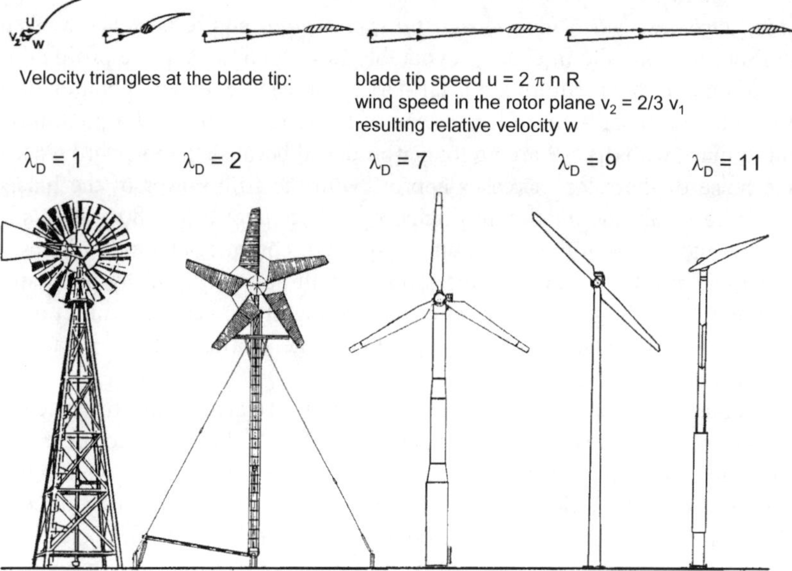

Fig. 3-3 Types of wind turbines, corresponding velocity triangles and tip-speed ratios

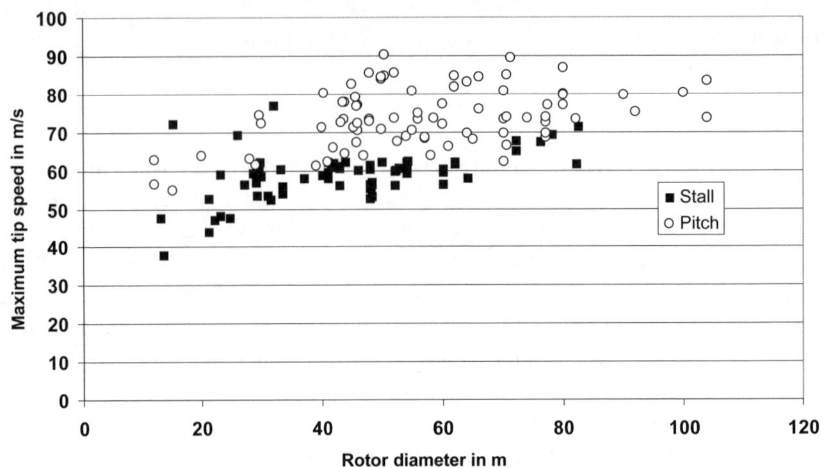

Fig. 3-4 Maximum tip speed for 3-bladed wind turbines with stall or pitch control

Wind turbines with a low design tip speed ratio provide a high start-up torque and require many blades for a high solidity of the swept rotor area (Fig. 3-3 and 5-15). Moreover, they have a high rotor thrust on the tower at low rotor speed (cf. section 6.4), so it is recommended that the rotor be turned out of the wind during a longer period of stand still, cf. chapter 12.

Wind turbines with a high design tip speed ratio require only a few, slender blades, but to attain the intended, favourable flow conditions at the profile during start-up some of them require a special start-up procedure: either operation of generator as motor at stall-controlled turbine or pitching the blade of a pitch turbine. Wind turbines with $\lambda_D > 9$ are no longer produced because the interfering aerodynamic noise of the rotor increases approx. with the fifth power of the blade tip speed. As a result, the maximum tip speed shall be kept below 80…90 m/s. Fig. 3-4 shows the values of the maximum tip speed of commercial wind turbines. The data is divided into two groups, the turbines with power control by pitch and by stall (see below). In general, the latter have a slightly lower maximum tip speed than the former because of their increased noise generation in the separated flow by the stall effect.

The *number of blades* of the rotor is indirectly related to the tip speed ratio (Fig. 3-1 and 3-3). Western mills, which need a high solidity due to their low tip speed ratio, often have 20 to 30 simple metal sheet rotor blades. Rotors with a high tip speed ratio for electricity generating wind turbines have mostly three rotor blades designed with high-quality aerodynamic profiles to reach a good efficiency (cf. chapter 5). A small number of blades is preferable due to the big investment cost share of the rotor, which is approx. 20 to 25% of the total wind turbine investment costs. Therefore, the first generation of Megawatt wind turbines were designed mostly with a two-bladed rotor, see Fig. 2-15, c, d and Fig 2-16, GROWIAN, cf. [2...5].

The distribution of mass and aerodynamic forces in the swept rotor area is more even at three-bladed rotors (cf. Fig. 3-3) with the result that their dynamic behaviour is calmer, thereby reducing the loads on all components. At the *two-bladed rotor*, the natural tilting frequency of the tower-nacelle system depends on the angular position of the blades. If the blades are in a vertical position, the natural frequency is smaller than when the blades are in the horizontal position. This causes very nervous dynamics (parameter excited system, cf. chapter 8).

The *three-bladed rotor* is the one with the smallest number of blades to be considered as dynamically disk-like and hence has a calm rotation. As well in terms of the visual impression the three-bladed rotor rotates in a calm way, in contrast to the nervous two-bladed rotor.

There is no physical reason for selecting one particular *sense of rotation*. Nowadays, it is common that the rotors turn clockwise when standing in front of it with the wind at one's back. If all the wind turbines in a wind farm have the same sense of rotation, then a more pleasant visual impression is achieved.

Apart from the main function of energy conversion, *safe operation* of the rotor has to be assured. Due to the increase in wind power with the cube of the wind speed, the power extracted from the wind has to be limited by the control. Furthermore, the total system has to be designed in such way that neither inadmissible loads and vibrations nor over-speeding occur, see chapters 8 and 12.

For the *power limitation*, two different aerodynamic concepts are applied. The more simple concept, already used in Denmark in the beginning of the 1980s, limits the power by the *stall effect, i.e. flow separation* at the rotor blade, see Fig. 3-5 (cf. chapter 12). The rotor turns at nearly constant rotor speed and circumferential speed u, because the wind turbine is rigidly linked to the grid frequency by the used asynchronous generator. At high wind speed v the angle of attack α_A between profile chord and relative velocity w becomes so large that the flow is no longer able to follow the profile contour and separates on the suction side. This principle of power limitation exhibits a somewhat stochastic behaviour, so the point of occurrence cannot be determined exactly. The stall effect may appear delayed, and there is a hysteresis when it disappears. Short-term gusts may cause torque peaks in the drive train because at constant rotor speed the power fluctuations result in torque fluctuations, cf. equation (3.1). In order to reduce this effect, some wind turbines of the MW class provoke the stall effect actively by pitching the blade (active stall control). The leading edge of the blade (or profile nose) is turned out of the wind, which is called '*pitch to stall*'.

The second concept of power limitation is also based on pitching the rotor blade around its axis, but here the leading edge of the blade is turned into the wind, which is called '*pitch to feather*', Fig. 3-6.

By pitching the rotor blade, the angle of attack α_A between profile chord and relative velocity is reduced in such way that the lift and the resulting power are limited. This power control is smoother than the active stall control because here the flow remains attached to the profile, however larger pitch angles are required.. Fig. 3-7 shows the comparison of typical power curves for these two concepts of power limitation. Both methods of pitching the blade – to stall or to feather - are extensively discussed in chapters 6 and 12.

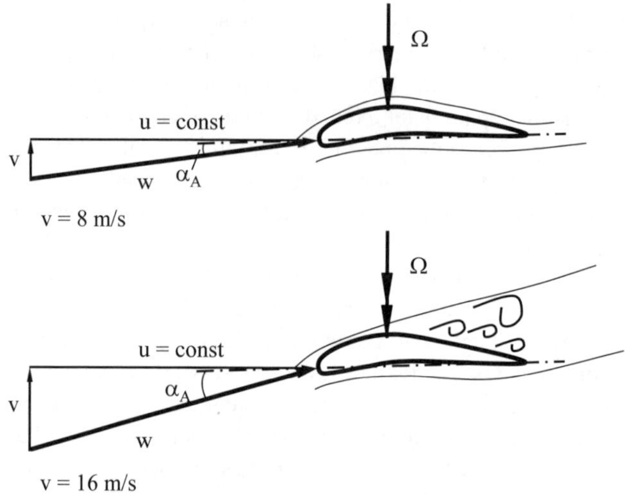

Fig. 3-5 Schematic view of the stall effect (limiting power at wind speed higher than rated wind speed)

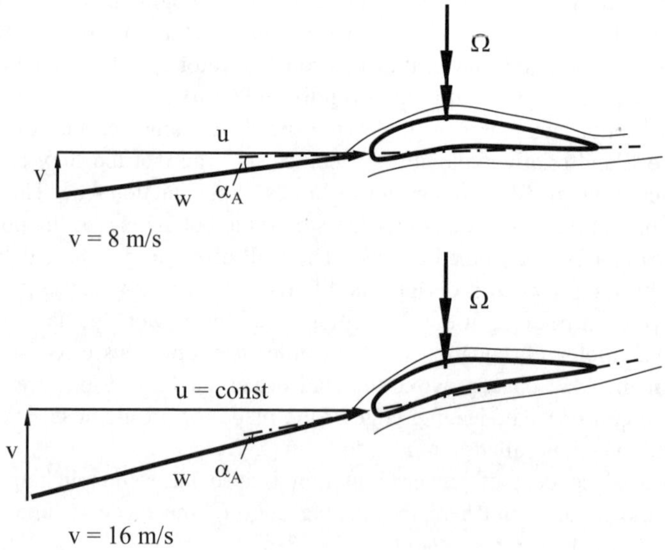

Fig. 3-6 Schematic view of "Pitching to feather" (limiting power at wind speed higher than rated wind speed)

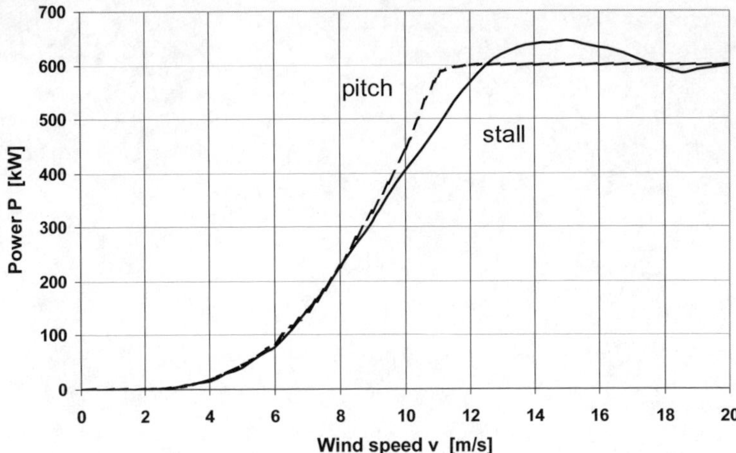

Fig. 3-7 Comparison of stall and pitch control: Power curves of two 600 kW wind turbines, data from type approval test reports; e.g. in [24]

3.1.1 Rotor blade

The design of the single rotor blade is determined by the selected aerodynamic profiles, the inner and outer geometry and as well as the chosen materials.

As it will be discussed in detail in chapter 5, the required quality of the aerodynamic profile depends on the chosen design tip speed ratio. Western mills (Fig. 3-3, left) only require the profile of a cambered plate, whereas wind turbines for electricity generation require high-quality profiles with a high ratio of lift and drag coefficient (i.e. high lift/drag ratio).

The aerodynamic profiles of the NACA44XX and NACA63XXX series are very common, but today there are also aerodynamic profiles developed specifically for wind turbine rotors, see Fig. 3-8.

Above all, a high lift/drag ratio is important in the area of the blade tip. The circumferential speed in the inner blade part is smaller, which leads to a smaller local relative velocity and local speed ratio. Therefore, the profile chord length has to be larger which allows thicker profiles to be applied in this rotor section, (cf. chapter 5). This is helpful for the reduction of the large material's stress, as the highest loads are found at the blade root. Hence, different aerodynamic profiles are used in the inner and outer part of the rotor blade. Fig. 3-8 shows a typical distribution of different profile geometries along the blade radius.

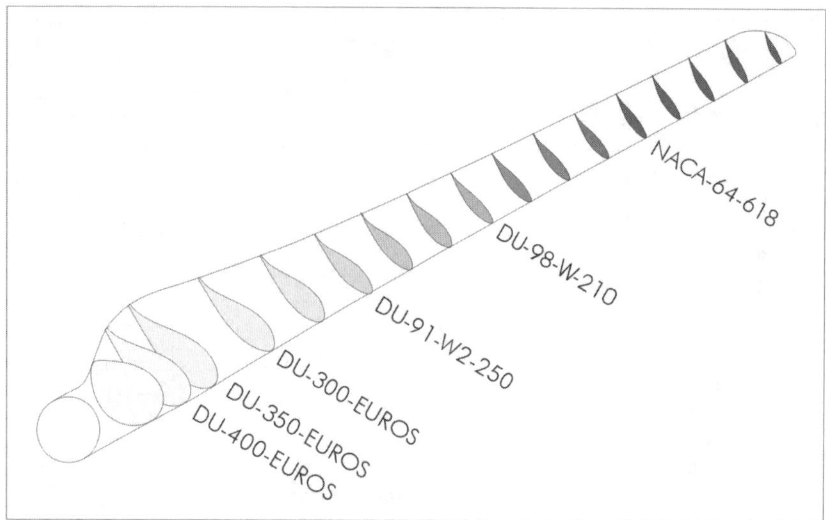

Fig. 3-8 Different blade profiles along the radius of a blade, (EUROS)

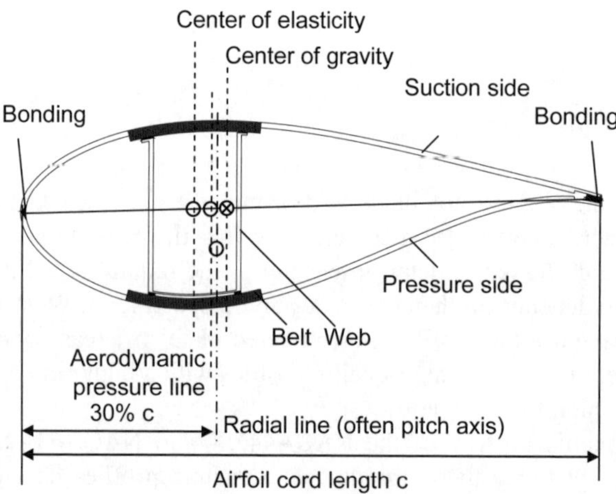

Fig. 3-9 Blade section - characteristic lines of the blade dynamics

In the design calculations of rotor blades (statics, vibrations, flutter etc.) the position of four lines, which pass radially along the blade, Fig. 3-9, is of great importance:

- The *radial line* is the axis of rotation of the blade pitch mechanism, resp. the line in the blade flange centre perpendicular to the rotor shaft axis
- The *elastic line* (centre of elasticity) is the position of the shear centre line in the supporting structure (approx. centre of gravity of the main

spar). The elastic deformations of the flap- and edge-wise movements are counted from here as well as the twist of the blade section caused by torsional moments

- The *line of the centres of gravity* represents the points of action of forces resulting from inertia and weight
- The *pressure line* consists of the points of attack of the lift and drag forces. When the flow is attached, it is approx. 30% of the chord length. If e.g. the stall effect occurs when exceeding the rated wind speed the pressure line will move which may cause the rotor blade to vibrate (stall flutter). These vibrations can be reduced by vibration dampers (containing liquids) in the blade tip. In order to provoke the stall effect along a defined line on the blade surface and prevent it from oscillating, some stall-regulated wind turbines are equipped with vortex generators (stall stripes) on their blade surface.

The necessary quality of the aerodynamic profile generates requirements in relation to the production process and the applied *materials of the rotor blade*. The simple profile of a Western mill, Fig. 3-3 left, is manufactured from cambered steel plates.

The rotor blades for electricity generating wind turbines, designed with a high tip speed ratio, Fig. 3-3 middle and right, have to meet higher demands. Their profiles are mostly laminated with glass fibre reinforced plastics (GFRP), and, most recently, with carbon fibre reinforced plastics (CFRP). The latter are more expensive, but their admissible material strength is up to three times higher than that of GFRP, Fig. 3-10. Their fatigue strength also tends to be higher which is ideal for lightweight design.

The separate moulds for the suction and pressure side of the blade (Fig. 3-11) are covered with the woven fibre fabrics (rovings), which are then soaked with polyester or epoxy resin. Today, this is mostly done automatically in a vacuum process in order to reduce the adverse health effects for the workers, prevent air bubbles which reduce the material strength and achieve a more defined material usage. After evacuating the mould sealed with a plastic film, the resin is pumped into the mould at defined points. Some rovings types are delivered already soaked with resin (so-called "pre-pregs"). Moreover, some manufacturers use so-called sandwich-constructions where balsa wood is located in between the inner and outer rovings. In a defined heating cycle the resin is hardened, and finally the two halves of the blade are bonded together. The spar of the GFRP blade, which provides defined geometry and material strength, is filled with foam and/or additionally stiffened by GFRP ribs, webs and belts (Fig. 3-9). The final coating has to be weather-proof and UV resistant.

Erosion protection film is attached to the leading edge to reduce the abrasive material removal under operation. Flow controlling elements are applied on the blades, e.g. vortex generators, to assure defined flow conditions and flow direction despite the wind fluctuating by time and once per revolution due to the wind profile.

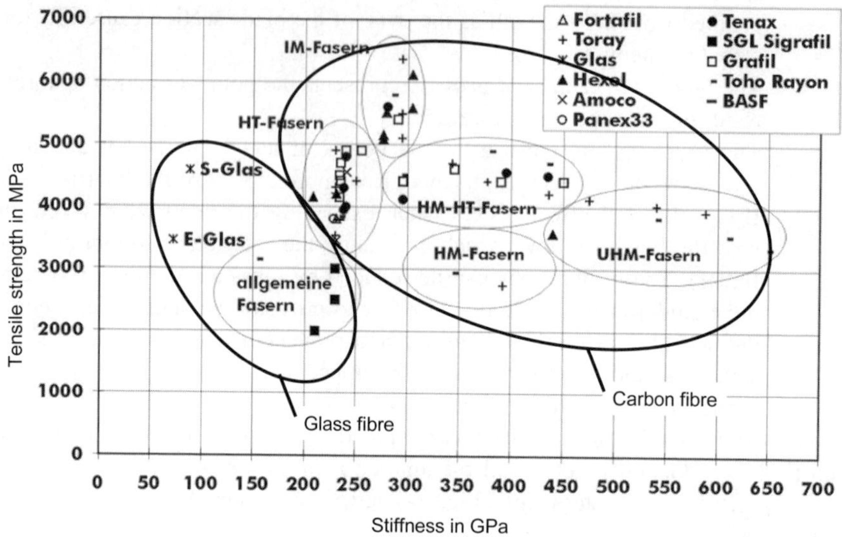

Fig. 3-10 Material strength values of glass and carbon fibres, (EUROS)

Fig. 3-11 Blade production, separate moulds for suction side and pressure side, mould in front: soaking of the laminate with epoxy resin by a vacuum process (NOI)

Fig. 3-12 Connection of rotor blade at hub: left "IKEA-bolts" (Bonus), right: bolt sleeves (Vestas)

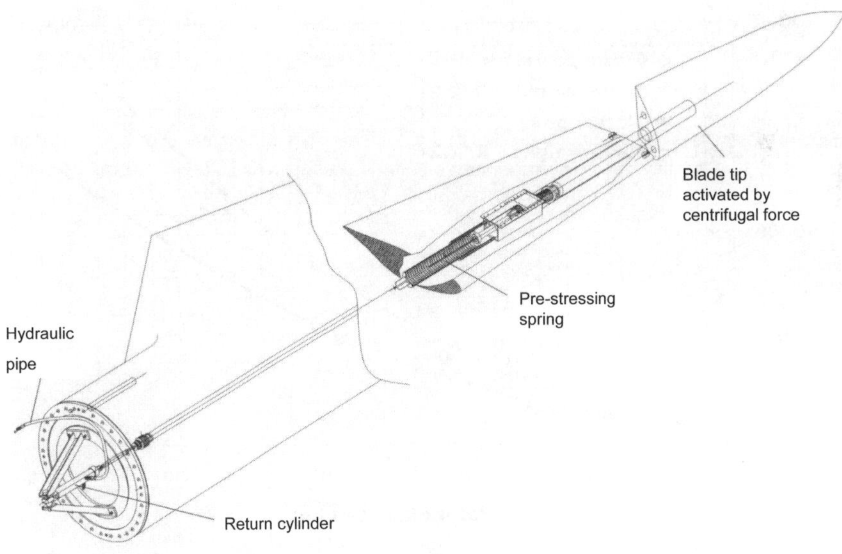

Fig. 3-13 Rotor blade with tip brake (turning the blade tip) (LM, [6])

The load transmission from the GFRP rotor blade to the metal flange of the hub is a tricky issue (Fig. 3-12). For this screw connection with the hub, there are used either sleeves for stud bolts laminated into the blade root, or the so-called "Ikea-connection" with a cross bolt

Another design detail which needs special attention is the *turnable blade tip* of stall-regulated rotors which serves as an aerodynamic brake. Activated by centrifugal forces, it deploys when the rotor is over-speeding (Fig. 3-2 and 3-13). For braking, a threaded guidance turns the blade tip by 90°, so it is nearly perpendicular to the relative velocity. Since the turnable blade tip is located at the maximum radius, the affected ring section, and as well the resulting braking force and torque, are very large.

The specific blade properties change with increasing size. Light weight design is an imperative, above all for the rotor blades of the Megawatt turbines. If the rotor blades were scaled up using the laws of similarity (cf. chapter 7), the weight of the rotor blades would increase with the cube of the blade radius (m ~ r³). The large weight would cause enormous bending forces and correspondingly high stress causing problems with the material strength. Fig. 3-14 shows the blade mass of commercial rotor blades versus the rotor diameter. The interpolation curves have an exponent of approx. 2.2 instead of 3.0, thanks to the lightweight design. Moreover, the diagram illustrates that when it comes to the rotor blade weight, it is advantageous to use epoxy instead of polyester resin.

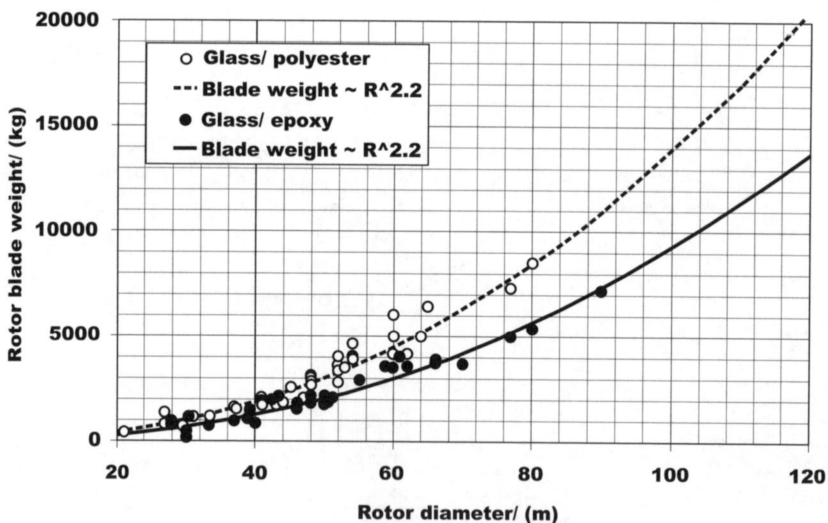

Fig. 3-14 Rotor blade weight depending on rotor diameter and material, (EUROS)

3.1.2 Hub

There are various possibilities for the design of the hub and the attachment of the blades. Most of them were tested in the 1980s at prototype turbines of different sizes. In the following, these variants are presented. Although the *rigid hub* is used nearly exclusively for commercial wind turbines (Figs. 3-15 and 3-23). Perhaps the other hub types will be considered again during future developments because they help to reduce stress and thus component weight.

The rotor blade connection to the hub can be made

- rigid or

- flexible by using a hinge (flapping), see Fig. 3-16. Another special hub type can be used at two-bladed rotors where the two blades are fixed together rigidly and have a common hinge in the hub, called the *teetering hinge*. All of these three types of blade-hub connection may be combined with a controlled pitching of the blade around its axis for power and rotor speed limitation. Theoretically, movement can also occur around the third axis at the blade-hub connection, the slewing axis. However, this is not used in practice. Instead, in order to compensate alternating loads (torque peaks) in this direction, additional components are included in the drive train (special couplings or the elastic support of the gearbox). Fig. 3-17 gives an overview of different hub types and the stress relief achieved in the blade root and rotor shaft.

Fig. 3-15 Nacelle with rigid hub of a three-bladed rotor (photo by company Zollern)

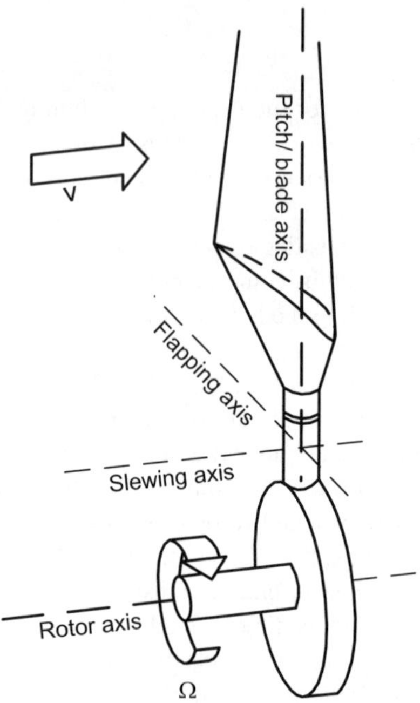

Fig. 3-16 Denomination of the áxes at the blade flange

The *flapping hinge rotor* is a characteristic of the SÜDWIND 1237 which has a downwind rotor (Fig. 3-18). The flapping hinge at each rotor blade relieves the blade root and the rotor shaft from all bending stress around the flapping axis. Such bending stress results from the 'wind pressure' (i.e. thrust) and the three-dimensional non-uniform wind velocity field (Fig. 3-19). At a hub with rigid blade connection the stochastic spatial fluctuations of the wind in the rotor swept area cause an eccentricity of the point of action of the resultant forces from the rotor axis leading to bending stress in the rotor shaft. This is also avoided by flapping hinges.

During operation, a balance is created between the centrifugal forces F_C and the thrust F_T at the rotor, Fig. 3-21. This causes a self-adjusting of the flapping angle and also its limitation, typically below 10°. Larger cone angles only occur if the centrifugal forces are too small due to low rotor speed, i.e. shortly before rotor standstill. Additional components in the rotor are therefore required - e.g. return springs (SÜDWIND), stoppers, gear or hydraulic components - to assure the start up of the wind turbine. These components are complicated and costly, hence the flapping hinge principle is seldom applied to larger rotors [7].

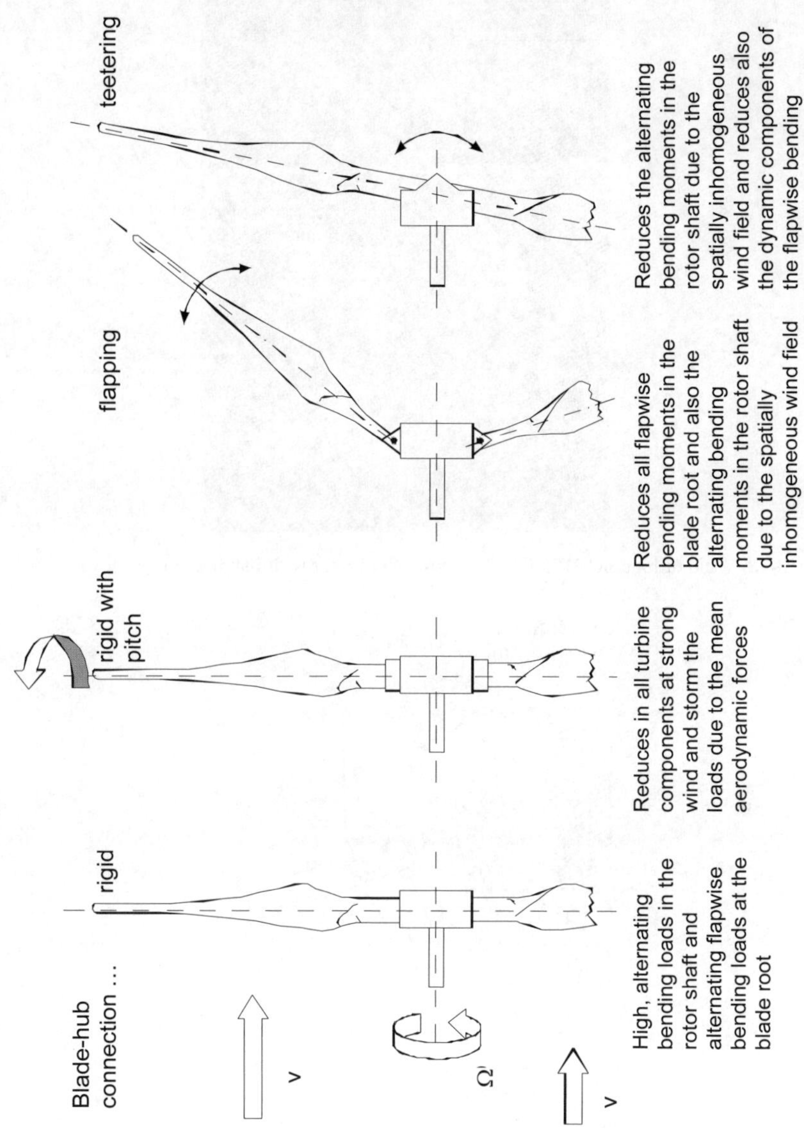

Fig. 3-17 Different hub design types

Fig. 3-18 Wind turbine SÜDWIND 1237, downwind rotor with flapping hinge

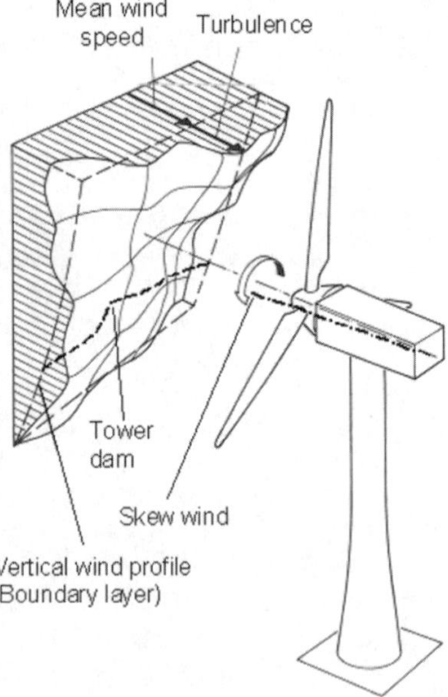

Fig. 3-19 Three-dimensional wind velocity field in front of the wind turbine, [6] modified

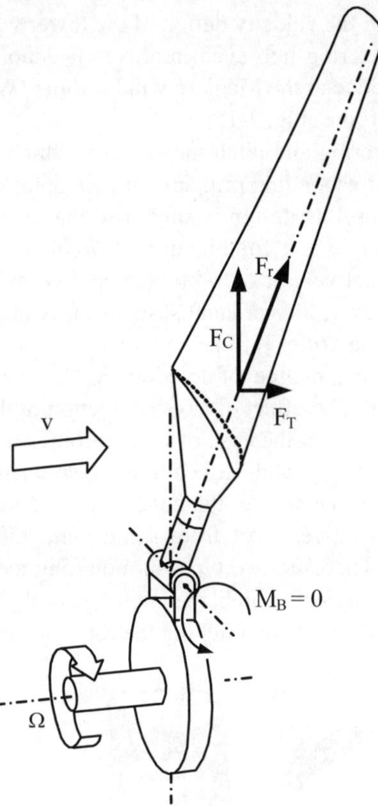

Fig. 3-20 Equilibrium of forces at the flapping hinge rotor

The flapping hinge reduces in the blade root the bending stress around the profile chord where the section modulus is small due to the slender profile sections. This way, the rotor blade weight may be reduced by up to 75%. At least for the design point, rotors with a rigid hub may also benefit from the effect of the partial compensation of thrust forces and the centrifugal forces, used at flapping hinge rotors. A fixed flapping angle may be introduced, which is then called cone angle.

A design which was developed specifically for to the two-bladed rotor is the *teetering hub* (Figs. 3-17 and 3-21). It reduces loads arising from spatial fluctuations of the wind. The rotor shaft is primarily relieved of the corresponding bending stress. At the rotor blade root, only the dynamic portion of the flapwise bending moment is reduced. The design principle of the teetering hub has been applied mainly for large downwind turbines, e.g. GROWIAN (Fig. 3-21) and WTS-3. Due to the large size of the turbine, the atmospheric boundary layer led to a significantly different wind speed at the top and bottom position of the blade thus changing flow conditions during the rotation. Due to the downwind position of the rotor, this asymmetry is even increased. This is primarily because, at its bottom position, the blade passes through the velocity deficit of the tower wake.

The effect of the teetering hub even improves if coupled with blade pitching. This was successfully done at the Maglarp wind turbine (WTS-3) and in the 1940s at the Smith-Putnam turbine (Fig. 2-15).

The coupling of teetering and pitch movement - which is called δ_3-coupling in accordance with the corresponding principle in helicopter design - is also used for hubs of one-bladed rotors. Its design is similar to that of the teetering hub. However, the physical principle is a combination of *flapping hub and pitch hub*. Fig. 3-22 shows that it is relatively easy to build such a complex hub by using a cardanic suspension. However, the cardanic suspension is imperative for this design, due to its dynamic particularities [8, 9].

Allowing an additional degree of freedom in the connection between rotor blade and hub has mainly the effect of stress reduction in the whole drive train, on the one hand. On the other, in the case of wind turbines in the MW range, the required construction is costly and susceptible to faults due to the heavy rotor weight and the big loads. Therefore, the concept of the *rigid hub* (Figs. 3-15 and 3-23) is preferred in practice, apart from some wind turbines for research and some other prototypes. There are two types of mounting the blades to a rigid hub:

- fixed blade angle, i.e. no blade pitching (stall concept) and
- variable blade angle by pitching the rotor blades (pitch concept).

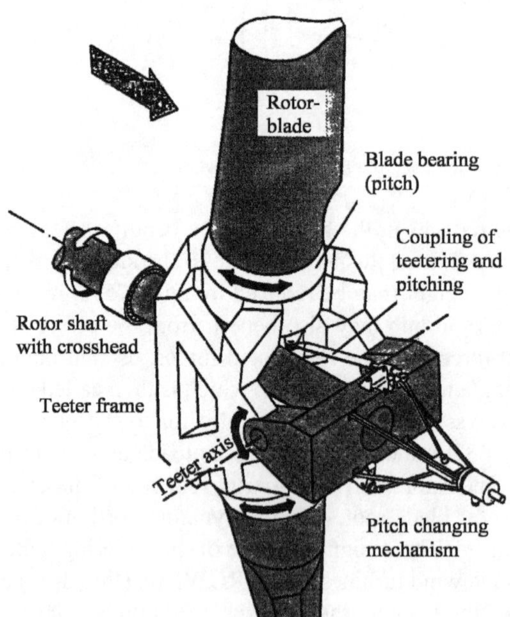

Fig. 3-21 Teetering hub of GROWIAN [26]

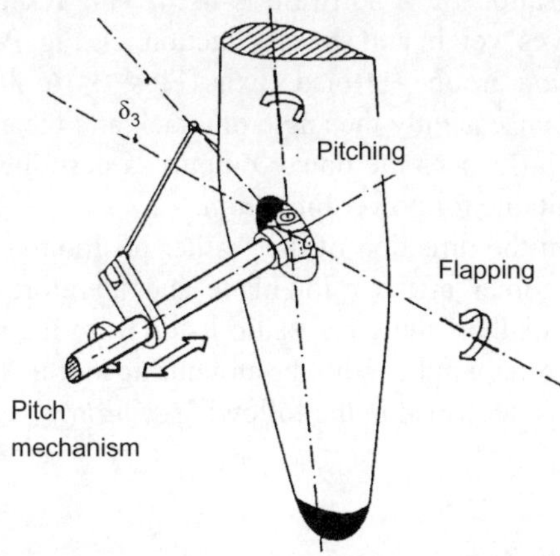

Fig. 3-22 Flapping pitch hub with cardanic suspension of FLAIR [8]

Fig. 3-23 350 kW wind turbine with rigid hub and integrated drive train (Suzlon)

The *weight of the hub* increases significantly with the growing size of the wind turbine. Due to their size, casting of the hub is limited to a few foundries. The material is nodular graphite cast iron (e.g. EN-GJS-400-18-LT, formerly known as GGG-40.3). A spherical hub has a comparatively small hub weight, but it is disadvantageous as far as stress-minimising design is concerned. Some hub designs include an extender to reach larger higher rotor diameters with the same hub and rotor blade.

For hub design optimisation FEM software is used. This results in topologically shaped hubs and achieves weight and stress reduction, cf. Fig. 9-9 in chapter 9.4.

By *pitching the blade* around its blade axis (Figs. 3-16, 3-17 and 3-22), the blade pitch angle and consequently the angle of attack and the aerodynamic forces are changed. Pitching influences the power output, as described in Figs. 3-6 and 3-7, and is therefore suitable for power limitation.

Pitching the blade in the direction of the feather position not only reduces the driving forces but all forces at the rotor blade and therefore also the resulting stress. By 'pitching to feather' the quasi static loads from the mean aerodynamic forces are reduced at strong wind conditions and during storm. The construction of the blade pitch system is described in the following section.

3.1.3 Blade pitch system

A significant effort is needed to build a rotor with a pitchable blade, regardless of whether it is to be pitched to feather (reduction of lift force) or to flow separation, i.e. stall ('active stall'). The blade bearing and the pitch drive have to be designed; moreover the blade has to be positioned precisely and firmly.

The *blade bearing*, Fig. 3-24, transfers the driving torque and the thrust as well as the centrifugal force and the bending moment from the blade weight. Single or double row four-point ball bearings are common.

The *blade pitch system* is driven by mechanical, hydraulic or electrical energy. In normal operation, a synchronous pitching has to be assured, no matter which drive concept is applied, else the aerodynamic forces become asymmetrically causing an aerodynamic unbalance.

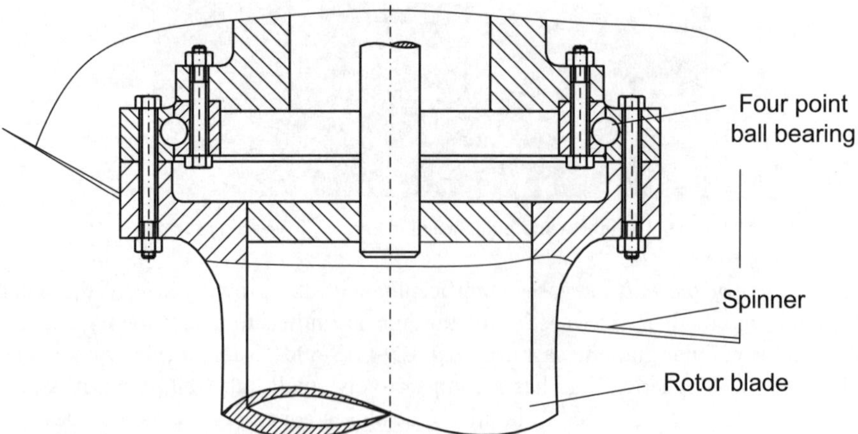

Fig. 3-24 Section of a blade pitch four point ball bearing (INA)

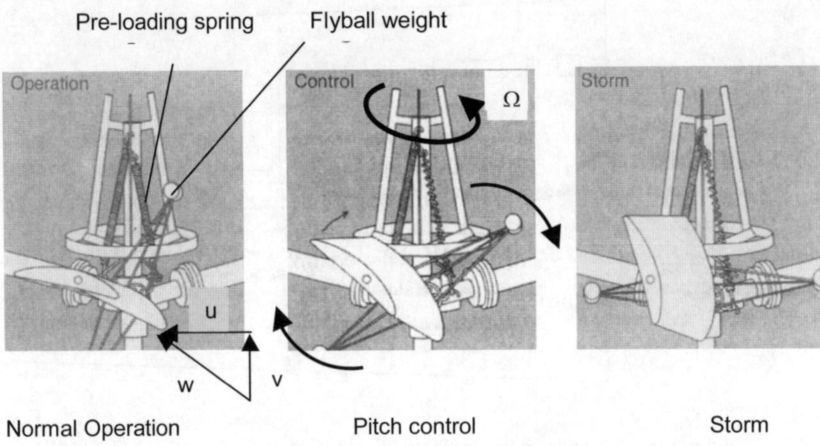

Fig. 3-25 Brümmer hub with blade pitch control by flyball weights

The synchronous pitch is achieved either with a central drive unit acting simultaneously on all three rotor blades (e.g. Adler 25, ENERCON E-32, DEwind D4, Vestas V44 or Zond 750), cf. Figs. 3-26 and 3-29. Or the pitch angles of the three blades are measured separately, and the controller adjusts the individual blade pitch, cf. Figs. 3-27 and 3-30.

Recently, more complex control systems were developed, and are now being applied at some Multi-MW wind turbines, which measure the loads at the blade and consider them in the individual blade pitch control. This reduces the dynamic loads significantly.

Moreover, the blade pitch system has to be designed as a *"fail-safe system"* since it is often one of the two required safety braking systems. When a dangerous operating state occurs (e.g. over-speeding or emergency stop), the blade pitch has to bring the rotor blade to the feather position immediately. In the case of shortage of the external energy, this is done by stored mechanical, electrical or hydraulic energy.

Using mechanical energy for the blade pitch is more suitable at smaller wind turbines (rated power less than 100 kW). Either the blade weight itself or additional fly weights are used to create the acting centrifugal forces. Fig. 3-25 shows the design of a *mechanically driven blade pitch with additional fly weights*. The fly ball weights are specially arranged in order to create the acting torque (propeller moment) around the blade axis. A pre-stressed spring creates a defined counter torque to determine the start of pitching. Since the acting centrifugal forces origin from the rotation of the fly ball weights, the requirement of being a fail-safe system is always fulfilled. The rotor is protected against over-speeding without any need for an external energy supply.

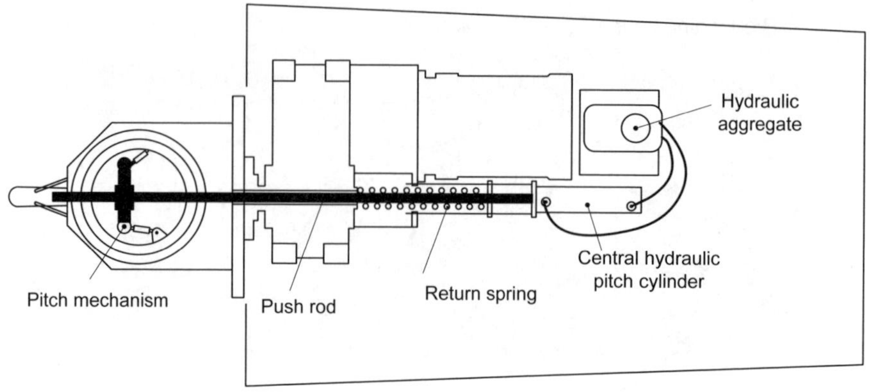

Fig. 3-26 Wind turbine Windmaster, central hydraulic pitch mechanism with push rod [6]

A *hydraulic blade pitch system* is applied at wind turbines with rated power from the 300 kW to the Multi-MW range. Fig. 3-26 shows a hydraulic central pitch system with a push rod transferring the pitch movement to the three rotor blades. Hydraulic drives are also used in some individual blade pitch systems (Südwind S.46, Vestas V52 to V90/ 3 MW). The weak points are the rotary joints and the leak-tightness in the rotating system. The required back-up energy for the fail-safe system is stored in hydraulic pressure accumulators in the hub. Additional safety measures have to be taken in order to prevent, in case of leakage or damage to the hydraulic system, the area and soil around the wind turbine suffering any ecologically problematic consequences.

An *electrical blade pitch system* is common to most of the wind turbine types. Nearly all manufacturers trust in this solution for the pitch, especially for larger wind turbines (rated power above 500 kW). A central pitch system is no longer suitable in this case due to the high acting forces and the space required. Therefore, the pitching of the three rotor blades is performed by three individual gear motors, Figs. 3-27 and 3-30. At wind turbines above 500 kW the rotor blade pitching speed is in the range of 5° to 10° per second. Therefore, the gear motors need a very high transmission ratio created by multi-stage planetary gears or worm gears.

For an exact *positioning of the rotor blades*, angular encoders are required to measure the actual blade position and report it to the pitch controller. The latter is often located in the rotating hub to prevent disturbances resulting from the signal transfer between the stationary nacelle and the rotating system of the rotor. For the fail-safe requirement of the electrical pitch system, the accumulators have to be integrated in the rotor hub. This location has disadvantages for the maintenance and the required periodical exchange of the accumulators due to their limited lifetime.

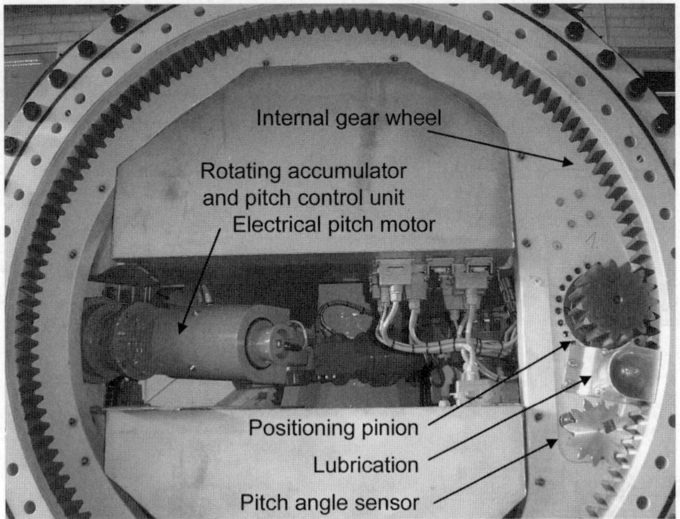

Fig. 3-27 Hub of a pitch-controlled 1,5 MW wind turbine with electrical gear motors (REpower)

Alternatively, the application of super capacitors is on the rise. The amount of stored energy has to serve for only one complete pitch movement to the feather position (park position). Another advantage of three individual pitch drives is that if one pitch drive fails the remaining two are still effective enough to protect the wind turbine against over-speeding and bring the rotor to stop.

3.2 Drive train

3.2.1 Concepts

There are different ways of *drive train component arrangements* (Fig. 3-28). Considering the development of commercial wind turbines from the 1980s on, there is no obvious "ultimate solution". However, the design engineers of different manufacturers seem to have different design philosophies. Several manufacturers even changed their drive train design with the growth in wind turbine size. The two basic concepts are
- the *integrated drive train* where different components with their different functions are fixed directly together and
- the *modular drive train* where most of the components are fixed separately on the nacelle frame.

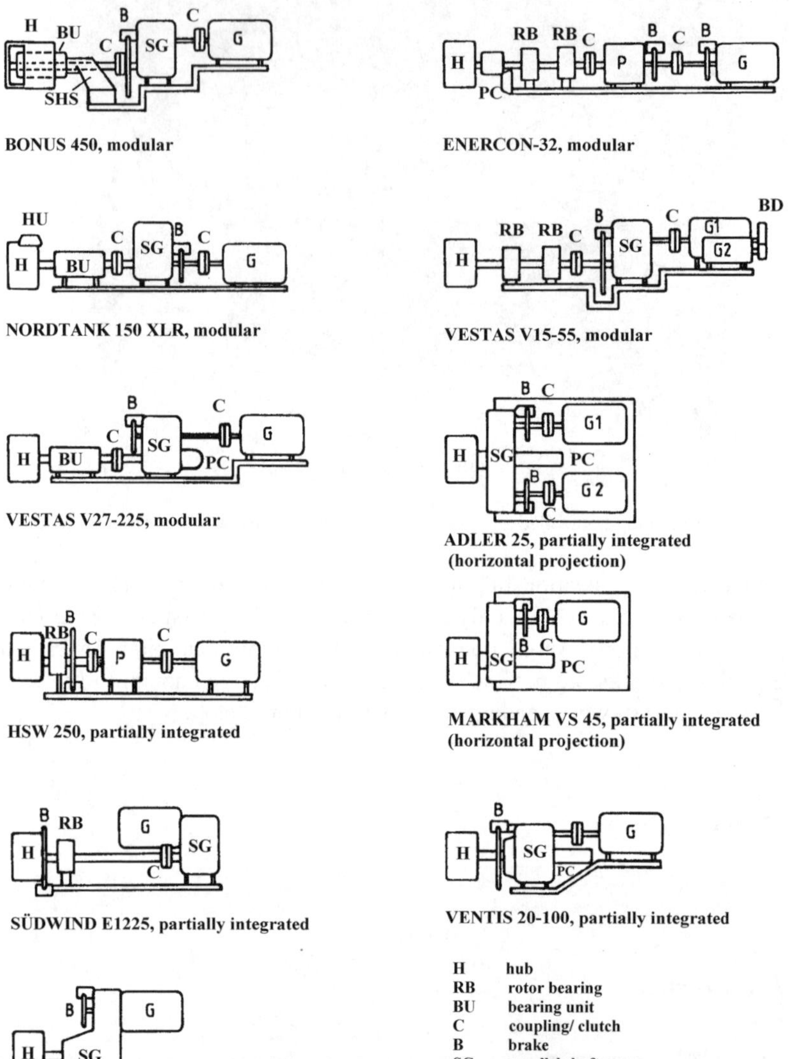

Fig. 3-28 Drive train concepts for various turbines

A mixed type is *the partially integrated drive train* where the rotor shaft support and the gearbox support are combined.

At the beginning of the 1990s, most of the Danish manufacturers trusted in the *modular drive train* design with a separate bearing unit for the rotor. Gearbox and generator were also fixed separately on the nacelle frame, Figs. 3-2 and 3-29. The advantages of this drive train type are the good accessibility of all components and the possibility of exchanging the gearbox without having to take down the rotor. The disadvantages include potential damages resulting from assembly faults: misalignment of the components and other assembly tolerances perhaps causing unexpected additional loads and accelerate wear.

Typical examples of the *integrated drive train* design are the wind turbines by ENERCON (Fig. 3-30). First presented in 1992 with the E-40 (rotor diameter 40m and rated power 500 kW), the gearless wind turbine type is nowadays the standard of the ENERCON design philosophy. Using a multi-pole ring generator there is no need for a gearbox. The nacelle is very compact in length with the hub directly fixed at the generator. So they are rotating together at low rotor speed on the conical axial pin.

The integrated design concept was also preferred in some wind turbines with a gearbox. Figs. 3-31 and 3-32 show two examples of this drive train design. The advantages, especially for wind turbine export and licensed manufacturing, are fewer transport problems due to the compact dimensions and easy assembly due to well defined positions of the drive train components. The disadvantages are that the gearbox requires a very stiff design because the complete rotor bearing support is integrated which causes correspondingly high loadings and, secondly, that the gearbox exchange requires dismantling the complete drive train.

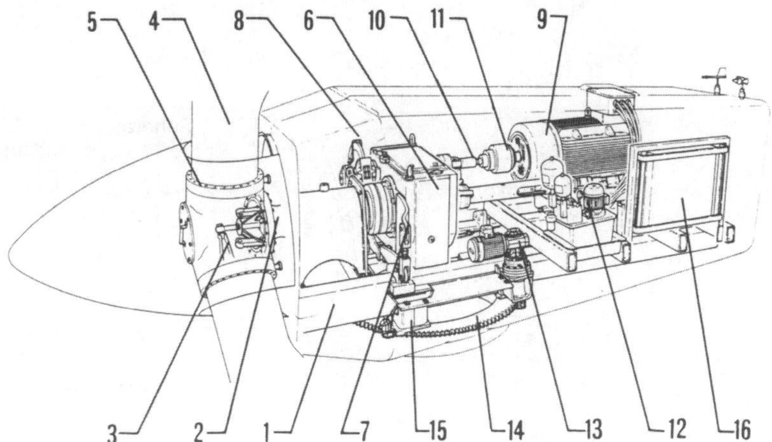

Fig. 3-29 Vestas V27-225, modular drive train; 1 nacelle frame, 2 main shaft, 3 pitch mechanism, 4 rotor blade, 5 cast steel hub, 6 spur gearbox, 7 torsionally elastic gear suspension, 8 brake, 9 pole-switchable asynchronous generator, 10 fast shaft with coupling, 11 sliding clutch, 12 hydraulic unit, 13 yaw drive, 14 yaw ring, 15 power cable twist control, 16 top control unit

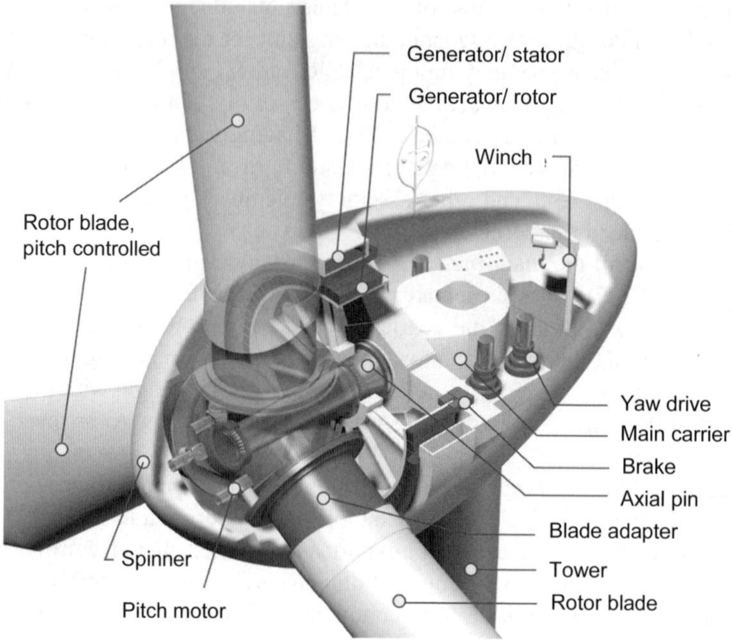

Fig. 3-30 E-66, gearless drive train (ENERCON)

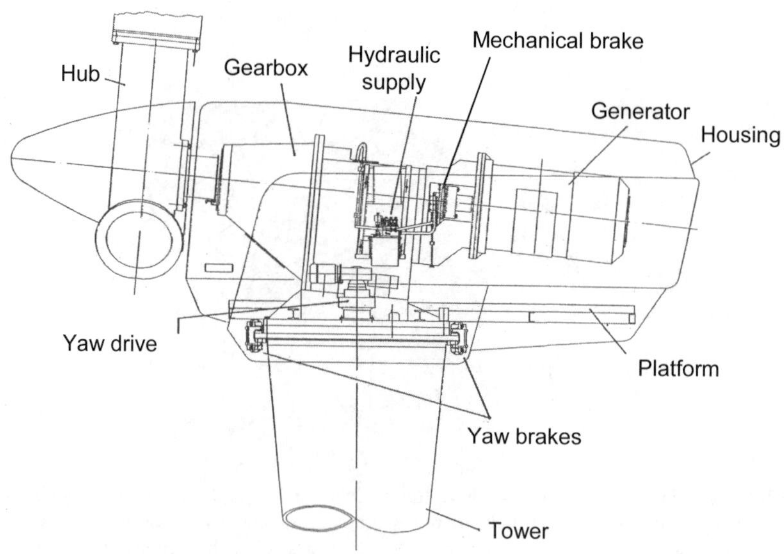

Fig. 3-31 N 3127, integrated design of the drive train (SÜDWIND, 1996 – today SUZLON)

As a mixed design type, the *partially integrated drive train* became established as wind turbine size increased in the mid 1990s, beginning with the 500 to 600 kW class wind turbines. Nearly all wind turbine manufacturers chose the so-called three-point support of the rotor shaft, Fig. 3-33. The main rotor bearing close to the rotor bears most of the gravitational loads of the rotor as well as the axial thrust. A second separate bearing does not exist, but the forces are transferred from the rotor shaft to the low-speed shaft of the gearbox with a hydraulic or mechanical clamping unit. The forces are then transferred to the nacelle frame via the torque support on each side of the gearbox which has a rubber-bonded-to-metal or elastomer mounting.

The type of power limitation is not relevant when choosing the *drive train concept*. All concepts are found in both stall and pitch controlled wind turbines. If one analyses the different drive train concepts in terms of their chronological development, one can see that there was a large variety at the beginning of the 1990s with partially integrated tending to dominate in the mid 1990s. However, current trends in the Multi-MW class once again reflect greater variety.

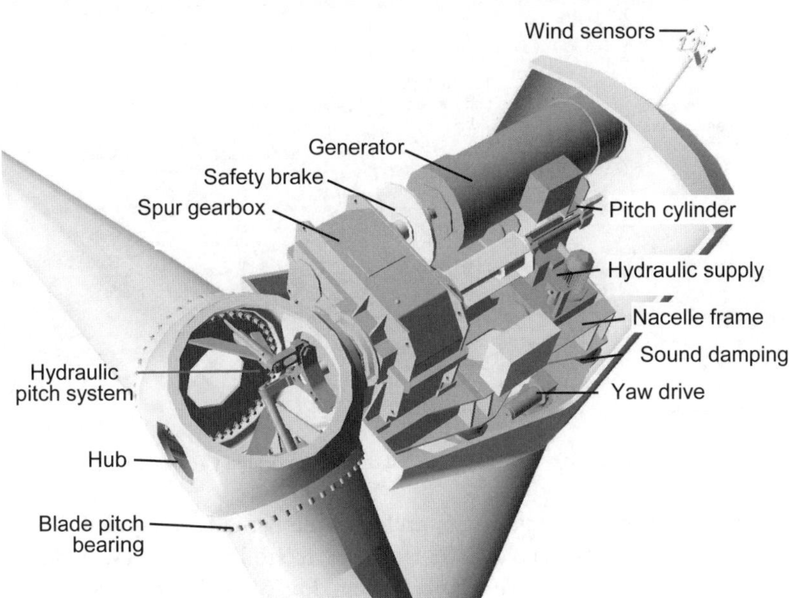

Fig. 3-32 D4, integrated design of the drive train (DeWind)

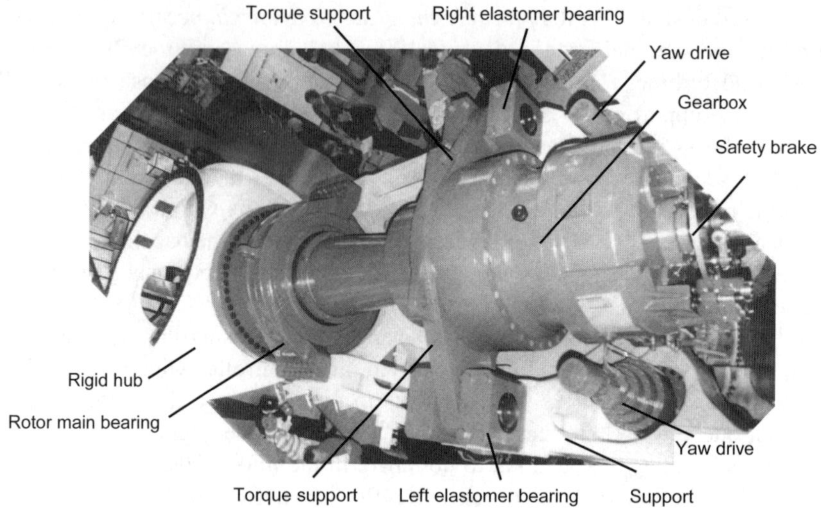

Fig. 3-33 N-60, partially integrated design with main shaft supported at three points (Nordex)

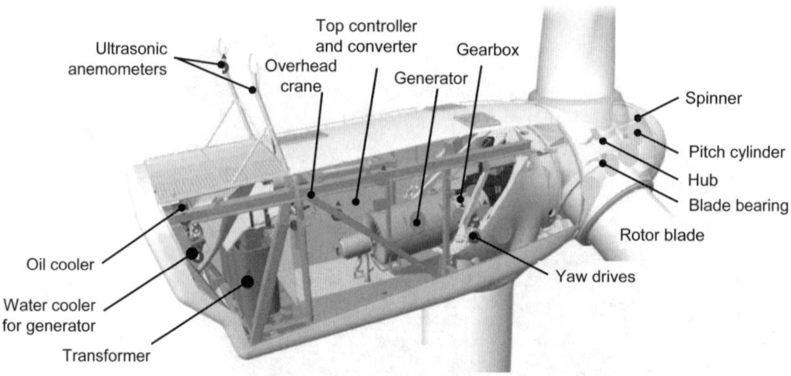

Fig. 3-34 V-90, D = 90 m, P = 3 MW, integrated drive train design (Vestas 2003)

For years the company Vestas, e.g., preferred the partially integrated drive train concept up to the 2 MW class (V-80). However, the following 3 MW wind turbine (V-90) shows an integrated drive train concept, Fig. 3-34. The company REpower Systems abandons the partially integrated drive train concept of their smaller Multi-MW wind turbines but introduces their 5MW wind turbine (5M) with a modular drive train, Fig. 3-35. The rotor shaft is supported by two bearings in order to reduce rotor loads entering the gearbox; moreover, the replacement of the gearbox without dismantling the rotor is facilitated.

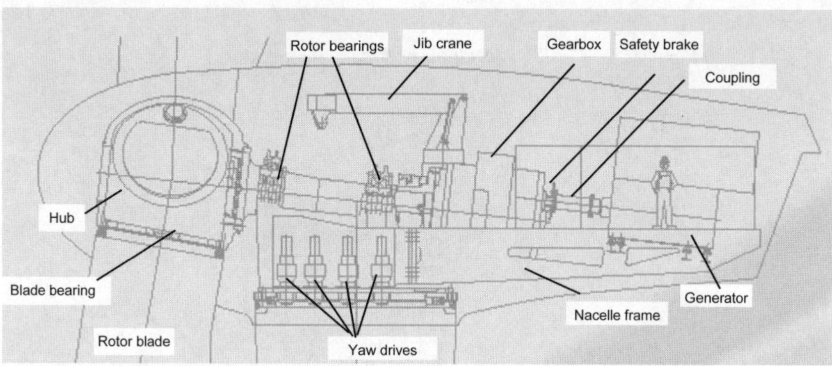

Fig. 3-35 5M, D = 126 m, P = 5 MW, modular drive train design (REpower 2004)

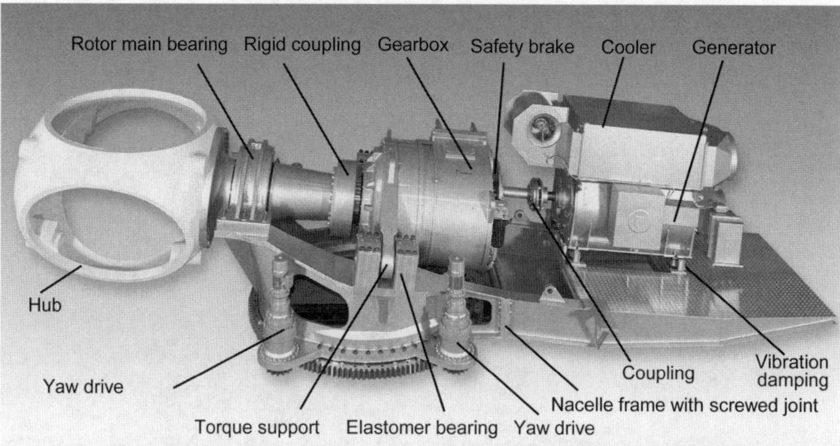

Fig. 3-36 D8, partially integrated drive train design with split, cast nacelle frame and main shaft supported at three points (DeWind 2002)

The *nacelles frames* of wind turbines are either casted or welded, Fig. 3-36. For the MW class wind turbines the nacelle frame is often split in two parts due to production requirements of the big part dimensions and high masses. The maximum capacity of foundries is reached if the weight of casted parts for Multi MW wind turbines reaches up to 100 t.

The wind turbines of the Finnish wind turbine manufacturer WinWind and the Multibrid M5000 of the company Multibrid GmbH (formerly Prokon Nord) represent a mixed type for the nacelle frame The latter operates with a medium speed multi-pole generator of up to 150 rpm, so there is no need for separate spur gear stages as required in most other wind turbine gearboxes. The drive train and the supporting nacelle frame are very compact thereby reducing the nacelle weight significantly compared to other wind turbine concepts both with and without gearbox.

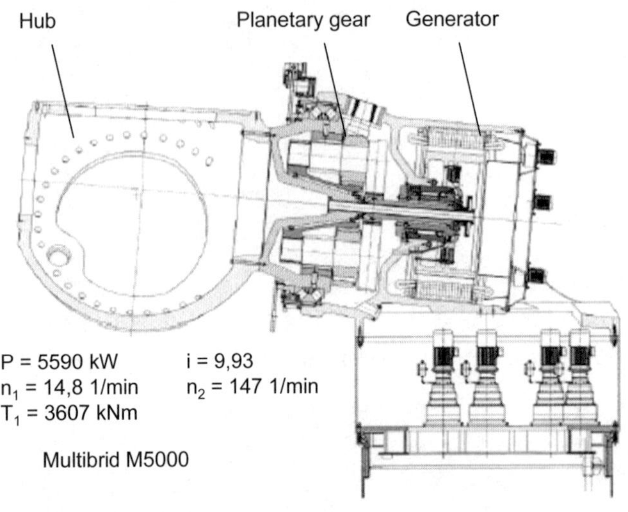

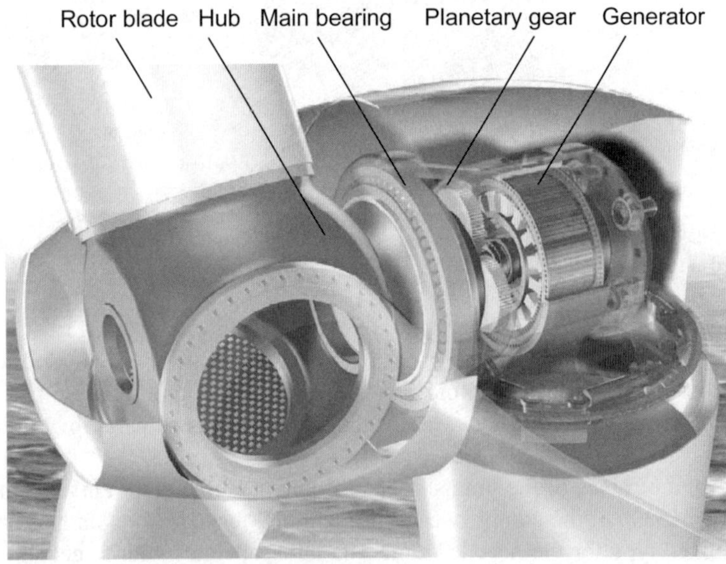

Fig. 3-37 Multibrid M5000, D = 116 m, P = 5 MW (Multibrid GmbH/Prokon Nord, 2004

3.2.2 Gearbox

Preliminary remarks

The power transmitting gearbox in wind turbines changes the rotational speed from rotor speed to the speed required by the driven work machine or generator. The torque changes correspondingly due to $P = T \cdot 2 \cdot \pi \cdot n$. As for the post wind mill, the low rotor speed was increased by the lantern gear to drive the faster running millstones, cf. Fig. 2-5b. The machine size determines the required transmission ratio. Since the maximum tip speed ratio and the corresponding blade tip speed is more or less given (cf. section 3.1), the rotor size determines the rotor speed, which, in most cases, is significantly lower than the speed of the driven work machine or generator. The speed of the generators' rotor is mainly determined by the grid frequency and the pole number, in particular for the directly grid connected asynchronous generators. For other work machines, the speed range results from the range of best efficiencies since the total system efficiency should be maximized.

Table 3.1 shows both gear types and other types of *torque and speed converters*. The *bevel gear* is required at wind pumping systems, cf. chapter 10, to change the orientation of the rotation axis from horizontal to vertical. A *belt drive* is then used to optimise the torque-speed characteristics of the driven centrifugal pump in relation to the wind turbine rotor characteristics. Early wind turbines for power generation based on the Danish concept with a direct grid connection were equipped with two generators. When the wind speed was small, a belt drive behind the gearbox was used to transmit power to the smaller generator. Later, this concept was replaced by pole switchable generators, cf. Fig. 13-2. A *chain drive* allows only a relatively small chain speed, so it is only found in historical wind turbines for power generation, e.g. the Gedser wind turbine, cf. chapter 2.

During the 1980s, *hydrodynamic converters and couplings* were used in several big prototype wind turbines (e.g. MOD-0A, WTS-3 and WWG-0600) for damping the load peaks and the "impacts" in synchronous generator (e.g. caused by the interaction of the rotor blade with the tower wake or by drive train oscillations). This is possible because the driving and the driven shaft are flexibly coupled by the fluid. But the disadvantages are smaller part load efficiencies and the requirement of additional oil coolers. The hydrodynamic converters were no longer needed when AC-DC-AC converters were introduced for speed-variable wind turbines allowing the rotor speed to be independent from the grid frequency. Due to technical progress, hydrodynamic converters are now being reused again in some modern wind turbines with a variable rotor speed but relatively constant generator speed (e.g. DEwind D8.2). Moreover, it should be mentioned that the part load efficiency of AC-DC-AC converters are also moderate to small.

Table 3.1 Torque and speed converters for wind turbines

Transmission type	Spur gear	Planetary gear	Bevel gear	Belt drive	Chain drive	Hydrodynamic converter (coupling)
Scheme	n_1, n_2	n_1, n_2	n_1, n_2	n_1, n_2	n_1, n_2	n_1, Pump, n_2, Turbine
Transmission ratio at wind turbine application	$i = n_1{:}n_2$ $\leq 1{:}5$	$\leq 1{:}7$	$\leq 1{:}5$	$\leq 1{:}3$	$\leq 1{:}5$	variable
Transmission principle	Form fit	Form fit	Form fit	Friction	Form fit	Hydraulic power transmission
Application	All applications	Electricity generation, > 500 kW	Wind pumping system	Wind pumping system, Danish concept	Historical wind pumps, do yourself	Prototype with synchronous generator
Overload safty machanism	No, to be realised externally	No, to be realised externally	No, to be realised externally	Yes, slip and slipping	No, to be realised externally	Yes, hydraulic slip in the fluid
Noise generation	high	smaller	high	small	high	small
Efficiency	high	highest	medium	medium	small	High at best point
Remark	Cost effective for smaller WT	Cost effective for big WT	90° angle between shafts	Slip reduces efficiency	Only for low rotational speed	Bad part load efficiency

In contrast to other common technical applications, the loads and operating states of the wind turbine and its gearbox vary considerably [10]. Respecting also the allowed sound pressure levels, this leads to high requirements for the gear teeth system, the bearing and the lubrication.

The size of the gearbox is determined by the required transmission ratio between rotor shaft and generator shaft. It is given by the rotor speed (calculated from the rotor diameter and the rated blade tip speed) and the generator speed. A stall-controlled Danish concept wind turbine with a four-pole asynchronous generator, e.g., has a generator speed of approx. 1,500 rpm. In reality it is slightly above 1,500 rpm depending on the slip of the generator. The maximum blade tip speed is, according to Fig. 3-4, around 70 m/s for common stall wind turbines. Table 3.2 shows the required transmission ratio for different rotor diameters. The values are quite high, as is the power to be transmitted. Hence only gearboxes with toothed wheels are suitable as torque and speed converters.

Table 3.2 Required transmission ratio for a stall-controlled Danish concept wind turbine (u_{Tip} = 70 m/s) with a four-pole asynchronous generator

Rotor diameter D in m	20	40	60	80
n_{Rotor} in rpm	66.8	33.4	22.3	16.7
$n_{\mathrm{Generator}}$ in rpm	1500 (slip not considered)			
Transmission ratio i	22.4	44.9	67.3	89.8
Power in kW (approx.)	100	600	1.300	2.300

Spur gears have parallel axes of the paired gear wheels, cf. table 3.1 and 3.4. In a helical spur gear, Fig. 3-38, there are always two teeth pairs in contact, so there is less noise emission, and the expected life time is higher due to a better load distribution [11]. So this gear type has become increasingly established despite higher manufacturing costs.

But the higher the transmission ratio, the larger the distance required between the two shafts. Gearboxes using only spur gears were still cost effective in wind turbines of the 500 kW class, whereas at least one *planetary gear* is suitable for larger wind turbines. For the same required transmission ratio, its dimensions, costs and noise emissions are smaller. Table 3.3 shows a comparison study of size, weight and costs of a 2.5 MW wind turbine showing the advantages of a multi-stage planetary gearbox. Table 3.4 characterizes the differences of spur gears and planetary gears.

In most cases, the first (slow speed) stage of the gearbox is a planetary gear with a hollow shaft, Figs. 3-39 and 3-40. This allows the rotary feedthrough of hydraulics, electrics and electronics to the hub.

Fig. 3-38 Small wind turbine (two-bladed downwind rotor) with two-stage helical spur gear

Planetary gears have, arranged concentrically around the sun wheel, three (or more) planet wheels with teeth contact at both the sun wheel and the hollow wheel [11…13]. The planet wheels are mounted in the planet carrier which itself is either immobile or also rotating, table 3.4. The transmission ratio differs slightly, depending on whether the hollow wheel or the planet carrier is fixed, Figs. 3-39, 3-40 and 3-41.

If the hollow wheel rotates and the planet carrier is fixed the transmission ratio is

$$i = n_{Sun} / n_{HW} = r_{HW} / r_{Sun}$$

If the hollow wheel is fixed and the planet carrier rotates, the situation is a little more complicated. Fig. 3-41 shows that the length of AA_1 equals BB_1, so the transmission ratio in this case is

$$i = n_{Sun} / n_{Pc} = 1 + r_{HW} / r_{Sun} \, .$$

Table 3.3 Comparison of different gearbox types for a 2.5 MW wind turbine with a gearbox ratio of 1:60, according to [6] in [18]

Number of stages and configuration	Scheme and dimensions	Weight in t	Relative costs in %
2 stages: spur gears	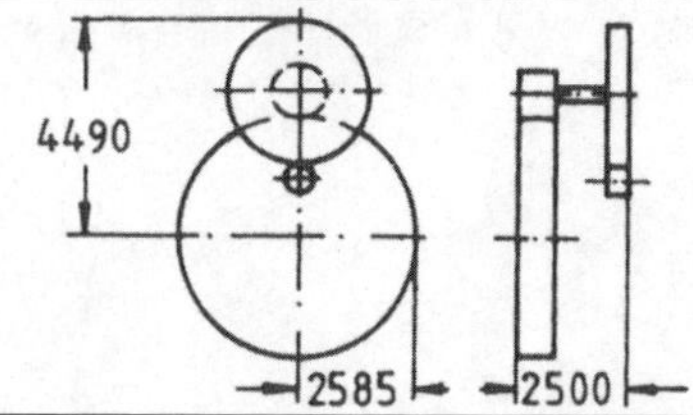 4490 / 2585 / 2500	70	180
3 stages: spur gears	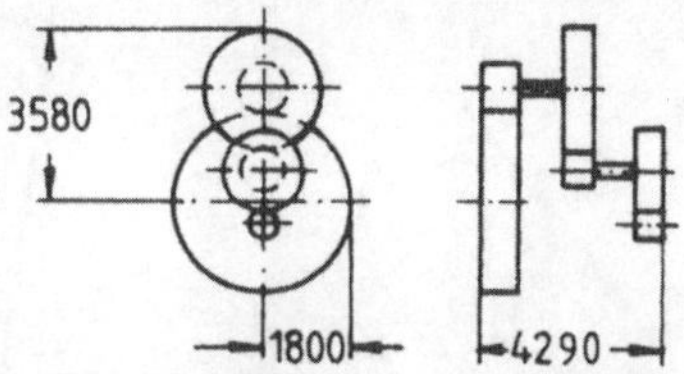 3580 / 1800 / 4290	77	164
2 stages: 1 Planetary gear 1 spur gear	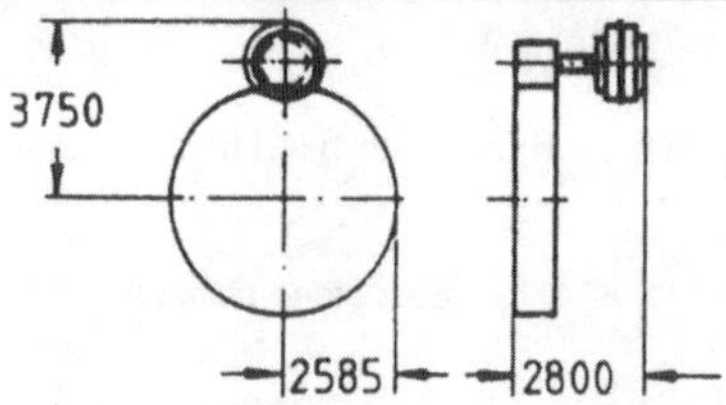 3750 / 2585 / 2800	41	169
3 stages: 2 Planetary gears 1 spur gear	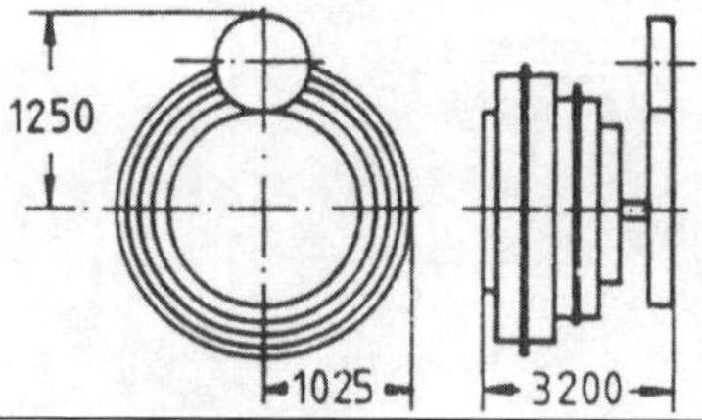 1250 / 1025 / 3200	17	110
3 stages: Planetary gears	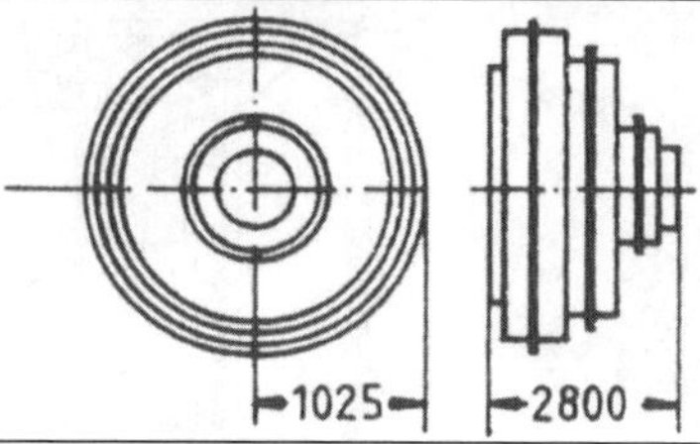 1025 / 2800	11	100

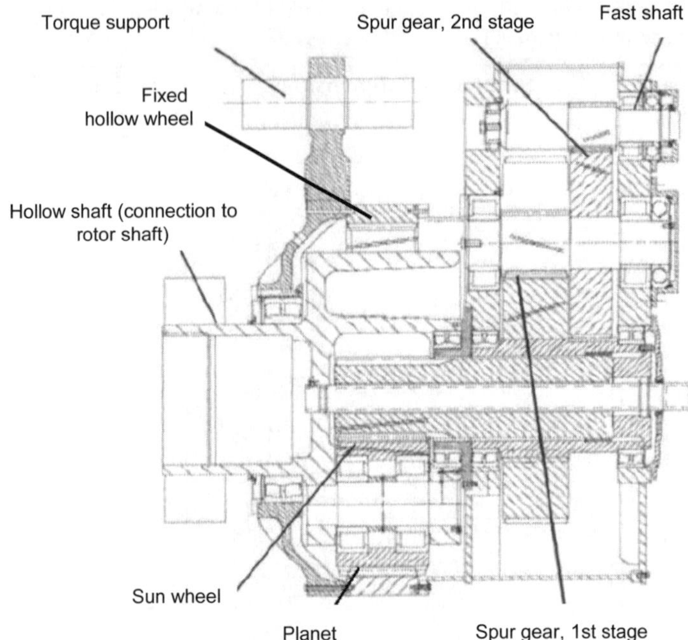

Fig. 3-39 Three-stage gearbox for wind turbines with fixed hollow wheel (Metso)

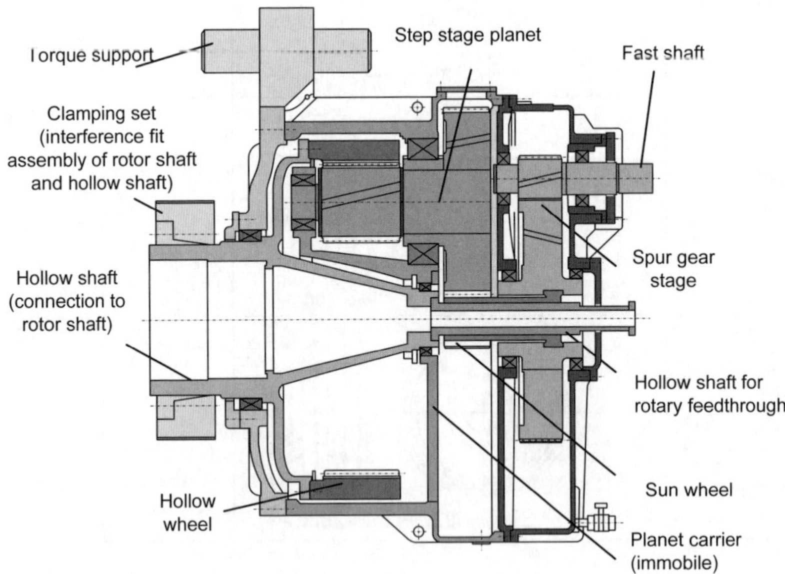

Fig. 3-40 Two-stage gearbox for wind turbines: one step stage epicyclic gear stage with rotating hollow wheel and one spur gear stage (Renk)

The advantage of the planetary gear is that the tangential force per teeth contact is reduced by the number of planets. With three planet wheels it is F/3 under 120° at the sun wheel and under 180° (sun and hollow wheel contact) at each planet wheel, table 3.4. The design with a rotating hollow wheel is more complicated, but less noise propagates directly into the casing than in the case of a fixed hollow wheel [14].

The gearbox has to meet various demands: it should certainly operate without any problems, have small weight and dimensions, have a low noise emission level (above all without tonality), survive even greater damages and also be maintenance-friendly [e.g. 15]. Moreover, the required lubrication has to be assured under very different operating states, also at very low speeds (idling, start-up) [10] and under difficult climatic conditions. These demands partially conflict in course of the design procedure and require a very high effort when calculating statics, dynamics, material strength and service life [15…17]. This is even increased by the insights gained in the course of operating experience and can only be performed by extensive PC based numerical calculations.

Table 3.4 Comparison of spur gears and planetary gears

Type	Spur gear	Planetary gear
Scheme	n_1 ⊙ n_2	Sun wheel / Hollow wheel / Planet wheel / Rotating planet carrier / Rotating hollow wheel Hollow wheel fixed Planet carrier fixed
Transmission ratio	$\leq 1{:}5$	$\leq 1{:}7$
Force per teeth contact	F	$F/3$ (with 3 planets)
Teeth contact frequency	$1 \cdot n$	$3 \cdot n$
Weight	high	low
Dimensions	big	small
Noise emission	high	low
Efficiency	approx. 98% per stage	approx. 99% per stage
Costs	Favourable for P < 500 kW	Minimum 1 planetary stage favourable for P > 500 kW

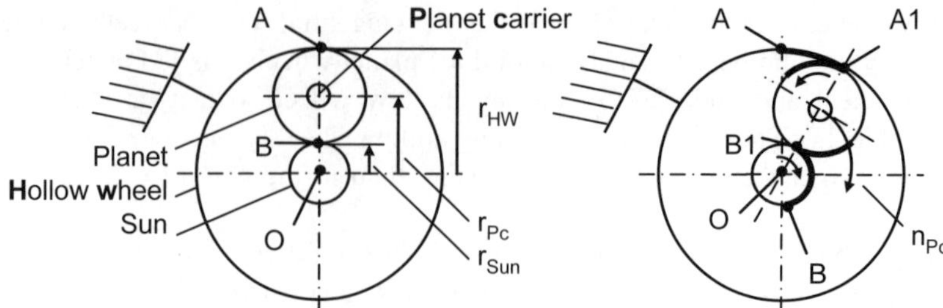

Fig. 3-41 Relations in a planetary gear with fixed hollow wheel and rotating planet carrier

3.2.3 Couplings and brakes

Due to the enormous torque, there is a *rigid coupling* between rotor shaft and slow gearbox shaft. In drive trains with a three-point support of the rotor shaft, the latter is fixed in the hollow shaft of the gearbox by an interference fit assembly (Fig. 3-40), either by shrinkage or force fitting which is easier to dismantle.

Between the fast shaft of the gearbox and the generator, only a slender shaft is required to transmit the smaller toque. But these two shafts require an *elastic coupling* since there may be a misalignment between the gearbox and the generator. Moreover, both drive train components are elastically mounted on noise and vibration absorbing damper elements. So the coupling used is torsion proof but elastic with respect to bending (multi-disk clutch, disks or steel bolts in rubber eyes), Figs. 3-42 and 3-43. In order to protect gearbox and generator, an overload protection is often integrated in the fast shaft coupling (slipping clutch or shear bolts).

Fig. 3-42 Coupling shaft between gearbox and generator with brake disc and two couplings, left: in leakage current isolating version, right: flexical GKG link type coupling (Winergy)

The certification guidelines of the German Lloyd [15] require two independent *braking systems*. At least one of them has to act on the aerodynamic side at the rotor. At stall-controlled wind turbines this is done by the tip brakes, i.e. the turnable blade tips (Figs. 3-2 and 3-13), whereas pitch-controlled wind turbines rotate the entire blade (cf. chapter 3.1). The second braking system tends to be a *mechanical disk brake*. At smaller wind turbines (< 600 kW) it is either found at the fast shaft or at the low speed shaft (Fig. 3-38) which has the advantage that the loads do not pass the gearbox when braking the rotor. But with increasing wind turbine size the braking torque increases strongly, as does the required disk diameter. One of the biggest commercial wind turbines with the mechanical disk brake on the low-speed side of the gearbox is the TW-600 manufactured by Tacke (today GE Wind).

Commercial wind turbines of more than 500 kW tend to have the disk brake on the fast shaft, Fig. 3-43 and Figs. 3-31 to 3-35. The brake is dimensioned for extreme load cases with emergency stops where the rotor has to be brought from full load operation or over speeding to standstill within seconds. During normal operation (no emergency stop) the braking procedure activates the aerodynamic brake at first and then the mechanical brake at small remaining torque, in order to stop the rotor completely.

In principle, wind turbines with individual blade pitch systems do not need a mechanical brake because the aerodynamic braking system is redundant due to autonomous pitch drives. Basically, pitching only one blade to feather is enough to brake the rotor from full load to standstill.

However, a mechanical brake and a rotor lock with fastening bolts are required for maintenance and repair at the rotor and in the nacelle. In all wind turbines, whenever staff are working at the rotor or in the hub, the rotor lock must be used for safety purposes.

Fig. 3-43 Disc brake at the fast shaft of a gearbox (Svenborg)

3.2.4 Generators

A detailed discussion of the generators and their electrical characteristics is found in chapter 11. So this section treats only aspects of the generator types relevant for the drive train design. The pole number, together with the grid frequency, determines the generator speed and whether a gearbox is required or not. *Asynchronous generators* which are coupled directly to the grid, employed in stall-controlled wind turbines usually have 4, 6 or 8 poles. From a grid frequency of 50 Hz then follows a super synchronous generator speed slightly above 1500, 1000 or 750 rpm (depending on the slip). Suitable gearbox types and their transmission ratio were discussed in chapter 3.2.2.

Doubly-fed asynchronous generators have a variable speed but the operating speed range is in a similar order. Only low-speed *multi-pole ring generators* (separately excited or permanently magnetic excited synchronous generators) may operate without a gearbox, Fig. 3-30.

A hybrid type is the medium-speed generator of the Multibrid M5000 with a moderate pole number driven by a planetary gearbox, with a generator speed of approx. 150 rpm, cf. Fig. 3-37.

With generator power of more than 1 MW, air cooling reaches its limits. So, water cooled generators are also applied. The ring generators of gearless wind turbines, Fig. 3-30, are large enough for effective air cooling, but the noise generation in the small gap between rotor and stator of the generator has to be minimized.

3.3 Auxiliary aggregates and other components

3.3.1 Yaw system

In the case of historical windmills, major effort was required for the *orientation of the rotor* perpendicular to the wind. The miller had the troublesome task of pushing the tailpole in order to adjust the rotor to the variable wind direction, see chapter 2. Only in the middle of the 18th century, did the invention of the fantail allow the yawing process to be automated. Even today the yaw system is a "non-trivial" functional subsystem of the wind turbine.

Wind turbines with a horizontal rotor axis orientation allow either

- *passive yaw systems* - e.g. autonomous yawing of a turbine with a downwind rotor or windvanes at upwind rotor turbines - or

- *active yaw systems* are applied - e.g. a fantail or yaw drives driven by external energy (also known as azimuth drives).

As already explained in section 3.1, the downwind rotor (leeward in relation to the tower, e.g. SÜDWIND; Fig. 3-18) is suitable for *autonomous passive yawing of the wind turbine* because if the wind direction is not parallel to the rotor axis, the force from the wind pressure on the rotor causes a yaw moment around the tower axis which adjusts the rotor to the wind direction, similar to a windvane. But for wind turbines with a high tip speed ratio, which have a relatively low solidity of the rotor area, this works only when the rotor is turning. Hence, for low wind speeds either the nacelle housing itself has a passive function as a kind of additional "windvane", or an active drive is required.

The *windvane* for passive yawing of upwind rotor turbines is a characteristic of the Western mill, cf. Figs. 2-10, 3-1 and chapter 12. Due to its simple system design and the fact that neither external energy nor control is required, it is also commonly used in other small wind turbines, especially battery chargers.

Passive yaw systems have to be designed in such a way that sudden changes in the wind direction do not provoke fast yaw movements producing strong additional loads due to gyroscopic forces. At single- and two-bladed rotors the situation is even worse. The inertia against yawing depends on the angular rotor blade position which increases the strong dynamic loads. Therefore, the application of passive yaw systems is limited in general to a rotor diameter of up to 10 m.

Active yaw systems position the nacelle using drive units and are applied in wind turbines of both upwind and downwind rotor configuration. No external energy is needed if the wind itself is driving the *fantail*, orientated perpendicular to the rotor as in the Dutch smock windmills, Figs. 2-8 and 2-12. The torque from this small auxiliary rotor (rosette), Fig. 3-44, is transmitted using a worm gear with a high transmission ratio (up to 4,000) to the rotating assembly of the yaw system.

A yaw system with one or more *electrical or hydraulic yaw drives* is most common for larger wind turbines. They are controlled using the signal from a small windvane on top of the nacelle, Figs. 3-30 and 3-32, and act on the spur gear of the big rotating assembly at the tower-nacelle connection, Figs. 3-44 right and 3-45 right. Multi-MW wind turbines may have up to eight yaw drives.

Since there is always an unavoidable clearance in this spur gear system, nacelle oscillations, which would increase wear of the teeth flanks, have to be prevented. For this purpose, the nacelle is either fixed by *yaw brakes* which are released only during a yawing movement, and/or additional friction brakes are permanently acting. The yaw drive then has to work against this friction force. If several electrical yaw drives are installed, another possibility is locking the yaw system "on the electrical side": when the yaw movement is completed, half of the drives get the signal to "try to turn" into the other direction. The drives work against each other, and the fixing torque assures in each yaw gear teeth contact on a defined flank. When designing the wind turbine, the fact that the active yaw system couples nacelle and tower rigidly has to be taken into account. *Torsional oscillations* of the tower are transferred into the nacelle.

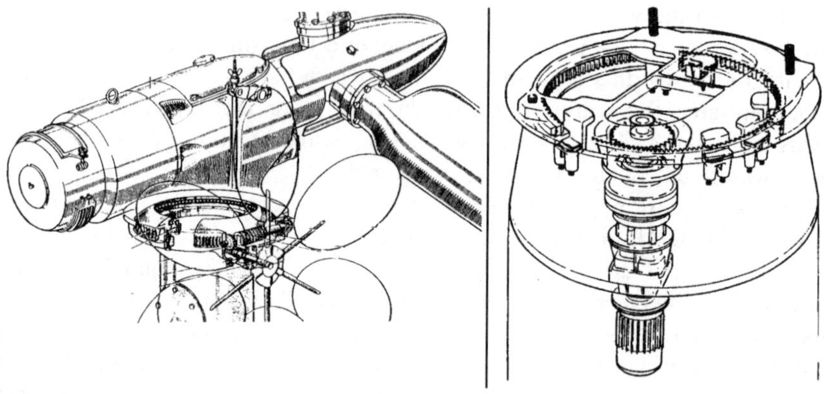

Fig. 3-44 Left: wind turbine with rosette for yawing (Allgaier) [19]; right: yaw bearing with yaw drive and yaw brakes (Wind World)

Fig. 3-45 Left: section of a yaw drive with multi-stage planetary gear, motor removed (Liebherr); right: electrical yaw drive, yaw brakes and yaw angle sensor (REpower)

3.3.2 Heating and cooling

Wind turbines have to operate within a large temperature range based on the climatic conditions on site. Moreover, the *heat losses* of gearbox and generator heat up the nacelle. The temperatures should not exceed the admissible operating range of the installed components, above all the sensible electronics. For a defined heat removal, the air flow is directed by special *ventilation systems*. Their design has to consider also the minimisation of noise emission and airborne sound propagation.

Apart from the *cooling system* for the nacelle itself, there are separate coolers e.g. for the drive train components gearbox and generator, Figs. 3-34 and 3-36. Since the admissible temperature level of the generator is often higher than that of the gearbox, a combined oil-water-cooling cycle may be used, Fig. 3-46. The water-air cooler is often located outside the nacelle, which may then be encapsulated completely, e.g. for air conditioning in offshore applications. If the gearbox oil is overheated, the lubrication properties may degrade due to the destruction of the additives, and the service life of the gearbox may be reduced significantly.

In the winter time, the ambient temperatures may fall below 0°C, even outside of cold climate sites. After a longer standstill, the gearbox oil also becomes cold and its viscosity increases, causing lubrication problems during the wind turbines' start-up. To assure good lubrication, *additional heaters* are then installed. Moreover, the temperatures in the electronic cabinets should not go below the allowed range. An encapsulation of the nacelle or entire wind turbine e.g. with shutters at all ventilation openings is suitable, but the effective cooling of the nacelle should not be disturbed in summertime.

Moreover, *rotor blade heating systems* are installed, if there is increased risk of icing. Ice on the blades changes the aerodynamic characteristics drastically and causes increased vibrations due to the additional mass unbalance of the rotor. Negative influences of condensation water in the blade are also reduced by the rotor blade heating, which is either done electrically with heating wires in the blade laminate or by blowing hot air from the nacelle into the hub and blades.

Heated anemometers and wind vanes are also useful under icing conditions. A frozen anemometer would cause e.g. yield loss because the wind turbine cannot start despite good wind speed. A frozen windvane may suggest, depending on the relative angle to the rotor axis, a permanently occurring deviation of the wind direction causing permanent yawing - or it may hide existing wind direction deviations causing yaw angle errors with additional loads. Both pose a potential dangerous to the wind turbine, see e.g. [20].

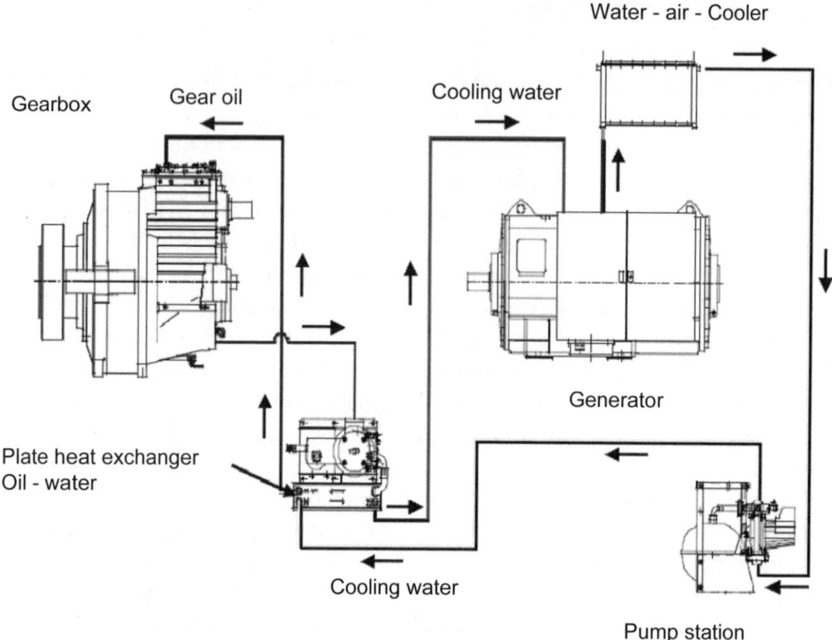

Fig. 3-46 Combined Oil - water - cooling cycle for gearbox and generator (Nordex)

3.3.3 Lightning protection

Lightning is a potential danger to wind turbines because of their large total height
and their installation at exposed sites. Accordingly, there are guidelines for the design of the lightning system, e.g. IEC 61400-24, see table 9.1. Statistically, a wind
turbine in Germany is only struck once every 10 years by lightning [21, 22], but at
exposed sites in the German highlands this number is significantly higher. Often,
the flash strikes the turbine at the highest point, i.e. the blade tip. However, in the
case of large wind turbines of the MW class, a flash may be propagate upwards
from the turbine into the clouds.

Rotor blades are often equipped with special *lightning receptors* in the blade tip
area. These are metal disks of approx. 5 cm diameter integrated into the blade surface. Other manufacturers install several receptors along the blade radius. Or the
blade tip as well as the leading and trailing edge are made from aluminium. The
high lightning currents (up to 1000 kA) are conducted in a defined way e.g. by
metal cables inside the rotor blade. If condensation water in the rotor blade were
to heat up and evaporate by lightning energy, the blade would explode. Many
blade manufacturers install lightning registration cards at each blade root to record

the currents. Later analysis provides the basis for detecting potential blade damage.

In order to protect the entire wind turbine against lightning damage the currents have to be conducted further down into the ground. It is important to protect the bearings as the high currents would point-weld together the balls and the running surface.

Fig. 3-47 Components of the lightning protection system for the protection of the bearings; left: sliding brush contact on the main shaft, right: on the yaw drive, combined with spark gaps (REpower)

To prevent this, spark gaps and/or sliding contacts with carbon brushes are installed, Fig. 3-47. The first crucial point in a wind turbine with a pitch system is the blade bearing between rotor blade and hub, the second is the rotor bearing between hub and nacelle frame, and the third is the yaw bearing between nacelle frame and tower. The scheme of the lightning protection system with the different lightning protection zones (LPZ) is shown in Fig. 3-48. LPZ 0, e.g., means that the object (rotor, nacelle top with wind sensors) may be directly hit by the strike. From inside the tower down to the earth rod, the required protection measures are the same as for technical buildings. Electronic components, the control, switchgear, transformers, etc. have additional surge and lightning protection.

Only approx. 30% of the lightning damages at wind turbines in Germany result from direct strikes, while approx. 60% are caused by strikes into the power and telecommunication lines. Therefore, an entire lightning protection concept according to the standards is required [21]. A periodic inspection of the grounding resistance is required to assure that the lightning protection system functions properly under operation from the rotor blade down to the ground.

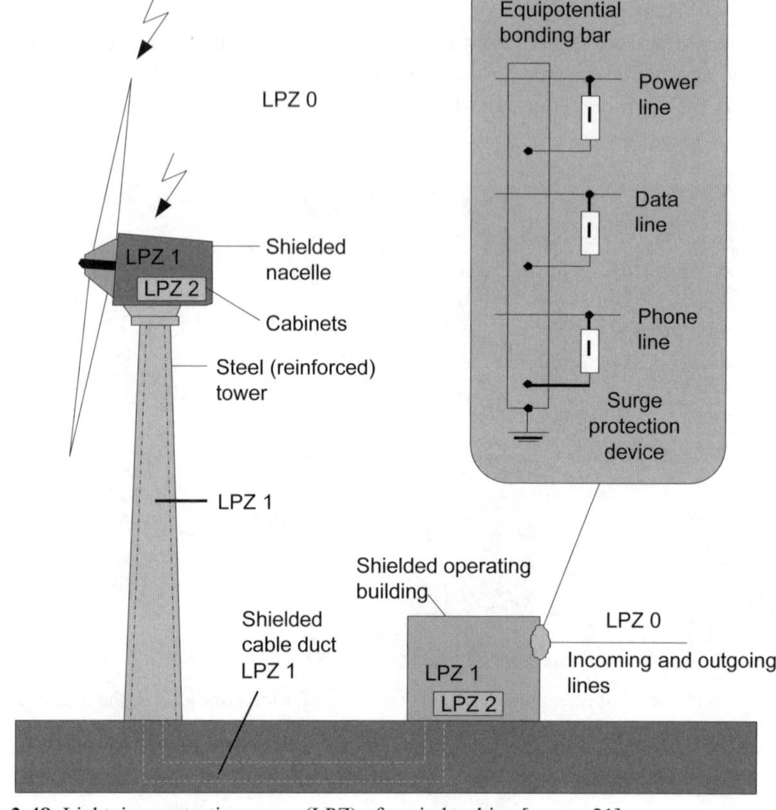

Fig. 3-48 Lightning protection zones (LPZ) of a wind turbine [acc. to 21]

3.3.4 Lifting devices

Most of the larger wind turbines are equipped with winches to lift tools and
smaller spare parts up into the nacelle. In Multi-MW wind turbines; there is often
also a slewing crane or overhead crane inside the nacelle to move heavier objects,
Figs. 3-30, 3-34 and 3-35. For the exchange of large components, mobile cranes
are needed, like for the erection of wind turbines. In the nacelles of offshore wind
turbines the integrated cranes are often larger since the costs for the (external)
crane ships are extremely high. Fig. 3-49 shows such a "nacelle crane" which is
suitable for the replacement of entire drive train components (rotor bearing, gearbox and generator).

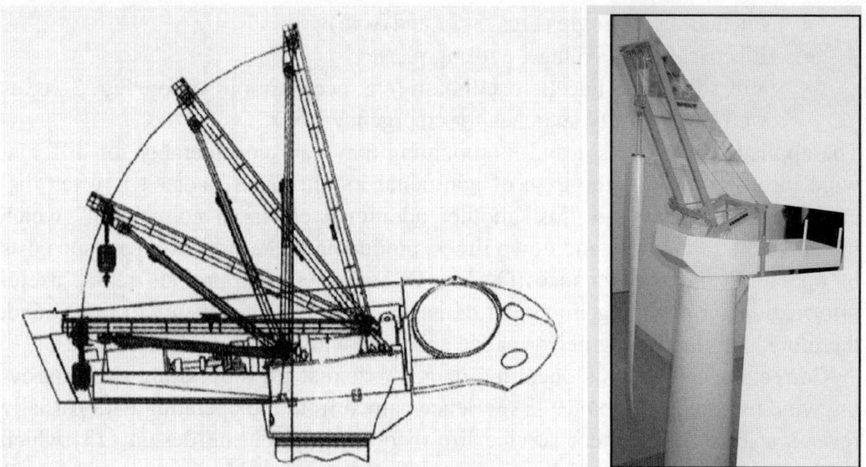

Fig. 3-49 Left: nacelle with additional crane, GE 3.6 (GE Wind); right: model (Liebherr)

Fig. 3-50 Left: Acceleration sensor (horizontal and vertical direction, company Mita Teknik) at the rubber bearing of the gearbox, right: revolution speed sensor at the generator back (REpower)

3.3.5 Sensors

The supervisory and control system of the wind turbine (see chapter 12) continually processes numerous data on the operating and ambient conditions which are provided by numerous sensors in the wind turbine and on the nacelle. The following *operating data* is acquired permanently:

- Wind speed and direction (Fig. 3-32)
- Rotor and generator speed (Fig. 3-50)
- Temperatures (ambient, bearings, gearbox, generator, nacelle)
- Pressure (gearbox oil, cooling system, pitch hydraulics)

- Pitch and yaw angle (Figs. 3-27 and 3-45)
- Electrical data (voltage, current, phase)
- Vibrations and nacelle oscillation (e.g. acceleration sensor, Fig. 3-50, as well as proximity switches for emergency stop)

The applied type of sensor and its mounting may vary considerably for different wind turbine types. In the case of additional monitoring sensors it is often discussed whether they are "just another additional electrical component" which increases the failure risk and down times, producing yield loss and increased costs of operation and maintenance (O&M). Or, whether they provide really useful information for detecting imminent damages, scheduling (preventive) O&M and, therefore, help increasing performance and yield and reducing O&M costs.

Correct and continuous operating data is essential for monitoring and improving wind turbine performance. Experience gained from the operating data partially reveals a lower component service life than assumed during design [23] which may present the operator and insurance with financial problems. Severe damages often cause consequent damages at other components, moreover e.g. a gearbox replacement requires the costly removal of the rotor with a mobile crane. Not only the maintenance and repair costs, but the yield loss due to the increased standstill time until repair (bad weather and accessibility) is also an issue. Therefore, at larger wind turbines and especially those with difficult accessibility like the offshore wind farms, the number of sensors is increased and e.g. *condition-monitoring systems* (CMS) are installed. These are certified systems [24] for surveying the condition of bearings and gearbox by frequency analysis of the measured vibrations. Remote monitoring and analysis then observes the evolution of potential damages, so scheduling the repair measures is easier, and severe damages with consequent damages are prevented. At offshore wind turbines extensive load monitoring is also needed for insurance reasons. Despite all sensors and monitoring systems it is important to inspect wind turbines periodically. Some damage types and their causes – e.g. the detection of avoidable additional mass unbalance or aerodynamic unbalance [25] as well as alignment errors of the drive train - have not been integrated in CMS to date and are revealed only by additional measurements.

3.4 Tower and foundation

3.4.1 Tower

When designing a horizontal axis wind turbine, the mechanical engineer often under-estimates the construction complexity of the *tower and foundation*. However, the stability of the entire wind turbine structure is the most important issue when applying for a construction permit from the building authority. Therefore, in the course of assembling the large amount of certification documents the manufacturer is "forced" to investigate the construction in detail (cf. section 9.1.4). Besides the static stability, the dynamic behaviour of the wind turbine's tower is of similar importance, see chapter 8.

Moreover, the *tower* is of great relevance for the *economic efficiency of a wind turbine* for several reasons: On one hand, it produces a significant portion of the wind turbine's initial costs, approx. 15 to 20%, and also determines the costs for transport and erection. On the other, the energy yield (i.e. the profit) significantly depends on the turbine's hub height. The wind speed increases with height logarithmically at most sites, and a hub height above the atmospheric surface boundary layer delivers a constantly higher energy yield (see chapter 4). Therefore, the choice of the optimum hub height (respectively tower height) has to be made for each site individually. To facilitate certification and selection, the manufacturers offer each wind turbine with several set tower heights. This helps calculating the *optimum cost-benefit ratio*.

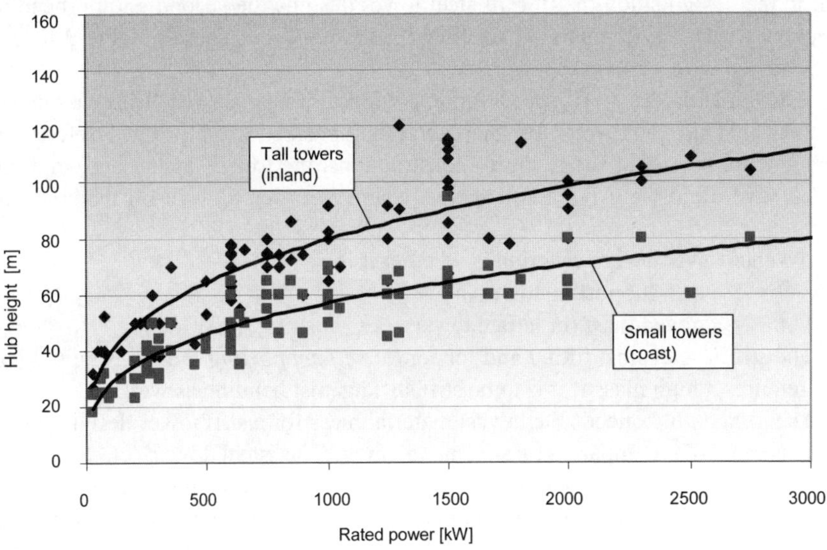

Fig. 3-51 Hub height versus rated power for commercial wind turbines

At the coast (short roughness length and lower turbulent intensity) the wind speed picks up fast with the height above ground, so mostly smaller towers are suitable, Fig. 3-51. The inland surface boundary layer (higher roughness length and turbulent intensity) is larger, which means that higher towers are more suitable. The ratio of hub height to rotor diameter is between 1.0 and 1.4 for wind turbines at coastal sites in Germany and between 1.2 and 1.8 for those at inland sites. The higher values are valid for smaller wind turbines, below approx. 300 kW, the larger values for the MW-class wind turbines.

The tower's structural design (see the Campbell diagrams in chapter 8) is either

- soft or
- stiff.

A *stiff tower design* means that first natural bending frequency of the tower is above the exciting rotor speed n, respectively the corresponding rotational frequency. A *soft tower design,* on the contrary, means that the first natural bending frequency of the tower is below the rotational frequency of the rated speed. At these wind turbines during start-up, the passage of the tower's natural frequency has to be controlled in order to prevent resonance effects with increasing vibrations of the system.

Fig. 3-52 shows the study of different tower designs for the wind turbine WKA-60. The rotational frequency of 0.3833 Hz is for all designs below the first tower's natural frequency, i.e. for all is the frequency ratio $f_{0.1} / n > 1$. But the blade passing frequency of 1.15 Hz (rotational frequency multiplied by the blade number) is always above the first tower's natural frequency. Moreover, it turned out in the design study that for all steel tower designs the second natural bending frequency was very close to this exciting blade frequency. Therefore, the WKA-60 was erected with a concrete tower.

The dynamic design of speed-variable wind turbines is a very tricky issue, see chapter 8. While the towers for small and medium sized wind turbines of less than 500 kW are mostly stiff (i.e. rigid) constructions, the towers of the big wind turbines have almost always a soft tower design in order to save on material and costs.

Another tower design criterion is whether it is

- a self-supporting tower or
- a guyed mast (or lattice tower).

The stiffness against tilting and torsion of *self-supporting towers* is quite high, but requires a high mass if it is to be stiff stiff against bending as well.

The *lattice tower* needs the lowest material mass for a stiff tower design: it may save up to 50% compared to an equivalent tubular steel tower, cf. Fig. 3-52. Moreover, due to having many connections, the structural damping is higher than that of the steel tower. This was the reason why a lot of Danish wind turbines of the first generation were designed with a lattice tower, cf. Fig. 3-2. Later, they were less frequently used due to their visual impact on the landscape and also because of costs, given the higher share of labour cost involved in manufacture and erection which is a disadvantage e.g. in Northern Europe. There is a higher degree of automation attainable in the manufacture tubular steel towers with bending and welding machines, Fig. 3-53, top.

Material:	Steel				Concrete		
Wind turbine: **WKA-60-II** Rotor: 3 blades, $\varnothing$ 60 m Rotor speed: n = 23 rpm = 0.3833 Hz Blade frequency: 3 x n = 1.15 Hz Head mass: 207 t hub height: 50 m	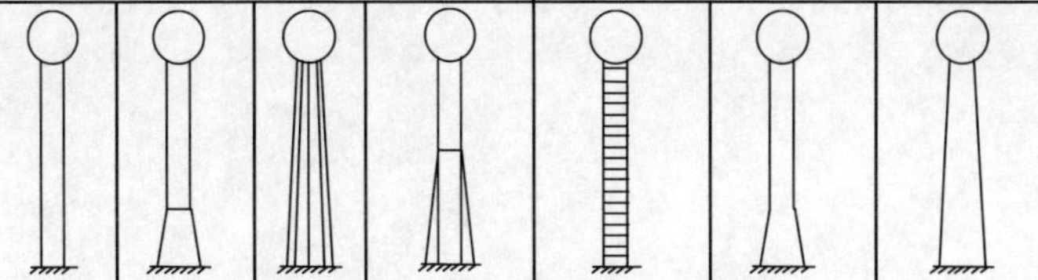 Cylindri-cal tower	Conical bottom	Lattice tower	Guyed lattice tower	Pre-fabricated segments	on-site concrete	on-site con-crete, conical
1st tower natural bending frequency $f_{0.1}$ [Hz] Frequency ratio $f_{0.1}$ / n [p]	0.55 1.44	0.56 1.46	0.55 1.44	0.55 1.44	0.65 1.70	0.96 2.50	0.96 2.50
$\varnothing$ top [m] $\varnothing$ bottom [m] Wall thickness [mm]	3.50 3.50 stepped 35–20	3.20 7.50 20	3.10 4.30 20	2.70 2.70 20	3.50 3.50 stepped 520/250	3.30 5.40 300	3.50 8.10 300
Tower mass [t]	114	90	87	63 + guy wires	430	455	540
Relative costs for supporting structure [%]	230	185	175	200	100	Not pre-stressed: 115 Prestressed: 160	Not pre-stressed: 135 prestressed: 185

Fig. 3-52 Various tower designs for WKA-60 [26]

However, the lattice towers are extensively used in countries which have a different structure of the production costs (lower labour costs compared to material costs). Moreover, they are an option for very tall towers (hub height above 100 m) at German inland sites where their low specific tower weight is advantageous. A lattice tower product line has lower certification costs. The on-site assembly of the tower segments, Fig. 3-53 bottom, reduces the component dimensions and therefore circumvents the transportation limits for the maximum tower diameter (clearance below bridges and road curve radius) existing for tubular steel towers and concrete towers, see below and chapter 15.

Fig. 3-53 Top: production of steel tube tower segments; left: bending of plate on 3-roll bending bench: right: welding together the plate of the tower segment (BWE); Bottom: FL 2500 in Laasow with 160 m lattice tower; left: assembling of the bottom tower segments; right: complete assembled wind turbine (Seeba, W2E, Fuhrländer)

Tubular towers are manufactured with a round or a polygonal section. The section area grows conically (or in cylindrical steps) from the top to the tower bottom taking into account the increasing bending moment, above all from growing lever of the rotor thrust. Moreover, at the tower bottom, a material saving tube design with large diameter and small wall thickness does not disturb the aerodynamics of the rotor. The already mentioned transportation limits for wind turbines of the Multi-MW class is the maximum clearance below bridges of 4.0 to 4.2 m. The maximum tower bottom diameter of a Vestas V-80 (2MW) is exactly 4.0 m. Tubular towers are mostly manufactured from steel. However, *spun concrete towers* are also applied which have lower production costs but higher transport and erection costs because a significantly larger mass is needed.

In order to simplify transport and erection there is the alternative of on-site production of the concrete tower using a jacking framework, Fig. 3-54. But in this case the quality control is difficult. Work has to be carried out very carefully, if the concrete is processed in 100 m height at low temperatures in the wintertime . Thus, another alternative was developed, e.g. by the manufacturer ENERCON, with its tower assembly from concrete segments produced in the workshop under defined conditions. This process has cost advantages compared to the on-site concrete method especially for a larger serial production. Concrete towers have a higher structural damping than steel towers, but they require tension anchors and steel ropes for pre-stressing the concrete which has only a limited capability of bearing tensile forces.

The *hybrid tower* is a tower type which combines the advantages of both the tower materials, concrete and steel. The tower bottom is made from concrete in order to avoid the transportation limits, and the upper tower part is from steel. Fig. 3-55 shows a hybrid tower with only a short concrete base. The tension anchors connecting the two tower parts and pre-stressing the concrete are inside the tower and freely accessible to allow the tension to be controlled and readjusted.

Fig. 3-54 Tower made of in-situ concrete with jacking formwork (Pfleiderer)

Fig. 3-55 Hybrid tower (GE Wind)

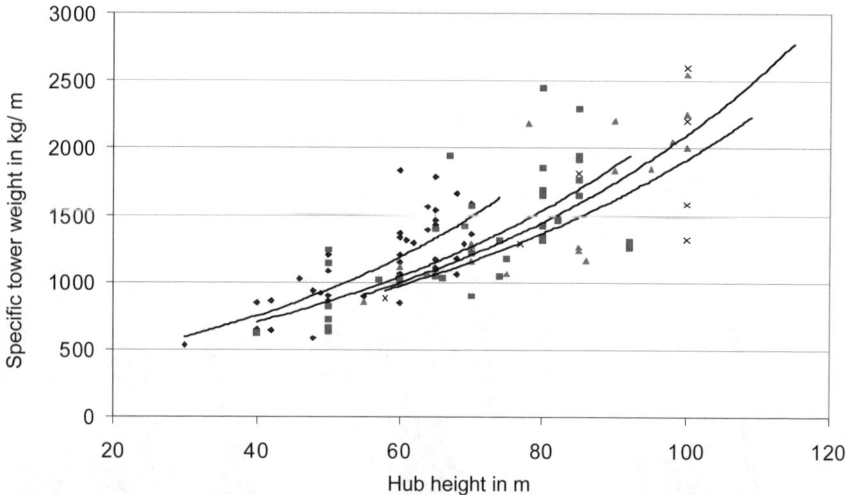

Fig. 3-56 Specific tower weight versus hub height [27]

Fig. 3-56 gives an overview of the typical specific tower weight versus the hub
height for different types of self-supporting towers. The large spreading of the
values for a certain hub height is attributable to different types and manufacturers.

Masts with guy wires are a common tower type of the smaller wind turbines,
e.g. Aerosmart 5 (Fig. 3-57 left) and SÜDWIND 1237 (Fig. 3-18). They are
lightweight towers and are suitable for erection with a winch, i.e. without any
auxiliary crane, which reduces significantly the costs for transport and erection,

Fig. 3-58, right. The guyed masts need a defined tension in the guy wires which has to be checked periodically. An interesting model of this mast was developed for a container hybrid system (comprising wind, solar, diesel, battery, etc.), Fig. 3-57b. The mast and guy wires are anchored to the container, so there is no requirement for an additional foundation and anchoring in the soil.

Other special tower types for small wind turbines also geared towards easy mounting and lowering of the turbine (for O&M or during storm) are the "A-shaped" mast. The latter needs only anchoring pegs instead of concrete foundations due to the big base area between its legs.

Fig. 3-57 Left: guyed pole (Aerosmart 5); right: mobile hybrid system (Terracon Energy Container)

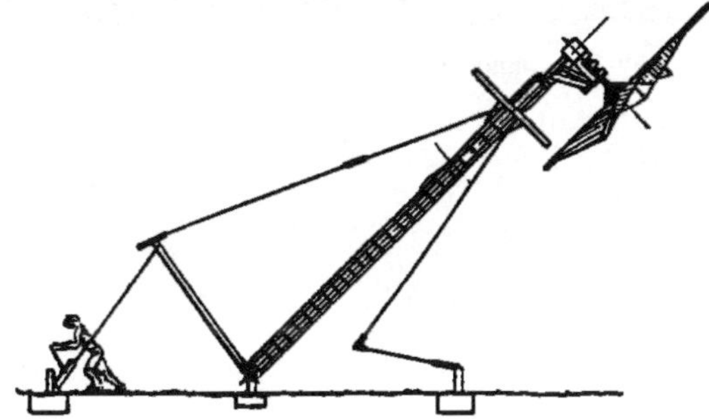

Fig. 3-58 Left: sail wind turbine with guyed A-pole; right: erection with a winch

3.4.2 Foundation

Wind turbines have mostly a *block-shaped foundation* made from concrete blocks Fig. 3-59. The flat centre foundation of self-supporting towers is dimensioned to prevent the turbine from tilting over (avoiding a gaping joint). In contrast to that, the guyed masts have a separated foundation design, (Figs. 3-57, 3-58 and 3-60). The main foundation of the mast prevents it from sinking into the soil (due to the weight and the vertical components of the tensile forces) whereas the anchoring foundations of the guying absorb the tensile forces. At lattice towers, the separated foundations have to bear a combination of this loading.

The basis of the flat foundation design is a *soil expertise* which proves that at the chosen site the soil provides at least the minimum bearing capacity assumed in course of the wind turbine design.

If the ground is very soft (e.g. the marsh lands at the North Sea coast) there is the requirement of an additional expensive pile foundation. Fig. 3-59 shows a typical flat foundation during its construction. Before the placing of the concrete it is important to install the cable feedthrough and the grounding. The foundation of a 600 kW wind turbine required e.g. approx. 165 m³ of concrete and about 25 tons of steel reinforcement.

Fig. 3-59 Flat foundation

Fig. 3-60 Foundation for guyed tubular steel pole

3.5 Assembly and Production

The nacelles of wind turbines are mostly assembled in the factory and transported as one unit to the site. In the factory, cranes of different capacity are needed for moving the components (nacelle frame, gearbox, generator, etc.), e.g. overhead cranes or jib cranes at the individual assembly areas. Fig. 3-61 shows a workshop of the company ENERCON. The maximum crane capacity, mostly an overhead crane, is required for loading the completed nacelle on the truck. Fig. 3-62 gives an overview of the nacelle and rotor mass of the different wind turbine sizes. The comparison of the actual curves with the theoretical curve obtained by scaling up the mass according to the similarity laws shows that with the increasing turbine size the lightweight design achieved a significant mass reduction.

Fig. 3-61 Nacelle assembly shop at company ENERCON

For smaller wind turbines, the hub may already be flanged to the rotor shaft in the workshop, Fig. 3-23, and transported together with the nacelle. But the hubs of Multi-MW wind turbines are very heavy and big, so they have to be transported separately. On site, the hub is assembled with the three rotor blades at the ground and then flanged to the rotor shaft of the already mounted nacelle, Fig. 3-55. Nevertheless, before the delivery, there is mostly a complete system test of the drive train including the hub with pitch system on a test stand in the workshop.

The *in-house production depth* (vertical integration) depends strongly on the manufacturer. The two main corporate philosophies are a high or low in-house production depth, see table 3.5.

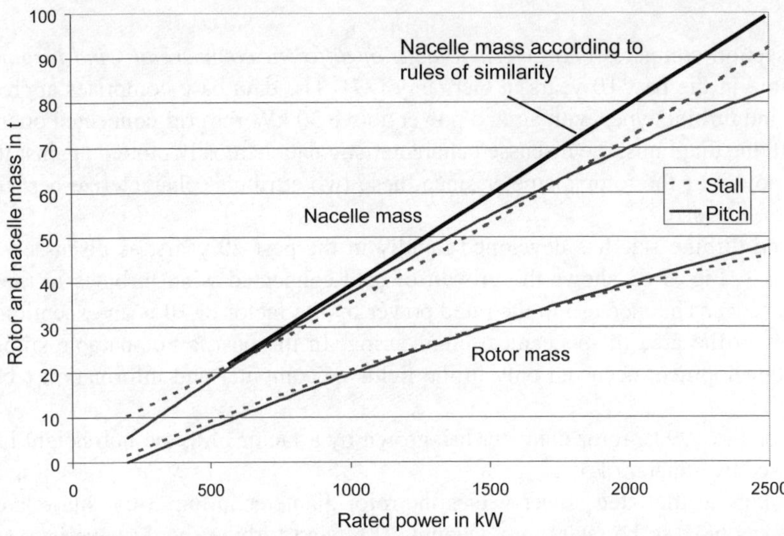

Fig. 3-62 Rotor- and nacelle mass versus size

Smaller manufacturers with a small market share tend more to a low in-house production depth. In-house production depth also grows in tandem with increasing company size. A specialty of the production of Multi-MW wind turbines is that most of the manufacturers do not have their own foundries, even if their in-house production depth is high, something which may turn out to be a bottle-neck. Only a few foundries are able to produce components with a weight of up to 100 t per single casting.

Table 3.5 Comparison of high and low in-house production depth

High in-house production depth		Low in-house production depth	
Advantage	Disadvantage	Advantage	Disadvantage
Low dependency on suppliers	High demand for capital	Low demand for capital	High dependency on suppliers concerning costs, quality and adherence to schedules
High value added	High risk of fluctuating production activity	High flexibility	Low value added
Easy quality control		Licensed manufacturing with local suppliers	

3.6 Characteristic wind turbine data

This section compares *characteristic data of different commercial wind turbines* available in the past 10 years in Germany [27]. The data base comprises approx. 300 wind turbine types with a rated power above 30 kW for grid-connected operation. In the diagrams, the discussed characteristic data is mostly plotted against the rated power or the rotor diameter since these two attributes characterize best the wind turbine size.

Wind turbine size has developed rapidly in the past 20 years, as discussed in chapter 1. Fig. 3-63 shows the growth of grid-connected wind turbines in these two decades. The increase in the rated power by the factor of 10 is an exceptional success in the area of mechanical engineering. In the past, a comparable stimulated development occurred only in the fields of computer and information technology.

Since the 1990s, rotor diameter has grown by a factor of 8, the hub height has increased by a factor of 5.

Looking at the rated power versus the rotor diameter in Fig. 3-64, site-related influences have to be taken into account. The wind turbine should have its best-efficiency point close to the wind speed of maximum energy density, cf. chapter 4. Wind turbines designed for coastal sites with a higher mean wind speed already achieve the same rated power with a smaller swept rotor area which results in a higher *area-specific power* (ratio of rated power and swept rotor area) of up to $520 \ W/m^2$ producing the upper limiting curve. The lower limiting curve of $290 \ W/m^2$ represents the design with a bigger rotor suitable for inland sites with a lower mean wind speed. For example, the 2 MW wind turbine MM82 of the company REpower with a rotor diameter of 82 m, designed for inland sites has an area-specific capacity of $378 \ W/m^2$, whereas its sister machine MM70 with the same rated power but designed for the windy coastal sites reaches $520 \ W/m^2$ due to the smaller rotor of 70 m.

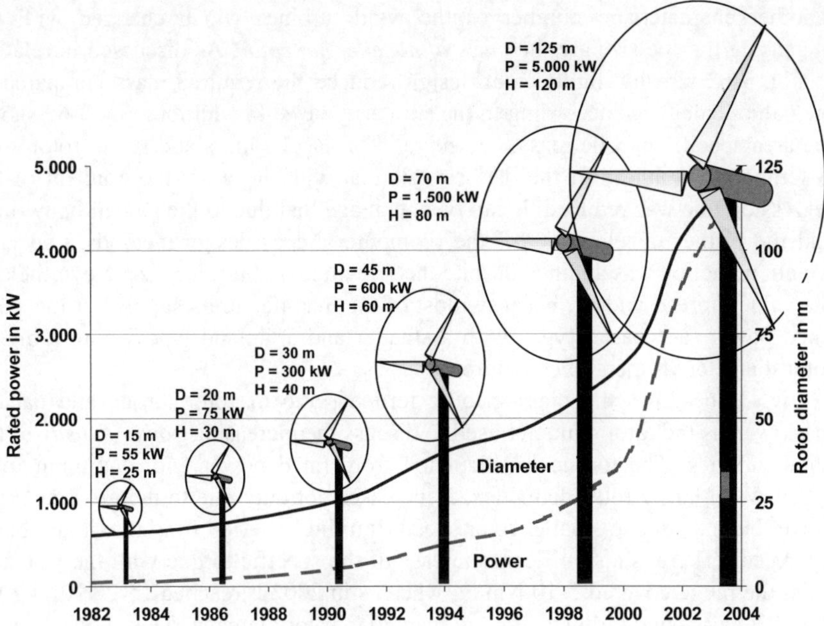

Fig. 3-63 Size and power of commercially produced wind turbines over time (cf. Fig. 1-1)

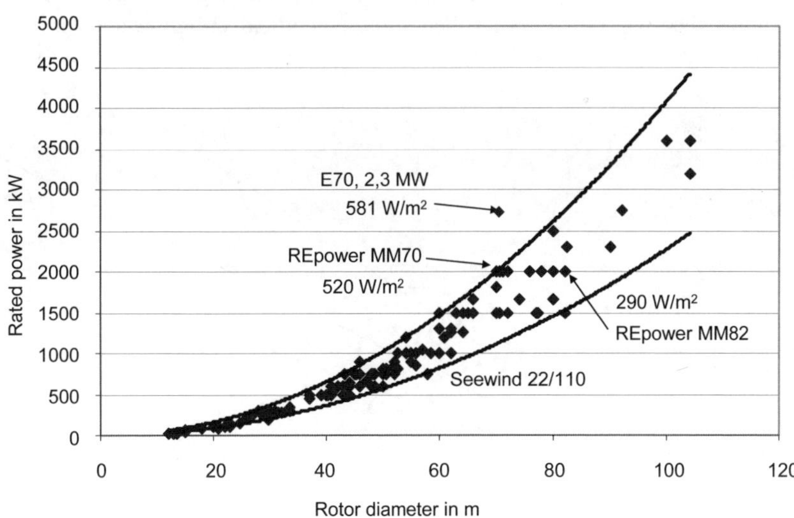

Fig. 3-64 Rated power versus rotor diameter

Another characteristic number of the wind turbines which changed with the increasing turbine size are the *mass of nacelle and rotor*. As discussed in relation to Fig. 3-62 already, lightweight design reduced the required mass compared to the values calculated according to the similarity laws. In addition, Fig. 3-65 shows the area-specific nacelle mass, i.e. the ratio of nacelle mass and swept rotor area. A temporal resolution of the data reveals that with the years the gradient of the regression line was reduced. It can be concluded that due to the gained know-how and the further development of the (computer-aided) design methods a specific weight reduction was attained despite the growing wind turbine size. Nevertheless, the values spread widely, because most of the manufacturers supply for the same rated power a "coastal type" with a smaller and a "inland type " with a bigger rotor diameter, as mentioned before.

Fig. 3-66 shows the mass-specific torque (ratio of rotor torque and nacelle mass) versus the rotor diameter used to discuss the increasing *power density* of the wind turbines. The torque is calculated from rated power and maximum rotor speed. With larger rotor diameters, it increases not only due to the growing rated power but also due to smaller rotor speed (limiting criterion from maximum blade tip speed). There is a significant increase of the specific torque with the years; in 1996 the range covered 5-10 Nm/kg whereas in 2002 it reached 15-20 Nm/kg for the MW class wind turbines of more than 60 m rotor diameter. This shows that the stress gets closer to the admissible material strength, the design is more stress-optimized (instead of high safety factors in the beginning) and also that new materials are applied. Moreover, the power density increases if generators and gearboxes are more compact because of e.g. better cooling and advanced power electronics. The wind turbines are becoming (specifically) lighter and more cost-effective.

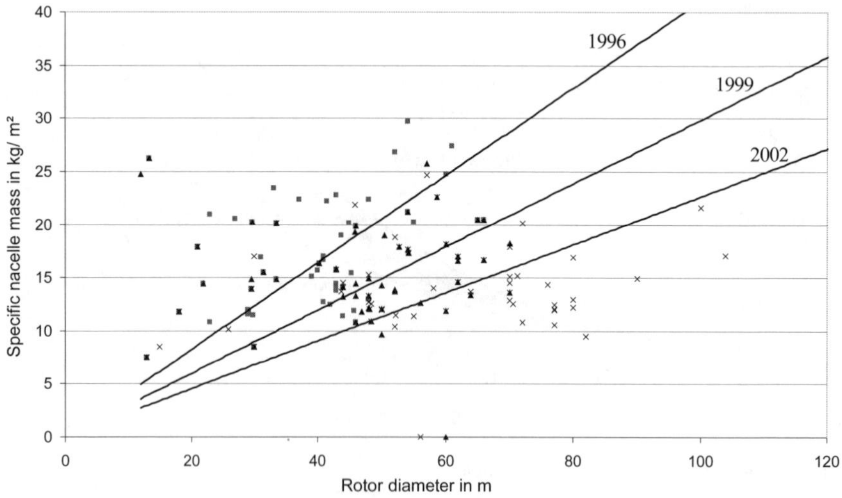

Fig. 3-65 Specific nacelle mass versus rotor diameter

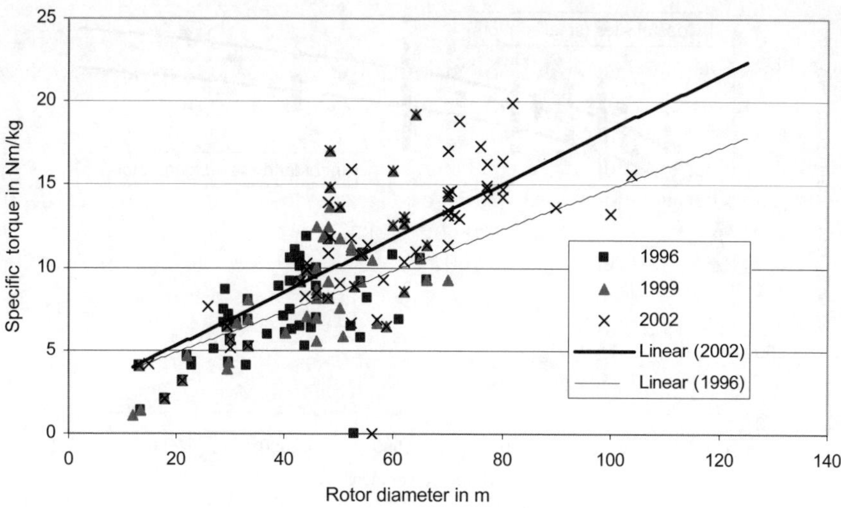

Fig. 3-66 Specific torque versus rotor diameter

The *energy yield of wind turbines* is basically site-specific. In order to get comparable values, the *reference yield* is introduced and defined by the wind conditions specified in the German Renewable Energy Sources Act (EEG) from April 2002 as follows [28]:

- Mean wind speed in 30 m height: 5.5 m/s
- Roughness length z_0:   0.1 m
- Wind frequency distribution according to Rayleigh: $k = 2$

The law establishes a virtual average German site with approx. 1,700 annual full load hours as the basis for wind turbine comparison. This is equivalent e.g. to moderately windy sites in the German federal state of Brandenburg. Fig. 3-67 shows that with the increasing turbine size (i.e. with the years) the area-specific reference yield (ratio of reference yield and swept rotor area) grew by more than 50%, from approx. 600 kWh_a/m^2 to approx. 1000 kWh_a/m^2. The spreading of the values depends not only on turbine type and manufacturer but also per type on the different available hub heights, see section 3.4. The achieved increase is mainly due to the growing hub heights providing better wind conditions for the turbine operation.

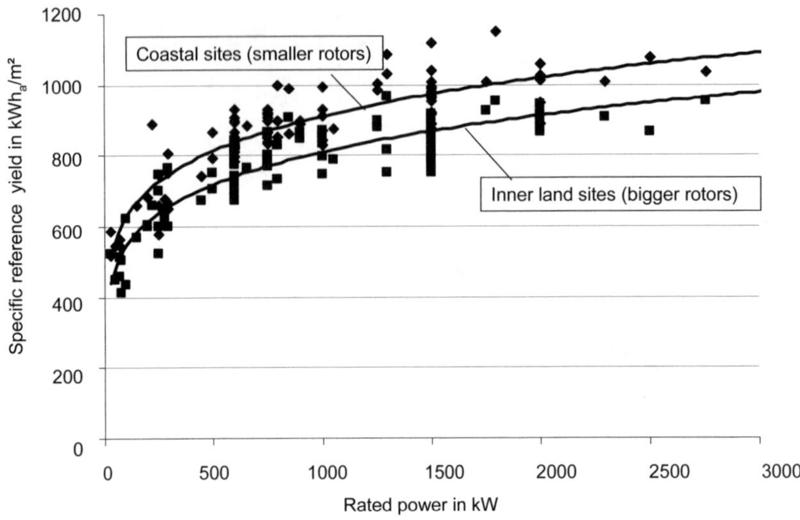

Fig. 3-67 Area specific annual reference yield versus rated power

Considering in Fig. 3-68 the best efficiency of the wind turbines, i.e. the power coefficient $c_{P.max}$, drawn from the power curves measured for type approval, a wide spreading of the values can be observed. The reason for this is the broad operating range of wind turbines. Hence, the manufacturers optimize the entire machine for a wide range of relatively high efficiency. This is needed especially for inland sites in order to maximize the annual energy yield: on one hand the very frequent weak winds have to be harvested efficiently, but on the other a good efficiency should also bring a high yield share from rarer strong winds ($P{\sim}v^3$!).
The final consideration, Fig. 3-69 discusses available commercial wind turbines concerning the two classical concepts of power limitation: stall and pitch. It shows a clear tendency to depart from a domination of stall-controlled turbines (without blade pitching) until the mid 1990s to a preference for pitch-controlled wind turbines evident in today's MW class turbines. In Germany, hardly any stall wind turbines have been recently installed, mainly due to strict grid codes, cf. chapter 14. In the first half of 2005, e.g. only pitch-controlled wind turbines were erected, of which 10% had an active stall control.

But in other countries with a different market situation and more restricted transport and erection conditions, there is a high demand for the robust and well-proven stall turbines. Even the company ENERCON, a classical manufacturer of gearless pitch-controlled wind turbines, now tests a 20 kW prototype with power limitation by stall. And also the 5 kW wind turbine Aerosmart 5, Fig. 3-57 left, recently developed for stand-alone systems and developing countries, has a downwind rotor with stall control.

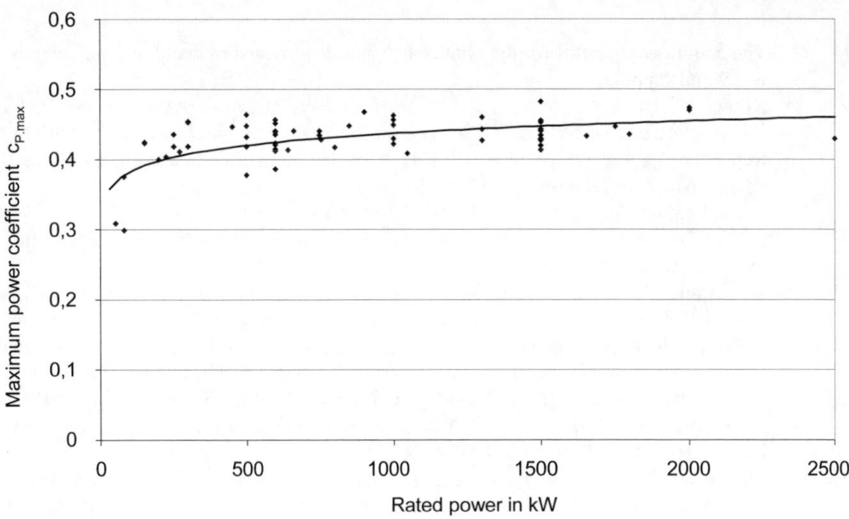

Fig. 3-68 Maximum power coefficient $c_{P.max}$ of a wind turbine versus rated power (from power curve measurements)

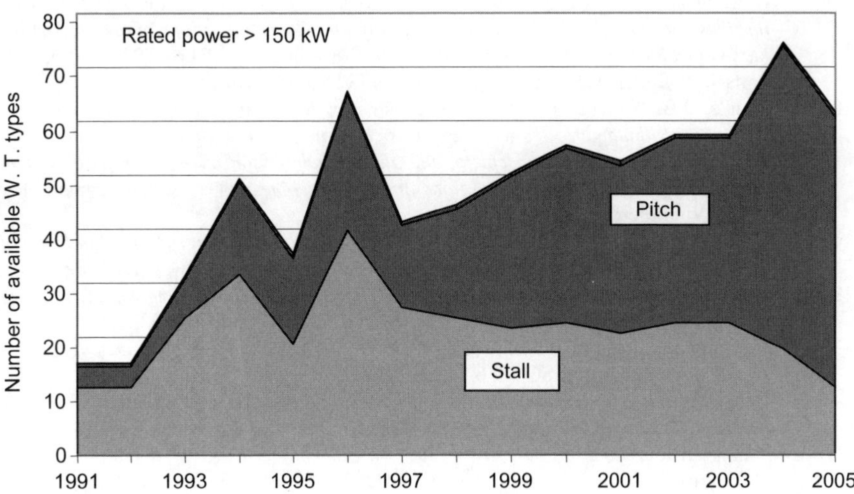

Fig. 3-69 Number of available stall and pitch wind turbine types on the German market [27]

References

[1]   The editors are grateful for the photos and figures provided by wind turbine manufac-
       turers and suppliers.

[2]   Körber F., Besel G. (MAN), Reinhold H. (HEW): *Messprogramm an der 3MW-
       Windkraftanlage GROWIAN (Test program on the 3 MW wind turbine GROWIAN)*,
       Report on the research project 03E-4512-A of the German Federal Ministry of Tech-
       nology, München, Hamburg 1988

[3]   According to information of Sydkraft; Malmö (S) 1991

[4]   N.N.: *What we have learnt about wind power at Maglarp and Näsudden*, Statens ener-
       gieverk, Stockholm (S) 1990

[5]   Wachsmuth R. (MBB*): Rotorblatt in Faserverbundbauweise für Windkraftanlage
       AEOLUS II (Rotor blade in fibre composite construction for the wind turbine
       AEOLUS II)*, Report on the research project 032-8819-A/B in the status report wind
       energy 1990 of the German Federal Ministry of Technology, KFA Jülich (Ed.) 1990

[6]   Hau E.: *Windkraftanlagen - Grundlagen, Technik, Einsatz, Wirtschaftlichkeit (Wind
       power plants – Basics, technology, application and economics)*, Springer-Verlag,
       Berlin, Heidelberg, New York 1988 / 2003

[7]   Maurer J.: *Wind turbinen mit Schlaggelenkrotoren, Baugrenzen und dynamisches
       Verhalten (Wind turbines with a flapping hinge rotor, design limits and dynamic
       characteristics)*, PhD thesis, TU Berlin; VDI Reports, series 11, No. 173, Düsseldorf
       1992

[8]   Wortmann F.X. (Institut für Aerodynamik und Gasdynamik der Uni Stuttgart): *Neue
       Wege zur Windenergie (New approaches to wind energy)*, Imprint of the DFG series:
       Forschung in der Bundesrepublik Deutschland - Beispiel, Kritik, Vorschläge, Verlag
       Chemie, Weinheim 1983

[9]   Person M.: *Zur Dynamik von Wind turbinen mit Gelenkflügeln - Stabilität und
       erzwungene Schwingungen von Ein- und Mehrflüglern (On the dynamics of wind
       turbines with flapping hinge blades – stability and forced vibrations of one and multi-
       bladed turbines)*, PhD thesis at the Institut für Luft- und Raumfahrt of TU Berlin, VDI
       Fortschritt-Berichte (Progress reports) Series 11, No. 104, Düsseldorf 1988

[10]   Franke, J.B.: *Erarbeitung eines Konzeptes zur Berechnung der Lebensdauer von
       Getriebeverzahnungen unter Berücksichtigung der Betriebszustände von
       Windenergieanlagen (A concept for the lifetime calculation of the gear tooth system
       considering the operating conditions in wind turbines)*, TU Berlin/Germanischer
       Lloyd WindEnergie GmbH, 2004

[11]   DUBBEL, Beitz W. und Küttner K.-H.: *Taschenbuch für den Ingenieur (Handbook of
       Mechanical engineering)*, Springer-Verlag Berlin, Heidelberg, New York, 2002

[12]   Deutsches Institut für Normung: *DIN 3990, Tragfähigkeitsberechnung von Stirnrädern
       (Calculation of load capacity of spur gears)*, Beuth-Verlag, Berlin, 1987

[13]   International Organization of Standardisation (ISO): *Calculation of load capacity of
       spur and helical gears*, ISO 6336, Genf, 1996

[14]   Boiger, P.: *Die Aerogear-Baureihe soll für mehr Sicherheit und Ruhe bei Getrieben
       sorgen (The Aerogear product line will improve safety and reduce noise of gearboxes)*,
       Windkraft Journal, 2002

[15]   Germanischer Lloyd WindEnergie GmbH: *Richtlinien für die Zertifizierung von
       Windkraftanlagen I...IV (Guideline for the Certification of Wind Turbines)*, Hamburg
       1993 to 2003

[16]   Haibach, E.: *Betriebsfestigkeit, Verfahren und Daten zur Bauteilberechnung (Fatigue
       analysis, procedure and data for component desig)*, VDI-Verlag, 1989

[17]   Schlecht, B., Schulze, T., Demtröder, J.: *Modelle zur Triebstrangsimulation von Multi-
       Megawatt Windenergieanlagen (Models for the drive train simulation of Multi-MW*

wind turbines), Tagungsband Dresdener Maschinenelemente Kolloquium 2003, S. 351-362, Verlagsgruppe Mainz, Aachen, 2003

[18] Thörnblad, P.: *Gears for Wind Power Plants*, 2nd Int. Symposium on Wind Energy Systems, Amsterdam, 1978

[19] N.N.: *Betriebsanleitung für die Allgaier Windkraftanlage (Operating manual for the Allgaier wind turbine) System Dr. Hütter, Type WE 10/G6*, Uhingen

[20] Institut für Solare Energieversorgungstechnik e. V. (ISET): *EU-Projekt NEW ICETOOL* (deutsches Teilprojekt), http://www.iset.uni-kassel.de/icetool/

[21] BINE Informationsdienst: *Blitzschutz für Windenergieanlagen (Lightning protection for wind turbines)*, Projektinfo 12/00, Fachinformationszentrum Karlsruhe, http://bine.info

[22] Fördergesellschaft Windenergie e.V. und Bundesministerium für Wirtschaft und Technologie (Hrsg.): *Blitzschutz von Windenergieanlagen(Lightning protection for wind turbines)*, Abschlussbericht, BMWi-Forschungsvorhaben 0329732, Juli 2000

[23] Deutsches Windenergie-Institut (DEWI): *Studie zur aktuellen Kostensituation 2002 der Windenergienutzung in Deutschland (Study on the costs of wind energy utilization in 2002)*, Wilhelmshaven 2002

[24] Germanischer Lloyd WindEnergie GmbH: *Richtlinie für die Zertifizierung von Condition Monitoring Systemen für Windenergieanlagen (Guideline for the Certification of Condition Monitoring Systems for Wind Turbines)*, Edition 2003

[25] Heilmann, C., Liersch, J., Melsheimer, M.: *Rotorunwuchten sind vermeidbar (Rotor unbalance is avoidable)*, Sonne, Wind und Wärme 4/2006

[26] Huß G. (MAN): *Modifizierung des Anlagenkonzepts WKA-60 im Hinblick auf eine Leistungssteigerung, Landaufstellung und eine Verbesserung der Wirtschaftlichkeit, (Report on the concept modifications of the WKA-60 wind turbine concerning performance improvement, onshore installation and improvement of cost-effectiveness)* Report on the research project 032-8824-A in the status report wind energy 1990 of the German Federal Ministry of Technology, KFA Jülich (Ed.) 1990

[27] Bundesverband WindEnergie e.V., BWE Service GmbH (Hrsg.): *Windenergie 2005 Marktübersicht (Wind energy market 2005)*, and previous editions starting 1996

[28] *Erneuerbare-Energien-Gesetz (EEG)* der Bundesregierung *(German Renewable Energy Sources Act (EEG))*, 2000, 2003, 2004 http://www.erneuerbare-energien.de/

[29] Körber, F.: *Baureife Unterlagen für GROWIAN (Documentation for the construction of GROWIAN)*, Final report on the research project of the German Federal Ministry of Technology, München, 1979

4 The wind

4.1 Origins of the wind

4.1.1 Global wind systems

The global atmosphere can be considered as a thermal engine in which the air masses are transported due to different thermal potentials. This thermal engine is powered by the sun. Water is the most important energy carrier in the atmosphere since it exists in the atmosphere in all the three states: vapour, droplets and ice. So, its latent heat when changing the state of aggregate from one phase to another is the dominant influence on the weather. The earth is a sphere, so getting from the equator closer to the poles the total irradiation by the sun declines more and more. Consequently, there is excess energy in the atmosphere in the equatorial zones and a deficit in the area of the poles. To equalize this imbalance, heat is transported by the air flow from the equator to the southern and northern hemisphere. This is done by the air mass exchange of the global wind systems.

Additionally, the Coriolis force from the rotation of the earth deflects the flow direction of the global winds. The global wind systems are shown in Fig. 4-1.

In each hemisphere there are three different zones: the tropical latitudes, the mid-latitudes and the polar latitudes. The tropical latitudes (Inter-Tropical Convergence Zone, ITCZ) on each side of the equator are between the low-pressure belt (tropical doldrums belt), and the subtropical high-pressure belt. This region is also called the Horse Latitudes and is found along the 30° latitude. Within this region forms the Hadley circulation (also Hadley cell): hot tropical air rises at the equator and flows in the higher atmospheric layers towards the poles while it is deflected more and more to the east due to the rotation of the earth. Around 30° latitude the air descends and flows back to the equator in the lower atmospheric layers causing the Trade Winds. The two Trade Wind systems meet in the ITCZ, closing the loop of the Hadley circulation.

The mid-latitudes (i.e. the Ferrel cell) are limited in the direction of the poles by the sub-polar through which is found along 60° latitude. Within the mid-latitudes the cyclonic western winds (the prevailing Westerlies) dominate. The polar zone is governed by the widely stretching low-pressure area situated above the polar region, with mainly eastward air flows.

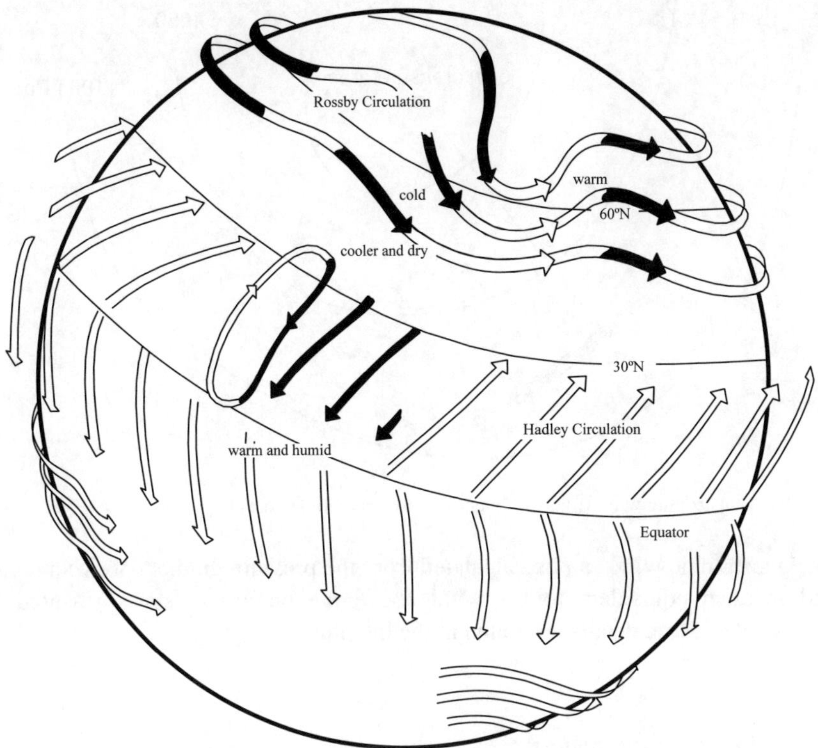

Fig. 4-1 Global wind circulation, black arrows: flow near the ground [2]

4.1.2 Geostrophic Wind

High above the ground, the friction of the ground has no influence on the air flow. This undisturbed flow is called the *geostrophic wind*; the pressure gradient force and the Coriolis force are in balance, Fig. 4-2. The gradient force results from the air pressure differences and points always towards the low-pressure region. If due to the pressure difference air mass moves, e.g., northwards along a Meridian, it is deflected by the Coriolis force to the right. If it moves from north to south, it is deflected to the left. So a balance between the two forces is found only if the geostrophic wind moves along the lines of constant pressure, the isobars. Hence, on the Northern hemisphere the low-pressure region is found always to the left of the geostrophic wind and the high-pressure to its right. On the southern hemisphere it works the other way round.

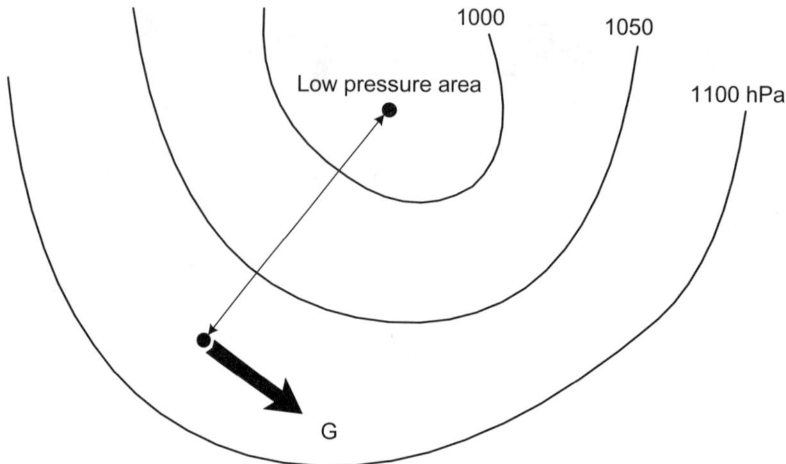

Fig. 4-2 Geostrophic wind *G* and isobars (on northern hemisphere) [2]

The geostrophic wind can be calculated from the pressure gradient above a region and is often equivalent to the wind above the boundary layer measured by radiosondes. More details are found in the literature, e.g. [5].

4.1.3 Local wind systems

Apart from the global winds there are also local wind systems which origin from potential differences. Again temperature differences are the driving cause. The most important of these local wind systems are the sea-land breeze and the mountain-valley breeze.

Sea-land circulation

The sea-land breeze is a diurnal wind system at the sea shore emerging under sunny weather conditions with a high temperature difference between the land and the water. In a weaker form it also occurs at the shore of bigger inland lakes. During the day, the onshore ground heats up stronger than the water. Correspondingly, the air above shows a temperature difference causing a pressure gradient from the sea to the land. The resulting "sea breeze" is the flow of cool, humid air to the shore, Fig. 4-3. High above the ground, there is a return flow from the land to the sea. In the late afternoon, the land is cooling down faster than the sea, so the flow conditions reverse. This causes the seawards pointing land breeze. Due to the friction at the ground it is weaker than the sea breeze.

Mountain-valley circulation

In the mountains, there are also thermal circulations from the superposition of two flow systems: the slope winds and the mountain-valley winds. High-pressure conditions with a high irradiation are required to form an intense circulation as long as there is no disturbing influence from the large-scale wind flow. After sunrise the slopes of the mountain are strongly heated by the sun, so are the air masses closely above it. Its thermal lift produces the rising slope wind. Later in course of the morning this flow is followed by the valley wind, an upwards flow through the valley. This flow replaces from below the air rising on the slopes. At night it is the other way round: due to the accelerated cooling of the ground, the air close to the slopes cools down faster than the air in free atmosphere at the same altitude. The gravitational force drives the falling downhill winds which meet in the valley and cause the mountain winds directed to the valley's exit, Fig. 4-4.

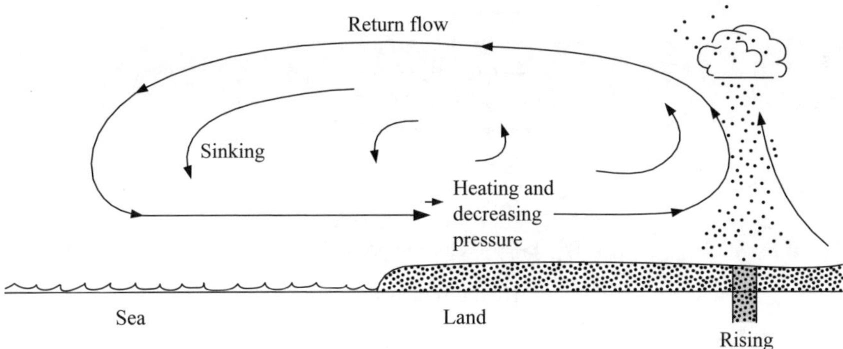

Fig. 4-3 Scheme of the sea-land breeze, example: day [2]

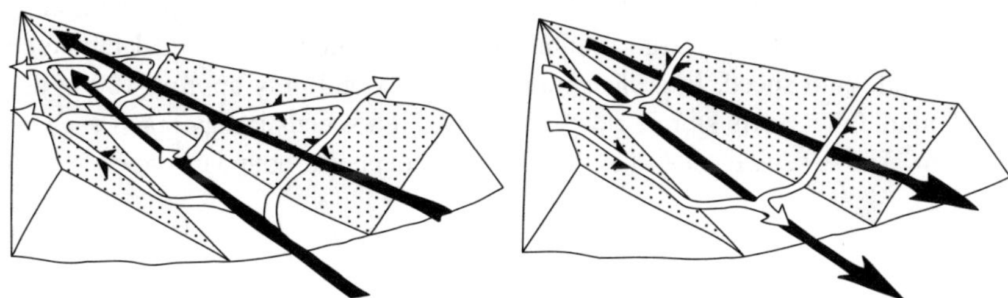

Fig. 4-4 Mountain-valley breeze (white: slope winds, black: valley wind). Left: valley wind at noon; right: mountain wind at night [10]

4.2 Atmospheric boundary layer

The lowest layer of the atmosphere is a turbulent layer called the *atmospheric boundary layer*. The air flow in this layer is influenced by the friction at the ground, the orography, the topography and as well by the vertical distribution of temperature and pressure. The geostrophic winds above the boundary layer are not influenced by the friction at the ground.

Thus, between this undisturbed geostrophic winds and the ground there is the layer where the wind speed is varying strongly with the height above ground. The friction at the surface roughness takes energy from the air flow causing a vertical gradient of the wind speed. This leads to a turbulent exchange of momentum and air masses with the air flowing in higher layers above the ground.

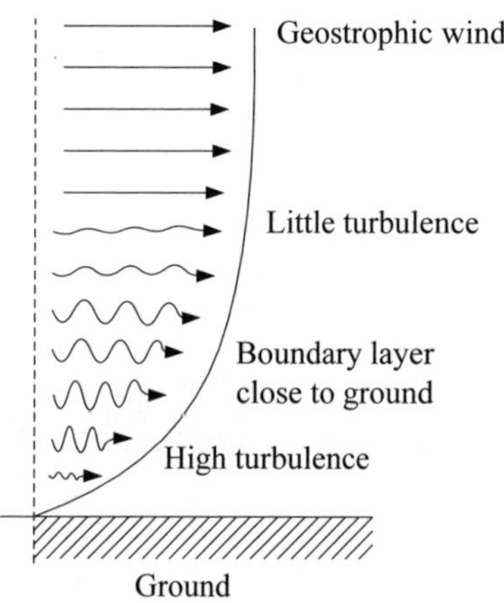

Fig. 4-5 Scheme of the atmospheric boundary layer

Wind turbines have to operate within this atmospheric boundary layer. The properties and the intensity of this air flow determine the amount of extractable energy and also the loads on the wind turbines. The height of the atmospheric boundary layer varies depending on the roughness of the ground, the vertical temperature profile and wind speed. In a clear night it may extend only approx. 100 m above the ground. On a warm summer day with low wind speeds it may reach even up to 2,000 m. The assumed mean height of the atmospheric boundary layer is around 1,000 m.

4.2.1 Surface boundary layer

The air mass flowing directly above the ground is called *surface boundary layer* (or Prandtl layer). Its height is often given as a fixed value of approx. 10% of the atmospheric boundary layer's height. In reality, it varies e.g. depending on the vertical temperature profile.

The variation of the wind speed with the height above ground is called the *vertical wind profile* (also called wind shear or height profile). Knowing its shape is very important for the determination of the energy yield of a wind turbine. Moreover, the vertical wind profile produces particular loads on the rotor and the entire structure of the wind turbine. It is influenced by the surface roughness, the nature of the terrain, the topography and also the *vertical temperature profile*. This thermal stratification is characterized by the three states of atmospheric stability: stable, neutral and unstable, Fig. 4-6.

If the thermal stratification of the atmosphere is *unstable*, the air close to the ground is warmer than the air above it. This situation is typical for the summer time when the sun heats up strongly the ground. The air close to the ground heat up as well, so its density reduces and it rises up. This causes a strong vertical mass transfer with increased turbulence. The stronger vertical mixing under unstable atmospheric conditions leads to a small gradient of the wind speed with increasing height above ground.

If the thermal stratification is *stable*, the temperature of the air close to the ground is smaller than in the layers above it. This situation is typical for the winter time when the ground cooled down strongly and correspondingly the air density close to the ground is higher than in the higher layers. This stable equilibrium leads to a small vertical mass transfer, turbulence is suppressed. So there is a strong gradient of the wind profile since the vertical differences in the (horizontal) wind speed are not equalized. In a stable atmosphere occur under certain circumstances also significant changes of the wind direction with increasing height.

If the atmosphere is *neutral* (neutrally stable) the surface boundary layer is neither heated up nor a cooled down resulting in an adiabatic temperature profile. The air temperature decreases by approx. 1°C per 100 m height. This situation often occurs at high wind speeds. In this case, the vertical wind profile depends no longer on the thermal mixing but only on the surface friction.

The influence of the temperature on the vertical wind profile is shown in Fig. 4-10.

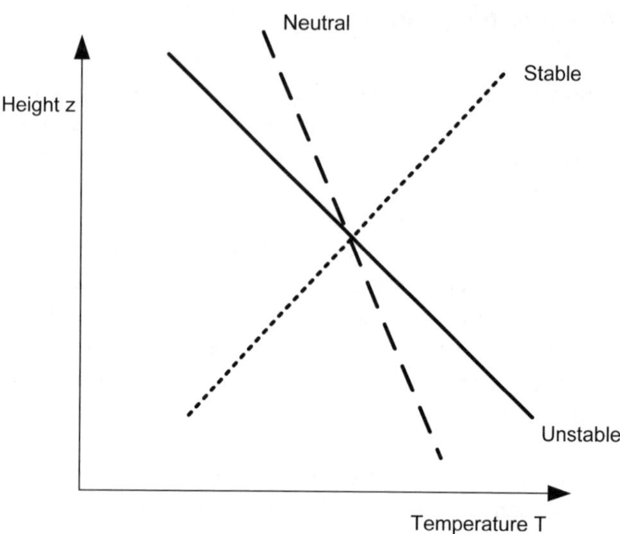

Fig. 4-6 Vertical temperature profiles: Change of temperature T with height z

4.2.2 *Vertical wind profile*

For the wind in a neutrally stable atmosphere above an idealised absolutely flat and unlimited landscape with a uniform roughness length there are analytic descriptions of the increase of the wind speed v (symbol V in the IEC standards) with the height z in the surface boundary layer.

Sometimes (e.g. also in [20]), the mean vertical wind profile is described by the *Hellmann power law*:

$$\frac{v(z_1)}{v(z_2)} = \left(\frac{z_1}{z_2}\right)^{\alpha},$$

(4.1)

where $v(z_1)$ resp. $v(z_2)$ is the wind speed in the height z_1 resp. z_2.

The power law exponent (also called empirical wind shear exponent) is $\alpha = 0.14$ for normal conditions [1]. But it varies with the height z, the roughness, the atmospheric stratification and the nature of the terrain and orography. There-fore, a measured power law exponent is valid only for the individual site and the used measuring heights z_1 and z_2. Its application to other heights is not permitted. Hence, the application of equation (4.1) is limited.

A description of the mean vertical wind profile for a neutrally stable atmosphere, based on the boundary layer physics of Prandtl, is the *logarithmic wind profile* which considers also the surface roughness:

$$v(z) = \frac{u_*}{\kappa} \ \ln\left(\frac{z}{z_0}\right) \ . \tag{4.2}$$

The logarithmic wind profile depends on several parameters: the shear stress velocity u_*, the height z above ground, the *roughness length* z_0 and the Kármán constant which is usually assumed to be $\kappa \approx 0.4$.

For a known roughness length at a site the application of equation (4.2) to two different heights gives the following very useful equation

$$v_2(z_2) = v_1(z_1) \cdot \frac{\ln\left(\dfrac{z_2}{z_0}\right)}{\ln\left(\dfrac{z_1}{z_0}\right)} \ . \tag{4.3}$$

Having measured a wind speed v_1 in the height z_1 (measuring mast) this equation allows to calculate the wind speed v_2 in the height z_2 (e.g. hub height) if the roughness length z_0 is given, table 4.1 shows typical values. The roughness length z_0 is a measure for the surface structure; the influence of z_0 on the vertical wind profile is shown in Fig. 4-7.

Table 4.1 Roughness length z_0 for different types of terrain

Type of terrain	z_0 in m
Calm water	0.0001 – 0.001
Farm land	0.03
Heather with few bushes and trees	0.1
forest	0.3 – 1.6
suburb, flat building	1.5
City centers	2.0

Having erected a measuring mast which records the wind speed at several heights of the wind profile, the roughness length z_0 may be determined for the close area around it. Assuming two measured wind speeds $v_1(z_1)$ and $v_2(z_2)$, and using $\ln(z_1/z_0) = \ln z_1 - \ln z_0$ the transformation of equation (4.3) gives

$$\frac{v_2}{\ln z_2 - \ln z_0} = \frac{v_1}{\ln z_1 - \ln z_0} \ . \tag{4.4}$$

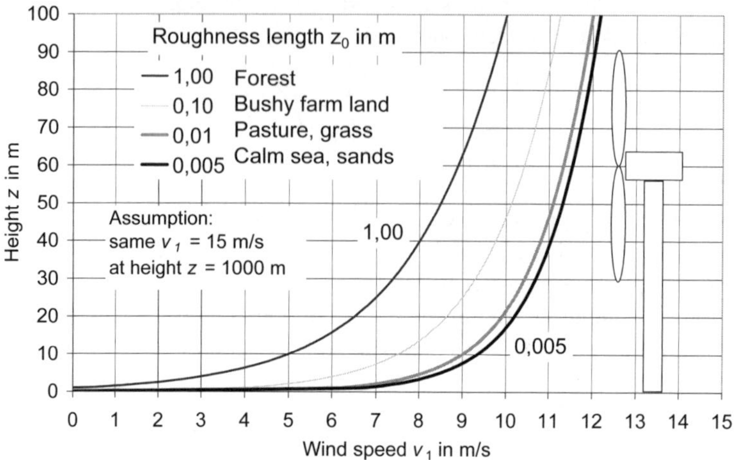

Fig. 4-7 Vertical wind profile for different roughness lengths z_0, assumed "geostrophic wind" of 15 m/s

The graphical interpretation of this equation (i.e. the intercept theorems) is shown in Fig. 4-8. The line drawn through the two measuring points hits the horizontal logarithmic axis ($\ln z$) exactly where the wind speed is zero. So Fig. 4-8 also shows the physical meaning of the roughness length, it is the height above ground where in the mean the wind speed is zero. In Fig. 4-7 the starting points of the wind profile curves at $v = 0$ m/s with the individual z_0 values arc subvisible; the vertical axis resolution is not fine enough.

If this procedure for the determination of the roughness length is applied, the height differences of the two anemometers must be sufficiently large, e.g. using the heights above ground of 20 m and 40 m [38]. Often the measuring mast is equipped with three or four anemometers with measuring heights up to 80 m. Then there are more points for fitting the interpolated line in Fig. 4-8.

In wind farm planning, the roughness length has to be estimated for a large area of up to several square kilometres since *changes of the roughness length* in the surrounding area, Fig. 4-9, have an influence on the vertical wind profile over a larger distance. These large-scale considerations for estimation of the roughness are based on the topography descriptions (site inspections, mappings, satellite images).

Offshore, the surface roughness is determined by the actual swell. Depending on the wind speed, the exposure time and the wind direction the swell develops characteristic wave heights and lengths which determine the roughness of the ocean surface. Recent investigations [15] indicate that offshore the wind profile is significantly influenced by the atmospheric stratification since the surface roughness is generally very small.

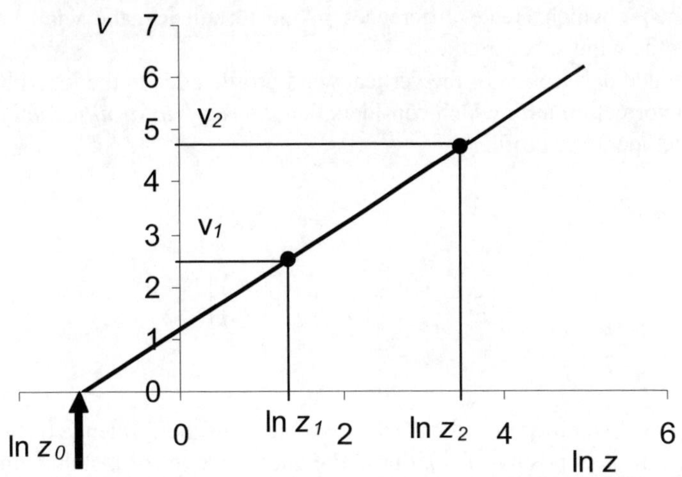

Fig. 4-8 Determination of roughness length z_0 from measurements at two different heights z_1 and z_2

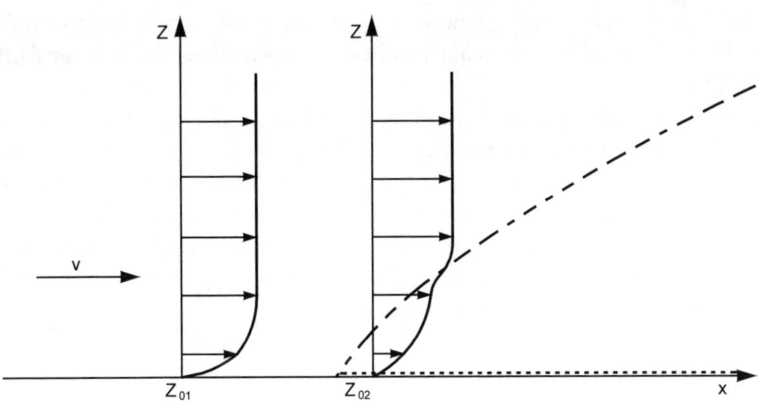

Fig. 4-9 Influence of roughness change on the vertical wind profile – transition zone

As mentioned, the above equations (4.1) to (4.4) are only valid for a neutrally stable atmosphere, an idealized absolutely flat surface with a uniform roughness length. A different *temperature profile* causes changes in the vertical heat and mass transfer which create differences in the turbulence, the wind profile and therefore the wind speed gradient.

The extended description of the vertical wind profile adds to the logarithmic wind profile a correction term which considers the *atmospheric stratification* due to the vertical temperature profile:

$$ v(z) = \frac{u^*}{\kappa}\left(\ln\left(\frac{z}{z_0}\right) - \Psi\left(\frac{z}{L}\right) \right) \ . \tag{4.5} $$

The empirical stability function Ψ takes into account the influence of the thermal stratification; it is positive for an unstable and negative for a stable atmosphere, and surely zero for a neutrally stable atmosphere.

The so-called Monin-Obukhov stability length L (its dimension is meter) is the parameter which describes the vertical mass exchange due to the ratio of friction forces and lift forces [31]. From measurements the Monin-Obukhov stability length can be determined directly or indirectly. A direct measurement is possible e.g. with an ultrasonic anemometer (cf. section 4.3.2) or by measuring the temperature difference of the air in two heights. For a first estimation, tables on the Monin-Obukhov stability length in dependence of the roughness length z_0 and the atmospheric stability are useful, e.g. in the German Technical Instructions on Air Quality (TA-Luft, dated 2002) [32].

Fig. 4-10 shows the already mentioned influence of the atmospheric stratification on the wind profile. The wind profiles are normalized for the measuring height of 30 m.

The logarithmic wind profile is valid for the infinitely flat and even terrain. But in reality, there may be considerable effects of the *surface contour* (i.e. the topography) on the wind profile and also on the wind speeds occurring in the rotor area of the wind turbine. The possibilities to describe these effects using fluid mechanics approaches are quite limited. Especially for complex and steep terrain as well as for strong atmospheric stratification the wind profiles obtained by extensive efforts in numerical simulation are still unsatisfactory.

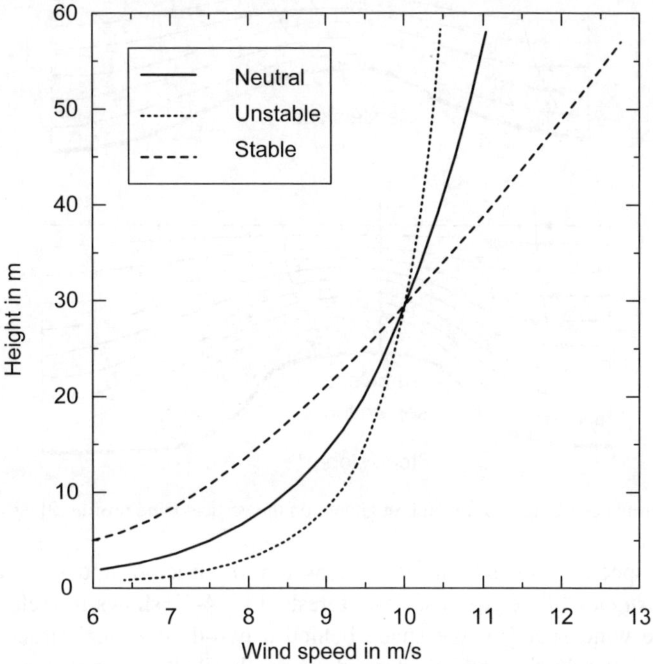

Fig. 4-10 Vertical wind profile – influence of atmospheric stability [1]

Surface formations lead to a deformation of the vertical wind profile. In general, the wind speed is accelerated when flowing over the top of hills and mountains, Fig. 4-11. The wind profile gets steeper. The wind profile shape on the summit depends on the surface roughness of the terrain. If the slope is very steep, with an inclination of more than 30%, flow separation occurs on the leeward side. Inside the turbulent separation bubble the wind is very intermittent. The size of the separation bubble depends on the inclination and the curvature of the elevation, the temperature profile and also on the surface roughness. In extreme cases there may be a decrease of the wind speed with height above ground.

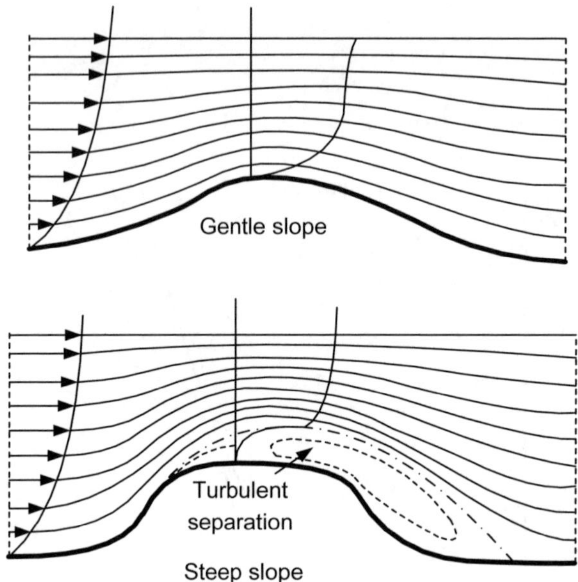

Fig. 4-11 Influence of topography and roughness on the vertical wind profile [9]

The wind speed is not only influenced by terrain structure and surface type but also by *obstacles* like e.g. houses and forests. Fig. 4-12 shows the relative reduction of the wind speed in the shade behind a two-dimensional structure. In the hatched area, the shading strongly depends on the detailed geometry of the obstacle.

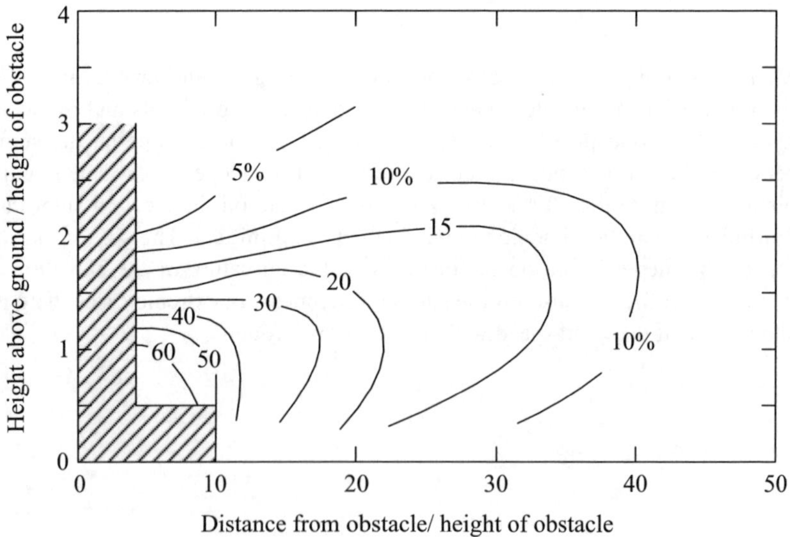

Fig. 4-12 Relative reduction of wind speed behind obstacles [5]

Above the surface boundary layer (Prandtl layer), there is the *Ekman layer*. Here, the wind profile is additionally influenced by the Coriolis force. This effect is considered by the following analytic expression [16] for the wind profile

$$v(z) = u_\mathrm{g} \sqrt{(1 - 2e^{-\gamma z}) \cdot \cos(\gamma z) + e^{-2\gamma z}} \qquad (4.6)$$

where u_g is the absolute value of the geostrophic wind and γ is the coefficient which describes the vertical turbulent mass transfer. It is assumed that it is constant, but its value is depending on the surface roughness and the atmospheric stratification and may be estimated only from measurements. Thus, there is also in equation (4.6) an uncertainty concerning the calculated vertical wind profile. But for large heights above ground it is more suitable than equation (4.5).

4.2.3 Turbulence intensity

The *fluctuations of the wind speed* in the atmosphere belong to processes with a time scale from less than one second to several days and a spatial structure from some millimetres to several kilometres, Fig.s 4-13 and 4-23.

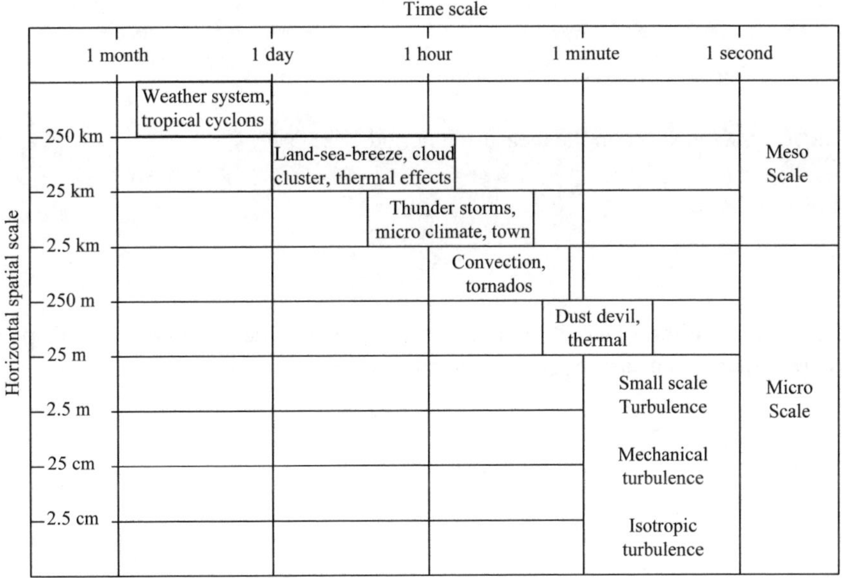

Fig. 4-13 Spatial and temporal scale of atmospheric phenomena [16]

The diurnal, synoptic and seasonal fluctuations of the wind are used together with the mean wind speed to predict the energy yield for a site, on one hand. On the other, the knowledge about turbulence and gustiness is required first of all for the load calculation of the wind turbine.

The fluctuating wind is described (roughly) by the mean wind speed $\bar{v}$ in the considered time period (mostly 10 min) and the turbulence intensity I_v, Fig. 4-14.

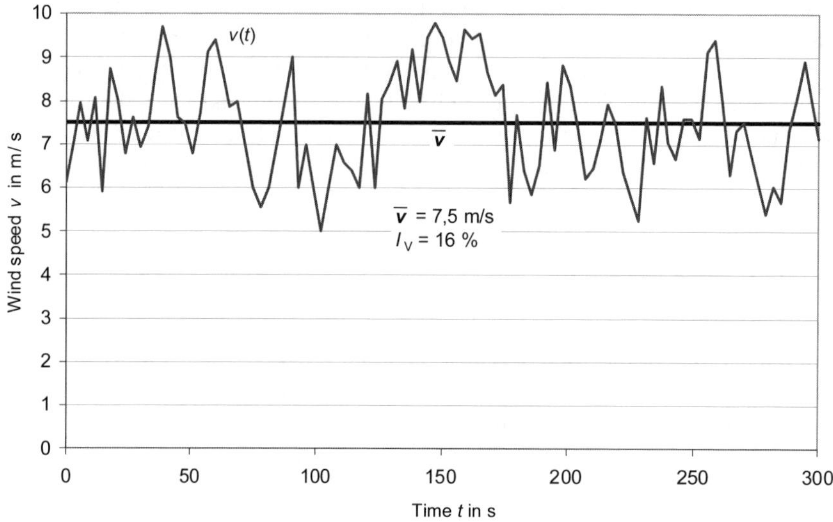

Fig. 4-14 Turbulent wind and mean wind speed $\bar{v}$

The *mean wind speed* in the measuring period T is

$$\bar{v} = \frac{1}{T}\int_0^T v(t)\,dt \; .$$

$$(4.7)$$

The average of the squared difference of the actual value to the mean wind speed is the *variance*, a measure for the "irregularity" of the wind

$$\overline{v^2} = \frac{1}{T}\int_0^T \left(v(t) - \bar{v}\right)^2 dt \; .$$

$$(4.8)$$

The square root of the variance is the standard deviation, $\sigma_v = \sqrt{\overline{v^2}}$, which has the same dimension as the mean wind speed $\bar{v}$. The ratio of standard deviation to the mean wind speed gives the *turbulence intensity* I_v

$$I_V = \frac{\sigma_v}{\overline{v}} = \frac{\sqrt{\overline{v'^2}}}{\overline{v}} \quad . \tag{4.9}$$

Roughly considered, the turbulent fluctuations have a Gaussian distribution around the mean value $\overline{v}$ with the standard deviation σ_V, cf. Fig. 4-30. But certainly extreme gusts like the 50-year extreme wind speed (see chapter 9, table 9.2) are not included in this range.

The turbulence intensity varies in a large range from $I_V = 0.05$ to 0.40 due to the natural fluctuations but also due to different averaging periods of the measurements and optionally the time response of the sensors.

There are two *fundamental types of atmospheric turbulence*: mechanical turbulence and thermally induced turbulence. *Mechanical turbulence* comes from the shear in the wind profile. It depends on wind flow from the large-scale pressure gradients and on the surface roughness. This share of the turbulence intensity is estimated based on fluid mechanics for a neutrally stable atmosphere in flat terrain with

$$I_v = 1 / \left(\ln\left(\frac{z}{z_0} \right) \right) . \tag{4.10}$$

Thermal turbulence is caused by heat convection and depends first of all on the temperature difference between the ground and the air masses above it.

Measurements show that the turbulence intensity decreases with increasing wind speed, Fig. 4-15. At low wind speeds, the turbulence intensity strongly depends on the atmospheric stability.

Above the ocean surface, these facts are no longer valid, since here the wind flow creates the roughness by itself since it forms the waves. The formation of the roughness is influenced not only by the actual wind speed, Fig. 4-16, but also by the time of action on the ocean surface meaning that the roughness is time-dependent. In general, the roughness increases, like the wave height, with higher wind speeds.

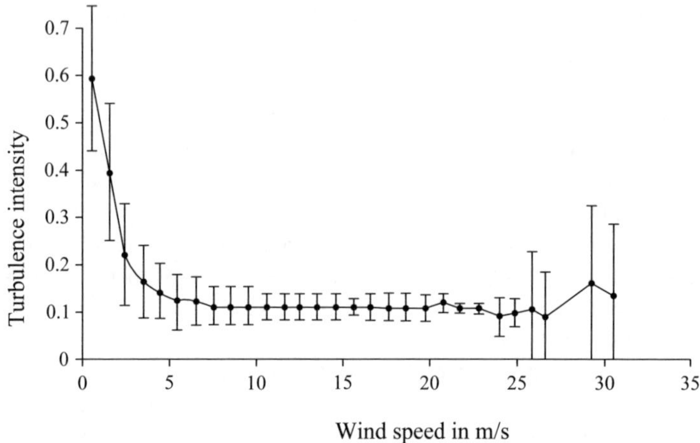

Fig. 4-15 Turbulence intensity versus wind speed, example of an onshore measurement

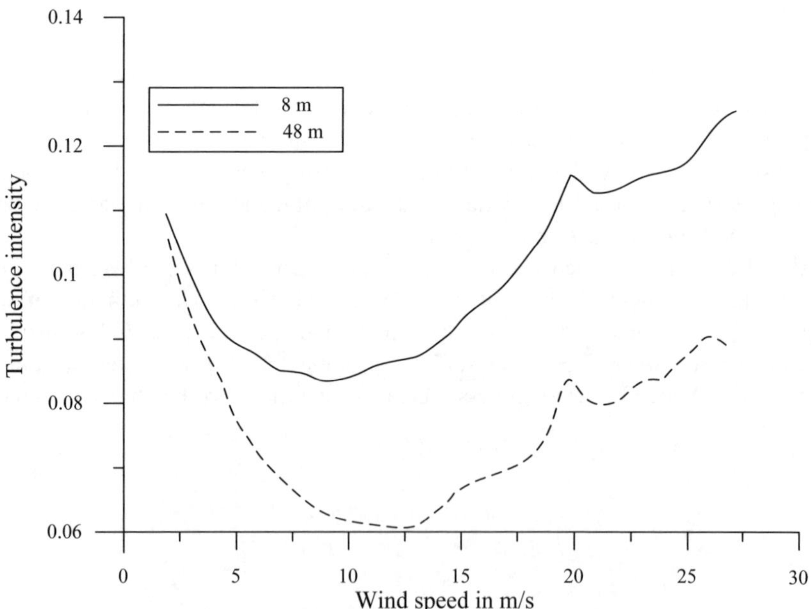

Fig. 4-16 Turbulence intensity versus wind speed, example of an offshore measurement at two different heights [26]

4.2.4 Representation of measured wind speeds in the time domain by frequency distribution and distribution functions

The determination of the annual energy yield and as well the assessment of the wind turbine loads are mostly based on measured 10-min averages of the wind speed; seldom is the averaging period a 1-min or 1-hour interval.

Additionally, the wind direction is measured during the whole measuring period using a wind vane, and all the values are recorded in time series.

Apart from the 10-min average wind speed, the maximum and minimum value as well as the standard deviation of the averaging period may be stored in the data series as well, cf. Fig. 4-30.

The measuring period should be long enough, especially for the calculation of the predicted energy yield at a site. Depending on the local climate, the wind speed may undergo strong seasonal fluctuations, so the measuring period should be an integer multiple of a year.

Investigations show that the annual energy content of the wind may vary in a range of up to ±25% [5]. Fig. 4-17 shows for a period of 100 years the energy content averaged over five-year periods and normalized with the 100-year average. The strong fluctuations are clearly visible.

On one hand, the long-term measuring data should be stored carefully, ("at best on a CD ROM in a safe"). On the other, it has to be processed using statistics.

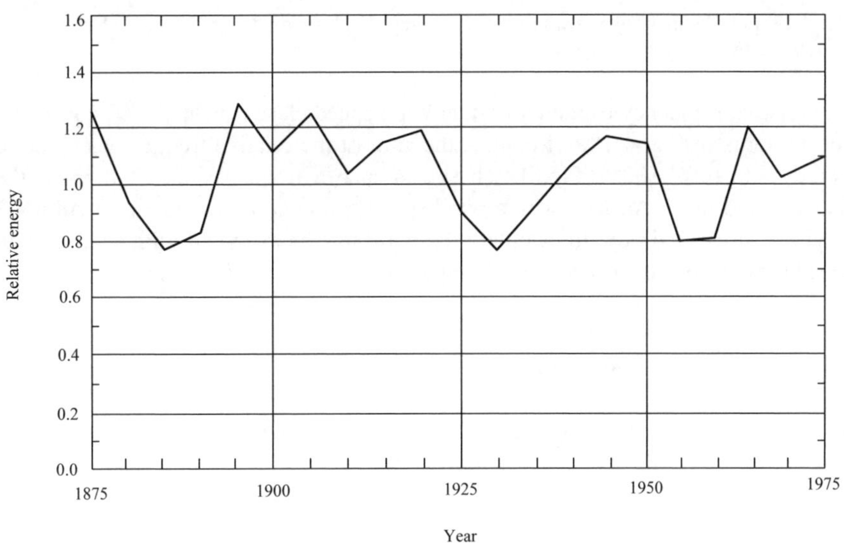

Fig. 4-17 Relative wind energy, averages over 5-year periods in Laeso, DK [5]

Frequency distribution – Histogram of wind speeds

For the wind turbine design as well as for the assessment of the expected energy yield, the pure time series of possibly several years of measurement are quite impractical. One possibility for attaining a compressed representation of the wind conditions is the generation of a *frequency distribution* of the wind speed. The wind speeds are sorted into classes and summed up, Fig. 4-18. This way, the temporal share of the individual wind speed class of the entire considered time period is determined. The width of the wind speed class is typically 1 m/s.

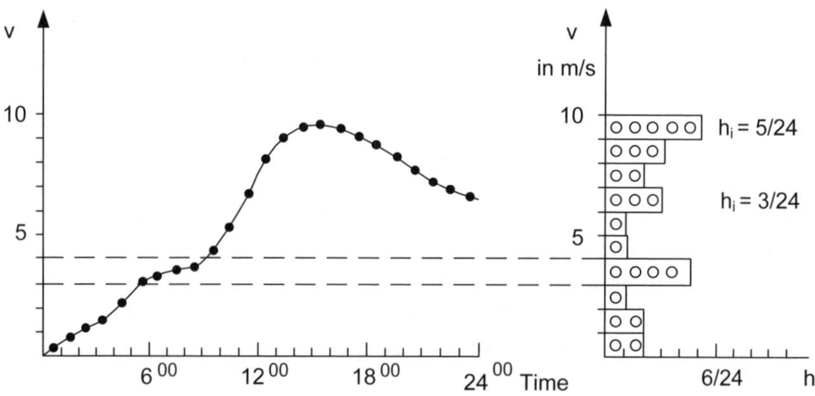

Fig. 4-18 Left: Diurnal time series of hourly averaged values; right: corresponding histogram of the relative frequency h_i of the respective day

The relative frequency of the individual wind speed class v_i is $h_i = t_i / T$, e.g. hours per 24 hours in Fig. 4-18. Of course, the sum of the relative frequencies must be exactly 1.0, resp. 100%. Fig. 4-21b shows such a frequency distribution of the Tauern wind farm (Austria) with a wind speed class width of 1 m/s. In section 4.3 it will be presented how to calculate the expected energy yield with such a wind speed histogram and a given wind turbine power curve, cf. Fig. 4-25.

Wind speed distribution function

The measured frequency distribution is mostly "compressed" into a mathematical description using the *Weibull distribution function* which is quite flexible due to its two parameters

$$h_W(v) = \frac{k}{A}\left(\frac{v}{A}\right)^{k-1} \exp\left(-\left(\frac{v}{A}\right)^k\right) . \tag{4.11}$$

The scaling factor A is a measure for the characteristic wind speed of the considered time series. The shape factor k describes the curve shape. It is in the range between 1 and 4, and the value is roughly characteristic for certain wind climate:

$k \approx 1$:   Arctic regions
$k \approx 2$:   Regions in Central Europe
$k \approx 3$ to 4: Trade wind regions

If there are small fluctuations around the mean wind speed $\bar{v}$ the value of k is high whereas large fluctuations give a smaller shape factor k, Fig. 4-19.

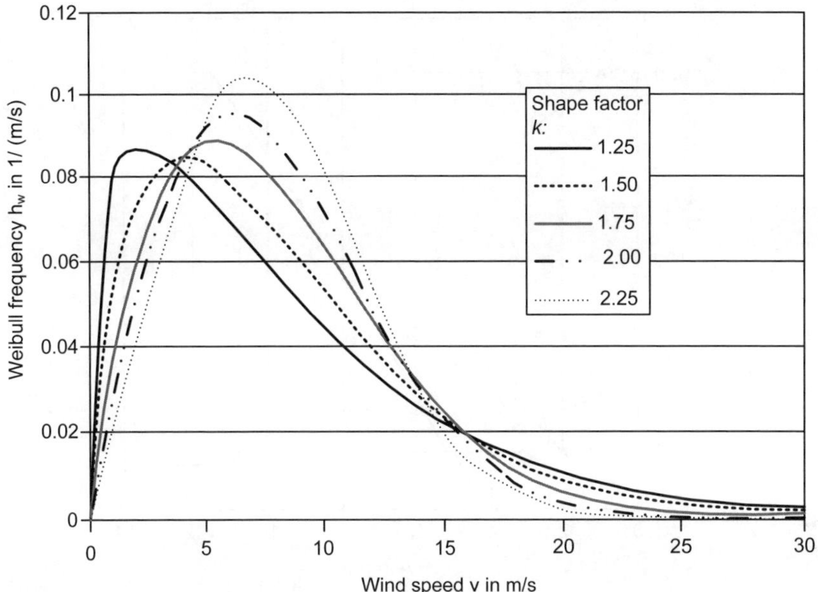

Fig. 4-19 Example of Weibull distributions for a mean wind speed $v = 8$ m/s and different shape factors k

Table 4.2 shows for some sites in Germany the Weibull factors A and k as well as the corresponding mean wind speed $\bar{v}$.

Given Weibull factors allow an estimation of the mean wind speed $\bar{v}$ [33]

$$\bar{v} \approx A\left(0{,}568 + \frac{0{,}434}{k} \right)^{1/k} .$$

(4.12)

Both Weibull factors A and k change with the height above ground, Fig. 4-20.

Table 4.2 Weibull factors for different sites in Germany, measuring height 10 m, from [36]

Site	k	A in m/s	$\bar{v}$ in m/s
Helgoland	2,13	8,0	7,1
Hamburg	1,87	4,6	4,1
Hannover	1,78	4,1	3,7
Wasserkuppe	1,98	6,8	6,0

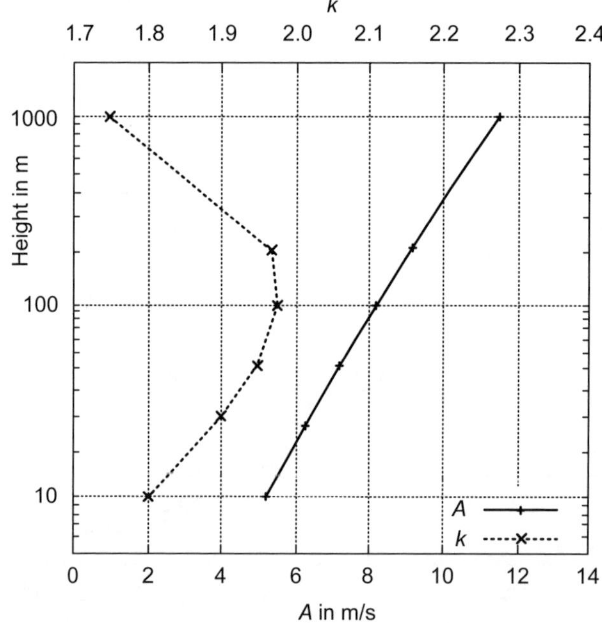

Fig. 4-20 Change of Weibull factors with height [1]

The *Rayleigh distribution function*, Fig. 4-21 (a), is a special, simplified case of the Weibull distribution function for the shape factor $k = 2$.

$$h_R(v) = \frac{\pi}{2}\frac{v}{(\bar{v})^2}\exp\left(-\frac{\pi}{4}\left(\frac{v}{\bar{v}}\right)^2\right). \tag{4.13}$$

It is very straightforward since it depends only one parameter: the mean wind speed $\bar{v}$, e.g. the annual mean wind speed which is roughly known for many sites. In general, the reference energy yield given by wind turbine manufacturers in the data sheets is calculated based on a Rayleigh distribution function.

Fig. 4-21 (b) shows for the site of the Tauern wind farm a measured wind speed frequency distribution and the fitted curves for both, the Weibull and the Rayleigh distribution function. The differences between the measurement results and the analytic description are strong.

If the analytic approximations by the distribution functions are used for the yield calculation the following criteria should be considered:

- The calculated wind energy of the analytic and measured frequency distribution should be approximately equal.
- For wind speed classes higher than the measured mean wind speed the frequencies should be identical.
- The sum of the frequencies should be used as checksum and always be 1.00, else the relative frequencies have to be weighted with the used wind speed class width (e.g. 0.5 m/s).

Moreover, the analytic description should represent correctly the classes which contain the maximum energy. The determination of the Weibull factors for a given measured frequency distribution is done e.g. by least square fitting, eventually it is beforehand necessary to take twice the logarithm [7, 34].

Note: If the measured frequency distribution, Fig.s 4-18 and 4-21, (discrete representation) is transformed into a distribution function $h(v)$ (continuous representation) or vice versa, it has to be considered that $h_i = h(v)\, dv \approx h(v_i)\, \Delta v_i$, since the distribution function has the dimension $1/(m/s)$. A different wind speed class width Δv_i causes other relative frequency values h_i , which is obvious: the larger the class width of the distribution function the more events are found within this class, $h_i = t_i /T$.

The Weibull distribution function with its two factors is quite useful to describe the histograms. But even if measuring the wind at a site for several years, extreme events like the 50-year wind speed are perhaps not recorded because they did not occur. So they have to be represented separately, cf. chapter 9.

Moreover, the analytic distribution functions are not very suitable to represent the range of calms, since the functions always start with the value $h\ (v = 0\ m/s) = 0$. Frequencies of less than 1%, i.e. 10-min average values with less than 525 events per year, may not be estimated with a Weibull distribution. Therefore, if the calms statistics are required, e.g. for wind pumping and stand-alone systems, they are recorded separately.

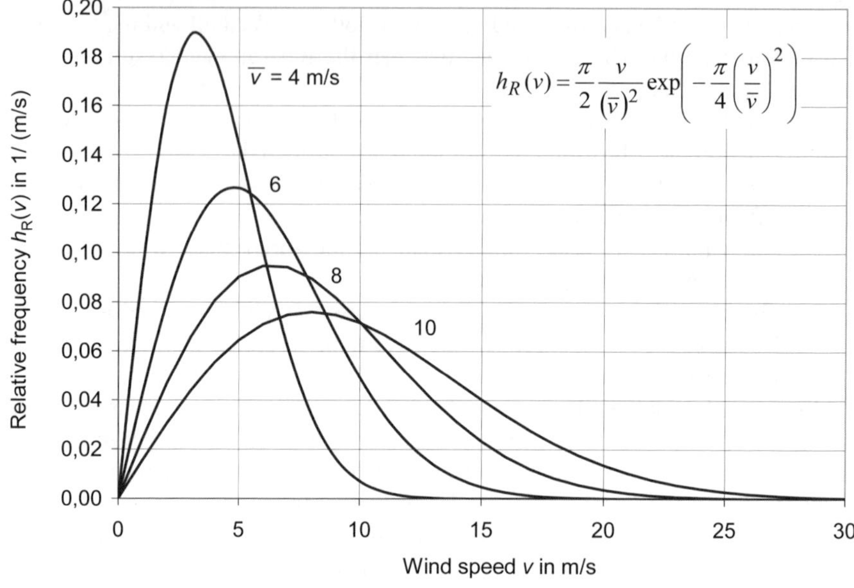

a)

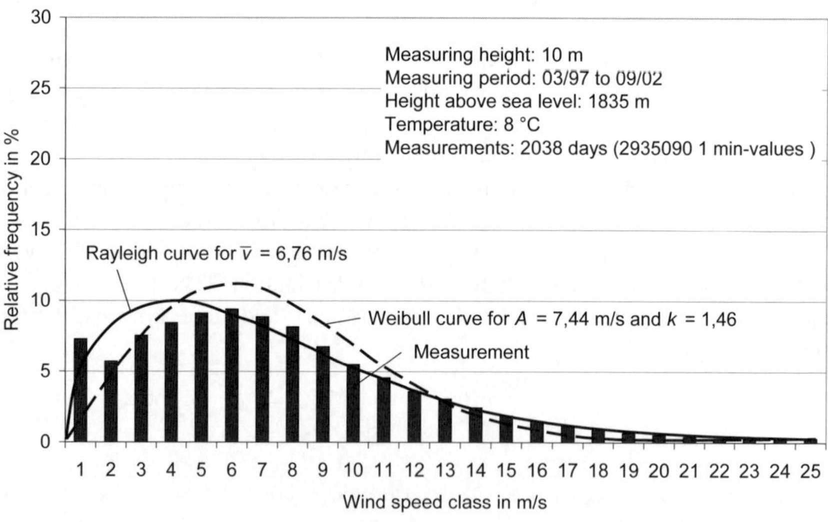

b)

Fig. 4-21 a) Frequency distribution according to Rayleigh for different mean wind speeds; b) measurements in the "Tauernwindpark" in Austria, wind speed histogram and the fitted Weibull and Rayleigh frequency distribution functions [www.tauernwindpark.com]

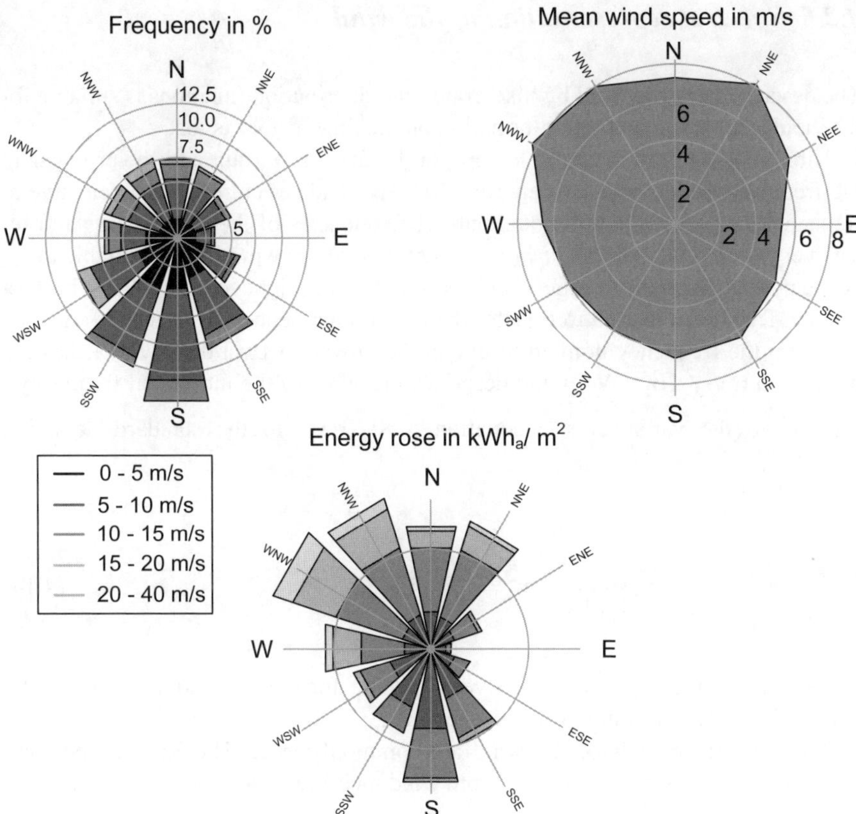

Fig. 4-22 Wind roses: Frequency rose, rose of mean wind speed and energy rose

For siting wind farms with more than one wind turbine it is important to know the frequency distributions of the wind speed for the different wind directions in order to arrange the wind turbines sophistically to prevent that the turbines shade each other, cf. section 4.3.5. This sectorised wind information is represented in a wind rose, Fig. 4-22. Three types of roses are important: the wind speed rose, the frequency rose and the energy rose.

The *wind speed rose* shows the mean wind speed in the individual wind direction sectors. In Fig. 4-22 the highest mean wind speeds are observed in the sectors NNE and WNW, while the Eastern wind is the smallest. In contrast, the *wind frequency rose*, shows a clear dominance of the Southern wind directions. But since the energy is proportional to the cube of the wind speed, and the Southern wind directions show relatively small mean wind speeds, the North-western sectors in the *energy rose* provide the largest energy share. These considerations serve for finding an optimum wind farm layout where, among others it is important to prevent that the turbines shade each other which reduces the energy yield and increases the loads due to wake effects, Fig.s 4-28, 4-29 and 4-39.

4.2.5 Spectral representation of the wind

The description of the wind by histograms and distribution functions is suitable for the yield calculation, but the information on the time history is lost.

Vibrations and the dynamic loading on blades, drive train and tower occur in the frequency range of 0.1 to approx. 30.0 Hz. The structure is very sensitive at frequencies close to the individual natural frequencies of the components and reacts very "nervous". The *spectral representation* of the wind is very suitable to investigate this frequency range which is found to the right in Fig. 4-23. The Fast Fourier Transform is used to transfer the time domain data (the measured time series) into the frequency domain to obtain the power spectral density S. It has the dimension $(m/s)^2/Hz = W/kg$ and describes the share of the individual frequency f (in Hz) to the variance $\overline{v^2}$, equation (4.8) (resp. to the standard deviation $\sigma_v = \sqrt{\overline{v^2}}$)

$$\overline{v^2} = \int_0^\infty S(f)df \ , \tag{4.14}$$

The spectral representation is also very suitable for the generation of "synthetic winds" for the digital simulation, cf. section 4.3.

Two spectral models of the wind are commonly used. The Kaimal spectrum model was determined empirically from wind measurements

Kaimal spectrum model: $\qquad S_v(f) = \sigma_v^2 \dfrac{4 L_{1v}/\overline{v}}{(1 + 6 f L_{1v}/\overline{v})^{5/3}} \ .$ $\qquad$ (4.15)

The given turbulence length scale parameters L_{1v} and L_{2v} (Fig. 4-13) vary slightly depending on the chosen guide lines.

The Kármán spectrum model is the second model and describes very well the turbulence in wind tunnels and pipes, but is also often applied to describe the wind. It allows quite easy to formulate the correlation between neighbouring points (e.g. hub centre – middle of blade).

von-Kármán spectrum model: $\quad S_v(f) = \sigma_v^2 \dfrac{4 L_{2v}/\overline{v}}{(1 + 70{,}8(f L_{2v}/\overline{v})^2)^{5/6}}$ $\qquad$ (4.16)

In general the power spectral density, Fig.s 4-23 and 4-24, is displayed versus the frequency, but in Fig. 4-23 the time scale is added to relate the spectrum to Fig. 4-13.

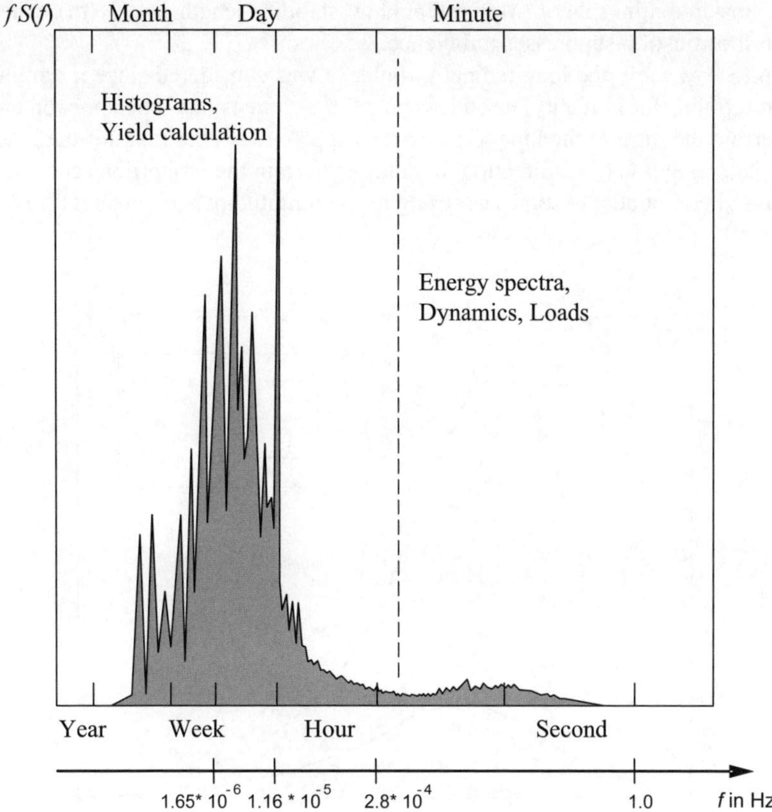

Fig. 4-23 Power spectrum of wind speed based on a continuous measurement in flat and homogeneous terrain [11]

The power spectrum in Fig. 4-23 is based on a one-year measurement with a sampling rate of 8 Hz. The fluctuations in the range of seconds and minutes, to the right in the spectrum, are caused by atmospheric turbulence. The diurnal cycle of the wind at the site creates the peak at the frequency ($1/86400$ s = $1.16*10^{-5}$ Hz) corresponding to the period of one day. The maxima in the spectrum in the range of several days reflect large-scale weather events, e.g. the passage of an Atlantic low-pressure system.

Fig. 4-24 shows schematically, for the case of flat, even terrain, the spectra for the three different states of atmospheric stability, cf. section 4.2.1. The area below the curve is proportional to the variance. Under neutrally stable conditions (L = infinite) the spectrum is dominated by a broad maximum. At higher frequencies f, it declines with the exponent $f^{-5/3}$. Low frequencies are characterized by a high variation and a high uncertainty. As explained in section 4.2.2 the temperature profile has a large influence on the vertical mass transfer and therefore on the turbulence, cf. Fig. 4-10. The Fig. 4-24 shows a strong increase of the turbulence in an unstable atmosphere (Monin-Obukhov stability length $L = -30$ m), whereas stable stratification suppresses turbulence.

Up to now, only the longitudinal turbulence was considered since it dominates in flat terrain. But in reality, turbulence is a three-dimensional phenomenon and in flat terrain the ratio of the kinetic energies is approx. 1.0 : 0.8 : 0.5 for the longitudinal, lateral and vertical direction. In complex terrain the proportions change, and the lateral component becomes as strong as the longitudinal component (1.0 : 1.0 : 0.8), [19].

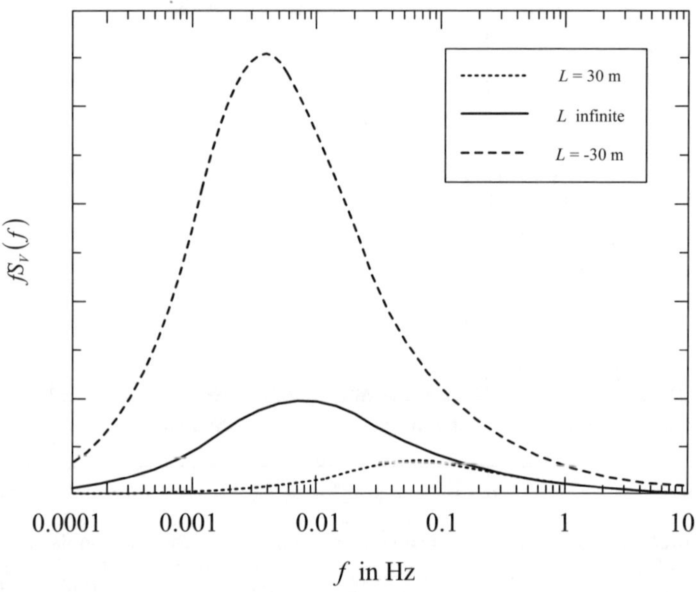

Fig. 4-24 Model spectra of the longitudinal wind speed component 50 m above ground in flat terrain for neutrally stable (L = infinite), stable (L = 30 m) and unstable (L = - 30 m) conditions; L stands for the Monin-Obukhov stability length [1]

Cross spectra and coherence functions

The turbulence spectra discussed above describe the temporal wind speed fluctuations of the turbulent components at a single point in the rotor swept area of the wind turbine. But since the turbine blades are moving through the turbulent wind field it is not enough to consider the spectra at one single point. The spatial change in lateral and vertical direction is also of importance because the blade is "collecting" all the spatial changes, cf. section 8.1.

In order to reflect these effects, the spectral models of the turbulence have to be expanded by the cross correlation of the turbulent fluctuations at two points of different lateral and vertical positions. The correlation is obviously decreasing with increasing radial difference Δr between the two points. Moreover, the correlation is for the high-frequency changes smaller than for low-frequency changes. The coherence $Coh(\Delta r, f)$ is measure for the relation between the turbulent fluctuations at the two individual points "1" and "2" in the rotor plane. The coherence is described depending on the frequency spectrum and the distance. It is defined as follows

$$Coh(\Delta r, f) = \frac{|S_{12}(f)|}{\sqrt{S_{11}(f)S_{22}(f)}} \qquad (4.17)$$

where $S_{12}(f)$ is the cross spectrum of the two points with the distance Δr, and $S_{11}(f)$ und $S_{22}(f)$ are the auto spectra of the individual points. More details on the spectral and coherence analysis is found in the literature [35, 37].

4.3 Determination of power, yield and loads

A prediction of the energy yield is possible using the measured wind speed and the derived histograms, resp. distribution functions, and the given power curve of a wind turbine. But since the wind is fluctuating strongly from year to year (cf. Fig. 4-17), it has to be verified whether the measuring period is representative enough for a long-term prediction. This is done by a comparing and correlating evaluation of wind data from neighbouring meteorological stations (also e.g. from airports) where long-term measurements are documented.

For the prediction of the energy yield of a wind farm it is moreover necessary to consider shading and disturbing effects at the site itself, and also from the neighbouring wind turbines. During the years of operation, the real yield will be surely smaller than the yield calculated in the first planning stages, because of standstill time due to failure, maintenance and repair, see chapter 15.

4.3.1 Yield calculation using wind speed histogram and turbine power curve

As discussed above, for a wind speed histogram with a given wind speed class width, the relative frequency $h_i = t_i / T$ of the individual wind speed class v_i follows from its temporal share t_i of the considered time period T, Fig. 4-18. If now the power curve $P(v)$ of the wind turbine is discretisised with the same wind speed class width, the energy yield E_i contributed by the individual wind speed class is

$$E_i = h_i \cdot P_i \cdot T .\qquad(4.18)$$

Summing up the individual energy yield of all the classes gives the total energy yield E_{total} in the considered time period T

$$E_{total} = \sum E_i = T \sum h_i \cdot P_i .\qquad(4.19)$$

If the wind regime is described by a Weibull or Rayleigh distribution function, $h_W(v)$ or $h_R(v)$, the relative frequency of the wind speed class is $h_i = h(v_i)\,\Delta\,v_i$, cf. note in section 4.2.4.

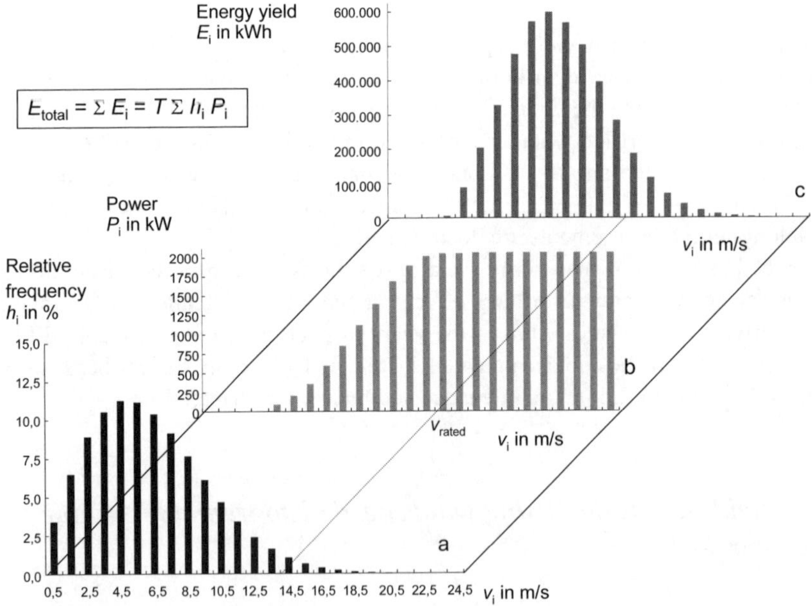

Fig. 4-25 Determination of the energy yield in kWh (c) in the period T of a year from the wind histogram (a) and the wind turbine's power curve (b)

4.3.2 *Yield calculation from distribution function and turbine power curve*

If both, the wind speed distribution $h(v)$ – to be inserted in 1/(m/s) - and the power curve $P(v)$ are given analytically, the yield is calculated in analogy to equation (4.19) as the integral

$$E = T \int_{0}^{\infty} h(v) \cdot P(v)\, dv \qquad\qquad (4.20)$$

In reality, the procedure is always a numerical integration (summation, see above) since the turbine's power curve is determined by measurement (cf. Fig. 4-26), and the sections between the measuring points of the power curve are linearised for the integration.

The specific annual energy yield, per square meter of the rotor swept area, for an idealized modern speed-variable wind turbine, calculated depending on the annual mean wind speed at hub height and using a Rayleigh distribution, is shown in Fig.s 15-23 and 15-24.

In general, the power curves are given for a standard air density of 1.225 kg/m³. The energy yield calculation for a particular site has to consider that the wind power is proportional to the air density, which itself depends according to the ideal gas equation from the ambient temperature and the atmospheric pressure. With increasing temperature (tropical zones) the air density decreases, as well with the increasing altitude of the site above sea level.

For the yield calculation of a single wind turbine the energy shares of all wind classes and wind directions are summed up, Fig. 4-25. But if a wind farm is planned, the yield calculation is done sector by sector, Fig. 4-22, in order to consider the shading by other wind turbines (wind farm effects).

4.3.3 *Power curve measurement*

The power curve of a wind turbine shows the output of electric power depending on the wind speed. Power curves are either calculated from the design data for rotor and drive train, or they are measured on the real machine standing in the wind. The *measured power curve* is determined from the simultaneous measurement of the wind speed at hub height and the produced power, Fig. 4-26. The averaging interval is usually 10 minutes [4]. It is assumed that the wind speed measured with the measuring mast is the same as the wind speed at hub height at the position of the wind turbine. To assure this, the distance between turbine and measuring mast should not be too large, on one hand. On the other, the mast should not be placed too close to the wind turbine else the blocking effects of the wind turbine influence the measured wind speed. Therefore, the wind speed measurement is installed at a relative upwind distance of 2 to 4 D (rotor diameter) to the turbine.

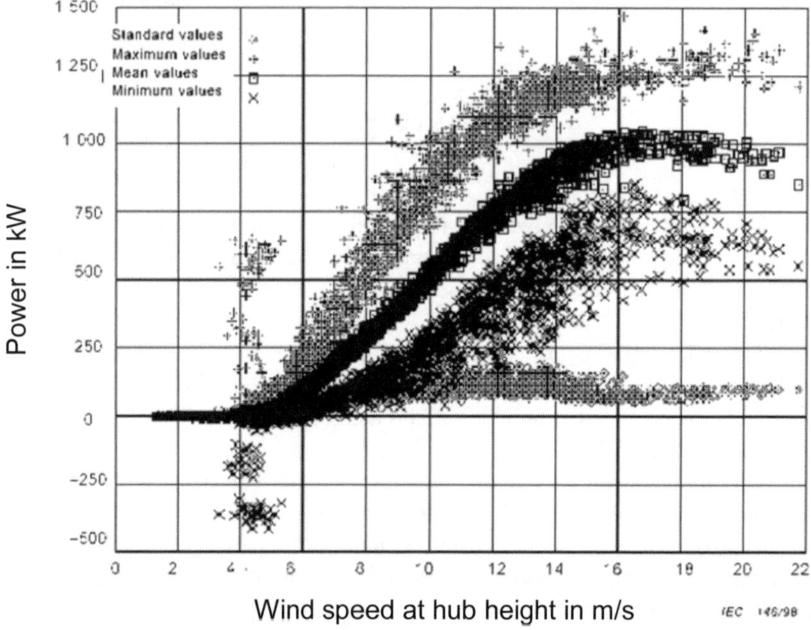

Fig. 4-26 Example of a measured power curve; 10-minute mean values, standard deviation and 1-second minimum and maximum values [4]

At this point it has to be remembered that small measuring errors of the wind speed have big effects on the determination of the power: If there is a measuring error of 3% of the wind speed, the energy in the wind may vary up to 9% (cube law).

The power curves of the wind turbines are essential for the planning and erection process, they belong to the scope of service of the turbine manufacturer.

The validity of the assumption that the hub height wind speeds at the measuring mast and the wind turbine are identical is quite limited to an ideal flat terrain without obstacles. But wind turbines are very often installed in complex terrain. If the site does not fulfil the requirements, an *on-site calibration* is necessary: before erecting the wind turbine a second measuring mast is temporarily installed at the foreseen position. Details are given in the international IEC guideline "Wind Turbine Performance Testing" [4] or in national guidelines, e.g. of the German FGW [18].

The performance and behaviour of the wind turbine is strongly influenced by site-specific parameters like turbulence, skew wind and vertical wind profile, so the application of a certified power curve obtained in flat terrain implies uncertainties.

The effects of the turbulence intensity on the power curve are shown exemplarily in Fig. 4-27. The curve part of smaller wind speeds is approximately a cube function. Here, the value of the averaged power is higher for increased turbulence. In contrast, increased turbulence lowers the averaged power output in the curve part close to rated power, since here the curvature of the power curve is negative.

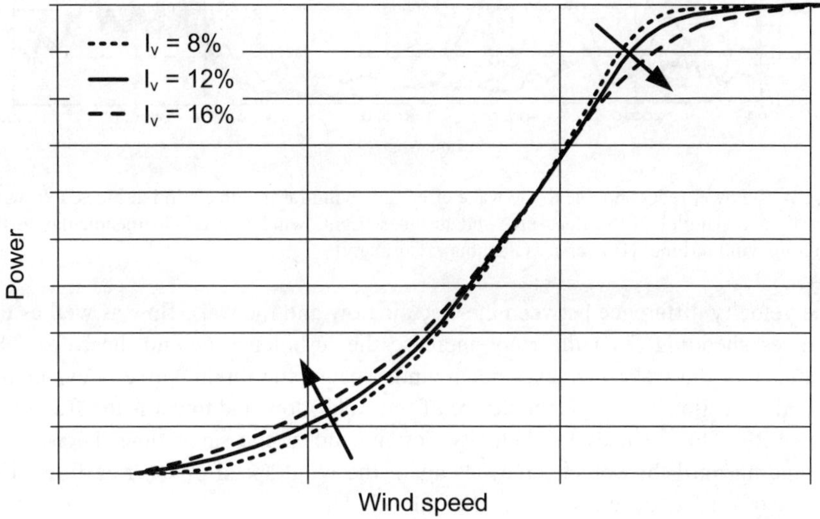

Fig. 4-27 Effect of the turbulence intensity I_v on the power curve [27]

4.3.4 Yield prediction of a wind farm

Since the wind turbine extracts energy from the free flowing air, the wind turbine wake is characterized by a reduced wind speed but also an increased turbulence. A wind turbine situated in the wake of one or several other wind turbines will therefore produce less electrical energy and experience higher loads than a wind turbine in an undisturbed flow, Fig. 4-28.

In the simplified consideration, the wake directly behind the wind turbine is a circular zone of reduced wind speed with a diameter slightly larger than the rotor diameter. The reduction of the wind speed is coupled to the thrust coefficient which describes the momentum change in the free air flow due to the wind turbine (cf. chapter 5).

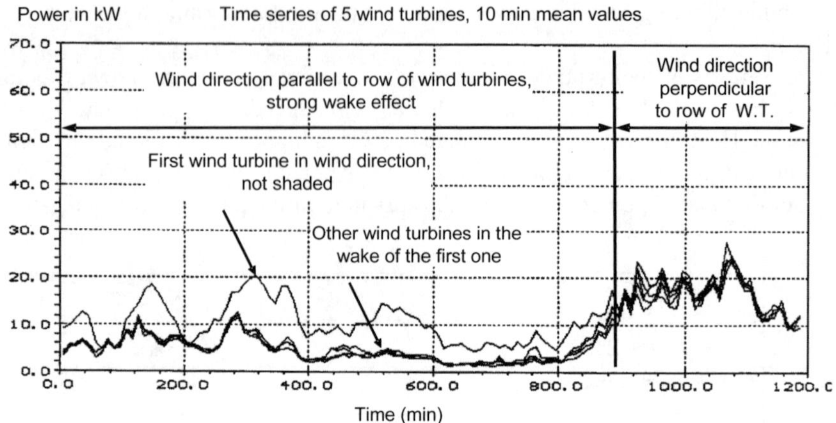

Fig. 4-28 Power reduction due to the wake effect in a wind farm with 5 wind turbines, left: wind direction in parallel to the lined-up wind turbines, right: wind direction perpendicular to the lined-up wind turbines [University Oldenburg, Germany]

The velocity difference between the free air flow and the wake flow as well as the vortices shedding from the rotor increase the turbulence behind the rotor. The diameter of the wake increases downwind because the mixing zone is expanding in both directions: towards the centre of the wake flow and towards the free flow, Fig. 4-29. This reduces the velocity deficit, and at the same time the wake is expanding until the velocity reaches again the wind speed of the free flow. The degree of mixing depends on the ambient turbulence.

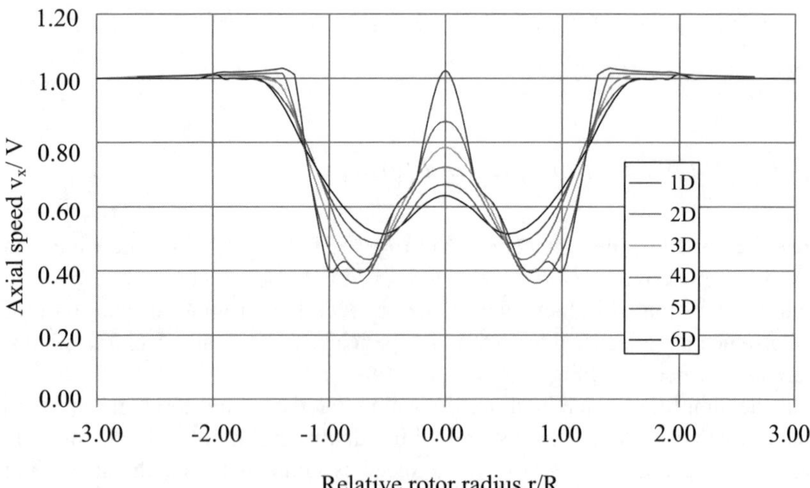

Fig. 4-29 Hub height velocity profiles downwind of a wind turbine. Distance shown as multiples of the rotor diameter D [29]

The wake flow model PARK [24] calculates the losses due to the wake based on the wind direction distribution, the position of the wind turbines and the turbine's thrust coefficient curve. Since it is a two-dimensional model there are several restrictions to the application. Accelerations of the wind caused by the topography are not considered, for example. In contrast to that, the software WAsP 7 (i.e. Wind Atlas Analysis and Application Programme 7 and higher versions) [14] allows the calculation of the wind farm efficiency even for a combination of different wind turbine types with different hub heights, rotor diameters and power curves.

4.3.5 Effects of wind and site on the wind turbine loading

Extreme wind speeds

As mentioned in section 4.2.4, the Weibull distribution function shows mostly a good agreement with the frequency distribution of the measured wind speed. In contrast to that, the distribution of extreme wind speeds follows other statistical laws and is not represented by the Weibull function. Extreme wind speeds are often expressed as n-year wind speed, i.e. the 10-min mean wind speed which is exceeded in the average once in the period of n years. In course of the wind turbine design, the 50-year wind is of interest. It is found that for a description of these extreme wind speeds the twice exponential Gumbel distribution is very suitable

$$F(V) = e^{\left(-\exp(-\alpha(V-\beta))\right)} \qquad (4.21)$$

where $F(V)$ is the cumulated probability that the mean wind speed V is exceeded. More details on this topic are found in the literature [1].

In practice, the extreme meteorological limiting conditions of a site are very important for the determination of the loads on the wind turbine and for the classification of the site, see table 9-2 (cf. IEC 61400-1).

There is a close relationship between the maximum gust wind speed and the maximum 10-min mean wind speed in the observed period: they are linked by the standard deviation σ_v, as shown in Fig. 4-30a. The measurement of 10-min averages on the Danish west coast in a height of 20 m comprised approx 1,500 values [33].

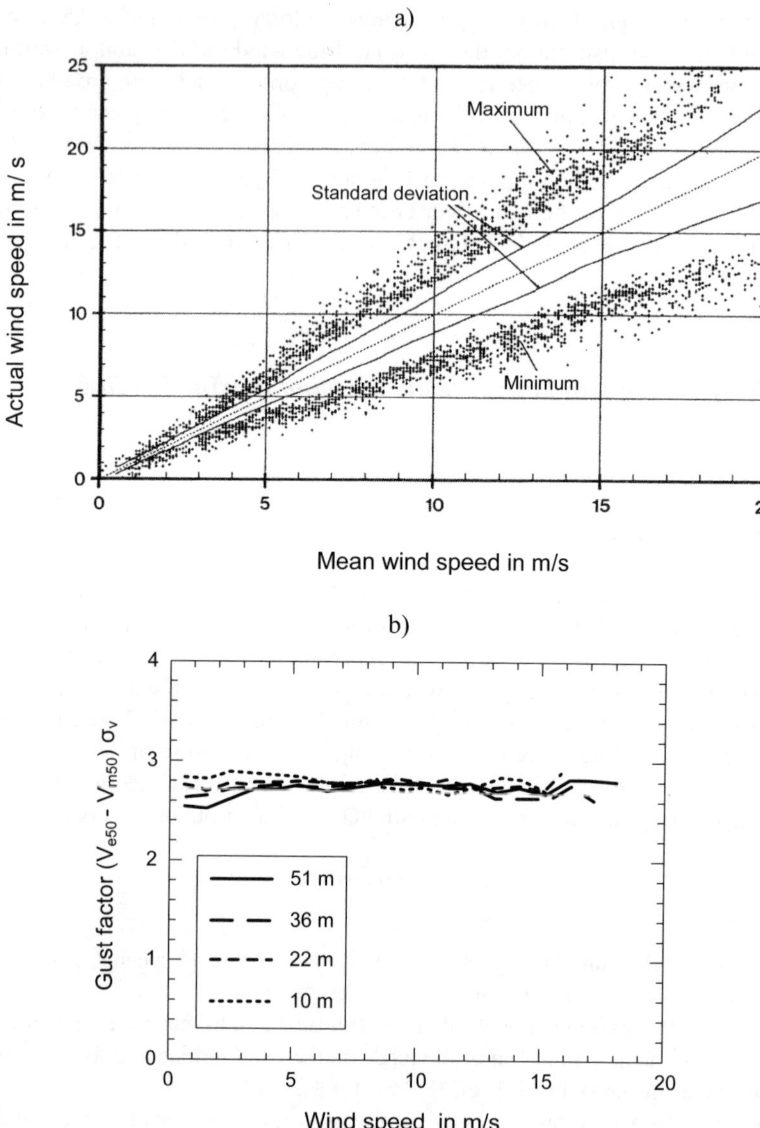

Fig. 4-30 a) Minima, maxima and standard deviations of the wind speed versus the 10-min average values [33]; b) Gust factor versus wind speed for different measuring heights [1]

The 3-sec-50-year extreme gust V_{e50} is therefore calculated from the maximum 10-min mean value V_{m50}, given in table 9-2 (cf. IEC 61400-1), and the estimated turbulence intensity I_v

$$V_{e50} = V_{m50}\,(1 + 2{,}8\,I_v)\,.\tag{4.22}$$

Remember in this context the equation (4.9) where $I_v = \sigma_v\,/\,\overline{v}$. The factor 2.8 in equation (4.22) is called gust factor and was confirmed by measurements for different heights, Fig. 4-30b. In the German guidelines of the German Lloyd dated 2003 [20] and the DIBt dated 2004 [21] the recommendations for the consideration of gusts are identical or similar.

Consequences of the turbulence

In section 4.2.3 the description of turbulence using the turbulence intensity I_v was already introduced, it is the ratio of standard deviation to the mean wind speed in the averaging period (mostly 10 minutes), equation (4.9). As discussed with Fig. 4-15, its value depends on the wind speed. This dependency is respected in the standards of IEC 61400 [19] and the guidelines of German Lloyd [20], table 9-2.

Apart from the blade weight, turbulence is a main cause of material fatigue. Producing *alternating loads* it stresses the whole blade, but above all the blade root, shown schematically in Fig. 4-31. Moreover, the turbulent wind field causes alternating torsion of the drive train, and furthermore the alternating thrust stresses the tower.

Fig. 4-32 shows exemplarily for two different turbulence intensities the operational loads at the blade root (flapwise bending) versus the wind speed.

High *natural turbulence intensities* I_v may be caused among other by obstacles (e.g. a near building), surface roughness (e.g. strong vegetation) and terrain inclination, according its origin named *ambient turbulence* (or environmental or natural turbulence.

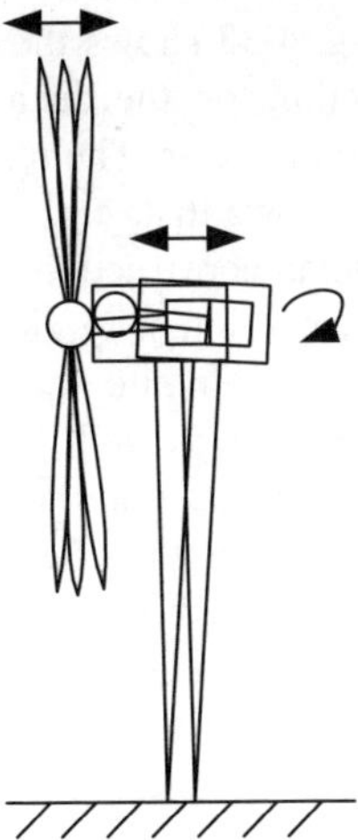

Fig. 4-31 Turbulences provoking dynamic loads and vibrations

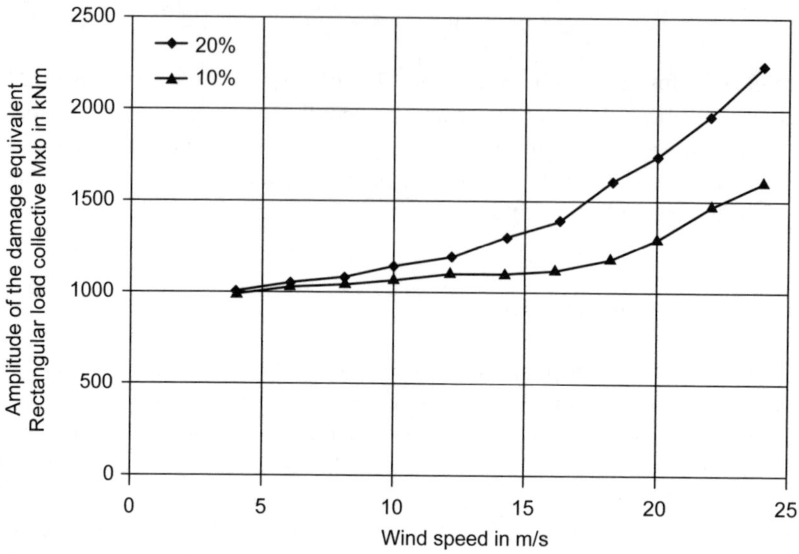

Fig. 4-32 Operational loads at the blade root of a 1,25 MW wind turbine for two different turbulence intensities [22]

But additionally, increased high turbulence intensity is caused in the wind turbine wake, named *induced turbulence intensity* I_W. It is problematic when there is a disadvantageous wind farm layout with too small distances between the wind turbines. Therefore, in course of the wind farm siting it has to be assured that the sum of natural turbulence I_v and turbulence I_W induced by the wake flow of neighbouring wind turbines does not exceed the limiting values applied in the design certification. Else, the service life may be reduced significantly (worst case: some months instead of 20 years). Fig. 4-33 shows the additionally induced turbulence intensity I_W in the wake depending on the distance from the wind turbine, expressed in multitude of the rotor diameter. The results are based on measurements at four different sites. The Fig. shows that, e.g., the limiting value of 20% turbulence intensity given in the German construction guideline of DIBt [21] is already exceeded by the induced turbulence intensity alone for a distance of less than 4 rotor diameters! Of course, for a certain site, the wind direction frequencies have to be included in a closer consideration as well.

It has to be mentioned in this context that the natural turbulence and the wake induced turbulence have a different length scale and therefore different effects on the wind turbine. The longitudinal length scale of the natural turbulence is in the range of 600 to 1.000 m, whereas the length scales for the wake flow turbulence are in the range of 1 to 2 rotor diameters. And at least it has to be considered that on one hand the turbulence intensity in the wake is increased, but the wind speed, and therefore the wind energy, is significantly reduced.

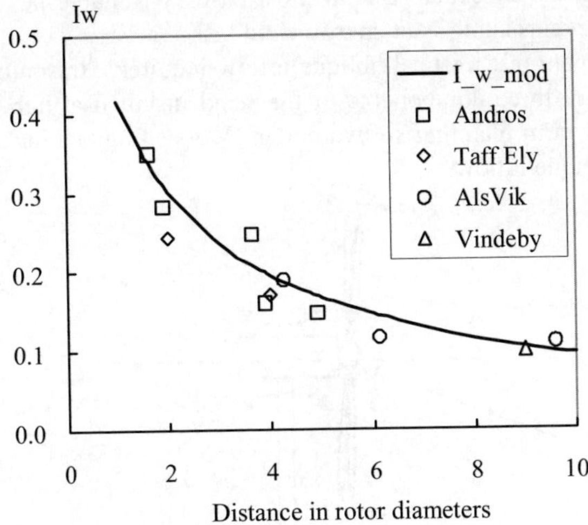

Fig. 4-33 Additionally induced turbulence intensity I_w in the wake behind wind turbines [23]

While the mean wind speed at the site is decisive for the occurring loads, the turbulence intensity is a measure for the frequency of the load cycles. Therefore, both have to be considered in course of the planning and the wind farm layout. For the load calculations, the standard deviation σ_v is of a higher importance than the turbulence intensity.

The data for wind speed and turbulence intensity are at best determined from measurements at the planning site. But apart from that, there are also planning tools, e.g. WAsP Engineering (Wind Atlas Analysis and Application Program) [14], which allow a consideration of the topography and the surface roughness in the siting calculations.

Consequences of oblique inflow

At sites on hills there is not only a deviation of the vertical wind profile from the logarithmic profile, but also a deviation of the inflow from the horizontal direction (i.e. an oblique inflow). Therefore, the blades are exposed permanently to an alternating inflow angle of the relative velocity leading to increased operational loads at the blade root. Additionally, the rotor shaft is stressed by bending, shown schematically in Fig. 4-34.

In course of the calculation of the loads for certification according to the guidelines of IEC and German Lloyd [20], a mean flow inclination angle of up to 8° is assumed. This deviation from a horizontal inflow is caused by an inclination of the terrain. The influence of the terrain decreases with increasing height above ground. Especially in very complex terrain, e.g. scarps and cliffs, the limiting value of 8° is easily exceeded. But in most cases, a suitable positioning of the wind turbines prevents significant energy yield losses.

The measurement of a vertical oblique inflow requires ultrasonic anemometers which measure all three components of the wind installed at hub height. Moreover, some wind farm planning software, e.g. WAsP Engineering, allow an estimation of the oblique inflow.

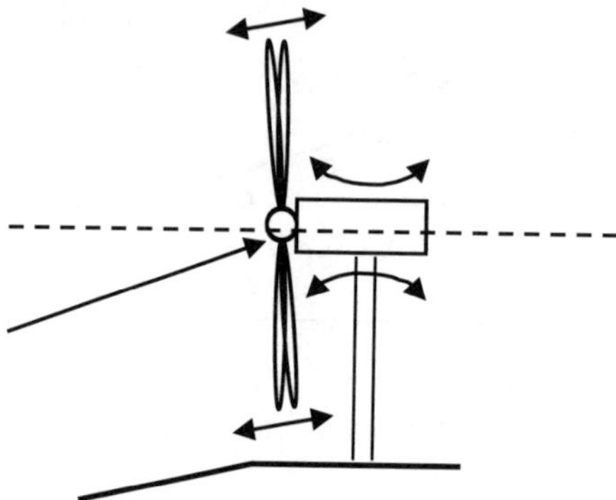

Fig. 4-34 Loads due to vertical oblique flow

Influence of wind shear

The wind shear is defined as difference of the horizontal wind speed at the top and bottom of the rotor. It is either given as the change of wind speed per meter of height, or as exponent of the power law, equation (4.1). In the certification guidelines according to IEC and German Lloyd [20], an exponent of 0.2 is applied in the load calculations.

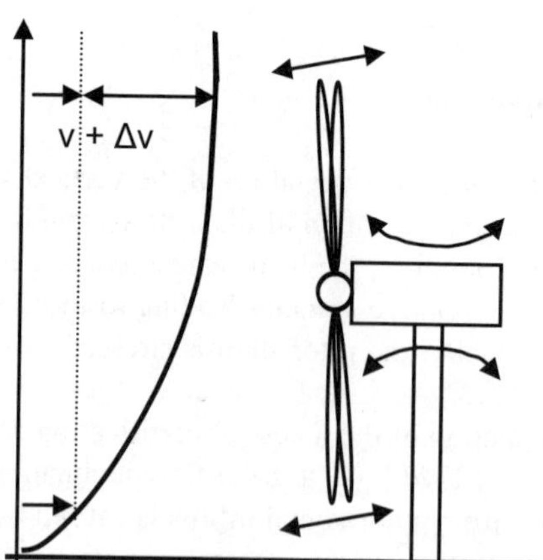

Fig. 4-35 Loads due to the wind profile (wind shear)

The wind shear causes alternating loads on the blades because the blade experiences during every revolution alternating inflow angles, Fig. 4-35. Therefore, the operational loads and also the bending stress of the rotor shaft increase as well.

On site the *wind shear* may be influenced by four phenomena:

- Terrain inclination

Strong inclinations of the terrain may cause deviations from the logarithmic wind profile. For smooth inclinations, the growth of the wind speed with height may be reduced. In some cases the wind even no longer increases with height, the gradient reduces to zero. This situation may be advantageous for the economic efficiency of the wind farm project since there is no requirement for very large hub heights for increasing the yield. But if the slopes are too steep and turbulent flow separation occurs, cf. Fig. 4-11, the wind profile may be deformed strongly. Parts of the rotor area may be exposed to a negative gradient. In other zones of the rotor area there may be very large gradients at the same time, Fig. 4-36.

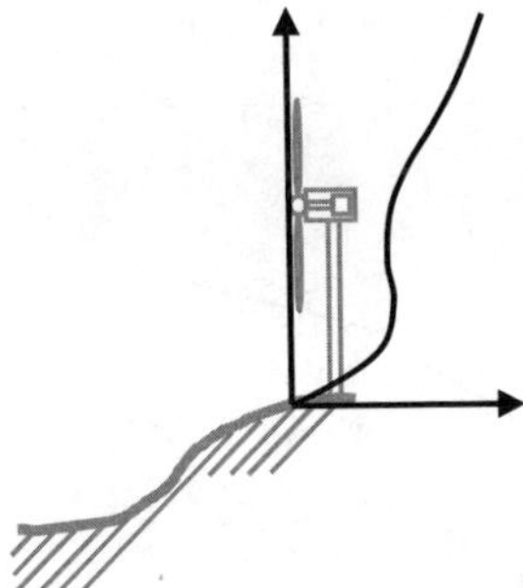

Fig. 4-36 Wind profile with strong wind shear due to terrain inclination (steep slope)

- Obstacles

If wind turbines are situated closely behind large obstacles, e.g. a forest, the wind speed at the bottom of the rotor may be decelerated heavily. The degree of deceleration depends on the dimensions of the obstacle, its porosity and the distance between obstacle and wind turbine. The deformation of the logarithmic wind profile is shown schematically in Fig. 4-37.

- Small distance between the wind turbines

As mentioned above, the expanding wake flow behind a wind turbine, cf. Fig. 4-29, induces additional turbulence and reduces the wind speed compared to the ambient free flow since the wind turbine extracts energy from the wind. This is called the velocity deficit in the wake flow. The continuous line in Fig. 4-38 shows the influence of the wake on the vertical wind profile at a distance of 5.3 rotor diameters behind the wind turbine. This deformed wind profile shows areas of a negative gradient, but also some with a strongly positive gradient. For comparison, the dashed line shows the wind profile in front of the wind turbine.

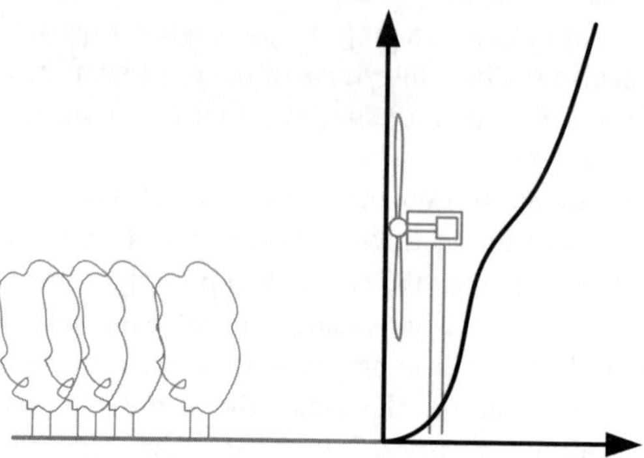

Fig. 4-37 Wind profile behind obstacles

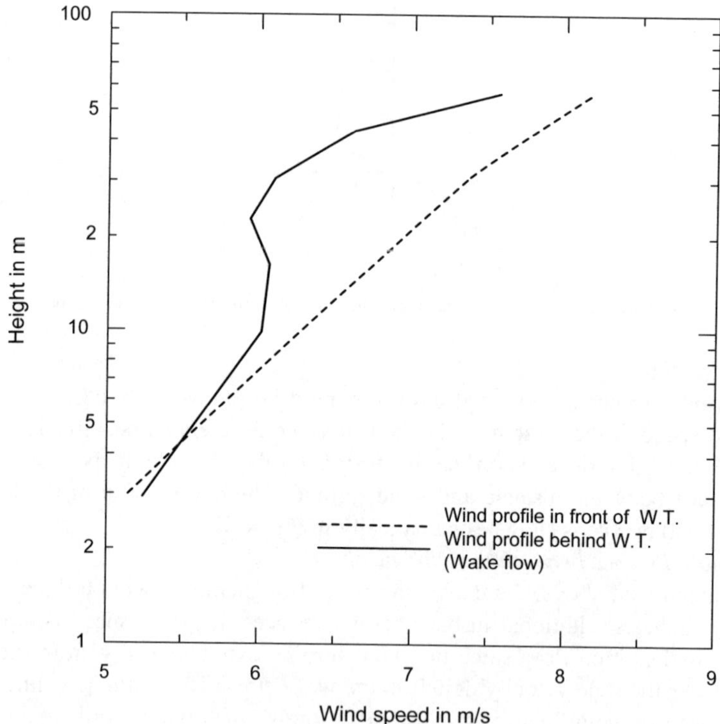

Fig. 4-38 Wind profile in front of a wind turbine and in the wake behind it at a distance of 5.3 rotor diameters [1]

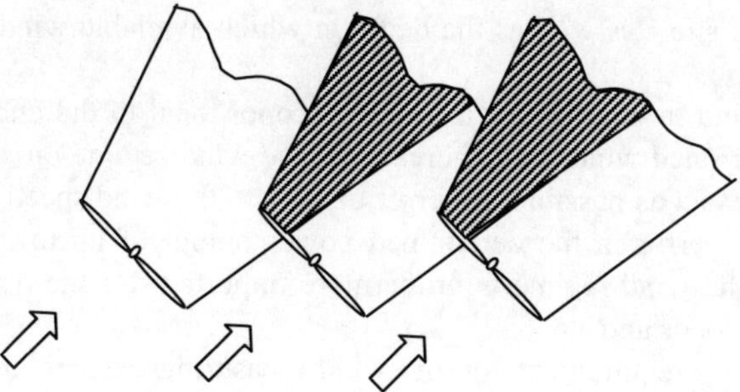

Fig. 4-39 Wind turbines partly shaded by the wake flow of the neighbouring wind turbine

The worst case for the loads on the wind turbine is not the total shading of the rotor by the wake, but partial shading, Fig. 4-39. Under these circumstances there is not only a vertical, but also a horizontal gradient, and additionally the turbulence is increased.

- Atmospheric stability

The different vertical temperature profiles produce different vertical wind profiles, as discussed in section 4.2.2. So at different heights there may be layers with strongly varying wind speeds. Fig. 4-10 showed clearly the vertical gradient of the wind profile increases with growing stability of the atmosphere (dashed line).

In the first three cases discussed above (influence of terrain inclination, obstacles and/or distances between the wind turbines), a suitable positioning and/or hub height of the turbines prevents that the stress due to strong wind shear exceeds the allowed material strength.

The wind shear at a site may be determined through wind speed measurements at different heights. But the problem is that the measured gradient is a function of the height itself, and therefore may not be simply extrapolated to other heights: The gradient determined from measurements at 10 m and 30 m height is not identical to the gradient observed between 30 m and 50 m height. Based on the terrain properties and data of good quality, siting software like the already mentioned WAsP allows a more precise determination of the gradient. Furthermore there are other programs like WAsP Engineering for the calculation of the gradient. But both of the mentioned programs are not able to predict directly the influence of atmospheric stability.

4.4 Wind measurement and evaluation

The magnitude, time history and direction of the wind are the most important parameters for the prediction of the expected energy yield – i.e. the quantitative assessment of a site - as well as the decision which available wind turbine is the most suitable one.

Since the wind turbine power is approx. proportional to the cube of the wind speed (until the rated wind speed is reached) the wind regime on a site has to be determined as exact as possible. An error of 10% in the wind speed measurements may produce an error in the determined power output of up to 33%. An exact knowledge of the wind regime is furthermore important for the determination of the mechanical loads and stress.

There are high requirements on the wind measuring devices, the sensors and the instrumentation for data recording. Apart from the accuracy of the wind speed sensor the devices have to be extremely robust in order to record data maintenance-free for long periods.

Moreover, faulty measurements due to wrong installation and sensor icing have to be prevented. In general, mechanical wind speed sensors like the *cup anemometer* are well proven and reliable sensors. Their limits of application and possible sources of errors are mostly known, so among other, they should be calibrated before and after a measuring campaign.

Sensors without moving parts, like e.g. the *ultrasonic anemometer* are up to now seldom applied for long-term measurements since firstly they are more expensive than the mechanical sensors, and secondly they are more susceptible to faults because they are generally more complex devices. The ultrasonic anemometer may measure, independent from the current wind direction, instantaneously all the three components v_x, v_y und v_z of the wind vector

$$v_{\text{Vector}}(t) = \sqrt{v_x^2(t) + v_y^2(t) + v_z^2(t)} \ . \tag{4.23}$$

Since a wind turbine is able to extract power only from horizontal wind speed component v_{horiz} which is perpendicular to the rotor swept

$$v_{horiz}(t) = \sqrt{v_x^2(t) + v_y^2(t)} \ , \tag{4.24}$$

a vertical component v_z, e.g. due to a slope (cf. Fig. 4-34), is of no use. The *cup anemometer* measures directly the horizontal wind speed component v_{horiz} but cannot resolve the vertical component v_z .

Propeller anemometers (also called windmill anemometers) measure as well only the horizontal wind speed component v_{horiz} and have an integrated wind vane. It is favourable that the wind direction is measured with the same sensor unit, whereas using a cup anemometer it has to be determined by a separate sensor. But unfortunately, the integrated wind vane of the propeller anemometer causes a severe problem: the propeller is wiggling in the wind which distorts the wind measurement.

4.4.1 Cup anemometer

International standards following IEC [4] and IEA [3] refer to the cup anemometer as the most suitable sensor type for wind speed measurements, Fig. 2-22 and 4-40. The cup anemometer is a small drag driven wind mill with a vertical axis of rotation. Cup-like drag bodies are fixed with a lever to the vertical shaft. Instead of hemispherical cups there are more and more conical shells applied since they have a more distinct edge for defined flow separation.

Anemometers produce either an analogue or a digital *signal output*. The analogue signal is the voltage, which is proportional to the rotational speed and the wind speed (cf. equation 2-12), produced by the rotation of a small generator. The digital signal, a defined number of impulses per revolution produced by reed contacts or an incremental encoder, is counted over in a certain interval and offers this way a measure proportional to the wind speed.

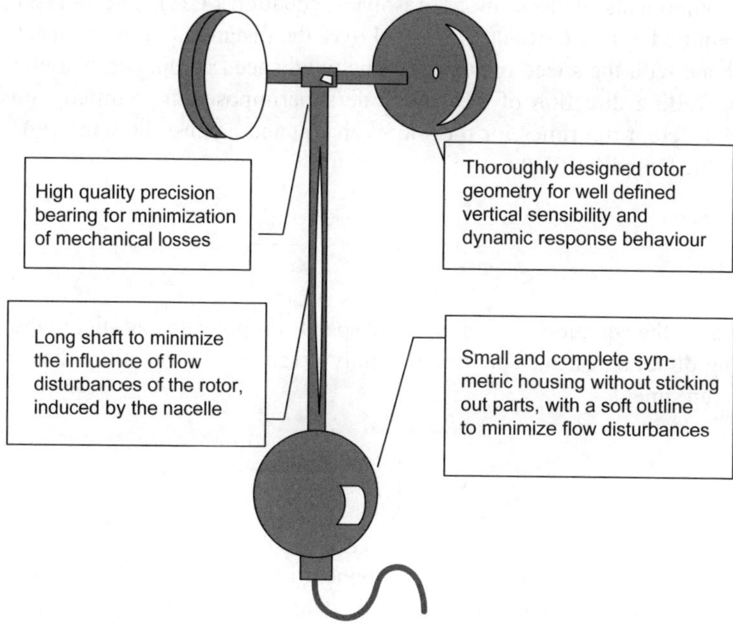

Fig. 4-40 Cup anemometer [3]

A special attribute of the cup anemometer is the so-called response length (also called distance constant, [13]) which results from signal delay occurring at rapid changes of the wind speed due to the inertia of the rotating cup star. If a virtual wind speed increases with a step function from v_0 to $v_0 + \Delta v$, the signal of the anemometer responds with an exponential curve (i.e. 1st order time delay).

There are several methods to determine the response length [13]. In the wind tunnel method, a constant wind speed is adjusted, and the acceleration of the anemometer from standstill to load-free idling is measured. The rotational speed increases rapidly, therefore, the required measurement instrumentation has to work fast (i.e. high sampling rate) and with a high resolution. Another method to determine the response length is the comparison of the cup anemometer with a high-resolution ultrasonic anemometer.

4.4.2 Ultrasonic anemometer

Ultrasonic anemometers, Fig. 4-41, were developed for the investigation of the turbulent field in the surface boundary layer. There are up to three pairs of sonotrodes (speaker-microphone combination) arranged in a way that the three spatial components of the flow are resolved, equation (4.23). The ultrasonic impulses, emitted at a rate of 100 Hz, travel over the distance s between speaker and microphone with the speed of sound c. The wind speed component parallel to the sound travelling direction of the sonotrode superimposes the emitted signal and leads to different run times for the way with (t_1) and against the wind (t_2), which is called the Doppler Effect.

$$t_1 = \frac{s}{c+v} \quad \text{and} \quad t_2 = \frac{s}{c-v} \tag{4.25}$$

Rearranging the equations yields the wind speed component v parallel to the sound travelling direction of this sonotrode, simply given by the travelling distance and the two run times.

$$v = \frac{s}{2} \cdot \left(\frac{1}{t_1} - \frac{1}{t_2} \right) \tag{4.26}$$

It is favourable that the wind speed component is determined independently from the speed of sound which varies with air density and humidity. As mentioned above, the combination of three sonotrodes gives a sensor which allows the instantaneous measurement of all three wind speed components, Fig. 4-41.

A problem inherent of the ultrasonic anemometer is the deflection of the flow by the sonotrodes heads, and the sensor arms as well. Therefore, the measurement in the direction of the arms is inaccurate. This is a potential source of measurement uncertainties of long-term measurements with theses sensors. Moreover, there are effects from the aging of the piezo-electric sonotrodes. At least, the sensor behaviour is temperature dependent.

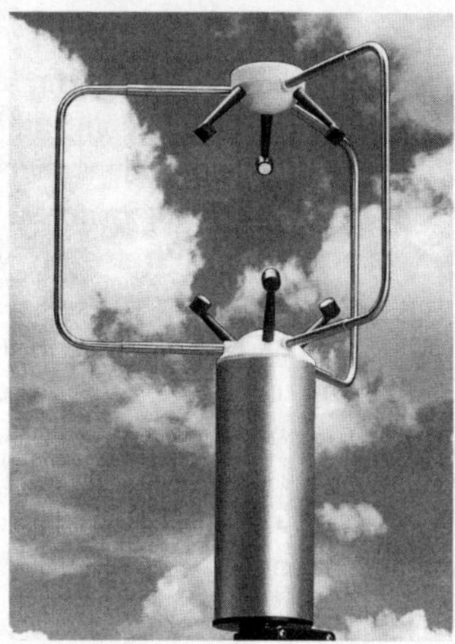

Fig. 4-41 Ultrasonic anemometer

4.4.3 SODAR

The abbreviation SODAR stands for **SO**nic **D**etecting **A**nd **R**anging. It is a remote sensing method for the exploration of the atmospheric boundary layer. The principle is similar to this of Radar and Sonar measurements. The SODAR technique is mainly applied for recording the vertical wind speed profile. The SODAR device emits small sharply bundled sound signals in the audible range. These are reflected at the boundaries between atmospheric layers of different refractive numbers which comes from differences in temperature and humidity. These boundaries between atmospheric layers move with the local speed of the wind. The frequency of the back-scattered signal is shifted with respect to the emitted signal due to the Doppler Effect. If the sender and the receptor are at the same location the system is called a monostatic SODAR. The measured Doppler shift is proportional to the wind speed component parallel to the beam direction of the sending antenna.

If several antennas or senders are applied which point in different directions the three-dimensional vector of the wind can be measured. The senders are mostly arranged in a way to receive a phase-shifted signal (phased array) which allows simulating the behaviour of several antennas pointing into different directions. Using the different run times between sender and scattering volume receptor parallel measurements of the wind speed in different heights are possible to determine the vertical wind speed profile.

The SODAR measures within a certain air volume, in contrast to the cup anemometer which is limited to the measurement at one single point of the wind field.

In order to measure in heights between 20 and 150 m above ground, so-called Mini-SODAR systems are applied. They operate in the frequency range between 4 and 6 kHz and reach a resolution of height between 5 and 10 m.

The biggest advantage of the SODAR systems is that they allow measurements of wind speed and wind profiles in large heights where a measuring mast would be too expensive. Especially in very complex terrain, e.g. where a forest is found in the main wind direction, the SODAR systems allow a comparison of measured and simulated wind profile. But it has to be considered that under stable atmospheric conditions there is only a small scattering, so there will be no valid measuring signals. Moreover, the signal-to-noise ratio decreases with increasing height, so the allocation of the signals to the related height above ground becomes more and more difficult. Thus, the installation and adjustment of the SODAR systems as well as the evaluation and interpretation of the SODAR measurements require extensive experience and routine.

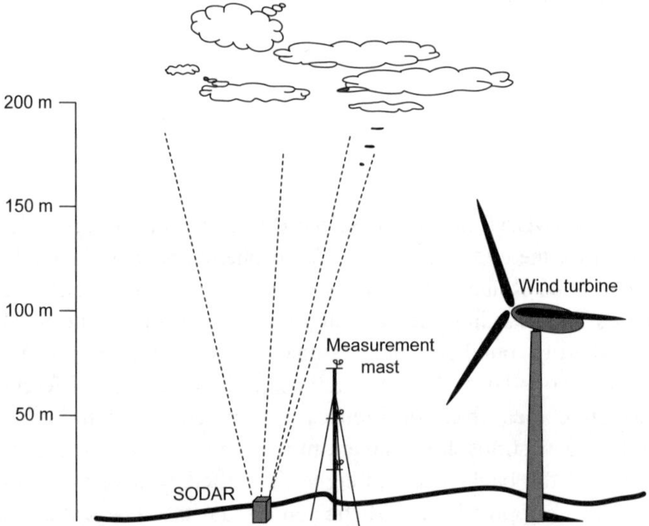

Fig. 4-42 Scheme of a SODAR measurement

Since the SODAR has a significantly higher power consumption than a cup anemometer the power supply may by a problem at remote sites. Moreover, the measurement quality is strongly dependent on measurement disturbances by obstacles close to the antennas which may disturb and reflect the signals. Since the emitted signals are in the audible frequency range the measurements are sensitive to background noise e.g. from cowbell, croaking frogs, the roar of the surf and more. It should be remembered that the wind profile measured with a SODAR system in a shorter measuring period is representative only for the specific prevailing atmospheric stratification. All in all, using a SODAR the determination of the mean wind speed with a desired precision is difficult, therefore it should be applied only in combination with conventional wind sensors installed on a measuring mast for additional investigations of the wind regime.

4.5 Prediction of the wind regime

4.5.1 Wind Atlas Analysis and Application Programme

The WAsP (Wind Atlas Analysis and Application Programme) was developed for the energy yield prediction of single wind turbines and wind farms [5].

With the years WAsP became a standard tool for the Micro-siting of wind farms, but its application has limits which are well known by now.

The software allows to analyze the wind regime at potential sites and to support the site selection. It is possible to transform the wind speed from one point A to an arbitrary point B, Fig. 4-43 [5]. For this purpose, in a first step all influences from orography, roughness and obstacles on the surface wind at point A are removed. The resulting wind represents the geostrophic wind which is then assumed to be uniform in a large-scale area. In the second step, the local influences at the point B are applied which gives the local wind at this site. The energy yield is then determined using the sectorised wind distributions and the power curve of the chosen wind turbine (cf. Fig.s 4-22 and 4-25).

The strongly simplified flow model applied in WAsP is based on the Navier-Stokes equations. The basic assumptions are:

- The atmospheric stratification is approximately neutrally stable.
- Thermal driven winds are neglected.
- The surface inclination is small, so there is no flow separation (inclination be low approx. 30%, i.e. 17°. But the critical inclination depends in reality e.g. on the surface roughness and the atmospheric stability).

These basic assumptions allow the linearization and the solution of the Navier-Stokes equations. This model is relatively simple and requires quite small computational resources.

The more complex the topography and climatology of a site, the higher are the uncertainties of this simplified calculation. Furthermore, the radius around point A in which this method is valid for application depends also on the quality of the input data. Own wind measurements close to the site are the favoured input data, so the radius for calculation results is a few hundred meters.

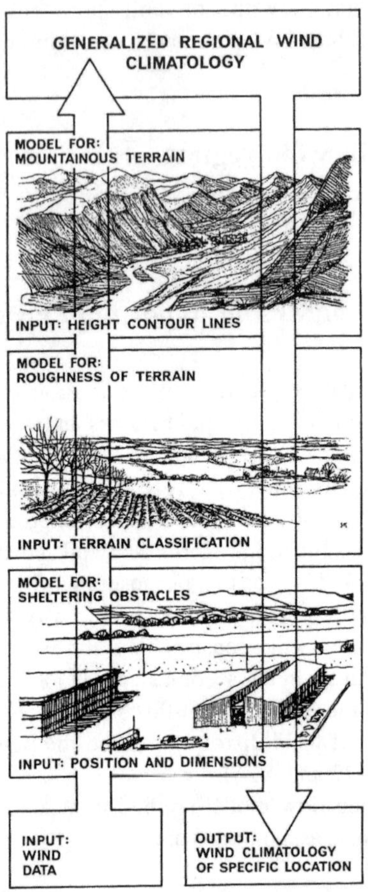

Fig. 4-43 Principle of conversion of the wind statistics between different locations according to WAsP [5]

One of the biggest problems of WAsP is that the large-scale dynamic effects in mountainous terrain are not considered and implemented up to now. These effects are among others the dependency of the vertical wind speed profile on the free convection. The heating and cooling of the ground leads to different lift forces which influence the dynamics of turbulence, cf. section 4.1.3. If on one hand the ground is cooling down at night (stable stratification) turbulence decreases, and therefore the wind gradient increases. On the other, the heating of the ground during the day leads to an increased turbulence and mixing of the air masses (unstable stratification), so the wind speed gradient decreases. In order to take account for this effect the model in WAsP is based on a simplified input of the climatologic mean and the standard deviation. If the annual cycle at the site differs from the basic assumptions of WAsP there may be a false estimation of the wind profile.

One way to quantify the complexity of a site is the ruggedness index (RIX) which describes the percentage of the area around a site which exceeds a critical inclination of e.g. 17°. Thus, the RIX serves for a rough estimation of the degree of flow separation and therefore to which extent the assumptions of the linearised model are violated.

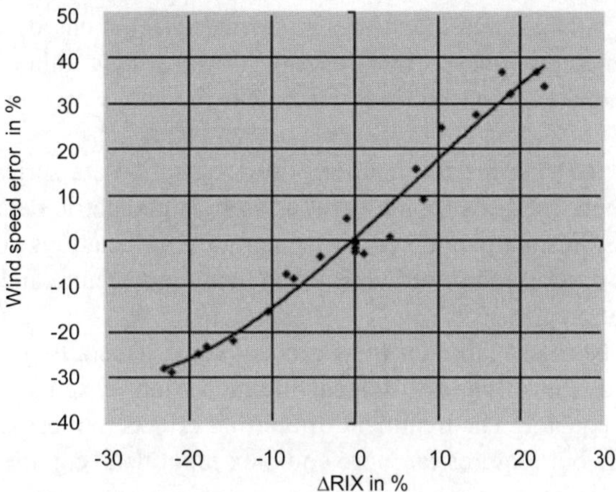

Fig. 4-44 Prediction errors of WAsP versus difference between ΔRIX of the investigated site and the reference site [6]

Comparing the RIX at the planning site (chosen wind turbine) with the RIX of the reference site (measuring mast), the relative difference ΔRIX may serve as an indicator for the prediction error of WAsP [6]. Fig. 4-44 illustrates the strong influence of flow separation on the prediction error: If the complexity of the terrain at the planning site and the reference site is quite similar (ΔRIX = 0), the prediction error is small. If the planning site is located in a more complex terrain than the reference site (ΔRIX > 0), WAsP tends to over-estimate the wind speed. On the contrary, if the planning site is less complex than the reference site (ΔRIX < 0), WAsP tends to under-estimate the wind speed.

The advantages and disadvantages of WAsP are briefly summarized as follows:

Advantages	**Disadvantages**
• Simple operation	• Valid only for a defined atmospheric stratification close to neutrally stable atmosphere
• Cost effective and fast	• Problems with turbulent flow separation in complex terrain
• Validated method, application limits are known	• Problems with the wind speed calculation outside the Prandtl layer (surface boundary layer)

4.5.2 *Meso-Scale models*

Meso-scale models describe the air flow with a spatial scale of up to 300 km and a time scale of the fluctuations ranging from hours (storms) to several days (passage of cyclones), cf. Fig. 4-13. Meso-scale models are applied to generate maps of the wind potential for large areas. The used systems of equations describe the fluid mechanics (momentum, mass flow, energy flow) among other based on the Navier-Stokes equations of motion. Different models vary strongly in their complexity.

Input data of the systems of equations are long-time data series for the wind speed and direction, the geostrophic wind as well as geometric data of the terrain and its roughness. Further parameters influencing the calculations are the temperature profiles, the description of turbulence and also the structure and density of the generated simulation grid.

While in the Northern latitudes the surface wind is generally dominated by the geostrophic wind, the influence of local thermal wind systems increases when approaching the equator. The treatment of thermal effects in meso-scale models is generally possible but requires far more complex input data, e.g. the albedo rate of the surface, which increases rapidly and heavily the computational efforts.

Since an adequately high spatial and temporal resolution requires high computational efforts, it is useful to combine meso-scale models with WAsP. A wind atlas is generated from the meso-scale calculations which are then the basis for the very detailed calculations with WAsP. But it is by all means recommendable to validate the simulation results with wind measurements at the site.

4.5.3 *Measure-Correlate-Predict-Methode*

For most planning sites wind measurements seldom exist for a sufficiently long period (minimum 20 years), on one hand. On the other, the wind regime at the site should be predicted as exact as possible for the expected operating period of 20 years because of the strong variations of the mean annual wind speed from year to year (cf. Fig. 4-17).

If a long-term time series of wind speed and direction is available for at least one reference site in the region of the planning site (minimum measuring period 20 years) then there is the possibility to generate a long-term wind prediction by correlation between the reference site and the planning site.

The purely statistical method *Measure-Correlate-Predict* (MCP) is based on the assumption that there is a linear correlation between simultaneous measurements at the planning site and the reference site. Using, e.g., the hourly mean values of the wind speed v_{Ri} at the reference site as x-coordinate and the simultaneous ones v_{Pi} from the planning site as y-coordinate, these pairs of values may be drawn in a Cartesian coordinate system. Under the assumption of a linear relation it is then allowed to draw a regression line through the points. The gradient of the line is a measure for the relation of wind speed v_P at the planning site to the wind speed of the reference site v_R. Calculating the correlation between the measuring data from reference site and planning site allows estimating statistically the relation between the wind regimes at the considered sites using the correlation coefficient R^2. It gives the linear correlation between the simultaneously recorded data as well as the scattering (variance). From the obtained averages and standard deviations of the sample it is possible to determine e.g. the Weibull distribution function, cf. section 4.2.2. If the correlation is sufficiently high, i.e. $R^2 > 0.70$, it is allowed to transfer the factors of the wind speed distribution at the reference site to the planning site.

Wind farm planning is in general based on the sectorised wind data, with usually 12 sectors, cf. Fig. 4-22. It follows that also the correlation between reference site and planning site has to be performed sector by sector: The distributions factors obtained for one individual wind direction sector of the wind rose at the reference site are transferred by the correlation to the corresponding sector of the planning site to achieve its individual distribution factors. In course of this procedure, many problems arise: If in the considered region a certain wind direction dominates the wind regime of the year or season there will be some sectors of the wind rose where there are less individual events than required for a correlation, meaning that the transformation from the reference site to the planning site is not possible. Moreover there may be a time delay between the reference and the planning site for the occurrences of distinct meteorological events, e.g. weather fronts. This leads to simultaneously recorded pairs of data which are not correlated. Due to the time delay it is moreover possible that in course of the passage of weather fronts there is a shift of the prevailing wind direction between the sites, and consequently the wind speeds of the meteorological event are located at the reference and planning site in different wind rose sectors.

All these problematic issues which reduce the quality of the correlation are reduced by choosing the sample length as long as possible, i.e. minimum one-year data sets. There are different ways for trying by methodological extensions to circumvent or reduce the effects of the mentioned deficits which is discussed extensively in the corresponding literature, e.g. [28].

Since the presented MCP method is a purely formal-statistical procedure, the physical conditions like orography, plant cover or building are not considered.

For the reliability of the correlation investigation it is important to choose either a suitable sample length to represent the annual wind cycle, or at best the correlation of an entire annual wind record of the reference and planning site.

The disadvantage of the method is that the calculated wind speed is valid only for the individual measuring height exactly at position of the measuring mast. A transformation of the long-term corrected wind regime to other points than the one measured (in measuring height at the mast position) is only possible by physical models which consider the effects of the local site conditions on the flow.

References

[1]   Petersen E.L., Mortensen N.G., Landberg L., Højstrup J., Frank H.P.: *Wind Power Meteorology*, Risø-I-1206, 1997

[2]   *Meteorological Aspects of The Utilisation of Wind as an Energy Resource*, WMO Rep. No. 575, Geneva, 1981

[3]   IEA: *Recommended Practices for Wind Turbine Testing, Part 11. Wind Speed Measurement and Use of Cup Anemometry*, 1999

[4]   IEC 61400-12-1: *Wind Turbine Generator Systems – Part 12: Wind Turbine Performance Testing*, 2005

[5]   Troen I., Petersen E.L.: *European Wind Atlas*, Risø National Laboratory, 1989

[6]   Mortensen N.G., Petersen E.L.: *Influence of Topographical Input Data on the Accuracy of Wind Flow Modeling in Complex Terrain*, Proceedings of the European Wind Energy Conference, Dublin, Ireland, 1997

[7]   Hübner H., Otte J.: *Windenergienutzung im Mittelgebirgsraum (Wind energy application in the low mountain range))*, University of Kassel, Germany, 1990

[8]   Antoniou I. et al: *On the Theory of SODAR Measurement Techniques*, Risø-R-1410(EN), 2003, Risø National Laboratory, Roskilde Denmark

[9]   Energia Eolica: *Le Gouriérès*, Masson, 1982

[10]   *Meyers kleines Lexikon der Meteorologie (Meyer's small encyclopaedia on meteorology)*, Meyers Lexikon-Verlag, Mannheim, 1987

[11]   Courtney M.S.: *An atmospheric turbulence data set for wind turbine research*, Proceedings of the 10th British Wind Energy Association Conference, London 22-24 March 1988

[12]   Burton T., Sharpe D., Jenkins N., Bossanyi E.: *Wind Energy Handbook*, John Wiley & Sons, 2001

[13]   Kristensen L., Hansen O.F.: *Distance Constant of Risø Cup Anemometer*, Risø-R-1320(EN), 2002, Risø National Laboratory, Roskilde Denmark

[14]   Mann J., Ott S., Jørgensen B.H., Frank H.P.: *WAsP Engineering 2000*, Risø-R-1356(EN), 2002, Risø National Laboratory, Roskilde Denmark

[15]   Lange B.: *Modelling the Marine Boundary Layer for Offshore Wind Power Utilisation*, PhD thesis, University of Oldenburg, Germany 2002

[16]   Stull R.B.: *An Introduction to Boundary Layer Meteorology*, 1988, Kluwer Acad. Publ. Dordrecht, 666pp

[17]   Petersen T.F., Gjerding S., Ingham P., Enevoldsen P., Hansen J.K., Jørgensen H.K.: *Wind Turbine Power Performance Verification in Complex Terrain and Wind Farms*, Risø-R-1330(EN), 2002, Risø National Laboratory, Roskilde Denmark

[18]   FGW (Federation of German Windpower): *Technische Richtlinien für Windenergieanlagen, Teil 2: Bestimmung von Leistungskurve und standardisierten Energieerträgen (Technical Guidelines for Wind Energy Converters, Part2: Determining the Power Performance and Standardised Energy Yields), Rev. 13*, 2000

[19]   IEC 61400-1, Ed.3: *Wind Turbine Generator Systems – Part 1: Safety Requirements*, 2005

[20]   Germanischer Lloyd: *Richtlinie für die Zertifizierung von Windenergieanlagen (Guideline for the Certification of Wind Turbines)*, Hamburg 2003

[21]   DIBt (Center of competence in civil engineering): *Richtlinie für Wind turbinen (Guideline for Wind Energy Plants)*, 1993, 1996, 2004

[22]   Kaiser K., Langreder W.: *Site Specific Wind Parameter and their Effect on Mechanical Loads*, Proceedings EWEC, Copenhagen, 2001

[23]   Frandsen S., Thøgersen L.: *Integrated Fatigue Loading for Wind Turbines in Wind Farms by Combining Ambient Turbulence and Wakes*; Wind Engineering, Vol. 23 No. 6, 1999

[24]   Katic I., Højstrup J., Jensen N.O.: *A Simple Model for Cluster Effeciency*, European Wind Energy Association Conference and Exhibition, 7-9 October 1986, Rome, Italy

[25]   Högström U.: *Non-dimensional wind and temperature profiles*, Bound. Layer Meteor. 42 (1988), 55-78

[26]   Barthelmie R., Hansen O.F., Enevoldsen K., Motta M., Pryor S., Højstrup J., Frandsen S., Larsen S., Sanderhoff P.: *Ten years of Measurements of offshore Wind farms – what have we learnt and where are Uncertainties?*, Proceedings The Art of making Torque of Wind EAWE, Delft, 2004

[27]   Kaiser K., Langreder W., Hohlen H.: *Turbulence Correction for Power Curves*, Proceedings EWEC 2003, Madrid 2003

[28]   Riedel V., Strack M.: *Entwicklung verbesserter MCP-Algorithmen mit Parameteroptimierung durch Verteilungsanpassung (Development of improved MCP algorithms with parameter optimization by distribution fitting)*, Deutsche Windenergie-Konferenz, Wilhelmshaven, 2002

[29]   Thomsen K., Madsen H.A.: *A new simulation method for turbines in wake – applied to extreme response during operation*, Proceedings: The Art of making Torque of Wind, EAWE, Delft, 2004

[30]   IEC 61400-121,88/163/CDV: *Wind Turbine Generator Systems – Part 121: Power Performance Measurements of Grid Connected Wind Turbines*

[31]   Obukhow A.M.: *Turbulence in an Atmosphere with non-uniform temperature.* Transl. in Boundary Layer Meteor., 1946

[32]   TA-Luft: *Technische Anleitung zur Reinhaltung der Luft (German Technical Instructions on Air Quality)*, Bundesministerium für Umweltschutz, Raumordnung und Reaktorsicherheit (German Federal Ministry of the environment, Nature Conservation and Nuclear Safety, Bonn, Juli 2002

[33]   Molly, J.P.: *Windenergie (Windenergy)*, C.F. Müller-Verlag, Karlsruhe, 1990

[34]   Manwell, J.F., u.a.: *Windenergy explained – Theory, Design and Application*, John Wiley & Sons Ltd., USA/UK, 2002

[35]   Thomsen, W.T.: *Theory of vibration*, 4th edition, Chapter 13: Random vibration, Prentice Hall, New Jersey, USA, 1993

[36]   Christoffer, J., Ulbricht-Eissing , M.: *Die bodennahen Windverhältnisse in der BR Deutschland (The surface wind regimes in Germany)*, Offenbach, Deutscher Wetterdienst (German Weather Service), Bericht Nr. 147, 1989

[37]   Hoffmann, R.: *Signalanalyse und –erkennung (signal analysis and recognition)*, Springer-Verlag, Berlin, 1998

[38]   Ammonit: *Messtechnik für Klimaforschung und Windenergie (Measuring equipment for climatic research and wind energy) 2006/07)*, Ammonit Gesellschaft für Messtechnik mbH, 2006

5 Blade geometry according to Betz and Schmitz

With the help of the Betz or the Schmitz (Glauert) theory [1, 2, 7], designing a wind turbine is relatively straightforward. These theories provide the blade chord and the blade twist relative to the radius, after the design tip speed ratio, the aerodynamic profile and the angle of attack or the lift coefficient have been specified.

The Betz theory only takes into account the axial downstream losses. Schmitz, additionally, takes into account the downstream swirl losses, due to the rotational wake. For rotors with a low tip speed ratio (design tip speed ratio of $\lambda_D < 2.5$), this results in a blade geometry significantly different from that determined by the Betz theory. Profile losses and losses resulting from the air flow around the blade tips are ignored in both theories. They have to be accounted for by an additional reduction in the turbine's power. The number of blades is not an issue as this only affects the tip losses.

5.1 How much power can be extracted from the wind?

In accordance with section 2.3.1 the kinetic energy of a mass in motion is

$$E = \frac{1}{2}\, m v^2 . \tag{5.1}$$

The power in the wind flowing through a control area A can be calculated as

$$P_{\text{wind}} = \dot{E} = \frac{1}{2}\, \dot{m}\, v^2 = \frac{1}{2}\rho\, A\, v_1^{\,3} , \tag{5.2}$$

since the mass flow rate is $\dot{m} = \rho\, A\, dx/dt = \rho\, A\, v_1$, see Fig. 5-1.

As already illustrated in chapter 2, the wind's power cannot be extracted completely. Betz [1] analysed a heavily idealised wind turbine that extracts the power from the wind at an 'active plane' by retarding its speed, without any losses. The physical process that extracts the kinetic energy from the air flow is not yet considered at this stage.

As shown, in Fig. 5.2, Betz assumed a homogeneous air flow v_1 that is retarded by the turbine to the velocity v_3 far downstream. That means that, for the reasons of continuity, he assumes a stream tube with divergent streamlines

$$\rho\, v_1\, A_1 = \rho\, v_2\, A = \rho\, v_3\, A_3 . \tag{5.3}$$

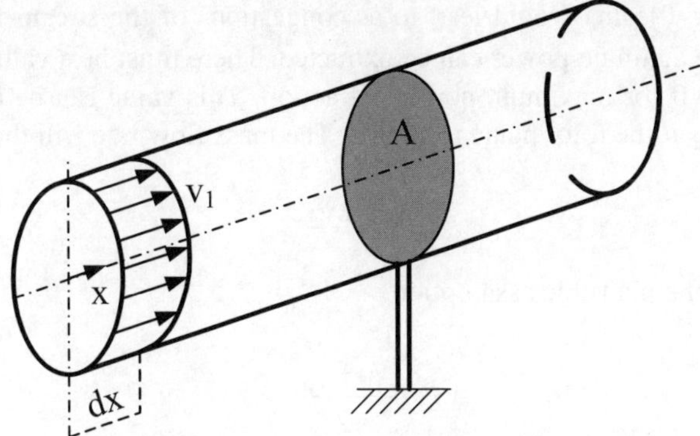

Fig. 5-1 Air flow in a stream tube of velocity v through a control area A (from chapter 2)

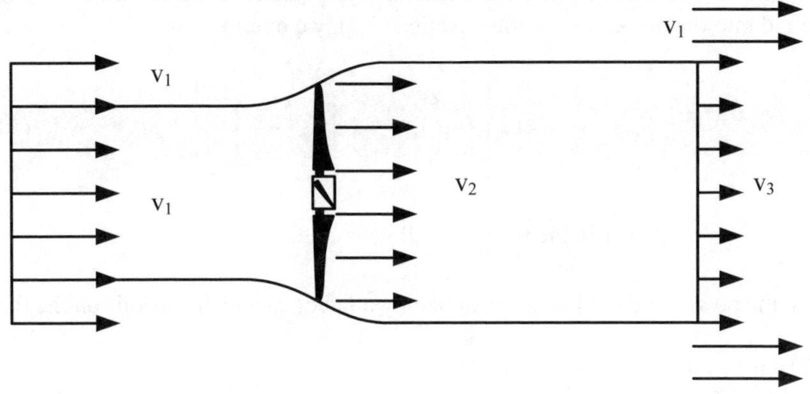

Fig. 5-2 Air flow through an ideal Betz wind turbine (from chapter 2)

As the changes in the air pressure are minimal, the air density ρ can be assumed to be constant.

The extracted kinetic energy E_{ex} is upstream energy minus downstream energy

$$E_{\mathrm{ex}} = \frac{1}{2}\, m\left(v_1^2 - v_3^2\right). \tag{5.4}$$

The power extracted from the wind is therefore

$$\dot{E}_{\mathrm{ex}} = \frac{1}{2}\, \dot{m}\left(v_1^2 - v_3^2\right). \tag{5.5}$$

If the wind were not retarded at all ($v_3 = v_1$), no power would be extracted. If the wind is retarded too much, the mass flow rate $\dot{m}$ becomes very small. Taken to the extreme ($\dot{m} = 0$), this would lead to a 'congestion' of the stream tube ($v_3 = 0$) such that once again no power can be extracted. There must be a value between $v_3 = v_1$ and $v_3 = 0$ for maximum power extraction. This value can be determined if the velocity v_2 at the rotor plane is known. The mass flow rate will then be

$$\dot{m} = \rho\, A\, v_2 \,. \tag{5.6}$$

At this point, the plausible assumption

$$v_2 = \frac{v_1 + v_3}{2} \tag{5.7}$$

is made, which will be proved later on (Froude-Rankine Theorem). If the mass flow rate of equation (5.6) and the rotor plane velocity v_2 of equation (5.7) are inserted into the power-extraction equation (5.5), we obtain

$$\dot{E}_{ex} = \frac{1}{2}\,\rho\, A\, v_1^3 \left[\frac{1}{2} \cdot \left(1 + \frac{v_3}{v_1} \right) \cdot \left(1 - \left(\frac{v_3}{v_1} \right)^2 \right) \right] \tag{5.8}$$

Power in the wind   Power coefficient c_{P}

Thus, the power in the wind is multiplied by a factor c_{P} which depends on the ratio v_3 / v_1 .

The maximum power coefficient is

$$c_{\mathrm{P,\,Betz}} = \frac{16}{27} = 0.59 \,. \tag{5.9}$$

It occurs when the wind velocity v_1 (upstream of the rotor) is retarded to $v_3 = (1/3)\, v_1$ (downstream). This can be found either by drawing the curve, Fig. 5-3, or, mathematically, by setting the first derivation of equation (5.8) to zero. Nearly 60% of the power in the wind is extractable by an ideal wind turbine! In this case, the velocity in the rotor plane is $2v_1/3$ and far downstream $v_1/3$.

The diagram, in Fig. 5.4, illustrates how much power can be extracted in Betz's ideal case of $c_{\mathrm{P,\,Betz}} = 0.59$. Of course, it depends on the wind velocity and the diameter of the wind turbine, Due to additional losses which will be analysed later on, the actual power of modern wind turbines is somewhat smaller. However, values up to $c_{\mathrm{P}} \approx 0.50$ are being achieved by some modern wind turbines.

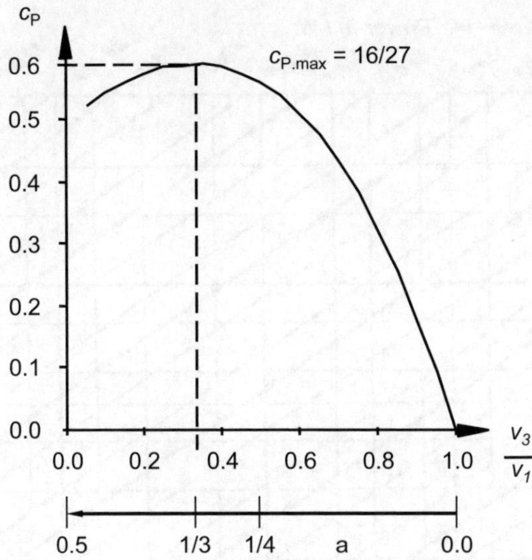

Fig. 5-3 Power coefficient c_P versus the ratio of wind velocity v_3 far downstream and the upstream wind velocity v_1 ; $c_{P,max} = 0{,}593$ at $v_3/v_1 = 1/3$, additional axis: induction factor a

Now, using the principle of linear momentum, we can determine the thrust acting on the wind turbine at the rotor plane when maximum power is extracted. The thrust equation

$$T = \dot{m}\,(v_1 - v_3) = \rho A\, \frac{v_1 + v_3}{2}\,(v_1 - v_3) \tag{5.10}$$

yields for the given ratio $v_3 / v_1 = 1/3$

$$T = c_T\left(\frac{1}{2}\,\rho v_1^2\right) A \quad ; \quad c_T = {}^8/_9 = 0.89\,. \tag{5.11}$$

The term in brackets is the dynamic pressure on the area A. Comparing this result with the drag D of a circular disk in the wind flow

$$D = c_D\left(\frac{1}{2}\,\rho v_1^2\right) A \quad ; \quad c_D = 1.11, \tag{5.12}$$

it is found that in the case of optimum power extraction the thrust is nearly 20% lower than that of a circular disk.

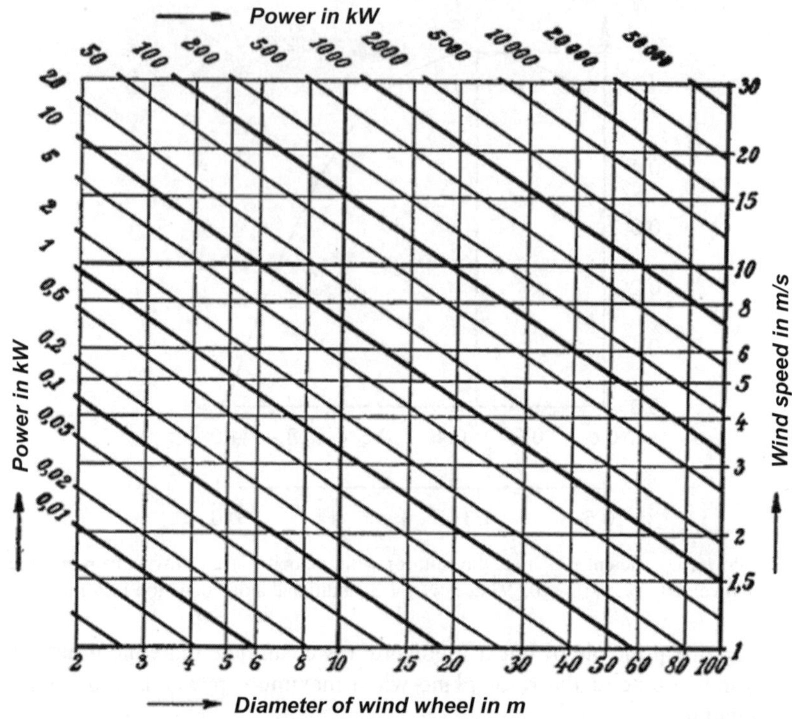

Fig. 5-4 "Betz power" of wind turbines as a function of wind velocity and diameter, Betz [1]

Remark: In the English literature, the power coefficient and the thrust are often shown versus the "induction factor a". This is based on the idea that the wind turbine superimposes a sort of headwind $a v_1$ on the wind velocity v_1. For the active rotor plane this gives

$$v_2 = v_1 (1 - a)$$

and far downstream of the rotor

$$v_3 = v_1 (1 - 2 \cdot a).$$

We show this induction factor on an additional scale in Fig. 5-3.

5.1.1 Froude-Rankine Theorem

In the following it will be proved that the velocity v_2 in the rotor plane according to Betz' theory is actually the arithmetic mean of the far-upstream and far-downstream velocities.

The thrust can be expressed using the principle of linear momentum in equation (5-10)

$$T = \dot{m}\left(v_1 - v_3\right).$$

Alternatively, it can be derived from the Bernoulli equation (energy balance), which we apply for both the plane left and the plane right of the rotor plane, Fig. 5-5.

$$p_1 + \frac{\rho}{2} \cdot v_1^2 = p_{-2} + \frac{\rho}{2} \cdot v_{-2}^2 \text{ and} \tag{5.13}$$

$$p_{+2} + \frac{\rho}{2} \cdot v_{+2}^2 = p_3 + \frac{\rho}{2} \cdot v_3^2 \ . \tag{5.14}$$

The subscript -2 denotes the plane immediately before the rotor, and +2 immediately after the rotor.

To preserve continuity, the velocity immediately left and right of the rotor must be equal, $v_{-2} = v_{+2}$. Moreover, the static pressure far upstream and far downstream are also equal, $p_1 = p_3$. Hence, the difference of equation (5.13) and (5.14) gives

$$\frac{\rho}{2}\left(v_1^2 - v_3^2\right) = p_{-2} - p_{+2} \tag{5.15}$$

According to these (energetic) considerations, the thrust at the tower results from the difference of the static pressure before and after the rotor plane

$$T = A \cdot \left(p_{-2} - p_{+2}\right) \ . \tag{5.16}$$

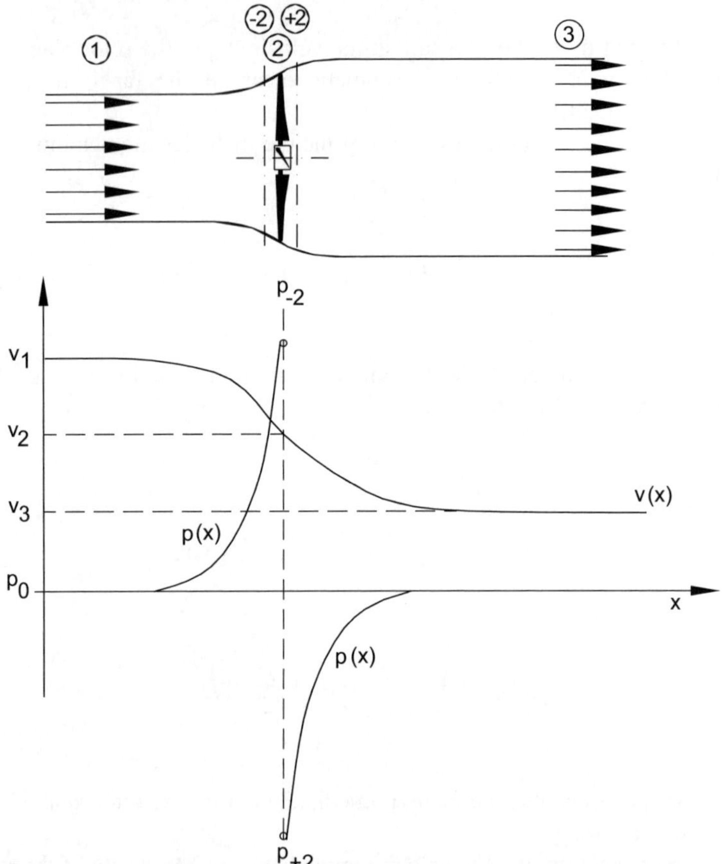

Fig. 5-5 Development of velocity and static pressure p along the stream tube

Putting the mass flow rate $\dot{m} = \rho\, A\, v_2$ into equation (5.10) and introducing this into the equations (5.16) and (5.15), we obtain the expression used in section 5.1 for the velocity v_2 in the rotor plane

$$v_2 = \frac{(v_1 + v_3)}{2} \quad .$$

5.2 The airfoil theory

So far, the method of power extraction at the rotor plane has been left open. Horizontal axis wind turbines use rotating blades designed to extract the maximum power as calculated by Betz. Before we start the discussion of blade dimensioning, the basic results of the airfoil theory will be presented.

For that purpose, we consider a *symmetric airfoil* profile attacked at the leading edge by a velocity w. When the angle of attack is $\alpha_A = 0°$ no lift occurs, only drag. For a streamlined airfoil, as shown in Fig. 5-6, it is small.

The positive lift force L results from an angle of attack greater than zero. It is proportional to the blade area (chord length c times width b), and to the square of the velocity w.

$$L = c_L(\alpha_A) \; \frac{\rho}{2} \; w^2 \; (c\ b) \tag{5.17}$$

$$D = c_D(\alpha_A) \; \frac{\rho}{2} \; w^2 \; (c\ b) \tag{5.18}$$

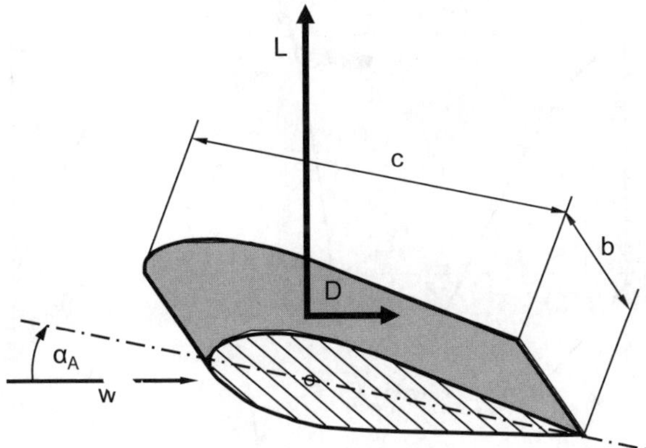

Fig. 5-6 Lift F_L and drag F_D at a blade element of the width b and chord length c

The lift coefficient $c_L = c_L\,(\alpha_A)$ denotes the dependence on the angle of attack α_A. Like the drag coefficient $c_D\,(\alpha_A)$, it is usually determined by means of wind tunnel experiments. Fig. 5-7 shows the measured curves of lift and drag coefficients for an asymmetric profile.

Initially, the increase in the lift coefficient c_L (and thus the lift force) increases linearly with the angle of attack (range $0° < \alpha_A < 10°$). Then, the curve flattens before reaching its maximum value. At even greater angles of attack, the flow separates from the profile: the lift is reduced in this range ($\alpha_A > 15°$), while c_D, and thus the drag force, increases rapidly with α_A, Fig. 5-7. This figure also shows that the thinner the airfoil profile is the smaller its drag coefficient for small angles of attack α_A.

The flow on the airfoil's upper side (i.e. suction side) is faster than on the lower side (i.e. pressure side) since it has to cover a larger distance. According to the Bernoulli equation, this results in a lower pressure on the upper side. The integration of $(p\,ds)$ along the profile contour gives the lift L and the drag D as components of the resulting force F, Fig. 5-9,

$$F = \sqrt{L^2 + D^2} \; . \tag{5.19}$$

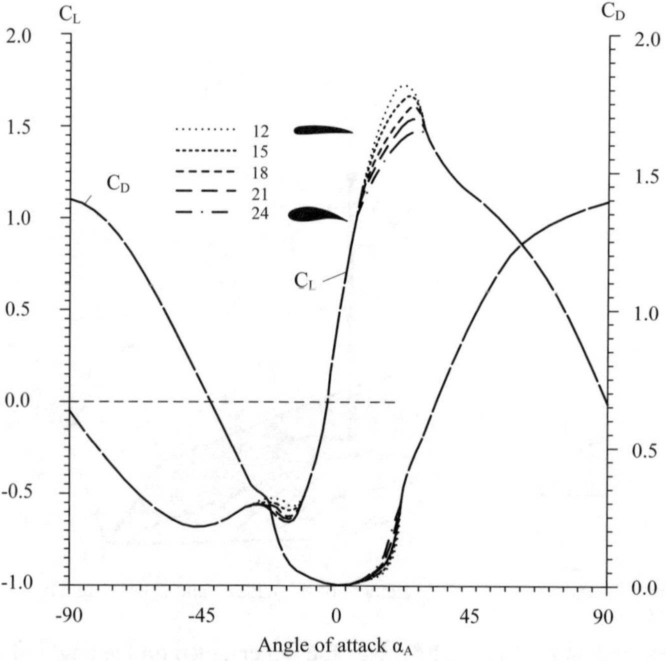

Fig. 5-7 Corrected lift and drag coefficients versus angle of attack α_A from wind tunnel measurements for the profiles NACA 4412 to 4424 [9]

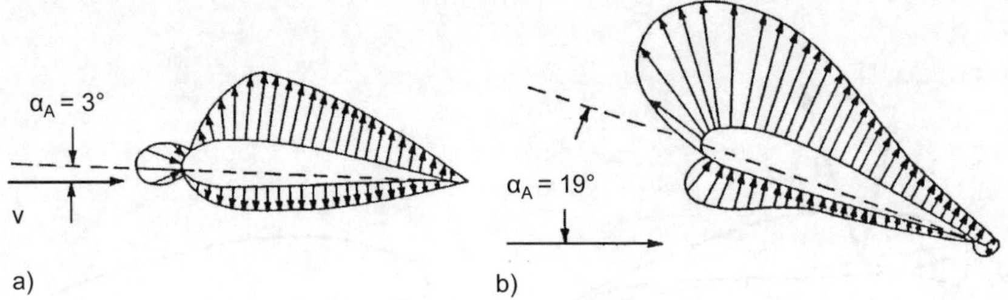

Fig. 5-8 Pressure distribution around an airfoil [13], a) for a small angle of attack ($\alpha_A = 3°$); b) for a large angle of attack ($\alpha_A = 19°$)

As long as the flow is attached, this force F attacks at a point between 25 and 30% of the profile's chord length c. If the flow is separated, this point moves further back; during heavy stall conditions it is found around $c/2$ which is immediately plausible for $\alpha_A = 90°$. For this case, the blade's surface is perpendicular to the attacking flow w, and the pattern of the flow around the blade is nearly symmetrical.

For the case of a flat plate with attached flow, the lift coefficient can be determined theoretically [6]

$$c_L(\alpha_A) = 2\ \pi\ \alpha_A .$$

(5.20)

For real airfoil profiles the lift coefficient c_A is slightly smaller

$$c_L(\alpha) = (5.1 \text{ to } 5.8)\,\alpha_A .$$

(5.21)

In catalogued measurements of asymmetric profiles [e.g. 3, 4, 5], it is important to check whether the angle of attack is measured from the resting edge (which will often be the case for profiles with a straight lower side), or from the chord line between nose centre of the leading edge and trailing edge, Fig. 5-9. In any case, the point of zero lift, $c_L = 0$, is found in the range of negative angles of attack. At $\alpha_A = 0°$ there is a lift due to the camber of the profile, see Fig. 5-10.

As for the symmetric profiles, the gradient c_L' in the rising part of the lift curve for the asymmetric profile is approximately $c_L' \cong 2\ \pi$.

In the next section the lift/drag ratio will be used which gives the ratio between lift force and drag force

$$\varepsilon(\alpha_A) = \frac{L}{D} = \frac{c_L(\alpha_A)}{c_D(\alpha_A)} .$$

(5.22)

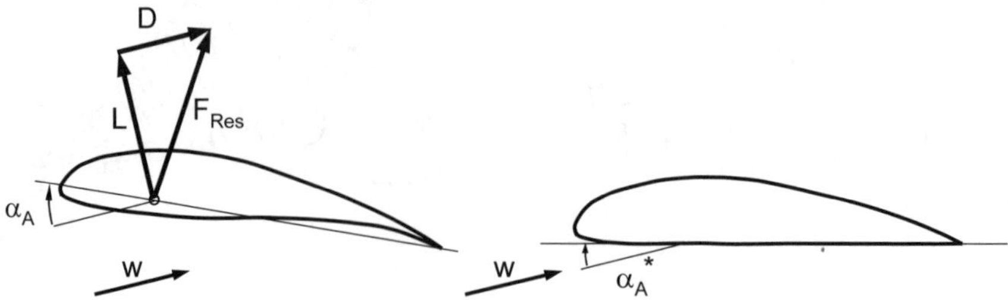

Fig. 5-9 Reference line for the definition of the angle of attack: Force F_{Res} resulting from lift force L and drag force D

Its maximum ε_{max} (usually found in the range $c_L = (0.8$ to $1.1)$, at moderate angles of attack) is a measure for the aerodynamic quality of the profile. High-quality profiles attain a lift/drag ratio of $\varepsilon_{max} = 60$ and higher; a flat plate may achieve a value of $\varepsilon_{max} = 10$. Be careful: there are different definitions to be found in the literature. Sometimes the inverse lift/drag ratio is applied, see Fig. 5-10.

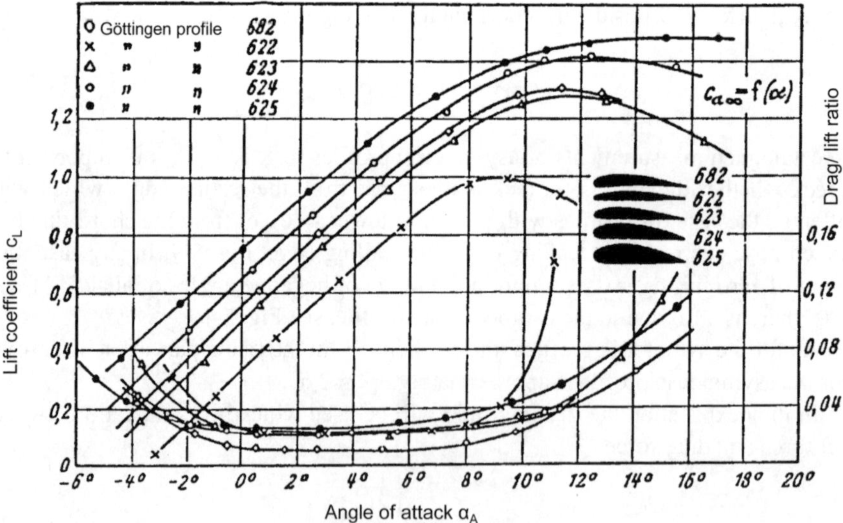

Fig. 5-10 Lift coefficients c_L and inverse lift/drag ratio of some Göttingen airfoil profiles [11]

5.3 Flow conditions and aerodynamic forces at the rotating blade

5.3.1 Triangles of velocities

There is a relative velocity w for each cross section r of the blade. This velocity consists of the retarded velocity in the rotor plane $v_2 = 2\,v_1/3$ according to Betz, and the local circumferential speed $u = \Omega\,r$ resulting from the blade's rotation with the angular speed Ω.

In Fig. 5-11 we see immediately that the relative velocity w is composed of the two components v_2 and $u(r)$

$$w^2\,(r) = \left(\frac{2}{3}\,v_1\right)^2 + \left(\Omega r\right)^2 \ . \tag{5.23}$$

Its direction ϕ in reference to the rotor plane is given by

$$\tan \phi\,(r) = v_2/(\Omega\,r) \tag{5.24}$$

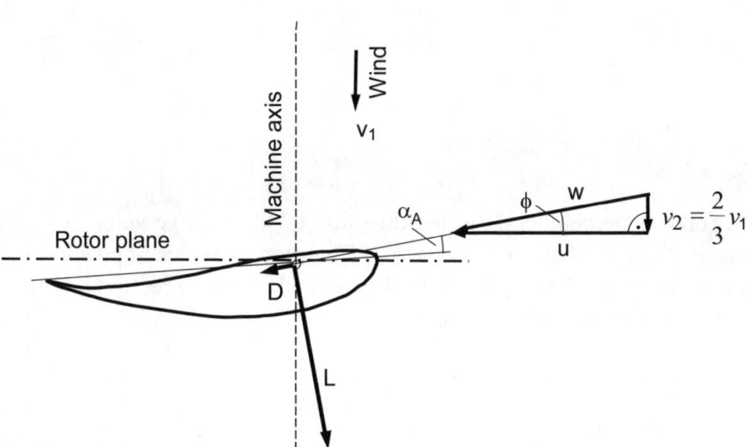

Fig. 5-11 Triangles of velocities, relative velocity w resulting from the geometrical superposition of the reduced wind velocity v_2 in axial direction, and the circumferential speed $u = \Omega r$, resulting from the blade's rotational speed

Introducing the design tip speed ratio λ_D which is the quotient of circumferential speed $\Omega\,R$ at the blade tip and the wind velocity v_1,

$$\lambda_D = \frac{\Omega R}{v_1} \tag{5.25}$$

we can modify equation (5.24) with $v_2 = 2\, v_1/3$ to

$$\tan \phi = \frac{2}{3}\,\frac{R}{\lambda_D\, r}\;. \tag{5.26}$$

Fig. 5-12 illustrates once more that the triangles of velocity vary for each blade section, as the circumferential component $u = \Omega\, r$ increases linearly with the radius.

5.3.2 Aerodynamic forces at the rotating blade

According to Fig. 5-13, the aerodynamic forces lift dL and drag dD pull and push (approx. at 0.25 of chord) at the blade element of length dr at the radius r,

$$\text{Lift:} \qquad dL = \frac{\rho}{2}\, w^2 c\; dr\; c_L(\alpha_A) \quad \text{and} \tag{5.27}$$

$$\text{Drag:} \qquad dD = \frac{\rho}{2}\, w^2 c\; dr\; c_D(\alpha_A)\;. \tag{5.28}$$

The lift force is perpendicular to the direction of the relative velocity w whereas the drag is parallel to it, cf. section 5.2. Therefore, the decomposition of these forces into the tangential force dU in the circumferential direction and the thrust dT in the axial direction yields (Fig. 5-13)

$$dU = \frac{\rho}{2}\, w^2 c\; dr\; [c_L \sin\phi - c_D \cos\phi] \quad \text{and} \tag{5.29}$$

$$dT = \frac{\rho}{2}\, w^2 c\; dr\; [c_L \cos\phi + c_D \sin\phi]\;. \tag{5.30}$$

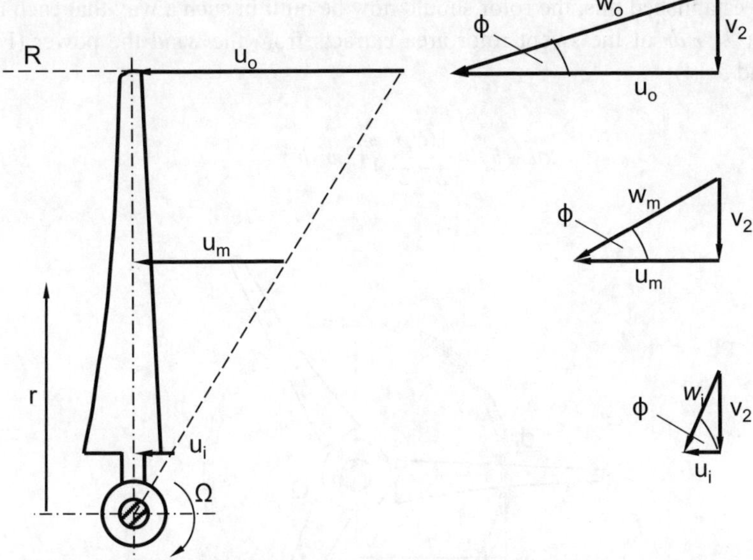

Fig. 5-12 Triangles of velocity at different blade sections: circumferential speed $u = \Omega r$, axial wind velocity at the plane of rotation $v_2 = 2\,v_1\,/3$

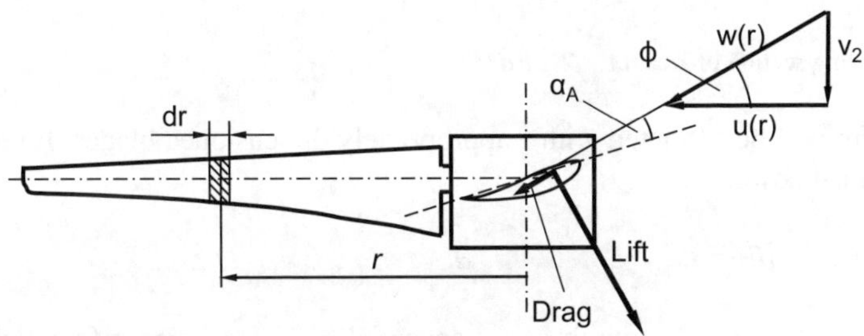

Fig. 5-13 Aerodynamic forces at the blade ring element dr

5.4 The Betz optimum blade dimensions

According to the general considerations in section 5.1, the maximum power that can be extracted from a circular area is

$$\dot{E}_{\text{Betz}} = \frac{16}{27}\frac{\rho}{2}\,v_1^3\left(\pi R^2\right).$$

Having established this, the rotor should now be built in such a way that each ring element $2\pi r\, dr$ of the swept rotor area extracts from the wind the power (Figs. 5-13 and 5-14)

$$d\dot{E}_{\text{Betz}} = \frac{16}{27}\frac{\rho}{2}v_1^3\left(2\pi r\, dr\right) .$$

(5.31)

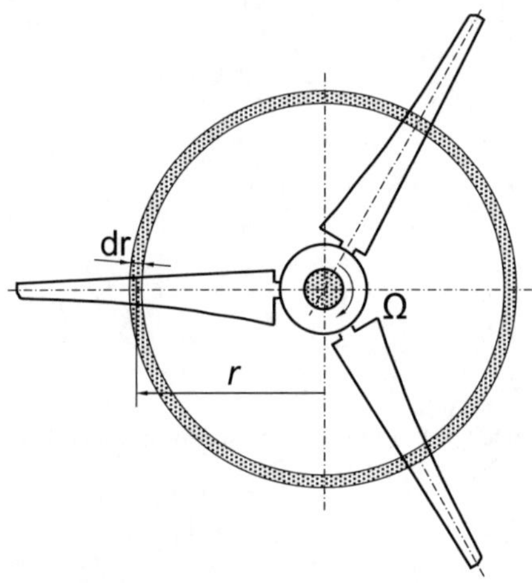

Fig. 5-14 Ring section of area $dA = 2\,\pi r\, dr$

This power will be extracted with z appropriately dimensioned blades. It results in a mechanical power of

$$dP = \quad z \quad \cdot \quad dU \quad \cdot \quad \Omega\, r \quad .$$

(5.32)

	number of blades	tangential component of the aerodynamic force	local circumferential velocity

at the ring section.

Since we assume that at the design point the aerodynamic profile is operated close to its best lift/drag ratio, the drag coefficient is very small, $c_D \ll c_L$. The lift is then the only significant contributor to the tangential force dU in equation (5.29)

$$dU \approx dL\ \sin\phi = \frac{\rho}{2}\ c_L\, w^2\ c\,(r)\, dr\, \sin\phi .$$

(5.33)

The mechanical power is now given by

$$dP \approx z\,\Omega\,r\,\frac{\rho}{2}\,c_L\,w^2\,c(r)\,dr\,\sin\phi\,.$$

(5.34)

Equating the mechanical power (5.34) and the Betz power (5.31), $dP = d\dot{E}_{Betz}$, the important formula for the blade chord $c(r)$ of an optimally designed blade can be obtained:

$$c(r) = \frac{1}{z}\,\frac{16}{27}\,\frac{2\,\pi\,r}{c_L}\,\frac{v_1^3}{w^2\,\Omega r\,\sin\phi}\,.$$

(5.35)

Using the relations given in the triangles of velocities,

$$v_1 = \frac{3}{2}\,w\,\sin\phi \qquad \text{and} \qquad u = \Omega\,r = w\,\cos\phi,$$

we can rewrite this as

$$c(r) = 2\,\pi\,R\,\frac{1}{z}\,\frac{8}{9c_L}\,\frac{1}{\lambda_D\sqrt{\lambda_D^2\,(\frac{r}{R})^2 + \frac{4}{9}}}\,.$$

(5.36)

In this equation, λ_D is the chosen design tip speed ratio and c_L the lift coefficient selected in the course of the design process. It can – but does not have to – be constant along the radius r. It is common practice to select a design value c_L close to the best lift/drag ratio, i.e.

$$\left.\begin{array}{l} c_L = 0.6 \text{ to } 1.2 \\[4pt] \alpha_A = 2° \text{ to } 6° \end{array}\right\} \qquad \varepsilon = c_L\,/\,c_D \approx \varepsilon_{max}\,.$$

In equation (5.36) which only distributes the required total blade chord over several blades there is no information on how to choose the number of blades. The blade number can be determined based on such issues as material strength, manufacturing or dynamic aspects.

The equation (5.36) for the blade chord becomes even more transparent in its simplified approximation

$$c(r) \approx 2 \cdot \pi \cdot R \cdot \frac{1}{z} \cdot \frac{8}{9 \cdot c_L} \cdot \frac{1}{\lambda_D^2 \cdot \left(\dfrac{r}{R}\right)}, \qquad (5.37)$$

This simplification can be used for turbines with a high tip speed ratio ($\lambda_D > 3$) and if we assume that the profiled begins at about 15% of the outer radius due to the space required for the hub. From the formula, it becomes clear that the blade chord required for the extraction of the Betz power actually decreases proportional to the square of the tip speed ratio λ_D .

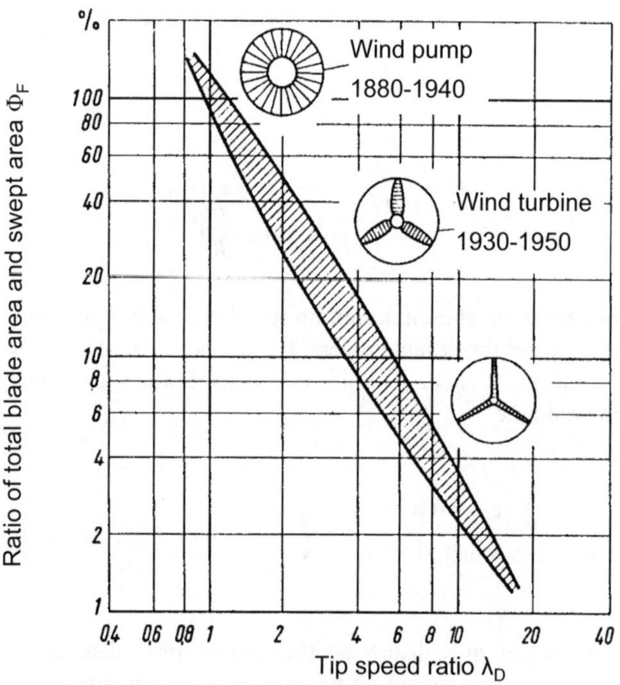

Fig. 5-15 Degree of solidity (ratio of total surface area of the blades to swept rotor area) versus design tip speed ratio λ_D [12]

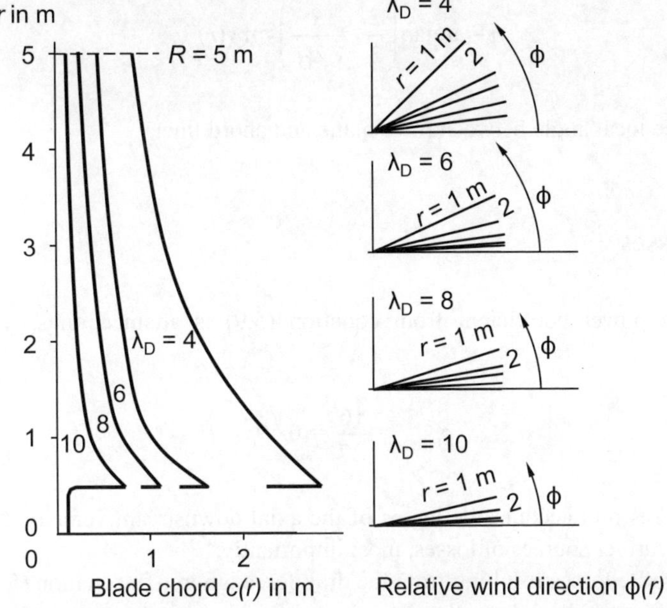

Fig. 5-16 Influence of design tip speed ratio λ_D on blade chord length $c(r)$ and direction of the relative velocity $\phi(r)$ in the rotor plane; three-bladed rotor of Betz design, $D = 10$ m

Hütter developed the diagram, Fig. 5-15, which shows the degree of solidity versus the chosen design tip speed ratio. The scatter results from the base range of c_L values close to 1.0.

In addition to the blade chord, the (total) blade twist angle $\beta(r)$ of the profile

$$\beta(r) = \phi(r) - \alpha_A(r) \tag{5.38}$$

must be determined, see Figs. 5-11 and 5-13. By selecting the tip speed ratio λ_D, we can determine the angle ϕ of the relative velocity w in the rotor plane contingent on the radius r, see equation (5.26) and Fig. 5-16

$$\phi(r) = \arctan\left(\frac{2}{3}\frac{R}{r\lambda_D}\right).$$

The angle ϕ of the relative velocity w is the first contribution to the blade twist. The blade chord has to be inclined in relation to this angle ϕ by the angle of attack α_A corresponding to the lift coefficient c_L taken as a basis for the profile chord length, Fig. 5-13. Therefore, the total blade twist angle is given by

$$\beta = \arctan\left(\frac{2}{3}\ \frac{R}{r\,\lambda_D}\right) - \alpha_A(r) \ . \tag{5.39}$$

This is the local angle between rotor plane and chord line.

5.5 Losses

The Betz power coefficient from equation (5.9) is attained only by an ideal machine

$$c_{P.Betz} = \frac{16}{27} = 0.59 \ .$$

It only takes into account the losses of the axial downstream velocity. Moreover, there are further sources of losses, most importantly:
-    the profile losses resulting from the drag force ignored in equation (5.34),
-    the losses resulting from the flow around the blade tip from the pressure to the suction side, the so-called tip losses, and
-    the wake losses due to the wake rotation downstream.

5.5.1 Profile losses

The profile losses are caused by the drag of the profile. These can be ignored when determining the ideal optimum blade geometry. However, they must be taken into account when calculating the power balance. Equation (5.32) together with equation (5.29) gives the real power at the blade element

$$dP = z\ \Omega\ r\ dU$$

$$= z\ \Omega\ r\left[\frac{\rho}{2}\ w^2\,c\,dr\,(c_L\ \sin\phi - c_D\ \cos\phi)\right]. \tag{5.40}$$

which takes into account the drag. For an ideal wind turbine, on the other hand, there is no drag ($c_D = 0$)

$$dP_{ideal} = z\ \Omega\ r\ \frac{\rho}{2}\ w^2\ c\ dr\ c_L\ \sin\phi .$$

The ratio of dP/dP_{ideal} gives the profile efficiency

$$\eta_{Profile} = 1 - \frac{c_D}{c_L}\frac{1}{\tan\phi}$$

$$= 1 - \frac{1}{\varepsilon}\frac{1}{\tan\phi} \tag{5.41}$$

$$= 1 - \frac{3}{2}\frac{r}{R}\frac{\lambda_D}{\varepsilon},$$

if $\tan\phi$ is expressed by equation (5.26). The losses at the ring section under consideration are proportional to the design tip speed ratio λ_D and the radius r

$$\zeta_{Profile} = \frac{3}{2}\frac{r}{R}\frac{\lambda_D}{\varepsilon}. \tag{5.42}$$

these losses increase with the proximity to the tip, but they are inversely proportional to the lift/drag ratio. Since the biggest portion of the power is extracted at the outer sections of the blade, wind turbines designed with a high tip speed ratio require high-quality profiles ($\varepsilon_{max} > 50$) for this part of the blade. In the inner rotor blade sections and for wind turbines designed with a low tip speed ratio (Western mill $\lambda_D \approx 1$, Dutch smock mill $\lambda_D \approx 2$) the quality of the profile is not an issue.

If a single airfoil profile type is used along the entire blade length with a fixed angle of attack α_A, the lift/drag ratio ε is independent of the local radius r. It is then possible to integrate explicitly the power (and the profile losses) along the radius

$$P = \frac{16}{27}\frac{\rho}{2}v_1^3\int_0^R \eta_{Profile}\, 2\pi r\, dr,$$

$$P = \frac{16}{27}\frac{\rho}{2}v_1^3\int_0^R\left(1 - \frac{3}{2}\frac{r}{R}\frac{\lambda_D}{\varepsilon}\right)2\pi r\, dr \quad \text{or} \tag{5.43}$$

$$P = \frac{16}{27}\,\frac{\rho}{2}\,v_1^3\,\pi R^2\left[1 - \frac{\lambda_D}{\varepsilon}\right].$$

In this case, the ratio of tip speed ratio and lift/drag ratio describes the total loss resulting from the profile drag.

Here, the overall losses due to profile drag are simply described by the tip speed ratio and the lift/drag ratio.

5.5.2 Tip losses

Another source of losses is the flow around the tip of the blade from the pressure side (lower side of the profile) to the suction side (upper side). This causes the blade lift to decreases toward the tip.

The superposition of the flow around the blade tip with the attacking wind creates an expanding vortex washed downstream by the wind, Fig. 5-17.

The more slender the blade, the more it resembles an infinitely long blade ($R/c = \infty$) for which the values of c_L and c_D from the profile catalogues hold true.

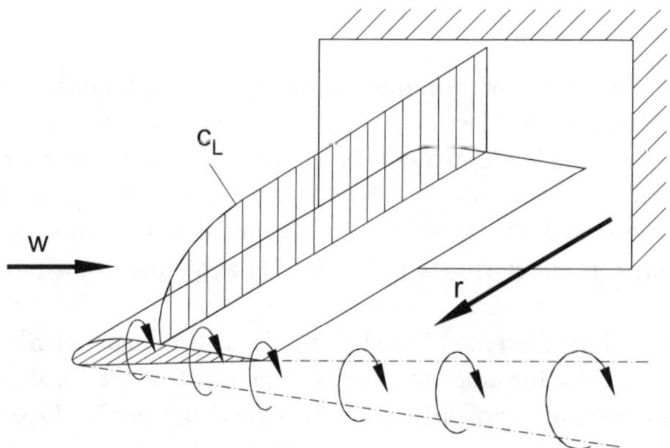

Fig. 5-17 Air flow around the blade tip from positive pressure side to suction side (tip vortex), lift distribution $c_L(r)$

In order to account for these losses at a blade of infinite length, Betz [1] introduces an effective diameter D' instead of the actual diameter D. Based on the Prandtl approximate method, this can be determined as follows

$$D' = D - 0.44\,b\,,\tag{5.44}$$

where b is the projection of the distance a between the tips of neighbouring blades (Fig. 5-18) onto a plane perpendicular to the direction of relative velocity w,

$$a = \frac{\pi D}{z}, \quad b = \frac{\pi D}{z} \sin \phi. \tag{5.45}$$

If one introduces the relationships given in the triangle of velocities at the blade tip

$$w \sin\phi = v_2 \quad ; \quad w^2 = (\Omega\, R)^2 + v_2^2,$$

and taking $v_2 = 2\, v_1/3$ for the design point, the reduced diameter D' is given by

$$D' = D \left(1 - 0.44 \, \frac{2\pi}{3\,z} \, \frac{1}{\sqrt{\lambda_D^2 + \frac{4}{9}}} \right). \tag{5.46}$$

Since the power is proportional to the diameter squared, the efficiency that takes into account the flow around the blade tip can be obtained from

$$\eta_{\mathrm{Tip}} = \left(\frac{P'}{P}\right) = \left(\frac{D'}{D}\right)^2 = \left(1 - \frac{0.92}{z\sqrt{\lambda_D^2 + \frac{4}{9}}} \right)^2. \tag{5.47}$$

For a design tip speed ratio $\lambda_D > 2$, this can be simplified to

$$\eta_{\mathrm{tip}} \approx 1 - \frac{1.84}{z\,\lambda_D}. \tag{5.48}$$

Roughly speaking, this loss is in inverse proportion to the product of blade number z and design tip speed ratio λ_D :

$$\xi_{\mathrm{tip}} \approx \frac{1.84}{z\,\lambda_D}.$$

Table 5.1 shows these losses and the ratio of effective to actual diameter, D'/D for some wind turbine types.

Table 5.1 Tip losses ξ_{tip} depending on the design tip speed ratio λ_D and the number of blades z. D' is the effective diameter

Wind turbine type	λ_D	z	$\lambda_D\, z$	ξ_{tip} in %	D'/D
Western mill	1	20	20	9	0.95
Dutch smock mill	2	4	8	22	0.88
Danish concept wind turbine	6	3	18	10	0.94
Single blade turbine (Monopteros)	12	1	12	15	0.92

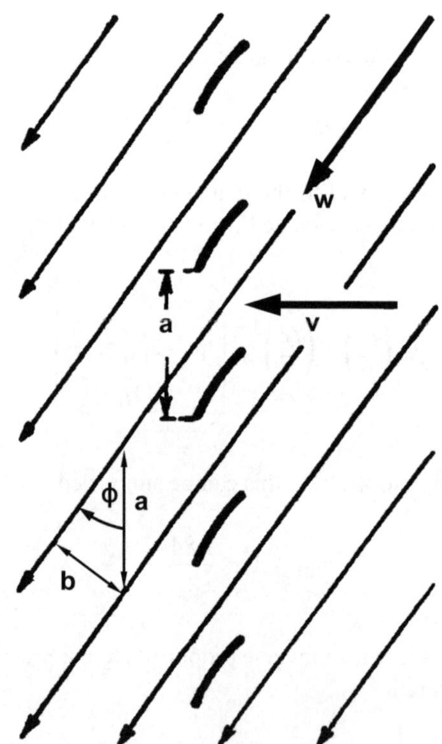

Fig. 5-18 Air distribution over the individual blades, Betz [1]

5.5.3 *Losses due to wake rotation*

These losses result from the extraction of torque at the active plane of rotation. The principle of action equals reaction means that the tangential force dU creates a counteracting torque on the downstream air flow via the lever r (Fig.   5-14). The smaller the design tip speed ratio of the wind turbine, the higher is this torque. This can be illustrated by equation (5.32) which gives the mechanical power dP in the ring section dr,

$$dP = \quad z \quad \cdot \quad dU \quad \cdot \quad \Omega\, r \quad .$$

| | number
of blades | circumferential
force | circumferential
velocity |

The wind turbine with a high design tip speed ratio extracts the power by a high angular speed Ω and a comparatively small torque $r\, dU$. The wind turbine with a low design tip speed ratio does it the other way round: its rotational speed is low and the aerodynamic torque $r\, dU$ high – consequently the wake rotation of the air flow downstream of the rotor is high.

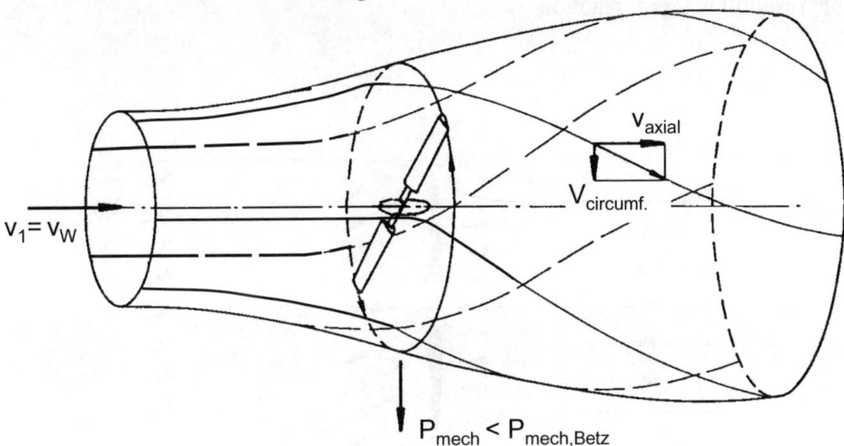

Fig. 5-19 Downstream wake rotation [10, modified]

Therefore, in contrast to Betz cf. equations (5.4) and (5.5), downstream losses are now not only due to the axial velocity v_3 far downstream but there is also a circumferential component $v_{circumf.}$ in the downstream flow causing additional losses, the wake rotation losses, Fig. 5-19.

For turbines with a high tip speed ratio, $\lambda_D > 3$, these losses are very low. But turbines with a low tip speed ratio, such as the Western mill with $\lambda_D \approx 1$, cannot reach the Betz power coefficient of $c_{P,Betz} = 0.59$. Due to unavoidable swirl losses, they can only reach a maximum coefficient of $c_{P,max} = 0.42$. And this value should be adjusted further for the profile and tip losses, see Fig. 5-25. The immense loss of 30% due to the torque reaction also affects the profile geometry of an optimally designed wind turbine. The blade chord and the blade twist along the blade will be different from the Betz dimensioning, as this is too simple for this case. As the losses due to rotational wake are, in any case, relevant for the optimum blade geometry, its calculation will be discussed in the next section.

5.6 The Schmitz dimensioning taking into account the rotational wake

Betz assumed that the air flow is retarded from the far upstream velocity v_1 via the velocity $v_2 = (v_1 + v_3)/2$ in the rotor plane, to the far downstream velocity $v_3 = v_1/3$, without any change to its exclusively linear axial direction.

However, Schmitz (and before him Glauert for his propeller calculation) takes into account the circumferential component Δu of the wake rotation which is an inevitable result of action = reaction.

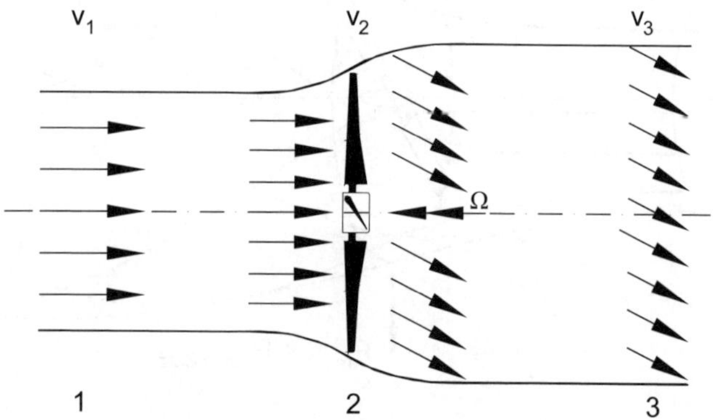

Fig. 5-20 Plan view of the air flow in a ring section, downstream flow with circumferential component Δu

This circumferential component is equal to zero far upstream of the rotor and Δu far downstream of the rotor, Fig. 5-20. It is created when the flow passes the blade.

Fig. 5-21 shows the relative velocity w at the airfoil, which is made up of Betz's axial velocity $v_{2.ax} = v_1 - \Delta v/2$ according to Betz, and the increased circumferential velocity

$$u = \Omega\, r + \frac{\Delta u}{2}.$$ (5.49)

Since the additional circumferential component Δu is created as the flow passes over the blade, the mean value $\Delta u/2$ of the value "before" and "after" is used (This assumption is an analogy to the Froude-Rankine theorem, section 5.1. A more profound theoretical consideration and proof is found in [2]).

The magnitude of Δu will depend on the design tip speed ratio λ_D of the wind turbine with optimum power extraction. The velocity triangle formed by v_1 and $\Omega\, r$, giving w_1 and ϕ_1, is that which would arise if the flow is not decelerated at all by the rotor, Fig. 5-21. The change Δw from the relative velocity w_1 far upstream the rotor to the relative velocity w_3 far downstream is caused by the airfoil effect. According to the principle of momentum "mass flow rate times change in velocity equals the force" this creates in the rotor ring section of the width dr the lift force dL which is perpendicular to w and in parallel to Δw,

$$dL = \Delta w\, d\dot{m}\ .$$ (5.50)

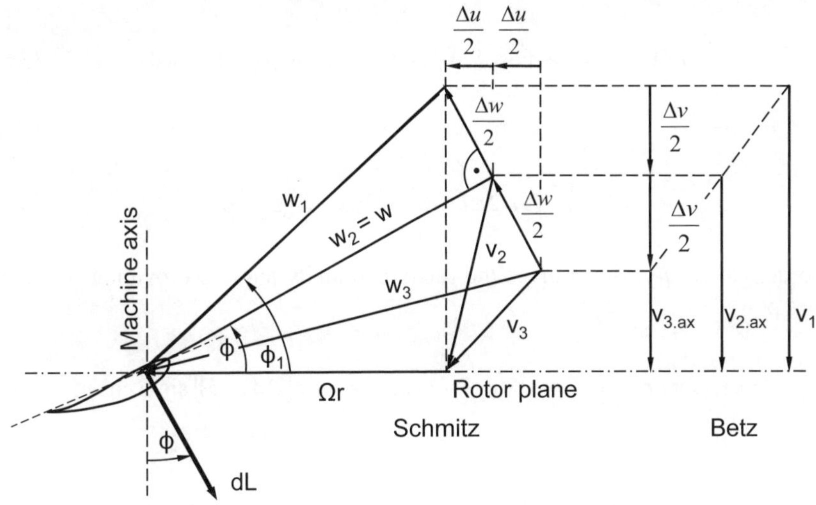

Fig. 5-21 Triangles of relative velocity upstream (subscript 1), in the rotor plane (no subscript), and downstream (subscript 3) of the rotor; angle ϕ of relative velocity direction in the rotor plane

The mass flow rate through the ring section is

$$d\dot{m} = 2\rho\pi r\, dr\, v_{2.ax}\ .$$ (5.51)

The power at the ring section, ignoring the drag, is obtained by

$$dP = dM\,\Omega$$

$$= dU\,r\Omega = dL\sin(\phi)r\Omega.\tag{5.52}$$

Fig. 5-21 gives the following geometrical relations:

Relative velocity in the rotor plane: $\quad w = w_1\cos(\phi_1 - \phi)$
Axial component of w in the rotor plane:

$$v_{2.\mathrm{ax}} = w\,\sin\phi$$
$$= w_1\,\cos(\phi_1 - \phi)\cdot\sin\phi$$

Velocity change Δw in the rotor plane: $\quad \Delta w = 2\,w_1\,\sin(\phi_1 - \phi)\tag{5.53}$

Hence, the power depends on the angle ϕ:

$$dP = r\,\Omega\,d\dot{m}\,\Delta w\sin\phi$$

$$= r\,\Omega\,\rho\,2\pi r\,dr\,w_1\cos(\phi_1 - \phi)\sin(\phi)\,2w_1\sin(\phi_1 - \phi)\sin(\phi)\tag{5.54}$$

or

$$dP = r^2\,\Omega\,\rho\,2\pi dr\,w_1^2\sin[2(\phi_1 - \phi)]\sin^2\phi\tag{5.55}$$

The derivation $dP/d\phi = 0$ gives the relative velocity angle ϕ providing the maximum power

$$\frac{dP}{d\phi} = \left(r^2\Omega\,\rho\,2\pi\,dr\,w_1^2\right)\left(-2\cos[2(\phi_1 - \phi)]\sin^2\phi + 2\sin[2(\phi_1 - \phi)]\sin\phi\,\cos\phi\right)$$

$$= \left(r^2\,\Omega\,\rho\,2\pi\,dr\,w_1^2\right)2\sin\phi\left[\sin[2(\phi_1 - \phi)]\cos\phi - \cos[2(\phi_1 - \phi)]\sin\phi\right]$$

$$= \left(r^2\,\Omega\,\rho\,2\pi\,dr\,w_1^2\right)2\sin\phi\left(\sin(2\phi_1 - 3\phi)\right)\tag{5.56}$$

For optimal power extraction this yields the angle

$$\phi = \frac{2}{3}\phi_1,\tag{5.57}$$

which is related to the design tip speed ratio by

$$\tan \phi_1 = \frac{v_1}{\Omega r} = \frac{R}{\lambda_D r} .$$

(5.58)

With this relative velocity angle $\phi = 2\phi_1/3$, the lift force dL in equation (5.50) can be obtained by

$$
\begin{aligned}
dL &= d\dot{m}\,\Delta w \\
&= \rho\, 2\pi r\, dr\, w_1\, \cos(\phi_1 - \phi)\sin\phi\, 2w_1 \sin(\phi_1 - \phi) \\
&= \rho\, 2\pi\, r\, dr\, w_1^2\, 4\sin^2\!\left(\frac{\phi_1}{3}\right)\cos^2\!\left(\frac{\phi_1}{3}\right)
\end{aligned}
$$

(5.59)

as $\quad \phi_1 - \phi = \phi_1/3$, and $\sin(2/3\,\phi_1) = 2\sin(\phi_1/3)\cos(\phi_1/3)$.

Now, the airfoil theory is required to determine the chord length necessary for the optimum power extraction via the lift force dL

$$dL = \frac{\rho}{2} w^2\, c_{\text{tot}} dr\, c_{\text{L}} .$$

(5.60)

After rewriting this as

$$dL = \frac{\rho}{2} w_1^2\, c_{\text{tot}}\, dr\, c_{\text{L}} \cos^2\!\left(\frac{1}{3}\phi_1\right)$$

(5.61)

and equating the expressions (5.55) and (5.52), we can derive the Schmitz blade chord length formula

$$c_{\text{tot}}(r) = \frac{16\pi r}{c_{\text{L}}} \sin^2\!\left(\frac{1}{3}\phi_1\right) .$$

(5.62)

Distributing the total chord length c_{tot} over z blades, we obtain the chord length per blade

$$c_{\text{Schmitz}}(r) = \frac{1}{z}\frac{16\pi r}{c_{\text{L}}} \sin^2\!\left(\frac{1}{3}\phi_1\right) .$$

(5.63)

For small angles ϕ_1, i.e. for high tip speed ratios, the values obtained from the Schmitz chord formula are equal to those of the Betz equation (5.36). This is illustrated in Fig. 5-22 which shows the dimensionless total blade chord length $\bar{c}$ for one-bladed turbines as per Betz and per Schmitz. This diagram was developed by J. Maurer, and we followed his concise presentation of the Schmitz dimensioning in this section [8].

$$\bar{c}_{\text{Betz}} = c_{\text{Betz}} \frac{z c_{\text{L}} \lambda_{\text{D}}}{R} = \frac{16\pi}{9} \frac{1}{\sqrt{\left(\lambda_{\text{D}} \dfrac{r}{R}\right)^2 + \dfrac{4}{9}}} \tag{5.64}$$

$$\bar{c}_{\text{Schmitz}} = c_{\text{Schmitz}} \frac{z c_{\text{L}} \lambda_{\text{D}}}{R} = \frac{16\pi \lambda_{\text{D}} r}{R} \sin^2\left(\frac{1}{3}\phi_1\right) \tag{5.65}$$

$$\text{with} \;\; \phi_1 = \arctan\left(\frac{R}{\lambda_{\text{D}} r}\right)$$

Both functions of the total blade chord length now only depend on one single parameter: $\lambda_{\text{D}}\, r/R$.

From Fig. 5-22, the development of the chord length $c(r)$ along the radius can be identified for wind turbines of any tip speed ratio λ_{D}. If, for instance, a turbine has a design tip speed ratio of $\lambda_{\text{D}} = 7$, and an inner radius of $r_{\text{i}} = 0.1\,R$, we obtain the (dimensionless) optimum blade chord length through the curve(s) in the range of values between

$$\lambda_{\text{D}}\, r/R = 0.7 \;\text{(at inner radius) and}\; \lambda_{\text{D}}\, r/R = 7.0 \;\text{(at blade tip).}$$

After determining a lift coefficient c_{L} and the number of blades z, the corresponding real blade chord length can be derived from the above definition

$$c = \frac{\bar{c} R}{\lambda_{\text{D}} z c_{\text{L}}}.$$

Taking into account the wake rotation, the smaller the local speed ratio $\lambda_{\text{D}}\, r/R$, the more the obtained optimum blade chord length differs from that of Betz. For turbines with a high tip speed ratio, this only affects the near hub area. This pleases the designers, since at this location it is difficult to implement a large chord length.

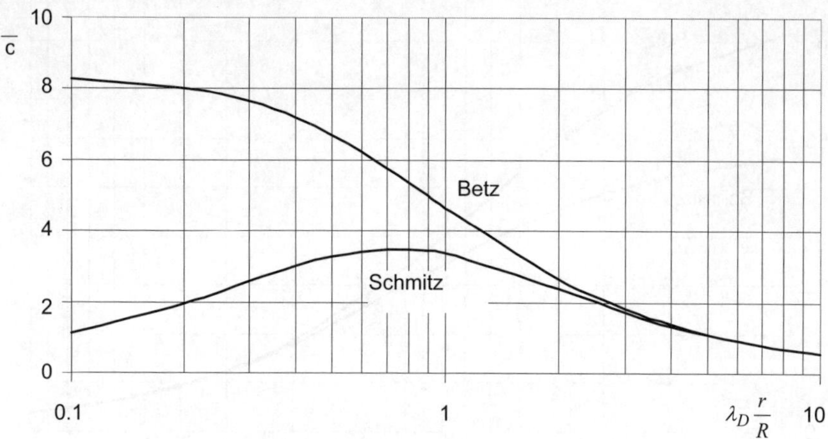

Fig. 5-22 Comparison of the dimensionless blade chord length $\bar{c}$ according to Betz and to Schmitz versus the local speed ratio $\lambda_D\ r/R$

For turbines with a low tip speed ratio, such as the American farm windmill with $\lambda_D \approx 1$, the chord is completely different if the wake rotation is taken into account: it tapers towards the hub. The manufacturers of Western mills seem to have had the right intuition!

Fig. 5-23 shows the curve of the relative velocity angle ϕ for the triangle of velocities in the rotor plane, taking into account the wake rotation (Schmitz theory), equation (5.57) and (5.58), and ignoring it (Betz theory). The graph illustrates that the Schmitz design results in less twist: for wind turbines with a high tip speed ratio in the inner area only, and for turbines with a low tip speed ratio along the entire blade length.

The total twist angle β of the blade results from the difference between the angle of the relative velocity ϕ and the angle of attack α_A used for obtaining the lift coefficient c_L in equations (5.64) and (5.65)

$$\beta = \phi - \alpha_A \ . \tag{5.66}$$

This is the local angle between the rotor plane and the chord line of the blade.

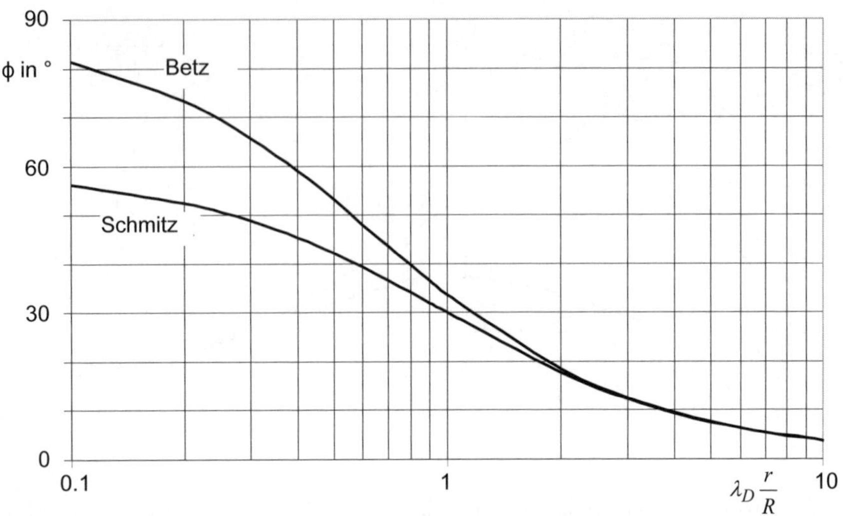

Fig. 5-23 Angle ϕ of relative velocity in rotor plane without (Betz) and with (Schmitz) wake rotation versus local tip speed ratio $\lambda_D\, r/R$

5.6.1 Losses due to wake rotation

According to Betz, without taking into account the downstream wake rotation, the maximum extracted power is

$$P_{\text{Betz}} = \frac{\rho}{2} v_1^3 \pi R^2 c_{\text{P, Betz}} \qquad c_{\text{P, Betz}} = \frac{16}{27}$$

Taking into account the downstream wake rotation, the maximum power dP at a ring section results from equation (5.55), provided that the optimum condition for the relative velocity angle $\phi = 2/3\ \phi_1$, equation (5.54), is inserted. Integrating all ring sections and taking into account the relations in equations (5.55), (5.57) and (5.58), the power P_{Schmitz} including the losses due to wake rotation is obtained with

$$\phi_1 = \arctan\left(\frac{R}{\lambda_D r}\right),$$

by

$$P_{\text{Schmitz}} = \frac{\rho}{2}\pi R^2\, v_1^3 \int\limits_0^1 4\lambda_D \left(\frac{r}{R}\right)^2 \frac{\sin^3\left(\frac{2}{3}\phi_1\right)}{\sin^2(\phi_1)}\, d\left(\frac{r}{R}\right). \qquad (5.67)$$

This integral can be solved analytically [2], although the solution is very complex and not at all transparent. Therefore, the result is given in a diagram, Fig. 5.24. It shows that the power coefficient $c_{P,\,Schmitz}$ is now strongly dependent on the tip speed ratio λ_D – in contrast to the constant Betz coefficient $c_{P,\,Betz}$.

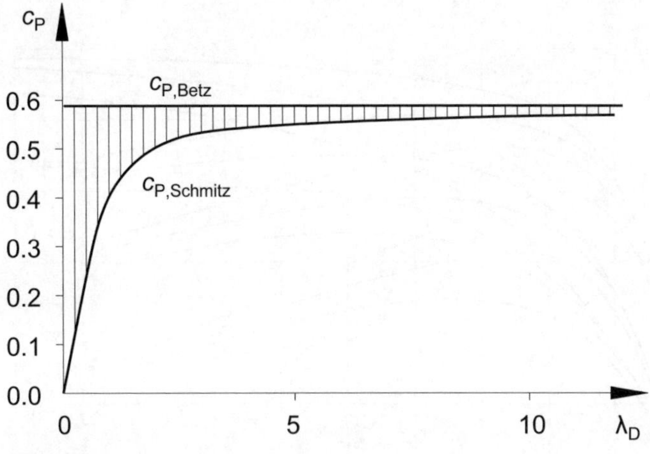

Fig. 5-24 Power coefficient according to Betz (without wake rotation) and Schmitz (with wake rotation). The hatched area shows the additional losses

5.7 Wind turbine design in practice

A first estimate of the maximum mechanical power to be expected for a wind turbine provides

$$P_{\text{Real}} = \frac{\rho}{2}\,\pi\,R^2\,v_1^3\,c_{P,\text{real}} \tag{5.68}$$

with

$$c_{P,\text{Real}} = c_{P,\text{Schmitz}}(\lambda_D)\,\eta_{\text{Profil}}(\lambda_D,\varepsilon)\,\eta_{\text{Tip}}(\lambda_D,z)\,,$$

where the power coefficient $c_{P,\,real}$ depends only on the design tip speed ratio λ_D, the lift-drag ratio $\varepsilon = c_L\,/\,c_D$ of the selected profile and the number of blades z, affecting the tip losses. Schmitz developed a diagram [2] showing $c_{P,\,real}$ versus the tip speed ratio. The parameters of the curves are the lift/drag ratio ε and the number of blades z, see Fig. 5.25. Without any calculation, this diagram provides the efficiency to be expected if the wake rotation, profile losses and tip losses are taken into account.

The estimation of the annual energy production in kWh is also now possible for a given wind velocity frequency distribution, see section 4.3.2 and 15.2.5.

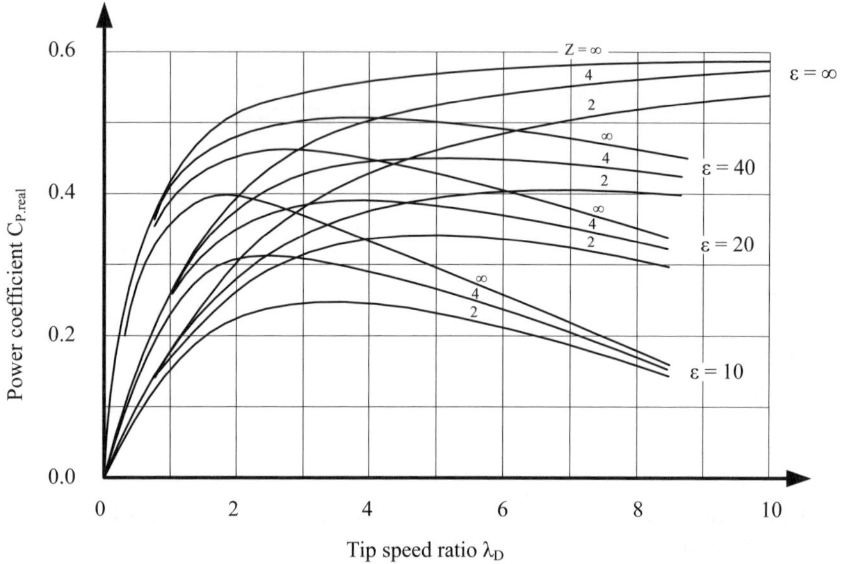

Fig. 5-25 Schmitz diagram: real power coefficient taking into account the profile, tip and wake rotation losses; number of blades z, lift/drag ratio ε [2, modified]

The Hütter diagram, Fig. 5-15, gives an idea of the required blade surface area. It shows that the required blade area diminishes substantially with increasing tip speed ratio. Whereas the Western mill with $\lambda_D \approx 1$ requires the full disk surface (100%), wind turbines with $\lambda_D = 6$ need only 4 to 6 percent, depending on the chosen lift coefficient c_L. The chord lengths and the relative velocity angles along the radius are shown in the Maurer diagram, Fig. 5.22 and 5.23, albeit in a logarithmically distorted form.

Having set the design tip speed ratio λ_D, the profile chord length $c(r)$ and the relative angles $\phi(r)$ will be drawn to visualise the twist along the radius. The total blade twist angle $\beta = \phi(r) - \alpha_A$ can now be obtained with the angle of attack α_A. For the Betz design, equations (5.36) and (5.39) are used, while the equations (5.63) and (5.66) together with (5.57) and (5.58) are applied for the Schmitz design. They give the blade twist

$$\beta(r) = \frac{2}{3}\arctan\left(\frac{R}{\lambda_D r}\right) - \alpha_A(r) \tag{5.69}$$

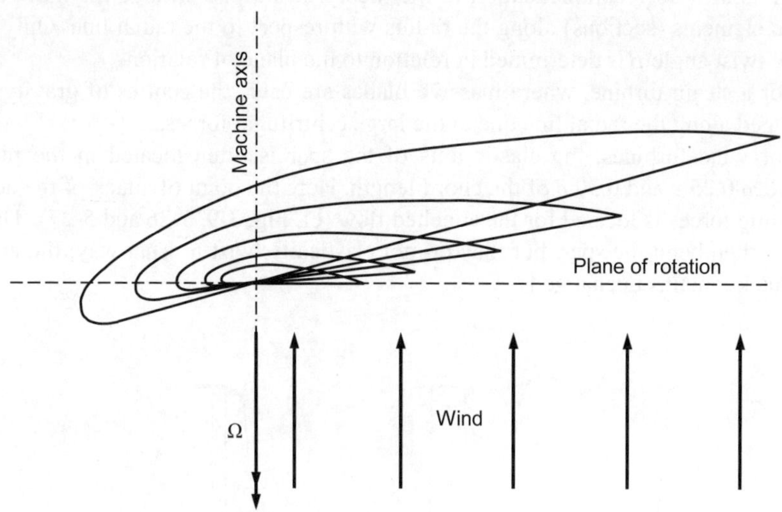

Fig. 5-26 Profile cross sections aligned at 0.25 of chord length c; wind turbine designed per Betz with a high tip speed ratio ($\lambda_D = 6$, Göttingen profile 797)

At some stage, the number of blades z must be determined. The dimensioning theories provide little help for this: z is a weak parameter, which only affects the tip losses. Therefore, manufacturing aspects (three blades are more expensive than two), dynamic considerations (dynamically speaking rotors with three or more blades run more smoothly, while two- and single-bladed rotors are jumpy and noisy) and strength issues are decisive.

For the hub with rigid blade attachments, it is theoretically better to use few blades because this reduces the bending moments at the blade root, caused by the thrust. However, rotors with individual flapping hinges (cf. Fig. 3-17 and 3-20) have zero bending moments at the root, so this is not an issue. But even for rotors with rigid blade attachments, the problem of a high bending moment at the blade root is reduced by using different aerodynamic profiles along the blade radius, see Fig. 3-8. At the blade tip, for high lift/drag ratios, thin profiles are used. For the middle part of the blade, the profiles are somewhat thicker, and at the root, the profile used is very thick. Here, the comparatively low lift/drag ratio (cf. equation 5.42) is hardly relevant. Thus, at the blade root, there is a sufficiently high blade thickness to provide high section modulus against the bending moment caused by the aerodynamic forces. Often, a family of profiles with different thicknesses is chosen, e.g. Delft University profiles DU… (cf. Fig. 3-8) or NACA 44… profiles (cf. Fig. 5-7).

Moreover, c_L need not remain constant along the entire blade length. Hence, the blade chord $c(r)$ can be manipulated, and it is, e.g., possible to obtain a linear increase of the chord length in the direction from the tip to the hub, when designing according to Betz or Schmitz.

There is also no recommendation in the theory on how to arrange the individual blade elements (sections) along the radius with respect to the radial line. Only the blade twist angle β is determined in relation to the plane of rotation.

For a steam turbine, where massive blades are used, the centres of gravity are arranged along the radial line due to the large centrifugal forces.

For wind turbines, the elastic axis of the spar is often located in the range between $0.25\,c$ and $0.30\,c$ of the chord length. Here the point of attack of the aerodynamic forces is located for the attached flow (cf. Fig. 3-9, 5-26 and 5-27). These forces then bend the spar, but they do not elastically twist it. That way, the angle of attack remains unchanged.

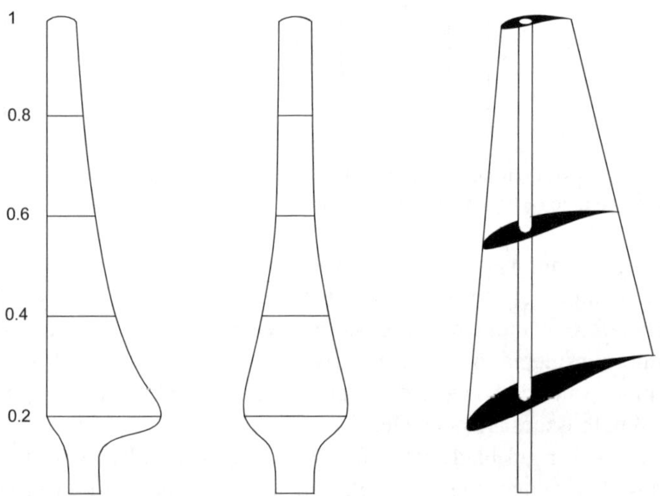

Fig. 5-27 Alignment of blade sections

The result of this design procedure is a rotor that has an optimum power coefficient when running at its design tip speed ratio. This is an advantage for variable-speed turbines, as their electronics allow the optimum rotational speed to be adjusted for the actual wind velocity, cf. chapters 11, 12, 13. For stall turbines, further design parameters for the rotor geometry should be included. In that case, the goal is a wide, flat c_P maximum, and a controlled flow separation (stall) at low tip speed ratios. Therefore, these rotor blades are to some extent different to the Betz/Schmitz design.

For wind turbines with a blade pitch mechanism, it is best to keep the torque around the pitch axis caused by the aerodynamic forces small, since it stresses the pitch mechanism. With common pitch-controlled wind turbines which turn their leading edge into the wind for power reduction, the air flow stays attached to the blade. Therefore, during control the point of action of the aerodynamic forces remains near to $c/3$ where the centre of elasticity of the spar was laid down during the design. Hence, the axis of rotation of the pitch mechanism is also located here.

However, if the pitch direction is set to "active-stall", i.e. to flow separation, the axis of rotation of the pitch mechanism will be located near to $c/2$ because the point of action of the aerodynamic forces is found here for separated flow (Fig. 5-28).

Further aspects regarding the choice of the design tip speed ratio, the number of blades and the profiles will be discussed at the end of the next chapter.

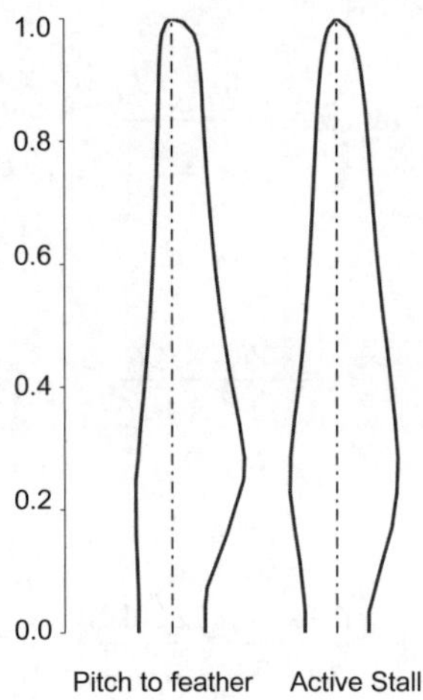

Fig. 5-28 Alignment of blade sections (blade pitch axis), left: pitch to feather, right: active-stall (LM34.0P and LM35.0P)

5.8 Final remark

The optimisation of the extracted power including the effects of the wake rotation in the downstream flow was presented here according to J. Maurer's description of Schmitz publication from 1956 [2]. It is easy to understand since only the relative velocity angle ϕ of the attacking flow is varied in order to find the maximum power with equation (5.56).

The Anglo-American literature [15, 16, 17] is mostly based on the earlier works of Glauert 1935 [7]. There, as already presented in Fig. 5-3, the axial deceleration a in the plane of rotation of the flow and the radial a' are considered separately as induction factors.

$$a = \frac{\Delta v / 2}{v_1} \qquad \text{and} \qquad a' = \frac{\Delta u / 2}{\Omega r} \; . \tag{5.70}$$

This complicates the algebra needed to find the optimum, but is comprehensive from the historical point of view: first came Betz who considered only the downstream losses due to the axial wind velocity deceleration Δv, and later this was followed by the more precise consideration of the wake rotation losses from Δu. However, both factors a and a' are interdependent, see Fig. 5-21. Here the flow angle ϕ of the relative velocity is, on the one hand

$$\tan \phi = \frac{v_{2.\mathrm{ax}}}{\Omega r + \dfrac{\Delta u}{2}} \tag{5.71}$$

$$= \frac{v_1}{\Omega r} \frac{1-a}{1+a'}$$

and on the other

$$\tan \phi = \frac{\Delta u / 2}{\Delta v / 2} = \frac{a'}{a} \frac{\Omega r}{v_1} \; . \tag{5.72}$$

So the relative velocity angle ϕ determines both a and a'. Equating the above expressions gives the relation between the induction factors (cf. equation (5–22) in Wilson, [15])

$$\left(\frac{\Omega r}{v_1} \right)^2 (1 + a')a' = (1 - a)a \; , \tag{5.73}$$

factors, however, which are not needed in the presentation of the theory by Schmitz.

Since the physical assumptions of Glauert and Schmitz are the same, the results of the optimisation are identical! The somewhat more complicated older presentation of Glauert blurs this fact a little. The easy to remember result optimisation of Schmitz (and Glauert) is visualised in the following table:

	Betz	**Glauert - Schmitz**
Far upstream of the rotor	$v_1 = v_{\text{wind}}$	ϕ_1
In the rotor plane	$v_2 = \dfrac{2}{3} v_1$	$\phi_2 = \phi = \dfrac{2}{3} \phi_1$
Far downstream of the rotor	$v_2 = \dfrac{1}{3} v_1$	$\phi_3 = \dfrac{1}{3} \phi_1$

The induction factors can be derived from Fig. 5-21 and equation (5.53) from the Schmitz and Maurer descriptions:

$$\Delta v = 2 \ w_1 \ \sin(\phi_1 - \phi) \cos\phi \tag{5.74}$$

$$\Delta u = 2 \ w_1 \ \sin(\phi_1 - \phi) \sin\phi. \tag{5.75}$$

They are the axial and radial components of Δw. Replacing w_1 with $w_1 = v_1 \ / \sin \phi_1 = \Omega r \ / \cos \phi_1$, a and a' are obtained by rearranging the definition, equation (5.70)

$$a = \frac{\sin(\phi_1 - \phi)}{\sin \phi_1} \ \cos\phi \quad \text{and} \tag{5.76}$$

$$a' = \frac{\sin(\phi_1 - \phi)}{\cos \phi_1} \ \sin \phi . \tag{5.77}$$

Design tip speed ratio λ_D at the maximum radius R and the local tip speed ratio λ_r in the ring section at the radius r and the relative velocity angle ϕ_1 are related

$$\tan \phi_1 = \frac{1}{\lambda_r} = \frac{R}{\lambda_D \ r} \ . \tag{5.78}$$

Inserting the result of the optimisation by Schmitz $\phi = 2 \ \phi_1/3$, the induction factors are determined:

$$a = \frac{\sin(\phi_1 / 3)}{\sin \phi_1} \ \cos(2 \phi_1 / 3) \quad \text{and} \quad a' = \frac{\sin(\phi_1 / 3)}{\cos \phi_1} \ \sin(2 \phi_1 / 3). \tag{5.79}$$

They are presented in Fig. 5-29 (cf. table 5-2 in [15]). The curve of the axial induction factor a clearly shows, on the one hand, that for higher tip speed ratios the consideration $a = 1/3$ of Betz is valid. On the other hand, it is obvious how strong the wake rotation is when $\lambda_\mathrm{D}\, r/R < 1$. That is the reason for the drastic increase of the wake rotation losses in this range, cf. Fig. 5-24, as well as the quite different blade shape near to the hub, cf. Fig. 5-22.

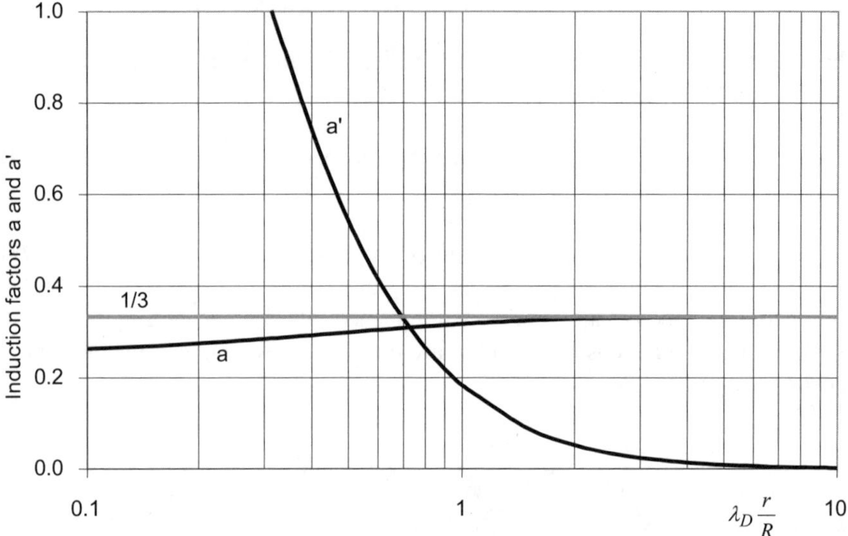

Fig. 5-29 Induction factors a and a' according to Glauert and Schmitz

Note:

In his original paper, [1], Betz counted the angle of the relative velocity w not in relation to the rotor plane (angle ϕ, e.g. in Fig. 5-13) but relative to the rotation axis of the machine (angle $\gamma = 90° - \phi$). For historical reasons, this nomenclature was used in the first three German editions and the first English edition of this book. But this caused a break in the description since, from chapter 5.6 on, the angle ϕ according to Schmitz is more suitable when including the effects from wake rotation in the calculations. For didactical reasons, we now use the relative velocity angle ϕ in the section on the simpler Betz theory. The main difference is that now in the equations (5.24) to (5.35) all the terms of $\cos \gamma$ are replaced by $\sin \phi$.

References

[1]   Betz, A.: *Wind-Energie und ihre Ausnutzung durch Windmühlen (Wind energy and its utilization through windmills)*, Vandenhoeck & Ruprecht, Göttingen 1926, reprint: Öko-Buchverlag Kassel 1982

[2]   Schmitz, G.: *Theorie und Entwurf von Windrädern optimaler Leistung (Theory and design of windwheels with an optimum performance)*, Wiss. Zeitschrift der Universität Rostock, 5. Jahrgang 1955/56

[3]   Riegels: *Aerodynamische Profile (Aerodynamic profiles)*, R. Oldenbourg Verlag, München 1958

[4]   Althaus, D.: *Profilpolaren für den Modellflug (Profile polar lines for aeromodelling)*, Neckar-Verlag, VS-Villingen 1980

[5]   Althaus, D.: *Stuttgarter Profilkatalog (Stuttgart profile catalogue)*, Vieweg-Verlagsgesellschaft, Braunschweig 1981

[6]   Betz, A.: *Einführung in die Theorie der Strömungsmaschinen (Introduction to the theory of turbo machinery)*, Verlag G. Braun, Karlsruhe 1959

[7]   Glauert, H.: Section "Airplane Propellers" in: Durand W.F.: *Aerodynamic Theory*, Springer Verlag, Berlin 1935, Reprint Verlag Peter Smith. Mass. US. 1976

[8]   Maurer, J.: *Wind turbinen mit Schlaggelenkrotoren - Baugrenzen und dynamisches Verhalten (Wind turbines with flapping hinge rotors - constructional limits and dynamic behaviour)*, VDI Verlag, Reihe 11, Nr. 173, Düsseldorf 1992

[9]   Paulsen, U.S.: *Aerodynamics of a full-scale, non rotating wind turbine blade under natural wind conditions*, Risø National Laboratory, Roskilde Denmark 1989

[10]   Hau, E.: *Windkraftanlagen (Wind turbines)*, Springer-Verlag, Stuttgart 1988

[11]   Sass, F. (Hrsg.): *Dubbels Taschenbuch für den Maschinenbau 11 Aufl. (Dubbel's paperback for mechanical engineering, 11th edition)*, Springer-Verlag, Berlin 1955

[12]   Miller, R.v (Hrsg.): *Energietechnik und Kraftmaschinen 6 (Power engineering and power engines 6, Technical encyclopedia)*, Techniklexikon
        Rowohlt Taschenbuch-Verlag, Hamburg 1972

[13]   Smith, H.: *The illustrated guide to aerodynamics*, TAB Books Inc, USA 1985

[14]   Le Gourierès, D.: *Wind Power Plants*, Pergamon Press, Oxford a. New York, 1982

[15]   Spera, D.A. (editor): *Wind Turbine Technology*, ASME Press, New York, 1994, 2nd printing 1995

[16]   Burton, T. et al.: *Wind Energy Handbook*, John Wiley & Sons, Chichester (UK), New York, 2001

[17]   Manwell, I.F. et al.: *Wind Energy Explained*, John Wiley & Sons, Chichester (UK), New York 2002

[18]   Hansen, Martin O.L.: *Aerodynamics of Wind Turbines*, James & James, London (UK), 2000

[19]   Althaus, D.: *Niedriggeschwindigkeitsprofile (Aerodynamic profiles for low Reynolds numbers)*, Vieweg Verlag, Wiesbaden 1996

6 Calculation of performance characteristics and partial load behaviour

6.1 Method of calculation (blade element momentum method)

Calculating the forces and the relative velocities at the rotor blades for tip speed ratios other than the design tip speed ratio requires a substantial effort. In the following a method is described which is comparatively simple to understand: the *blade element momentum method.*

When dimensioning the blade geometry according to Schmitz-Glauert (cf. section 5.6), the relative velocity angle ϕ at the plane of rotation for a given design tip speed ratio λ_D was determined. At this ϕ the maximum power can be extracted from the wind using the optimum angle of attack α_A. Afterwards, the blade chord c and the twist (the total blade twist angle β) were dimensioned so that the optimum flow conditions are reached during operation at design tip speed ratio.

Now blade chord and twist are given. For tip speed ratios other than λ_D, there are different relative velocity angles ϕ in the plane of rotation, Fig. 6-1. For the calculation of these relative velocity angles ϕ, we use the same equations as for the dimensioning of the blade chord: the lift force formula obtained using airfoil theory and the principle of linear momentum.

We use the airfoil theory to obtain the lift force at a blade section:

$$dL = \frac{\rho}{2} w^2 c\, dr\, c_L(\alpha_A) \tag{6.1}$$

with $w = w_1 \cos(\phi_1 - \phi)$ c = blade chord length

                                       dr = width of the blade section

and $\alpha_A = \phi - \beta$ c_L = lift coefficient

                                         ρ = air density

The lift force at a blade section can also be obtained by using the principle of linear momentum:

$$dL = d\dot{m}\, \Delta w \tag{6.2}$$

with $\qquad\qquad\qquad d\dot{m} = \rho \dfrac{2\pi r}{z} dr\, w \sin\phi$

$d\dot{m}$ = mass flow rate, r = local radius of the blade section,

z = number of rotor blades and $\Delta w = 2 w_1 \sin(\phi_1 - \phi)$

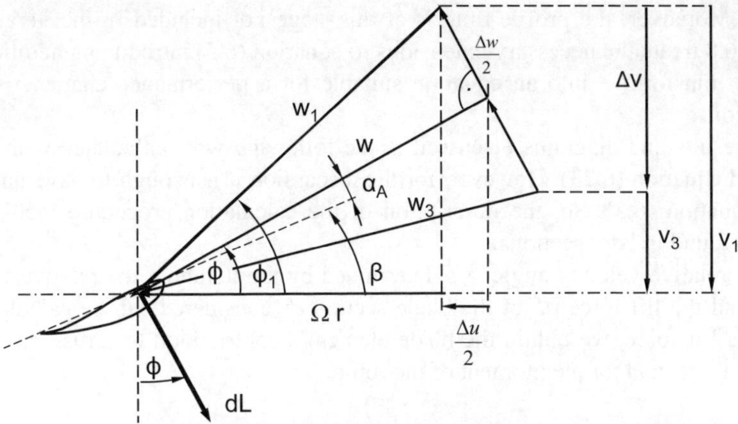

Fig. 6-1 Triangles of velocity upstream (subscript 1), in-plane (no subscript), and downstream (subscript 3) of the rotor, ϕ and w at the plane of rotation

By equating the two lift forces, we obtained an equation that was used to calculate the blade chord length c for the given relative velocity angle ϕ. Now the blade chord c is given, and the relative velocity angle ϕ is unknown.

$$\frac{\rho}{2} w^2 c \, dr \, c_L(\alpha_A) - d\dot{m} \, \Delta w = 0 \tag{6.3}$$

Inserting all values of equations (6.2) and (6.3), we obtain

$$\frac{\rho}{2} w_1^2 \cos^2(\phi_1 - \phi) c \, dr \, c_L(\alpha_A) - \rho \frac{2\pi r}{z} dr \, w_1 \cos(\phi_1 - \phi) \sin(\phi) 2 w_1 \sin(\phi_1 - \phi) = 0 \,. \tag{6.4}$$

This equation can be reduced and simplified. The result is an equation without the air density ρ, the width dr of the ring section and the undisturbed relative velocity w_1. Using the formula of the angle of attack α_A with the known twist angle β, we obtain an equation with only one unknown variable left, the relative velocity angle ϕ in the rotor plane.

$$c \, c_L(\phi - \beta) - \frac{8\pi r}{z} \sin(\phi) \tan(\phi_1 - \phi) = 0 \tag{6.5}$$

Unfortunately, this equation cannot be solved directly for the unknown relative velocity angle ϕ, but must be determined by iteration. Moreover, there are still some defects in this equation: reasonable calculation results are only obtained in the range around the design point λ_A. During start-up ($\lambda \ll \lambda_A$) and close to load-free idling there are additional aerodynamic effects which have not been considered up to now. Moreover, the profile drag is at this stage not included in the iteration. Section 6.8 treats the necessary extensions to equation (6.5) introducing additional terms to transform it into an equation suitable for a performance characteristics calculation.

The results and diagrams discussed in the following were calculated with this extended equation (6.23). However, for the discussion it is enough to have understood equation (6.5). So, the description of the calculation procedure including losses is found in later sections.

If the relative velocity angle ϕ is determined by the iteration, the relative velocity w and the lift force dL at the blade section dr considered can be calculated. From the lift force, we obtain the blade element's contribution to thrust, circumferential force, and torque moment of the rotor:

$$w = w_1 \cos(\phi_1 - \phi)$$

$$dL = \frac{\rho}{2} w^2 \, c \, dr \, c_L(\alpha_A)$$

$$\left. \begin{array}{lll} \text{thrust force:} & dT(r) = dL \, \cos\phi \\[2ex] \text{circumferential force:} & dU(r) = dL \, \sin\phi \\[2ex] \text{torque moment:} & dM(r) = dU \, r \end{array} \right\} \qquad (6.6)$$

After that, we repeat the procedure for the next blade element, calculate the relative velocity angle which gives then the forces. The iteration must be repeated for each blade element! A computer is used for this complex calculation.

The forces and torque of the entire rotor result from summing up the results of all blade elements:

$$\left. \begin{array}{lll} \text{circumferential force of a blade:} & U = \sum_r dU(r) \\[2ex] \text{thrust force of the rotor:} & T = z \sum_r dT(r) \\[2ex] \text{torque moment of the rotor:} & M = z \sum_r r \, dU(r) \end{array} \right\} \qquad (6.7)$$

$$\text{power of the rotor:} \qquad P = \Omega \, M \qquad\qquad (6.8)$$

If the losses resulting from the air flow around the blade tip and the profile drag, cf. section 6.8, and others are to be included, only equations (6.4) and (6.5) have to be extended. The calculation procedure for the iteration remains the same.

6.2 Dimensionless presentation of the characteristic curves

As illustrated above, the most demanding calculation is the iterative determination of the relative velocity angle ϕ. The equation (6.4) of the lift contains neither wind speed nor rotational speed. As a first step, dimensionless values for forces, moments and power can be calculated for a set undisturbed relative velocity angle ϕ_1. These dimensionless factors can be used to calculate the thrust force, torque moment and power of any combination of rotational speed and wind speed by simply multiplying them with the reference values.

Two procedures are commonly used for the representation by dimensionless characteristics:

For the *wind turbine characteristics*, the selected reference speed is the wind speed v_1 far upstream, and the dimensionless characteristic curve is plotted versus the tip speed ratio $\lambda = \Omega R/v_1$ (where R is the tip radius of the rotor). The *characteristics of propellers and helicopter* rotors are calculated analogously, but the chosen reference speed is the circumferential speed ΩR, and the dimensionless curve is plotted versus the inverse tip speed ratio $1/\lambda = v_1/\Omega R$ (degree of propagation).

In the following, we consider only the representation commonly used for wind turbines. The chosen *reference force* F_{dyn} is the product of dynamic pressure $v_1{}^2\,\rho/2$ and the swept rotor area πR^2.

$$
\begin{aligned}
F_{\mathrm{dyn}} &= \frac{\rho}{2}\pi R^2 v_1^2 && \text{reference force}\\[2mm]
T &= F_{\mathrm{dyn}} c_{\mathrm{T}}(\lambda) = \frac{\rho}{2}\pi R^2 v_1^2 c_{\mathrm{T}}(\lambda) && \text{thrust coefficient: } c_{\mathrm{T}}(\lambda)\\[2mm]
M &= R F_{\mathrm{dyn}} c_{\mathrm{M}}(\lambda) = \frac{\rho}{2}\pi R^3 v_1^2 c_{\mathrm{M}}(\lambda) && \text{moment coefficient: } c_{\mathrm{M}}(\lambda)\\[2mm]
P &= v_1 F_{\mathrm{dyn}} c_{\mathrm{P}}(\lambda) = \frac{\rho}{2}\pi R^2 v_1^3 c_{\mathrm{P}}(\lambda) && \text{power coefficient: } c_{\mathrm{P}}(\lambda)
\end{aligned}
\tag{6.9}
$$

The power $P = \Omega M$ is calculated with equation (6.8) from torque and rotational speed. Therefore, equation (6.9) gives for the power coefficient

$$
c_P = \lambda c_M .
\tag{6.10}
$$

In order to calculate the *dimensionless coefficients*, equation (6.5) is used once more to determine the relative velocity angle ϕ for each blade element. Then, it follows from the triangles of velocities in Fig. 6-1 that

$$
w = w_1 \cos(\phi_1 - \phi) \quad \text{and} \quad v_1 = w_1 \sin\phi_1 .
$$

Thus we find

$$\overline{w} = \frac{w}{v_1} = \frac{\cos(\phi_1 - \phi)}{\sin(\phi_1)} \quad , \quad d\overline{L} = \frac{dL}{F_{\mathrm{dyn}}} = \overline{w}^2 \frac{c\,dr}{\pi R^2} c_{\mathrm{L}}(\alpha_{\mathrm{A}})$$

and

thrust coefficient: $\qquad\qquad dc_{\mathrm{T}}\left(\dfrac{r}{R}\right) = d\overline{L}\,\cos\phi\,,$

torque moment coefficient: $\qquad dc_{\mathrm{M}}\left(\dfrac{r}{R}\right) = \dfrac{r}{R}\,d\overline{L}\,\sin\phi\,,$

power coefficient: $\qquad\qquad dc_{\mathrm{P}}\left(\dfrac{r}{R}\right) = \lambda\,dc_{\mathrm{M}}\left(\dfrac{r}{R}\right)\,.$

Summing up the coefficients of all blade elements along the radius, one point of the dimensionless characteristic curve is obtained.

6.3 Dimensionless characteristic curves of a turbine with a high tip speed ratio

The coefficients for power, torque and thrust were calculated for a wind turbine with a high design tip speed ratio $\lambda_{\mathrm{D}} = 7$, using the method described in the previous section. The three blades of the rotor are equipped with the profile FX 63-173 (e.g. in [1]), and they were designed according to Schmitz. In the calculation of the characteristics, tip and profile losses are taken into account, although they will only be discussed in section 6.8.

The maximum power coefficient of wind turbines with a high tip speed ratio is found at the optimum tip speed ratio λ_{opt} close to λ_{D} and is clearly lower than the ideal value $c_{\mathrm{P}} = 16/27$, Fig. 6-2 gives $\lambda_{\mathrm{opt}} = 6.5$ and $c_{\mathrm{P.opt}} = 0.52$. This is due to the actually occurring profile drag and tip losses which was not considered during the choice of the design tip speed ratio $\lambda_{\mathrm{D}} = 7.0$. For turbines with a high tip speed ratio, the profile drag is nearly parallel to the circumferential speed but has the opposite orientation; therefore it reduces the extracted power.

Typically, turbines with a high tip speed ratio have a low torque coefficient during start-up at $\lambda = 0$ (rotational speed $n = 0$ rpm), Fig. 6-3. This is the reason for their poor self-starting behaviour. Due to $c_{\mathrm{M}} = c_{\mathrm{P}} / \lambda$, the torque moment coefficient can be read directly from the c_{P}-λ curve. For $\lambda = 0$, the torque moment coefficient is the slope of the c_{P}-λ curve.

The maximum torque moment coefficient is obtained by drawing the tangent starting from the origin of coordinates against the c_P-λ curve. This torque moment coefficient is important for the dimensioning of an emergency brake.

As the tip speed ratio increases, the power and torque moment coefficient become smaller again. At approximately twice the design tip speed ratio both coefficients are zero. This (load-free) idling tip speed ratio of the turbine is reached if the turbine is over speeding when it runs without load.

The thrust coefficient increases steadily with higher tip speed ratios, Fig. 6-4. Turbines with high tip speed ratios have few and slender blades. During start-up, almost all the wind can pass undisturbed through the plane of rotation, hence c_T is very small. However, as the tip speed ratio increases, the swept rotor area is progressively blocked. At the design tip speed ratio $\lambda_{opt} = 7$, the thrust coefficient is $c_T = 8/9$ according to Betz, even after taking into account the profile and tip losses.

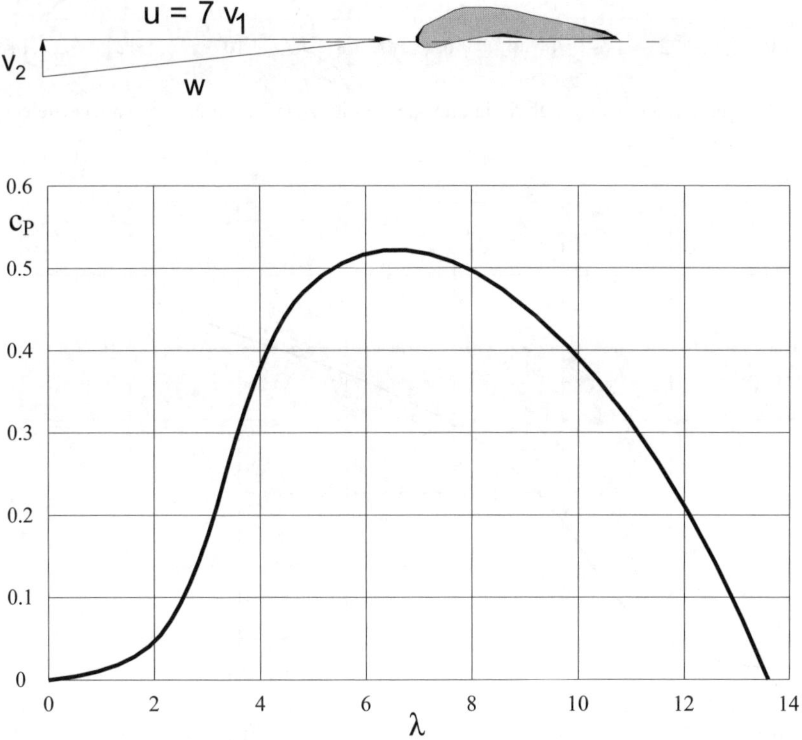

Fig. 6-2 Wind turbine with a high design tip speed ratio ($\lambda_D = 7$); power coefficient c_P versus tip speed ratio

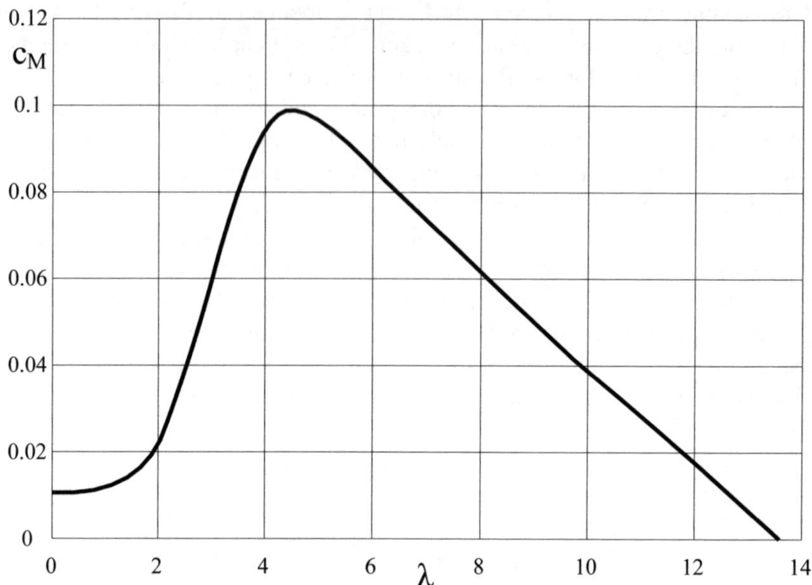

Fig. 6-3 Wind turbine with a high design tip speed ratio ($\lambda_D = 7$); torque moment coefficient c_M versus tip speed ratio

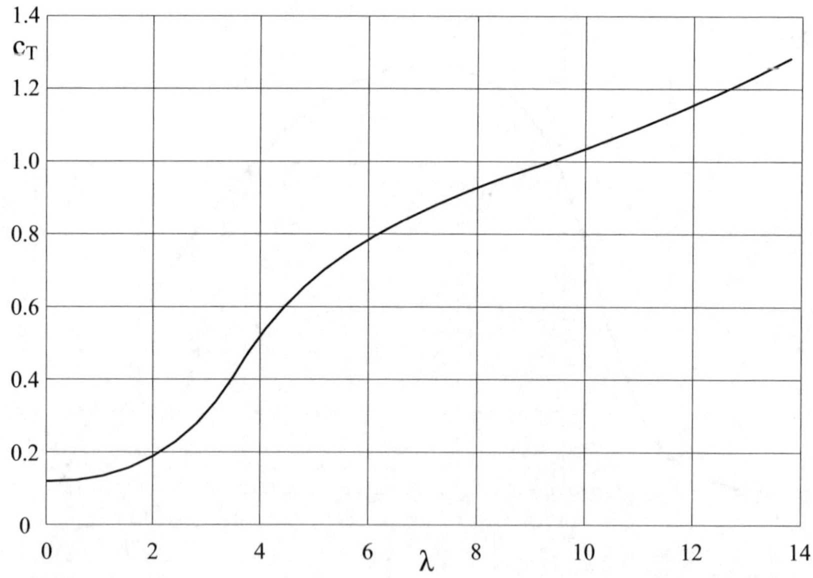

Fig. 6-4 Wind turbine with a high design tip speed ratio ($\lambda_D = 7$); thrust coefficient c_T versus tip speed ratio

When idling, the thrust coefficient is $c_T = 1.25$ which is roughly similar to the drag coefficient of a solid disk. It should be emphasized that for turbines with a high tip speed ratio the large thrust force at idling is not caused by the profile drag but by the lift which acts in this case almost parallel to the wind direction, i.e. perpendicular to the swept rotor area.

6.4 Dimensionless characteristic curves of a turbine with a low tip speed ratio

In the following, the power, torque moment and thrust coefficients for a turbine with a low tip speed ratio of $\lambda_D = 1$ will be discussed. The 15 blades of the rotor are also designed with the high-quality Wortmann profile FX 63-137. Though this profile would never be used for the design of a turbine with a low tip speed ratio, it is applied in our case to allow performance comparison and to illustrate the general differences to turbines with a high tip speed ratio.

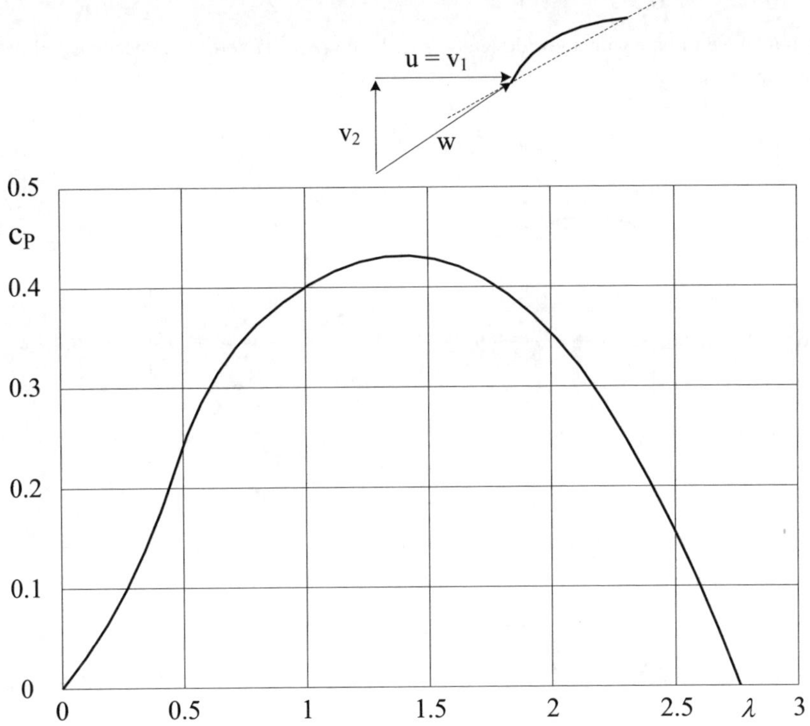

Fig. 6-5 Wind turbine with a low design tip speed ratio ($\lambda_D = 1$); power coefficient c_P versus tip speed ratio

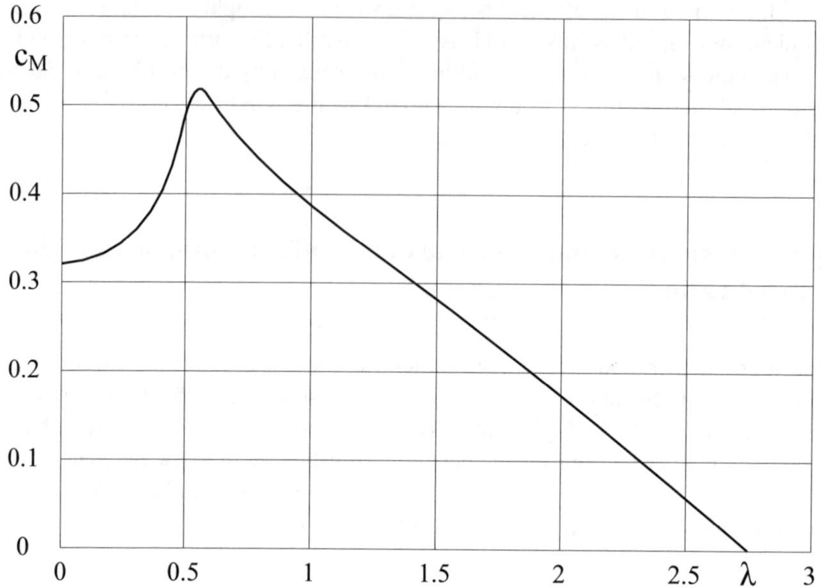

Fig. 6-6 Wind turbine with a low design tip speed ratio ($\lambda_D = 1$); torque moment coefficient c_M versus tip speed ratio

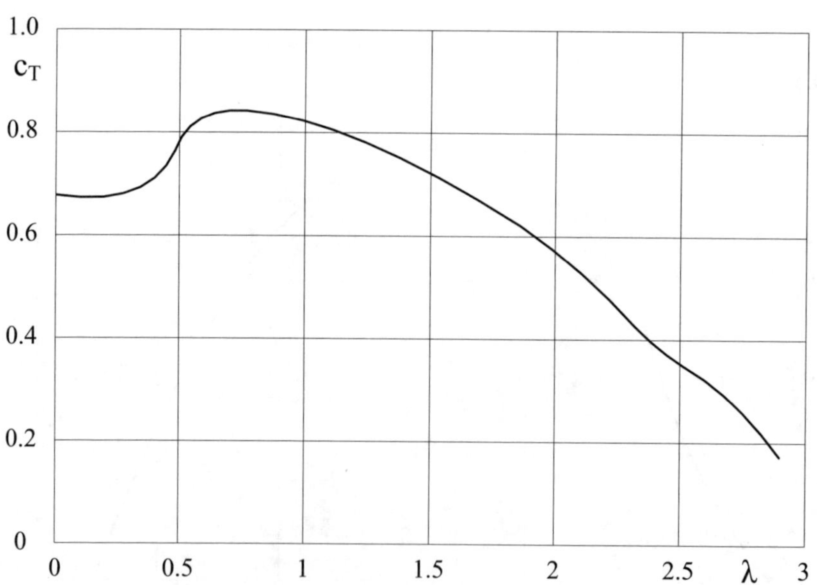

Fig. 6-7 Wind turbine with a low design tip speed ratio ($\lambda_D = 1$); thrust coefficient c_T versus tip speed ratio

The maximum power coefficient of 0.43 (Fig. 6-5) is well below $c_P = 16/27$ and much lower than for the turbine with a high tip speed ratio. This time, it occurs at a tip speed ratio of $\lambda_{opt} > \lambda_D$ and is due to the wake rotation losses considered in the calculations which decrease significantly for tip speed ratios of $\lambda > 1$, cf. Fig. 5-24. The profile drag has only little influence on these characteristics.

The high torque moment coefficient during start-up at $\lambda = 0$ (Fig. 6-6) is a typical feature of turbines with a low tip speed ratio. This results from the large blade twist angle β and from the high solidity. The large twist angle causes large lift forces at the profile, even during start-up. Turbines with a low tip speed ratio, therefore, are suitable for driving piston pumps (cf. chapter 10) or other machines which require a high starting torque.

The high solidity also causes high thrust coefficients during start-up, Fig. 6-7. The reduction of the thrust coefficients of turbines with a low tip speed ratio during idling is due to the fact that some regions of the blades with a disadvantageous inflow show negative angles of attack and therefore also negative lift coefficients when idling. These regions then accelerate the air flow (i.e. "acting as a ventilator") instead of decelerating them (i.e. "acting as a turbine"), cf. section 6.6.2. A summarizing overview of the characteristic features of wind turbines with high and low tip speed ratio is given in section 6.6.1.

6.5 Turbine performance characteristics

How can we now use the dimensionless curves presented to determine power, torque moment and thrust for given rotational speeds and wind speeds?
-    First, the tip radius R of the rotor is selected.
-    The wind speed v_1 is also given, e.g. the wind speed with the highest contribution to the energy yield, cf. chapter 4. Now the reference force F_{dyn} is calculated

$$F_{dyn} = \frac{\rho}{2}\pi R^2 v_1^2 \ . \tag{6.10}$$

-    Then, a choice is made of the rotational speed n at which the turbine should run, e.g. according to the chosen generator and gearbox. From the rotational speed n, the angular speed Ω is obtained which is used to determine the tip speed ratio:

$$\lambda = \frac{\Omega R}{v_1} \quad \text{with} \quad \Omega = \frac{n\pi}{30} \quad \text{in 1/s or rad/s and } n \text{ in rpm.}$$

- The characteristic curves give us the dimensionless coefficients for the determined tip speed ratio, and thrust force, torque moment and power can be calculated:

Thrust : $T = F_{dyn}\, c_T\,(\lambda)$

Torque moment: $M = R\, F_{dyn}\, c_M\,(\lambda)$

Power: $P = v_1\, F_{dyn}\, c_P\,(\lambda)$

If for the same wind speed the values of T, M or P are now calculated for different rotational speed, the same reference force F_{dyn} can be used. If the wind speed is changed, the reference force F_{dyn} must be recalculated.

Using this method, the power curve, power versus rotational speed for different constant wind speeds in Fig. 6-8, of a wind turbine with a high tip speed ratio and a rotor diameter of $D = 4$ m was determined. The used dimensionless c_P-λ curve of the turbine with a high tip speed ratio is the one shown in Fig. 6-2, section 6.3. For the same turbine Fig. 6-9 shows the power versus the wind speed for different constant rotational speeds.

P in kW

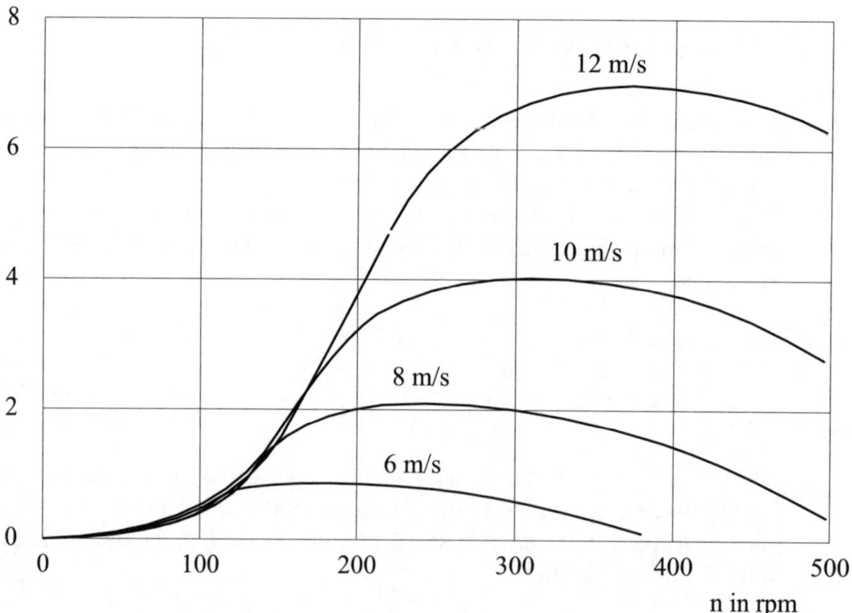

Fig. 6-8 Power versus rotational speed at different wind speeds for a wind turbine with $D = 4$ m, c_P-λ characteristics from Fig. 6-2

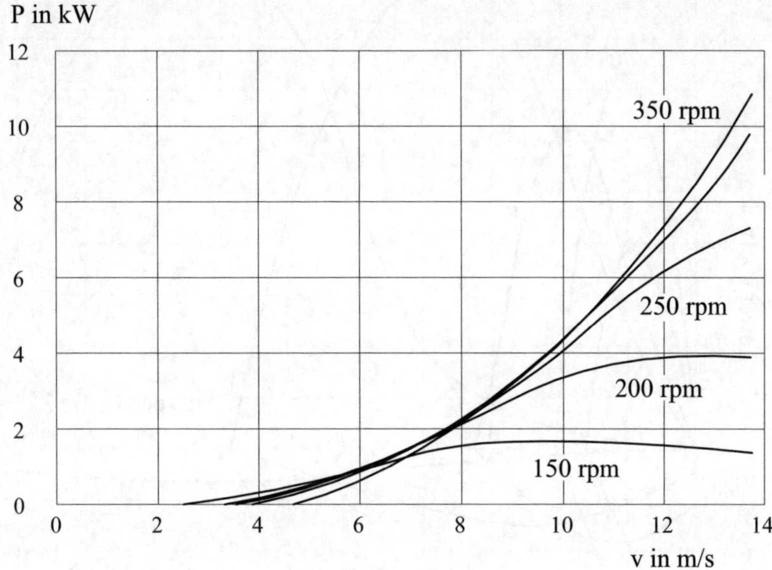

Fig. 6-9 Power versus wind speed at different rotational speeds for a wind turbine with $D = 4$ m, c_P-λ characteristics from Fig. 6-2

6.6 Flow conditions

6.6.1 Turbines with high and low tip speed ratio: a summary

This section summarizes once again the findings of chapters 5 and 6, and highlights the differences between turbines with a high tip speed ratio and those with a low tip speed ratio. The power and torque moment coefficients of wind turbines with different design tip speed ratios are given in Fig. 6-10.

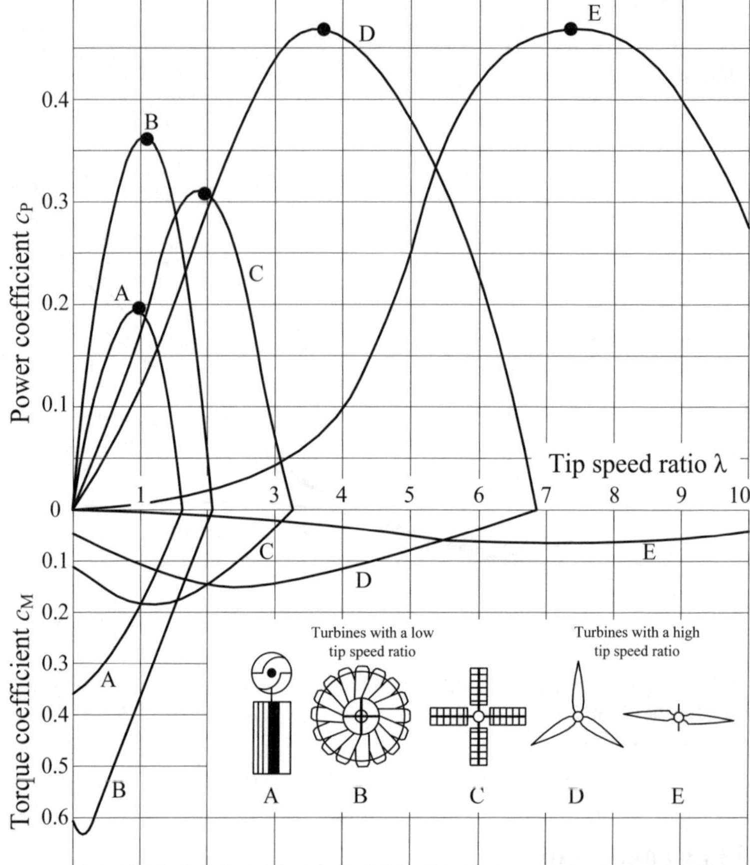

Fig. 6-10 Power and torque moment coefficients of different wind turbines varying in design and in tip speed ratio (Fateev [6], modified)

Power

- The power of a wind turbine increases proportionally to $v^3 R^2$, equation (6.9).
- The maximum power coefficient c_P is always smaller than 16/27.
- For turbines with a high tip speed ratio, $c_{P\text{max}}$ is basically only affected by the profile drag. The optimum tip speed ratio reached is always slightly smaller than the design tip speed ratio, $\lambda_{\text{opt}} < \lambda_D$.
- For turbines with a low tip speed ratio, $c_{P\text{max}}$ is mainly reduced by the wake rotation of the air; hence in this case λ_{opt} is always slightly larger than λ_D.

Torque moment

- The torque moment of a wind turbine increases proportionally to $v^2 R^3$.
- The maximum moment coefficient is always found at tip speed ratios of $\lambda < \lambda_D$, based on $c_P = c_M \lambda$.
- Turbines with a high tip speed ratio have small torque moment coefficients during start-up at $\lambda = 0$.
- Turbines with a low tip speed ratio have large torque moment coefficients during start-up.

Thrust force

- The thrust force of a wind turbine increases proportionally to $v^2 R^2$.
- Turbines with a high and a low tip speed ratio both have approximately the same thrust coefficient at the design point, $c_T \approx 8/9$ (according to Betz).
- Turbines with a high tip speed ratio have small thrust coefficients at start-up, and the thrust coefficient increases for higher tip speed ratios (c_T is at load-free idling nearly as high as for a solid disk).
- Turbines with a low tip speed ratio have large thrust coefficients at start-up; these decrease with larger tip speed ratios (at load-free idling partly "acting as a ventilator").

Idling

- In general, the idling tip speed ratio is approximately twice the design tip speed ratio.
- For turbines with a high tip speed ratio, the idling tip speed ratio is limited by the profile drag.
- For turbines with a low tip speed ratio, the profile drag has little effect on the idling tip speed ratio.

Blades

- Turbines with a high tip speed ratio have few and slender blades, which are smaller at the outer radius than at the inner radius. The blades require a high-quality surface; airfoil profiles with a high lift/drag ratio have to be used. There is heavy stress on the few blades due to aerodynamic and iner-tia forces.
- Turbines with a low tip speed ratio have a larger number of blades that have constant width, or are even wider at the outer radius than at the inner radius. There are no special requirements for the blade surface and the

profile. The stress on the single blade is low compared to turbines with a high tip speed ratio.

Application

- Turbines with a high tip speed ratio are used for electricity generation. Due to the higher rotational speed of the rotor, a smaller gear transmission ratio is required compared to turbines with a low tip speed ratio. The small start-up torque of turbines with a high tip speed ratio is not a problem as the generator only starts working at higher rotational speeds.
- Turbines with a low tip speed ratio deliver a high torque at start-up and are suitable for machines, such as piston or heat pumps, sawmills or corn mills.

6.6.2 Flow conditions in a turbine with a low design tip speed ratio

In order to reach a better understanding of the characteristic curves of a turbine with a low tip speed ratio, the flow conditions at the blade are discussed with the following Figs. 6-11 and 6-12. Again, a turbine with a tip speed ratio of $\lambda_D = 1$ is used as an example. The turbine has 21 blades dimensioned according to Schmitz theory.

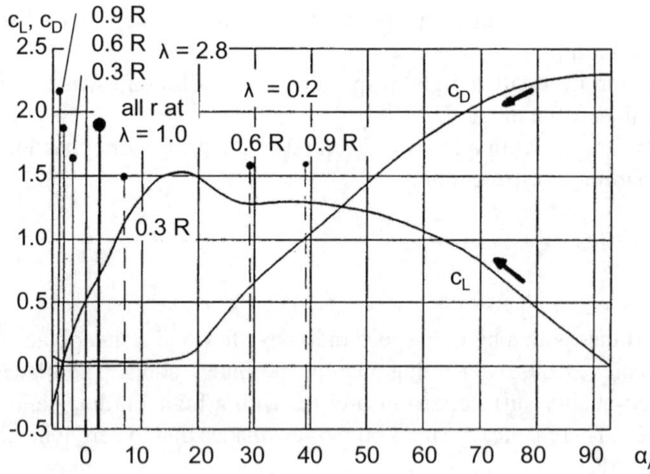

Fig. 6-11 Change of operating points on the profile characteristics during start-up (turbine with a low design tip speed ratio, $\lambda_D = 1$)

Fig. 6-12 shows three cross sections of the blade: the section above is close to the outer radius of $r = 0.9\ R$, the one in the middle at $r = 0.6\ R$ and the one below close to the inner radius at $r = 0.3\ R$. The upper half of each displayed section shows the magnitude and direction of the resulting forces (i.e. sum of lift and drag forces) for three different tip speed ratios, while the lower half shows the corresponding relative velocity w at the blade section. In addition, Fig. 6-11 shows the profile characteristics of the lift and drag coefficient. The operating points that are considered are marked for each blade section.

The blades were analyzed for three tip speed ratios: $\lambda = 0.2$ (start-up, dashed lines), $\lambda = 1$ (design tip speed ratio, thick line), and $\lambda = 2.8$ (load-free idling, thin lines). The dimensionless characteristic curves of this turbine were discussed in section 6.4, Figs. 6-5 to 6-7. They were obtained with the extended calculation method of section 6.8, where consideration was given to fluid dynamic effects (blocking of the stream tube) and losses (e.g. profile drag and reduced flow deflection). The calculation was performed for the case that the wind speed is constant and the rotor accelerates from standstill to load-free idling.

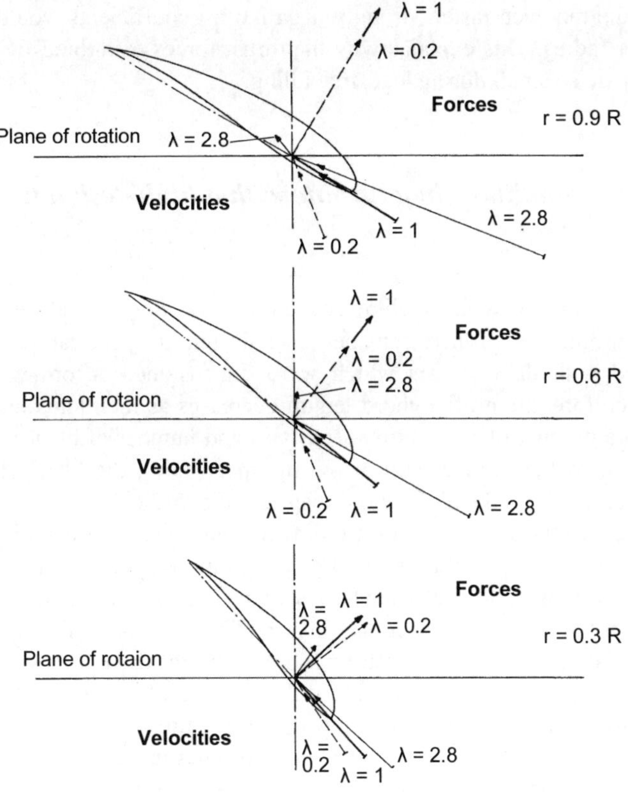

Fig. 6-12 Forces, relative velocity angles and relative velocities at three radial cross sections of a blade for different tip speed ratios (turbine with a low design tip speed ratio, $\lambda_D = 1$)

During *start-up* (following the curves from the right along the arrows in Fig. 6-11), the inner blade section has already a favourable lift/drag ratio, whereas at the outer sections the flow is still separated. Nevertheless, the circumferential forces across the entire blade are almost constant due to the increasing profile chord length with a larger radius. This produces the high starting torque moment coefficient of the turbine with a low tip speed ratio in Fig. 6-6.

For the *design point*, design tip speed ratio of $\lambda_D = 1$, there is along the entire blade an angle of attack of $\alpha_A = 2°$ with maximum lift/drag ratio as the blade was designed for this operating point. The circumferential force is somewhat larger at the outer sections than close to the inner radius due to the larger local circumferential speed and profile chord length.

When *idling*, the flow at the blade shows negative angles of attack ($\alpha_A < 0$) due to the increased circumferential speed. As a result, the lift coefficient becomes very small or is even negative. This means that the power extracted in the blade sections of positive lift (which "acts as a turbine") is used in the sections with negative lift coefficient (which "acts as a ventilator") to actively accelerate the air. If the turbine ran even faster, strongly negative lift coefficients would be reached at the outer radius. This explains why the thrust forces of turbines with a low tip speed ratio are so small during load-free idling.

6.6.3 *Flow conditions in a turbine with a high design tip speed ratio*

The blade of a turbine with a high tip speed ratio can now be analyzed in the same way. The turbine we will here consider has a design tip speed ratio of $\lambda_D = 7$ and is equipped with three blades which were dimensioned according to Schmitz theory. Therefore, the profile chord length decreases as the radius increases, Fig. 6-14. As the product of design tip speed ratio and number of blades $\lambda_D z = 21$ is equal to that of the turbine with a low tip speed ratio, the blade chord can be directly obtained from the blade chord diagram, Fig. 5-22.

The blade sections in Fig. 6-14 are drawn using the same scaling factor as for the blade of the turbine with the low tip speed ratio (Fig. 6-12). The upper blade section is again close to the outer radius of $r = 0.9\,R$, the blade section in the middle at $r = 0.6\,R$, and the lower at $r = 0.3\,R$. The upper half of the section shows the values and directions of the resulting forces, but for these forces, as well as for the relative velocities in the lower half of the section, a larger scaling factor had to be chosen than for the turbine with a low tip speed ratio.

The forces were calculated for both wind turbines using the same wind speed. The profile characteristics, Fig. 6-13, are the same as for to that of the turbine with the low tip speed ratio since the same aerodynamic profile is used, which makes more sense for the turbine with the high tip speed ratio. The dimensionless characteristic curves of this turbine were discussed in section 6.3, Figs. 6-2 to 6-4.

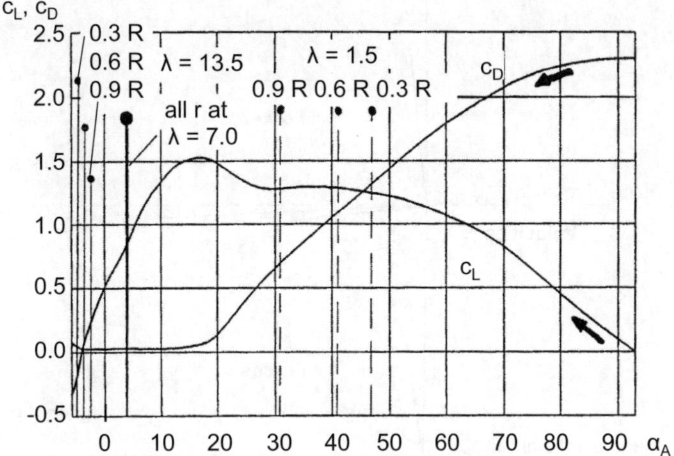

Fig. 6-13 Change of operating points along the profile characteristics during start-up (turbine with a high design tip speed ratio, $\lambda_D = 7$)

The blade sections are analyzed now at constant wind speed for the three tip speed ratios: $\lambda = 1.5$ (start-up, dashed lines), $\lambda = 7$ (design tip speed ratio, thick line), and $\lambda = 13.5$ (load-free idling, thin lines), Figs. 6-13 and 6-14.

During *start-up*, we again follow the arrows from the right in Fig. 6-13. Large parts of the blade are in the same range (i.e. the region of separated flow) of the profile curve as for the blade of the turbine with a low tip speed ratio, but now the inner section is also operating here. The values of the forces, Fig. 6-14, are nearly equal to those of the turbine with a low tip speed ratio, but their direction is less favourable so it produces only small start-up torque, cf. Fig. 6-3. As the turbine with a low tip speed ratio has seven times more blades, the thrust forces are substantially higher for the same blade loading, cf. Figs. 6-7 and 6-4.

When operating at the *design tip speed ratio*, the angle of attack is again at 2° along the entire radius as the blade was designed for this operation point. The circumferential force along the blade is nearly constant. The thrust force at the design point is equal to the one of the turbine with a low tip speed ratio, approx. 8/9, cf. Figs. 6-7 and 6-4.

At load-free *idling*, the angles of attack are slightly negative, Fig. 6-13. The lift coefficient at the outer radius does not diminish as much as that of a turbine with a low tip speed ratio. Due to the very high relative velocities which are nearly parallel to the plane of rotation, cf. Fig. 6-14, the forces increase substantially and act nearly parallel to the rotor axis producing very high thrust, cf. Fig. 6-4. The wind speed v_2 in the plane of rotation is strongly decelerated compared to the cases of start-up and design point where it has the same value as the undisturbed wind speed v_1.

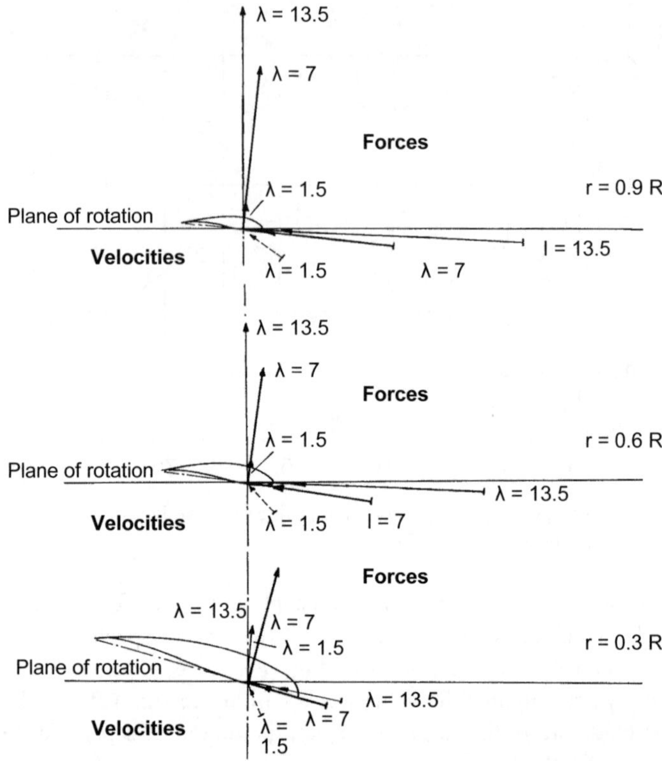

Fig. 6-14 Forces, relative velocity direction and velocities at three radial cross sections of a blade for three different tip speed ratios (turbine with a high design tip speed ratio, $\lambda_D = 7$)

6.7 Behaviour of turbines with high tip speed ratio and blade pitching

Blade pitching changes the blade twist angle globally over the blade length. In the following, "pitching" means pitching the blade to feather position, i.e. the trailing edge turns out of the wind and the nose of the profile into the wind. The inflow conditions at a turbine with a high tip speed ratio then become e.g. for the start-up, more similar to the inflow at a turbine with a low tip speed ratio. However, the high driving forces of a turbine with a low tip speed ratio are not achieved, since the blade chords of the turbine with a high tip speed ratio are designed for higher relative velocities. Nevertheless, the turbine now has an improved self-starting behaviour, see Fig. 6-16.

Pitching the blade is also useful when the rated power is reached since it reduces the angle of attack and consequently the lift coefficients, leading to a power limitation, see Fig. 6-15.

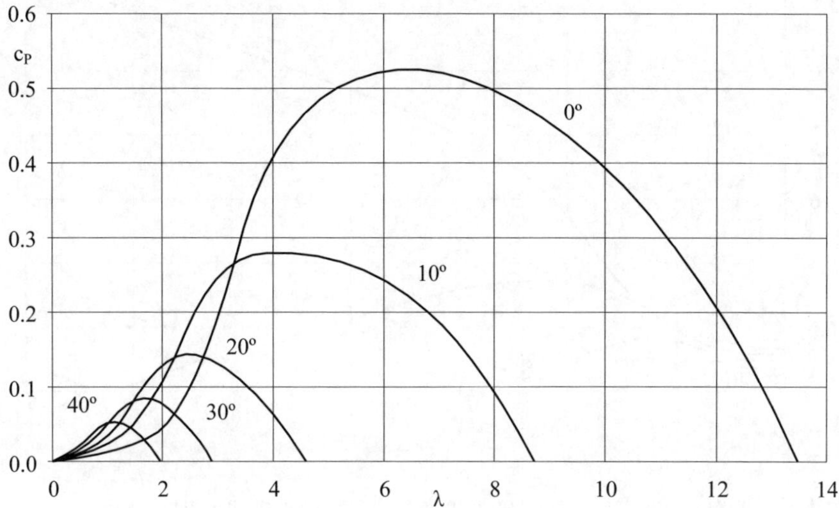

Fig. 6-15 Power coefficient c_P versus tip speed ratio for different pitch angles ($\lambda_D = 7$)

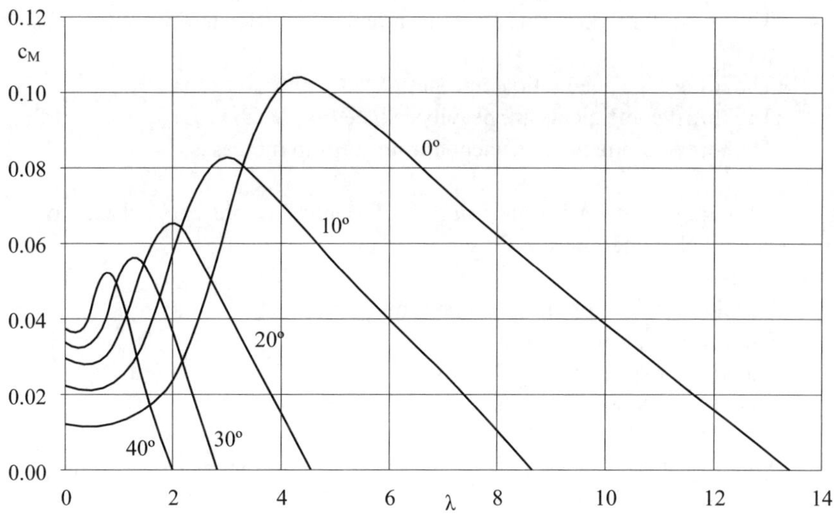

Fig. 6-16 Torque moment coefficient c_M versus tip speed ratio for different pitch angles

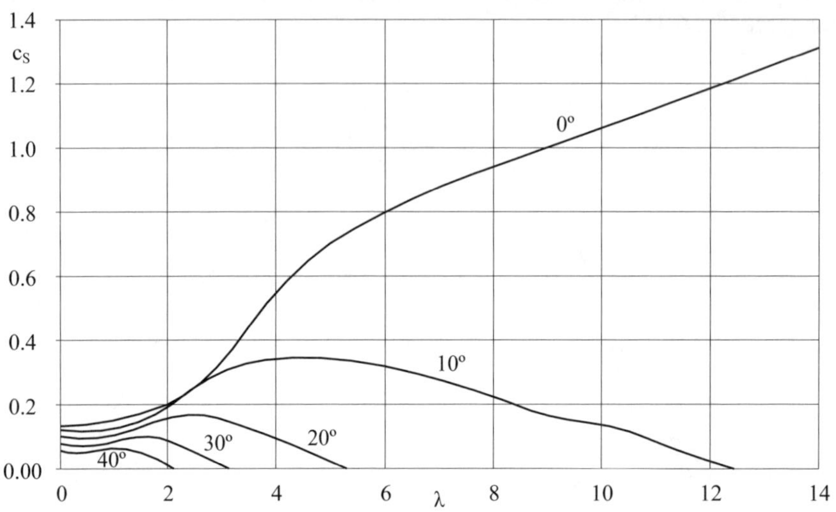

Fig. 6-17 Thrust coefficient c_T versus tip speed ratio for different pitch angles

So, in turbines with a high tip speed ratio, blade pitching is used for improved start-up, for power control and also serves as an emergency brake. This will be discussed in mire detail in chapter 12.

Figs. 6-15 to 6-17 show that with an increasing pitch angle:

- The maximum power and torque moment coefficients decrease substantially.
- The idling tip speed ratio decreases.
- The thrust coefficients are heavily reduced.
- The torque moment coefficient during start-up increases.

In the following an *example for using pitch control* is presented, based on the dimensionless characteristics of Figs. 6-15 to 6-17.

Given data:

- Turbine with a high tip speed ratio of $\lambda_D = 7$,
- Rotor diameter of $D = 4$ m,
- Rated rotor speed of $n_R = 300$ rpm and
- Rated rotor power $P_R = 4$ kW.

By pitching the blade, the rotor speed shall be kept constant at $n_R = 300$ rpm. The rated power of the variable speed generator is 4 kW.

We start by determining the factor k_{dyn} which allows us to make a simplified calculation of the dynamic pressure:

$$k_{dyn} = \frac{\rho}{2}\pi R^2 = 7.85 \; kg/m$$

a) wind speed $v_1 = 10$ m/s.

Tip speed ratio at $n = 300$ rpm:

$$\lambda = \Omega\frac{R}{v_1} = n\frac{\pi R}{30 v_1} = 6.3$$

From the power coefficient curve, Fig. 6-15, we obtain the power coefficient for the pitch angle of $0°$:

$$c_P\,(\lambda = 6.3) = 0.52$$

Power:

$$P = A\;v_1^3 c_P = 7.85 \cdot 1000 \cdot 0.52 W = 4\,kW$$

At this wind speed, action of the pitch control is unnecessary.

b) The wind speed increases to 12 m/s.

Tip speed ratio at $n = 300$ rpm:

$$\lambda = \Omega\frac{R}{v_1} = n\frac{\pi R}{30 v_1} = 5.3$$

For the pitch angle of $0°$ this leads to a new c_P of:

$$c_P\,(\lambda = 5.3) = 0.50$$

Power:

$$P = k_{dyn} v_1^3 c_P = 7.85 \cdot 1728 \cdot 0.5 W = 6.3\,kW$$

However, the generator should draw only 4 kW of power from the rotor to prevent overload. What does this mean for the required c_P?

$$c_{P.demand} = \frac{4\,kW}{k_{dyn} v_1^3} = 0.29$$

If there is no pitch control but we leave the pitch angle at 0°, the rotor would accelerate and therefore increase the tip speed ratio until the power coefficient of $c_{P.demand} = 0.29$ is reached. This would occur at $\lambda = 11$. What would then be the corresponding rotational speed?

$$n = \lambda \frac{30}{\pi} \frac{v_1}{R} = 630\,rpm > 2n_R$$

At this rotational speed, the generator would be destroyed (it will shed its windings and then burn out), that is if the blades have not already been torn off (the centrifugal forces for blades turning at 630 rpm are 4.4 times higher than at 300 rpm).

But when pitching the blade to feather by 10°, we reduce the angle of attack and the corresponding lift coefficient. Consequently, we can attain the $c_{p.demand} = 0.29$ for $\lambda = 5.3$, i.e. with the rated rotational speed.

c) The wind speed continues increasing to $v_1 = 14$ m/s; the pitch angle is 10°.
   Tip speed ratio at $n = 300$ rpm:

$$\lambda = \Omega \frac{R}{v_1} = n \frac{\pi R}{30 v_1} = 4.5$$

$$c_P\,(\lambda = 4.5) = 0.29$$

Power:

$$P = k_{dyn} v_1^3 c_P = 7.85 \cdot 2744 \cdot 0.29\,W = 6.2\,kW$$

Required c_P for $P = P_R$:

$$c_{P.demand} = \frac{4\,kW}{k_{dyn} v_1^3} = 0.18$$

This means the pitch angle must be further increased to approximately 15°.

The pitch angle must be adjusted quickly otherwise the rotor may overspeed and the generator burn. Moreover, all blades need to be pitched simultaneously to prevent aerodynamic imbalances.

Fig. 6-18 shows the characteristics of a pitch-controlled wind turbine for the pitch angles 0° (i.e. design blade twist angle), 22.5° and 39.8°. The generator load curve of the speed-variable wind turbine is shown as well. In the normal wind range between 4 and approx. 10 m/s the pitch angle remains constant at the design value of $\gamma = 0°$ (cf. Fig. 12-2) and the wind turbine extracts always the optimum (i.e. maximum possible) power from the wind. The rotor speed is variable and the optimum is adjusted according to the current wind speed.

Above the rated wind speed $v \geq 10.3$ m/s, i.e. the strong wind range, blade pitching starts to limit the power to rated power (as indicated by the circle drawn at the rated operating point (n_R, P_R)). It also keeps constant the rotational speed n_R of the turbine. The curves for constant wind speeds of 22 m/s (pitch angle 22.5°) and 30 m/s (pitch angle 39.8°) are shown here; these pass through the rated operating point of (n_R, P_R). In general the wind turbines are shut down for wind speeds exceeding 25 m/s. More details on the blade pitching are found in chapters 12 and 13.

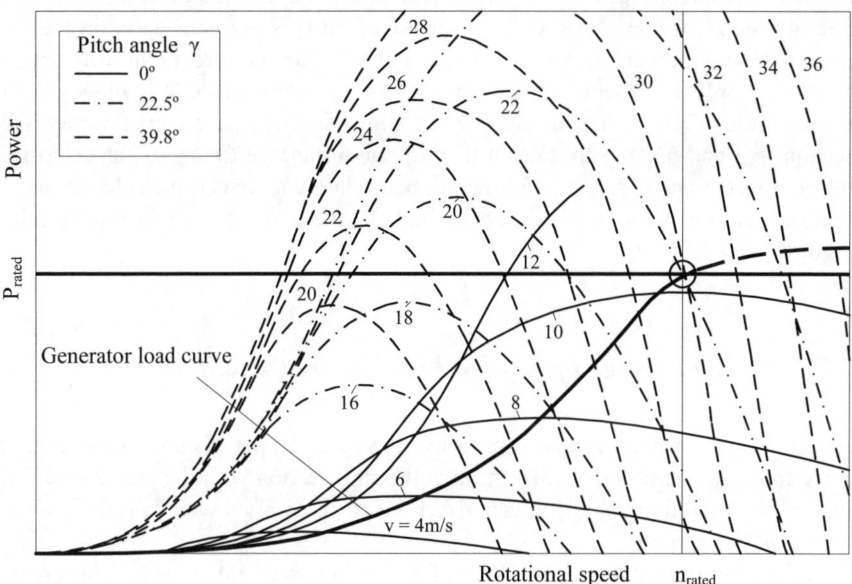

Fig. 6-18 Rotor power versus rotational speed (P-n characteristics) for different pitch angles and wind speeds of a pitch-controlled wind turbine with a speed variable generator load

6.8 Extending the calculation method

In section 6.1, by equating aerodynamic lift force and aerodynamic force calculated from the principal of linear momentum, the relative velocity angle at a blade section was determined. These two forces can be equated, when the following assumptions hold true:

- The forces at the blade section cause a steady and homogeneous change in the velocity of the air mass that flows through the ring section with the area $dA = 2\pi r\, dr$.
- The air mass is only affected by the aerodynamic forces in the plane of rotation. The stream filaments do not excert any forces on each other.
- The profile drag is very small and can thus be neglected.

These assumptions are approximately valid as long as the rotor runs near the design tip speed ratio. During start-up ($\lambda \ll \lambda_D$) of a turbine with a high tip speed ratio, air flows through the rotor without being influenced by the blades. However, during load-free idling ($\lambda > \lambda_D$), the deceleration of the air is so heavy ($v_2 < v_1/2$) that some of the retarded air mass flows rather around the outside of the rotor than through the rotor area. This loss of utilized air mass is not considered in the formula of linear momentum we used in section 6.1. Furthermore, the profile drag is negligible only for operation with the design tip speed ratio. We already mentioned these effects when discussing the dimensionless characteristic curves in sections 6.3 and 6.4 and considered them in the turbine performance characteristics of sections 6.5 and 6.7. In the following, better late than never, the calculations of section 6.1 will now be extended to include these influences in the iteration algorithm.

6.8.1 Start-up range of $\lambda < \lambda_D$ (high lift coefficients)

A turbine with a high tip speed ratio has few and slender blades. These apply a force to the air masses that flow through the ring section with the area $2\pi r\, dr$, cf. Fig. 5-14. An individual airfoil can affect the air mass in its vicinity only in a certain *effective width b**.

If the rotor runs at the design point of λ_D, the relative velocity w has the angle ϕ with respect to the rotor plane, Fig. 6-19. The width b of the region the blade should affect is comparatively small, $b = (2\pi\ r \sin\phi)/z$, because, due to the small angle of attack, the incident relative velocity is nearly parallel to the profile chord.

In contrast, at standstill the individual blade should affect a width of $a = 2\pi r/z$ (i.e. the circumferential distance between the blades) which is usually significantly larger than its actual width of influence $b*$. Therefore, during standstill (and start-up) of a turbine with a high tip speed ratio, some of the air flows through the plane of rotation without being disturbed and utilized by the blades.

How large is the effective width $b*$ of the region where a blade affects the air masses? Strictly speaking, its influence reaches into infinity - it only decreases rapidly with growing distance. From the *Prandtl airfoil theory* [4], we obtain the maximum of air mass for which a blade can cause a constant change in speed.

From that, we can estimate the effective width b^*. For the airfoil of an airplane, Prandtl proved that for an *elliptic lift distribution* (Fig. 6-20), the air flow at each blade section is directed downwards at an identical angle.

Far downstream of the airfoil, the air moves downwards with a speed of $2\delta v$ which depends on the lift coefficient c_L, the blade chord c and the relative velocity w.

$$2\delta v = 2wc_L \frac{c}{\pi R} \qquad (6.11)$$

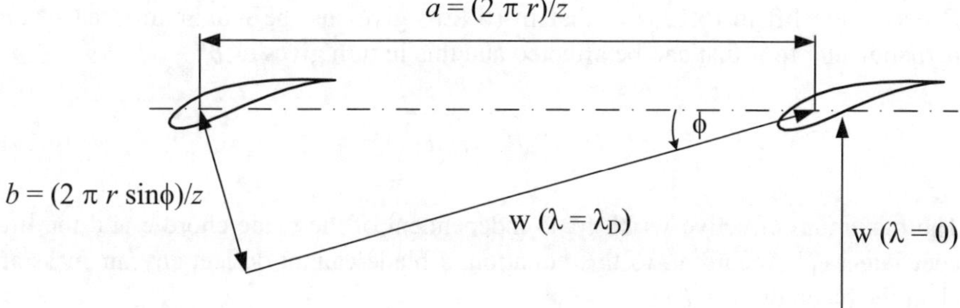

Fig. 6-19 Width of area in which the rotor blades should influence the flow: width a at standstill ($\lambda = 0$) and b at the design point ($\lambda = \lambda_D$)

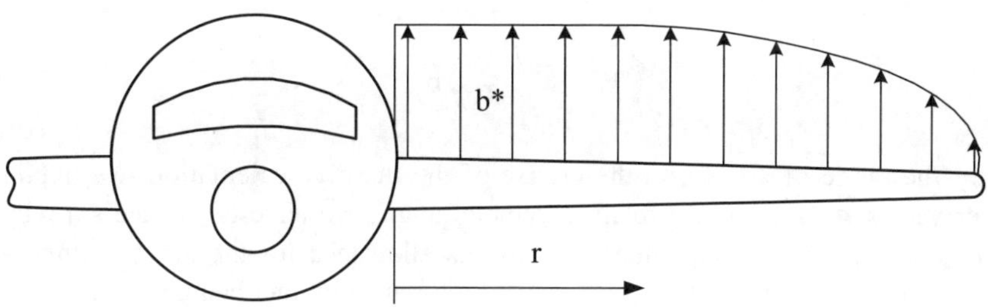

Fig. 6-20 Area in which an airfoil affects the air flow with reduced influence close to the tip

If a similar elliptic lift distribution is assumed for the rotor blade, the lift of each blade section can be calculated by

$$dL = \frac{\rho}{2} w^2 c\, dr\, c_L \frac{4}{\pi} \sqrt{1 - \left(\frac{r}{R}\right)^2} \, . \qquad (6.12a)$$

The mass flow is now the product of air density ρ, relative velocity w and the still unknown area $b^*\,dr$ around the blade where it affects the flow. The lift obtained from the principle of linear momentum with a change in speed of $2\delta v$, equation (6.11), is therefore

$$dL = d\dot{m}\; 2\; \delta v$$

$$dL = \rho\; w\; b^*\; dr\; 2w\, c_{\mathrm{L}}\, \frac{c}{\pi R} \qquad\qquad (6.12b)$$

Equating the lift in (6.12a) to that in (6.12b) gives us the maximum area $b^*\,dr$ perpendicular to w that can be affected and this in turn gives us b^*

$$b^*\,dr = R\sqrt{1-(r/R)^2}\; dr \quad . \qquad\qquad (6.13)$$

The maximum effective width b^* is independent of the blade chord c and the lift coefficient c_{L}! According to this equation, a blade cannot deflect any air mass at all at the blade tip ($r = R$).

If in Fig. 6-19 we take $b = b^*$ and use the width a between the blades, we obtain the maximum relative velocity angle ϕ_{max} up to which the air mass in the ring section of $2\pi r\,dr$ is affected and utilized. The maximum relative velocity angle ϕ is then dependent on the number of blades z

$$\sin\phi_{\mathrm{max}} = \frac{z\sqrt{1-(r/R)^2}}{2\pi(r/R)} \qquad\qquad (6.14)$$

If, for the range of $\lambda < \lambda_{\mathrm{A}}$, in the course of the iterative calculation, we obtain larger values $\phi > \phi_{\mathrm{max}}$ of the relative velocity angle, we have to replace $\sin\phi$ by $\sin\phi_{\mathrm{max}}$ in the iteration equation. This makes allowance for the fact that then a fraction of air mass flows through the swept rotor area without being utilized.

6.8.2 Idling range of $\lambda > \lambda_{\mathrm{D}}$ (Glauert's empirical formula)

If the rotor is idling load-free the thrust coefficient of a turbine with a high tip speed ratio, cf. Fig. 6-4, approaches the drag coefficient of a solid circular disk, cf. Fig. 2-19.

If therefore the wind speed v_2 in the plane of rotation is slowed to zero, the relative velocity angle and consequently the mass flow approaches zero, too. Using the principle of linear momentum, we would then obtain the thrust force

$$dT = d\dot{m}\,\Delta v = \rho\,w\left(\frac{2\pi r}{z}\,dr\sin\phi\right)\Delta v = 0 \qquad (6.15)$$

i.e. the thrust becomes zero, which does not seem altogether logical. Formally, it is correct that if no air flows through the plane of rotation, we do not need any force to retard it. However, in reality, the point is that the retarded air no longer flows through the rotor plane as assumed in the formula of linear momentum, but passes outside around the circular rotor swept area as it does with a solid circular disk. This kind of flow is not taken into account in the stream tube theory that has been applied so far.

In order to consider this fluid dynamic effect in the calculation of rotor characteristics we will now apply an approximated approach, without which it would be necessary to compute the entire three-dimensional flow in the area from $-\infty$ to $+\infty$.

The thrust forces of this operating state were measured and investigated by Glauert (1926) and Naumann (1940), [2, 3]. The purpose of the measurements was to determine how efficiently a propeller airplane could be slowed down if the propeller blades were pitched so that they act contrary to the direction of flight. Glauert discovered that the thrust force can be calculated using our formula of linear momentum until approximately $v_2 \geq 2/3\ v_1$.

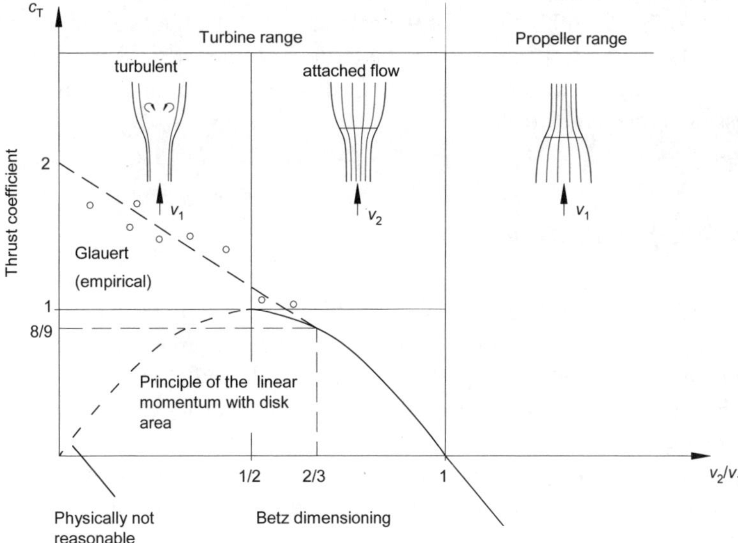

Fig. 6-21 Thrust coefficient according to the principle of linear momentum and according to Glauert's empirical approach in the range of $0 < v_2/v_1 < 0.5$, [2,3]

For smaller v_2, i.e. larger λ, the thrust force, and correspondingly the thrust coefficient, increases steadily and reaches a value of twice the dynamic pressure at $v_2 = 0$ m/s (i.e. $\phi = 0°$), see Fig. 6-21. Given this finding, the air mass that is retarded by the forces of the blade can be approximated depending on the relative velocity angle using a continuous function, for which the following conditions apply:

- For $v_2 = 0$, twice the dynamic pressure is obtained
- For $v_2 = 2/3\ v_1$, the value of the function and the slope must agree with the Betz optimum dimensioning.

Using the substitution:

$$y = \frac{\sin\phi}{\sin(2\phi_1/3)}$$

the mass flow is described as follows:

$$\dot{dm} = \rho\,\frac{2\pi r}{z}\,dr\,w\left[\frac{1}{4}\sin\left(\frac{2}{3}\phi_1\right)\sqrt{9 - 2y^2 + 9y^4}\right] \tag{6.16}$$

This approximation is only valid for small relative velocity angles ϕ_1 of the undisturbed flow! The angle ϕ_1 has to be so small that the approximation $\sin\phi_1 \approx \phi_1$ is valid. For turbines with a high tip speed ratio, this operational state is reached, for instance, during load-free idling.

Equation (6.5) can still be used for the calculation of the relative velocity angle ϕ, however, for $\phi \leq 2/3\ \phi_1$ (i.e. $\lambda \geq \lambda_D$) the term of the air mass ($8\ \pi r/z\ \sin\phi$) has to be substituted by the larger air mass of equation (6.16) (i.e. $\sin\phi$ is replaced by the expression in the large brackets).

6.8.3 The profile drag

For an airfoil, the profile drag acts in parallel to the relative velocity. The profile drag reduces the relative velocity w by the velocity difference δw which denotes the component parallel to the profile drag of the change of relative velocity, Fig. 6-22. For the sake of clarity, the component of the velocity change parallel to the lift is denoted Δw. The velocity difference δw is obtained again from the force equations of the airfoil theory and the principle of linear momentum.

The calculation of the drag force at a blade section using the airfoil theory was derived in equation (5.28) of chapter 5. The principle of linear momentum can also be used to obtain the drag force at a blade section:

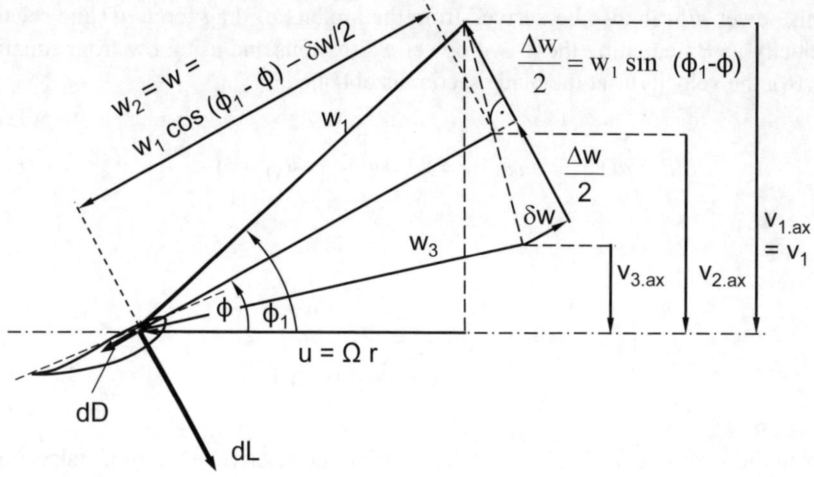

Fig. 6-22 Flow conditions at the blade section

$$dD = d\dot{m}\ \delta w = \rho w \frac{2\pi\, r}{z}\, dr \sin(\phi)\, \delta w \quad . \tag{6.17}$$

By equating both forces, the velocity difference δw is obtained depending on the relative velocity w at the blade section

$$\delta w = w \frac{z\, c\, c_\mathrm{D}(\alpha_\mathrm{A})}{4\pi\, r \sin\phi} \quad . \tag{6.18}$$

Observing the flow from the rotating blade as the reference (rotating frame of reference), the far downstream wind has the velocity w_3. As w_3 is smaller than the undisturbed upstream relative velocity w_1, the rotating observer will note that the power dP is extracted from the air:

$$\left.\begin{aligned}
w_3 &= \sqrt{[w_1 \sin(\phi_1 - \phi)]^2 + [w_1 \cos(\phi_1 - \phi) - \delta w]^2} \\[2ex]
w_3 &= \sqrt{w_1^2 - 2\delta w\left(w_1 \cos(\phi_1 - \phi) - \frac{\delta w}{2}\right)} \\[2ex]
dP &= \frac{1}{2}\, d\dot{m}\left(w_1^2 - w_3^2\right) = d\dot{m}\ \delta w\left[w_1 \cos(\phi_1 - \phi) - \frac{\delta w}{2}\right]
\end{aligned}\right\} \tag{6.19}$$

This power dP can also be derived from the product of drag force dD and relative velocity w. By equating these two power expressions and using δw from equation (6.18), the velocity w at the blade section is obtained:

$$dP = dD\, w = d\dot{m}\, \delta w\, w = d\dot{m}\, \delta w \left[w_1 \cos(\phi_1 - \phi) - \frac{\delta w}{2} \right] \tag{6.20}$$

$$w = w_1 \cos(\phi_1 - \phi) \frac{\dfrac{8\pi r}{z} \sin\phi}{\dfrac{8\pi r}{z} \sin\phi + c\, c_D(\alpha_A)} \tag{6.21}$$

Since the lift force dA is also calculated using the relative velocity w, taking into account the profile drag changes the basic equation (6.5) of the iteration as follows:

$$dL = \frac{\rho}{2} w^2\, c\, dr\, c_L(\alpha_A) = d\dot{m}\, \Delta w$$

$$c\, c_L(\alpha_A) - \left[\frac{8\pi r}{z} \sin\phi + c\, c_D(\alpha_A) \right] \tan(\phi_1 - \phi) = 0 \tag{6.22}$$

6.8.4 The extended iteration algorithm

Considering the flow of unused air around the swept rotor area (Prandtl), the Glauert empirical formula for the flow near to the tip and the profile drag, the *iteration algorithm* for every blade section is as follows:

1) Starting value: $\phi = \phi_1$,

2) Limiting angles of the relative velocity angle:

$$\sin\phi_{max} = \frac{z\sqrt{1-(r/R)^2}}{2\pi(r/R)}$$

$$\sin\phi_{min} = \sin\frac{2}{3}\phi_1$$

3) With this ϕ: $\alpha_A = \phi - \beta(r) \rightarrow c_L(\alpha_A)$ and $c_D(\alpha_A)$ from the airfoil

profile characteristics

and substitution: $x = \sin\phi$

4) Verification:

    If $x < \sin\phi_{min}$, then modified according to Glauert:

$$\text{Set } x = \frac{1}{4}\sin\left(\frac{2}{3}\phi_1\right)\sqrt{9 - 2y^2 + 9y^4} \text{ with } y = \frac{\sin\phi}{\sin(\phi_1 \cdot 2/3)}$$

    If $x > \sin\phi_{max}$ then modified according to Prandtl:

    Set $x = \sin\phi_{max}$

5) Solution of

$$f = c(r)c_L(\alpha_A) - \left(\frac{8\pi r}{z}x + c(r)c_D(\alpha_A)\right)\tan(\phi_1 - \phi) \qquad (6.22/\ 6.23)$$

6) Verification:

        If $f > 0$ then reduce ϕ and go back to step 3).

        If $f < 0$ then increase ϕ and go back to step 3).

With this procedure we zero in on the relative velocity angle ϕ until the residual reaches $f = 0$ m. Aerodynamics is once more reduced to searching the functions zero.

To compute the iteration it is quite reasonable to select a suitable loop termination criterion for the residual f. Due to the profile chord length and the blade radius in equation (6.23) f is dimensionful (meter); moreover for wind turbines with a high tip speed ratio the chord length decreases strongly with increasing radius. Therefore, it is not useful to apply one and the same termination criterion (e.g. $f \leq 0.002$ m) for the entire blade: it is perhaps suitable for the small blade tip, but unreasonable for the range of the maximum chord length of up to 4 m close to the blade root, where it is smaller than the manufacturing tolerances. In the course of programming the iteration this problem is solved by transforming equation (6.22) into a dimensionless equation, e.g. dividing by the chord length c. Then the percentage termination criterion for the residual f/c is applicable along the entire blade radius.

7) The relative velocity w is calculated with the determined relative velocity angle ϕ and equation (6.21). Here the $\sin\phi$ should be replaced by the term x from step 4) of the iteration algorithm.

8) The aerodynamic forces now obtained for the blade section are:

$$dL = \frac{\rho}{2} w^2 c(r)\,dr\,c_L(\alpha_A) \quad \text{and} \quad dD = \frac{\rho}{2} w^2 c(r)\,dr\,c_D(\alpha_A)$$

and subsequently:

Thrust force:	$dT(r) = dL\,\cos\phi + dD\,\sin\phi$	
Circumferential force:	$dU(r) = dL\,\sin\phi - dD\,\cos\phi$	(6.24)
Torque moment:	$dM(r) = dU\,r$	

With the presented extended iteration algorithm of the blade element momentum method it is possible to determine, with a fairly close approximation, the forces and moments produced by the aerodynamics as well as the power of the wind turbine rotor for the entire range of its performance characteristics.

6.9 Limits of the blade element theory and three-dimensional calculation methods

The blade element method discussed in this chapter has the advantage of providing a comparatively accurate calculation of the performance characteristics without requiring the solution of extensive systems of equations. However, it ignores the fact that there are interactions between the flow in the different radial blade sections. For the simplified approach, optimum rotor geometry and ideal lift distribution were assumed. In the range of low tip speed ratios ($\lambda < \lambda_D$) and for non-optimum rotor geometry, there may occur considerable discrepancies between the calculated and the actual measured power. In particular, the beginning of the flow separation (stall) cannot be predicted accurately. There are in the main three effects which produce the divergence between calculation and actual values:
- Ignoring the real lift distribution
- Change of the lift curves by the 3-D effect
- Dynamic effects

In the following, various approaches for addressing these effects are discussed.

6.9.1 Lift distribution and three-dimensional effects

In the literature, there are different effects referred to as the "3-D effect":
- Induced velocities produced by the tip vortices,
- Modified profile characteristics due to the finite length of the blade,
- Modified profile characteristics due to Coriolis and centrifugal forces.

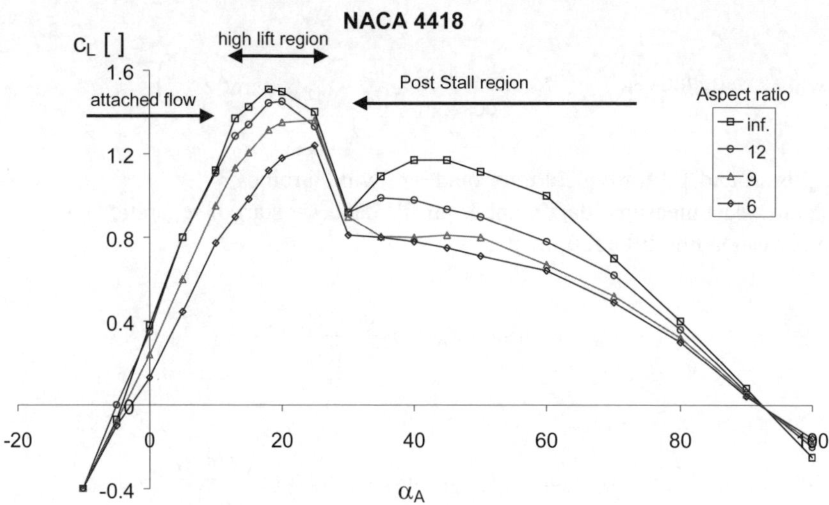

Fig. 6-23 Lift coefficient versus angle of attack for blades with different aspect ratio, [7, modified]

For *attached flow* the lift can be calculated with the Prandtl correction. Our version, equation (6.22), assumes an elliptical lift distribution. More extensive calculation procedures based on Panel methods (method of singularities) or Euler equations are, however, only applicable for attached flow.

Generally the measured airfoil characteristics are available for the region with attached flow up to the *high lift region*. Airplanes do not have separated flow across the wing under normal flight conditions. Therefore, it is very difficult to get accurate values from wind tunnel measurements for the region of separated flow. So the database for post-stall values is very small. The last measured value for lift and drag is usually for the angle of attack $\alpha_{A.s}$ where the flow starts to separate.

The lift and drag values in the *post-stall region* with fully separated flow seem to be quite similar for all airfoils. Here, it is possible to fall back on the aerodynamic characteristics of a flat plate where there are approximations for the region of separated flow [8]. Based on this, Viterna and Corrigan formulated the description of the passage to the measurement data [e.g. in 26]. In the following we will present their equations for the range of the angle of attack between $15 < \alpha_A < 90°$.

For separated flow, the drag coefficient may be described as follows:

$$c_D = B_1 \sin^2 \alpha_A + B_2 \cos \alpha_A \tag{6.25}$$

where $\qquad B_1 = c_{D.max} \; ; \; B_2 = \dfrac{1}{\cos \alpha_{A.s}} \left(c_{D.s} - c_{D.max} \sin^2 \alpha_{A.s} \right);$

$c_{D.max}$ is around 1.3 for typical rotor blades of wind turbines

$\alpha_{A.s}$, $c_{D.s}$ – last measured data point of profile data (s = start of separation)

The corresponding lift coefficient is

$$c_L = A_1 \sin 2\alpha_A + A_2 \frac{\cos^2 \alpha_A}{\sin \alpha_A} \tag{6.26}$$

where $\qquad A_1 = \dfrac{B_1}{2} \; ; \; A_2 = \left(c_{L.s} - c_{D.max} \sin \alpha_{A.s} \cos \alpha_{A.s} \right) \dfrac{\sin \alpha_{A.s}}{\cos^2 \alpha_{A.s}}$

$c_{L.s}$ – last measured data point of profile data (s = start of separation)

After testing a prototype these values should be replaced with values determined from the prototype measurements.

If the rotor blades rotate, further 3D-effects act on the flow. Measurements indicate higher lift values in the near-hub region of the blade. *Coriolis and centrifugal forces* influence the flow. Snel published an estimate to calculate the additional lift [9]. He found that the local solidity (cf. Fig. 5-15) influences the lift coefficients.

$$c_{L.3D.rot} = c_{L.2D} + \Delta c_L \tag{6.27}$$

The additional lift coefficient Δc_L depends on the local radius r and the local chord length c, the value $c_{L.inv}$ is from equation (6.30):

$$\Delta c_L = 3 \left(\frac{c}{r} \right)^2 \left(c_{L.inv} - c_{L.2D} \right) \; . \tag{6.28}$$

Provided that
-    the lift does not exceed the lift for attached flow,
-    the lift is zero for very high angles of attack ($\alpha_A = 90°$)
-    it is practical to modify equation (6.27) to

$$\Delta c_{\mathrm{L}} = \tanh\left(3\left(\frac{c}{r}\right)^2 \right)\left(c_{\mathrm{L.inv}} - c_{\mathrm{L.2D}}\right)\cos^2 \alpha_{\mathrm{A}} \qquad (6.29)$$

The correction term is, however, limited to positive lift values with this modification. And it is only valid for rotating blades. Its influence is limited to the near-hub region of rotor blades with a large maximum chord length.

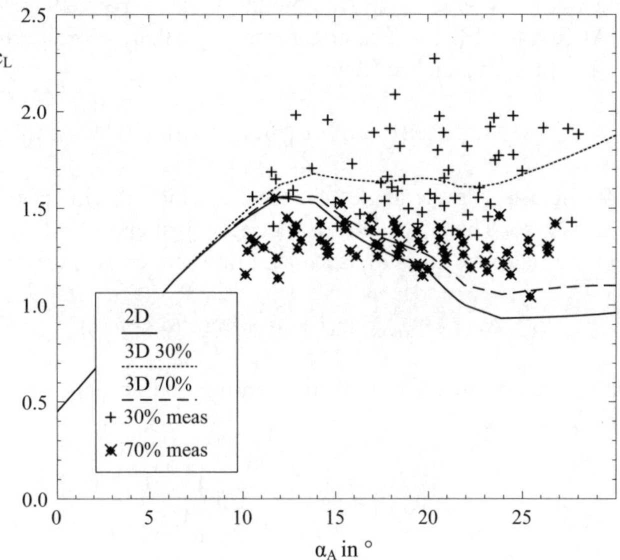

Fig. 6-24 Influence of 3D-effects on the lift coefficient c_{L}. 2D-measurement, 3D-measurement and 3D-corrected lift coefficient c_{L} for 30% and 70% of the blade length [9]

There are different theories about the drag under these conditions. Usually the changes are ignored because they only have a small impact on the turbine. Of course, all this is only a rough approximation to describe the start of flow separation which, in reality, involves very complex physics.

Nowadays, all these "3D-effects" can be calculated by Computational Fluid Dynamics (CFD) which also takes into account the viscous effects. The results depend strongly on the chosen turbulence model and the Navier-Stokes equation solver [10], cf. section 6.9.5. The main problem is that the calculation with these programs is very time-consuming. So it seems to be expedient to back-calculate the 2D-lift curves from these models and then use them in conventional BEM-programs to calculate the dynamic loads.

6.9.2 *Dynamic flow separation (Dynamic stall)*

A model for the dynamic flow separation is needed to calculate the aerodynamic forces at *elastic blades in turbulent flow*. Otherwise the calculated dynamic loads can be at an unrealistic high level. The real physics is very complex, especially for leading edge stall. On rotor blades of wind turbines with large Reynolds numbers and thick leading edges we expect mostly trailing edge stall. The Beddoes-Leishman model was developed for helicopters, and this was further simplified for wind turbine rotors by Øye [11]. This model works very well for trailing edge separation. At first the lift coefficient curve is modelled according to the characteristics of a flat plate at attached flow:

$$c_{L.inv} = 2\pi(\alpha_A - \alpha_{A.0}) \quad \text{(for thin airfoils)} \tag{6.30}$$

The zero-lift angle $\alpha_{A.0}$ is the angle of attack at which the lift coefficient of the chosen airfoil is zero. The curve of separated flow is then modelled. According to Hoerner [8] for a flat plate and fully separated flow (index *sep*) we can say that

$$c_{L.sep}(\alpha_A) = c_{D.max} \sin(\alpha_A - \alpha_{A.0}) \ \cos(\alpha_A - \alpha_{A.0}) \tag{6.31}$$

The lift coefficients of an existing profile can now be described as a function of the degree of separation f

$$f(\alpha_A) = \frac{\left(c_A - c_{A.sep}\right)}{\left(c_{A.inv} - c_{A.sep}\right)} \tag{6.32}$$

For each angle of attack α_A the steady-state degree of separation f can now be determined, and for each degree of separation f there is now a corresponding lift coefficient c_A. The degree of separation varies between zero and one and may be interpreted as the location of the separation point on the blade (cf. Fig. 6-27). If there is a quick change in the angle of attack, the separation point needs a certain time to move to the new steady-state point. The time response is described in the model of Øye by a simple PT1 behavior

$$\frac{df}{dt} = \frac{\left(f_{stat} - f\right)}{\tau_{fak}} \tag{6.33}$$

The time delay constant τ_{fak} depends on the chord length c and the relative velocity w:

$$\tau_{fak} = a\frac{c}{w} \tag{6.34}$$

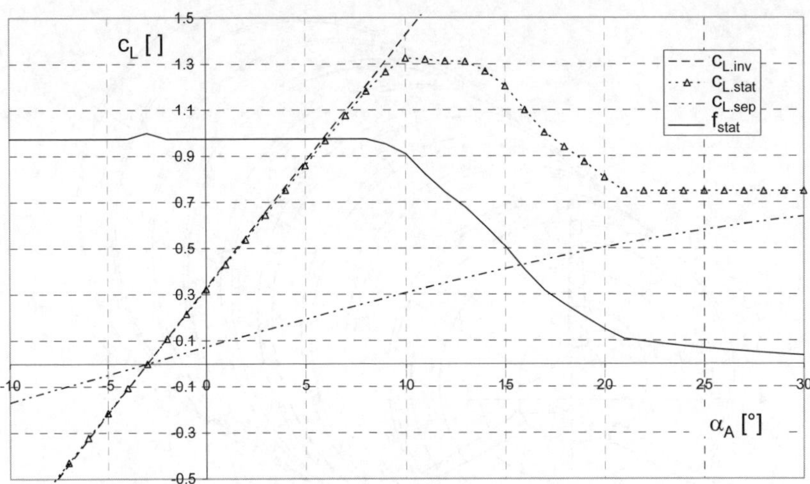

Fig. 6-25 Lift coefficient c_L composed of auxiliary curves for attached flow $c_{L.inv}$ and fully separated flow $c_{L.sep}$ using the degree of separation f_{stat} depending on the angle of attack α_A

Measurements to determine the factor a are needed. For one wind turbine it was found that $a = 4$ [11, 12]. This means for a simulation in the time domain

$$f_i(t) = f_{i-1} + \left(f_{stat} - f_{i-1}\right)\frac{\Delta t}{\tau_{fak}} \tag{6.35}$$

The lift coefficient can now be obtained by

$$c_L(\alpha_A, t) = f(t)c_{A.inv}(\alpha_A) + \left(1 - f(t)\right)c_{A.sep}(\alpha_A) \ . \tag{6.36}$$

6.9.3 *Method of singularities*

An older method based on the Laplace equation which describes the inviscid flow (potential flow theory) is the method of singularities. Objects in the flow, together, if possible, with the known fluid dynamic effects at the rotor, are described by vortices or by sources and sinks which are superimposed on the main flow. It is difficult to describe the wake flow behind the rotor. This is either assumed to be static, or its shape is determined by iteration. Despite the Laplace equation being only valid for inviscid flow, the method of singularities is from time to time still applied in combination with the airfoil characteristics, although good results are only achieved in the range of low profile drag.

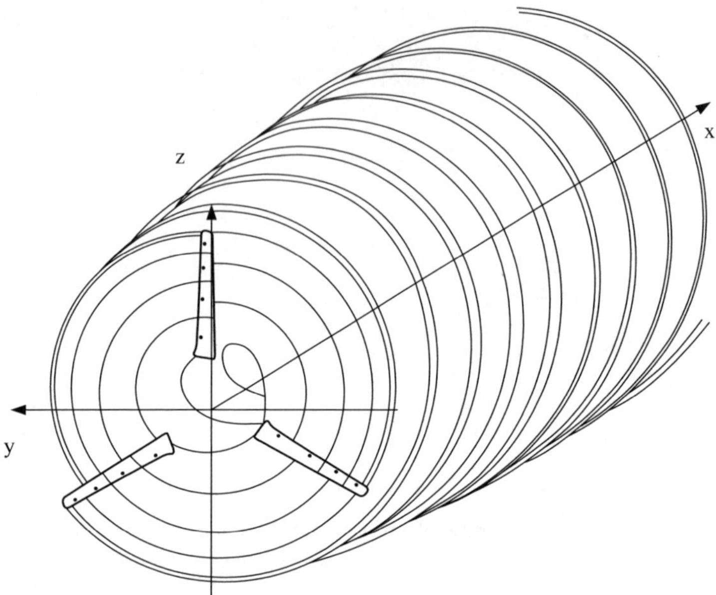

Fig. 6-26 Model for a free wake rotation

6.9.4 Computational fluid dynamics applied to wind turbines

Preliminary remark on computational fluid dynamics

The aim of numerical flow simulation by *Computational Fluid Dynamics* (CFD) is the calculation and visualization of flow fields with a spatial and temporal resolution. It provides a deeper understanding of the flow phenomena and saves cost-intensive experimental investigations which, for wind turbines and other large rotordynamic machines, are only possible using significantly smaller turbine models or the already erected prototype. By choosing an appropriate physical model of the flow it is hoped that the so-called "numerical wind tunnel" will represent the occurring flow effects better than the blade element momentum method. The aero-dynamic simulation of wind turbine components (like blades, spinner and nacelle housing as well as the ventilation and cooling system) helps to optimize the distri-bution of forces, to minimize flow losses, to improve acoustics, and so on. CFD is based on the Euler equations of motion for inviscid fluid, or the Navier-Stokes equations of motion if viscosity is considered [e.g. 13, 14]. Using a time-step re-solved simulation, CFD allows the investigation of unsteady processes. Moreover, it is possible to choose any section of the flow volume for visualisation thus facilitating the investigation of areas where access with measuring instruments is difficult.

If CFD is applied by an experienced expert using reasonable boundary conditions, the results for steady flow obtained in an acceptable period of time are quite close to reality, e.g. close to the design point of the rotor. The computational efforts required for unsteady flow, e.g. when stalling, are significant, and there is still intense research needed to improve the modelling [10].

The *basic equations of fluid dynamics* applied in CFD can only be solved as coupled partial differential equations and by iteration. Therefore, there is always a residual deviation from the exact solution.

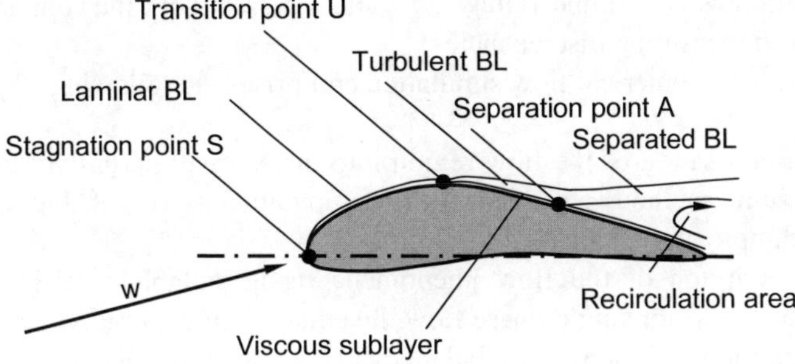

Fig. 6-27 Scheme of the development of the boundary layer (*BL*) and separation along the profile section

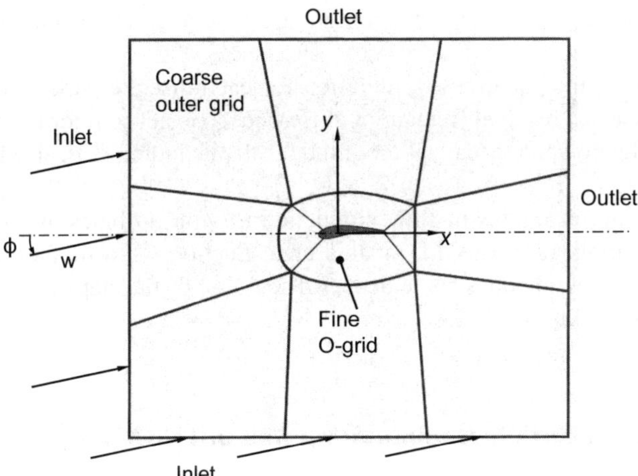

Fig. 6-28 Scheme of CFD-grid with a block-structured 2D-grid including an O-grid around a blade profile [15, modified]

The choice of flow model, which includes e.g. the approaches for modelling turbulence, has a significant influence on the result. For example, a simple turbulence model with two equations may be able to perform calculations quickly but can neither able to represent the development of the boundary layer (and the resulting force distribution on the blade surface) nor compute the transition from laminar to turbulent flow, Fig. 6-27. Complex flow models, in contrast, yield more exact results but with a significant increase in computing time.

Besides displaying the flow properties of interest (visualization as well as integral values), the *evaluation of CFD results* always includes a critical assessment of the chosen flow model and boundary conditions. If the CFD results are validated with existing measurements, both, the results of simulation and measurement should be viewed critically. It should be remembered that every measurement on real wind turbines has an uncertainty. In some research work the comparison with CFD revealed measuring discrepancies.

The procedure of numerical flow simulation comprises the following steps:

1. Discretisation of the flow region into many small partial volumes (finite volume method) which form the computational grid, cf. Fig. 6-28 as an example of the blocks of a 2D grid.
2. Description of the flow phenomena using suitable partial differential equations (for single-phase flow the equation of continuity, i.e. mass conservation and momentum balance) and, for example, the ideal gas law for consideration of the temperature.
3. Choice of a suitable simplified physical flow model for the solver and selection of the boundary conditions (e.g. inflow and outflow velocity and angle, as well as other conditions at the inlet, outlet and the walls, cf. Fig. 6-28)
4. Solving the equations by iteration for each discrete partial volume until the chosen residual (remaining difference to exact solution) is reached.
5. Evaluation and visualization of the results including critical assessment.

In the following, examples of flow simulation in wind turbines are presented to illustrate the application of CFD in this area. Section 6.9.6 of the 4th German edition of this book includes some deeper discussion of the steps in the CFD procedure mentioned above.

6.9.5 *Examples of CFD application to wind turbines*

The motivation for using CFD on wind turbines arises, at least in part, from the fact that in the real wind turbines the observation of flow phenomena by means of measurement techniques is difficult. Moreover, there is the constant attempt to describe ahead of time the flow effects relevant for the efficiency and structure which result from the interaction of blade, nacelle and tower and the wind farm effects as well.

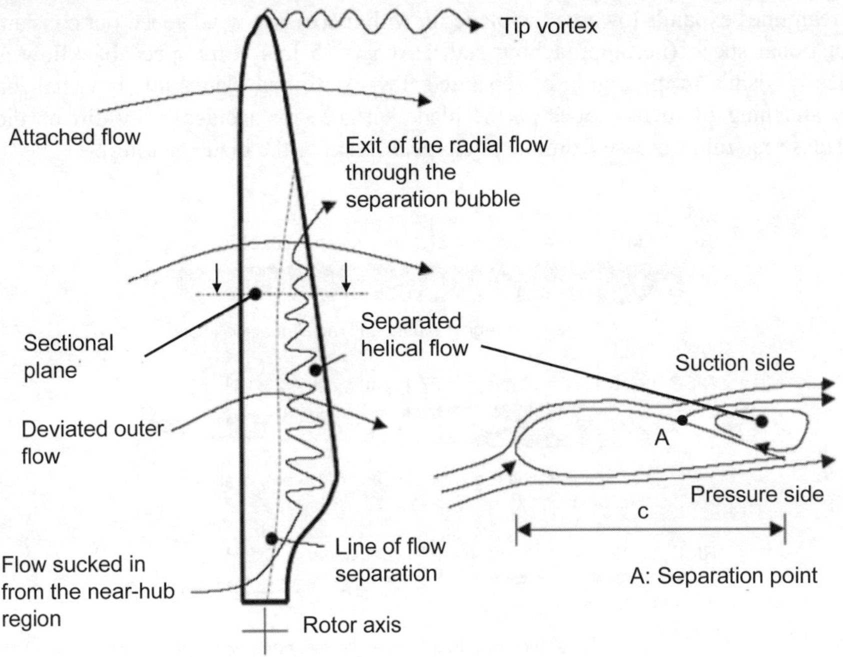

Fig. 6-29 Schematic representation of the flow in the separation area on the suction side of a blade at higher wind speed [15, modified]

Early experimental works on propellers revealed that´, due to *3D effects*, there are higher lift coefficients in the near-hub area compared to the blade tip, and that there are other lift curves than those measured for a 2D airfoil in the wind tunnel [16]. It was found by means of experiments [e.g. 17] and numerical simulation [18] that on the suction side, especially in the *near-hub region*, there are large areas of separated flow due to the high blade twist angle, the contour change from the circular blade root to the profiled blade and the unfavourable inflow. Fig. 6-29 provides the schematic representation of the effects described in the literature.

In the region of separated flow there are reduced flow velocities and therefore the centrifugal and Coriolis forces have a stronger effect. A helical flow forms in the separation bubble heading radially towards the blade tip.

In the European project VISCVIND, [10], various aspects of CFD were investigated and compared to measurement data. The aim was to compare the suitability of the CFD software of different institutions for flow simulations at wind turbines, to make improvements and to bundle the knowledge for further understanding of 3D effects.

Fig. 6-30 shows the shear stress lines (near-wall stream lines) on the blade surface obtained by a 3D simulation of the *flow around the rotating blade* LM 19.1. The region of flow separation on the suction side of the profile with the nearly radial streamlines expands towards the blade tip with increasing wind speed but constant rotational speed (i.e. approaching stall). At v_1 =15 m/s wind speed backflow is clearly visible in the region of separated flow. With real blades this is visualized by attaching filament probes on the blade surface - or accidently by dirt on the blades (e.g. oil or grease from the pitch mechanism or the blade bearing).

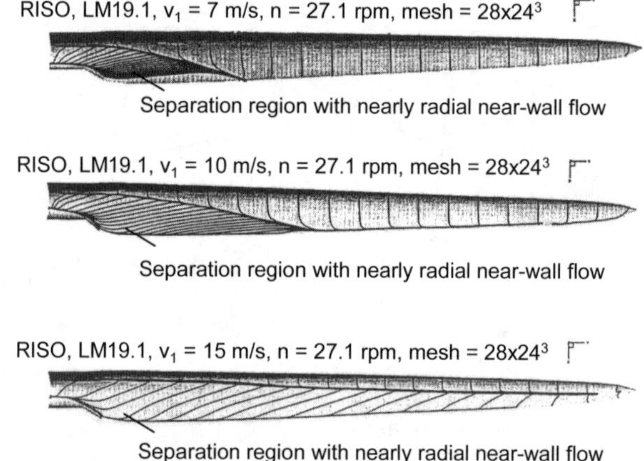

Fig. 6-30 Streamlines (wall shear stress lines) on the suction side surface of an LM-blade, from a CFD calculation at different wind speeds; rotor blade length 19.1 m [10]

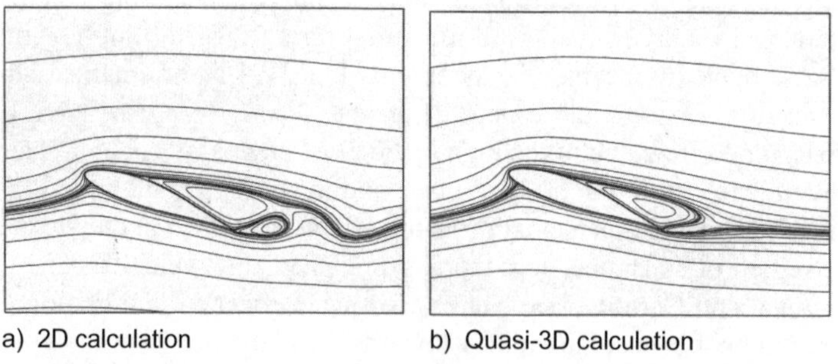

Fig. 6-31 Streamlines a) from a 2D-CFD calculation of the flow around an NACA 63_2-415 air foil at an angle of attack $\alpha_A = 20°$ and $Re = 1.5*10^6$, b) Reduction of the separation area in a quasi-3D calculation by introduction of radial force influences (centrifugal forces)

It is possible to transform 2D flow simulations around the airfoil into *quasi-3D simulations* by introducing additional terms for the radial centrifugal forces. For the Quasi-3D simulation, which considers centrifugal forces as 3D effect, Fig. 6-31 right, simulation of the flow around a rotating airfoil at an angle of attack of 20° results in a reduced size of the separation bubble on the suction side compared to the pure 2D simulation, Fig. 6-31 left. This improves the pressure distribution which reflects the increased lift values observed in the near-hub area of rotor blades.

These comparisons helped in the development of new correction equations, based on existing approaches [9, 19], for transforming wind tunnel measured 2D-profile characteristics to be then suitable as well for the tip and near-hub region of the rotating blade. These transformations consider not only the ratio c/r of the local chord length c to the local blade radius r (e.g. the values of 30 and 70% in Fig. 6-24), but also the blade twist angle β [10].

If simulations are computed for a *fully turbulent flow around airfoils* the drag coefficient is over-estimated in most cases, compared to measured profile characteristics (as there are only a few specially designed full turbulent profiles). The reasons for this may lie, among other, in the faster and stronger growth of the simulated fully turbulent boundary layer compared to the real boundary layer which is in its first section a laminar boundary layer, cf. Fig. 6-27. The effective profile shape results from the profile shape itself and the boundary layer around it, cf. Fig. 6-31. Therefore the fully turbulent approach may cause a displacement of the stagnation point (cf. Fig. 6-27) as well as a "deformation" of the effective profile.

Comparison studies of measured and CFD-simulated *wind turbine power curves* are performed at present mostly by simulating a rotor section of 120° around one blade, followed by a transformation of the result to the real wind turbine with three blades. In most cases the rotor-tower-nacelle interactions as well as the influence of the wind shear have been ignored so far to save computing time.

Such a comparison study was performed e.g. for the stall-controlled wind turbine Nordtank N500/41 [10] , Fig. 6-32. As long as the flow is attached, simpler turbulence models pretty much agree on the measurement as well. However, the SST model is obviously better for the more complex range with stall effect (strongly separated flow).

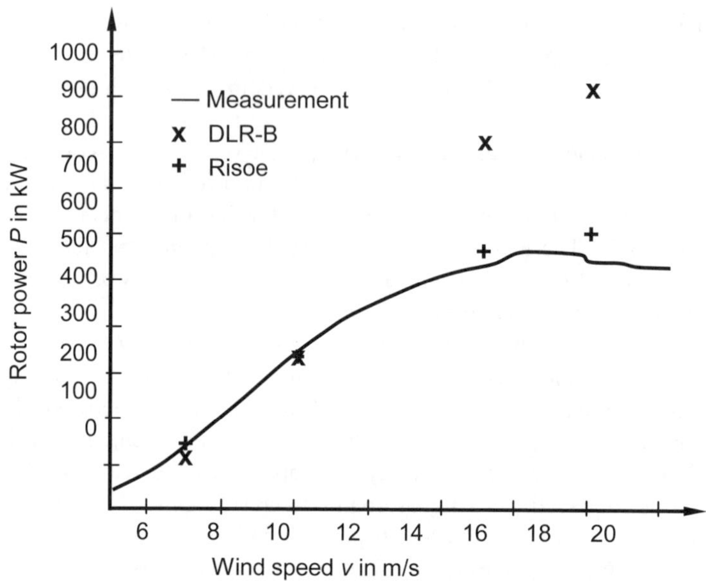

Fig. 6-32 Comparison of the measured and calculated rotor power of a Nordtank 500/541, DLR-B: Baldwin-Lomax turbulence model, Risoe: SST-k-ω turbulence model [7, modified]

The SST model uses different turbulence models for the calculations of the free flow and of the near-wall flow. For 2D simulations it is found too that the more simple turbulence models give fairly good results only for an angle of attack of less than 25°.

Fig. 6-33 shows stream lines from a CFD simulation for operation at the design point for selected regions of a rotor with a diameter of 60 m. The tip vortex (source of noise and tip losses) is clearly visible, as is the region of turbulent flow separation at the transition from cylindrical blade root to the profiled blade.

In the near-hub region one can also clearly see the centrifugal forces drive the air towards the outer radius, as was shown schematically in Fig. 6-29.

The accuracy of CFD computations is validated with experimental studies in wind tunnels. These investigations aim to improve of the modelling of transition (i.e. the point where the boundary layer changes from laminar to turbulent, Fig. 6-27), of the influence of roughness effects on the transition and flow separation, of changes of the profile contour and last but not least of the influence of the Reynolds number on the profile characteristics [20].

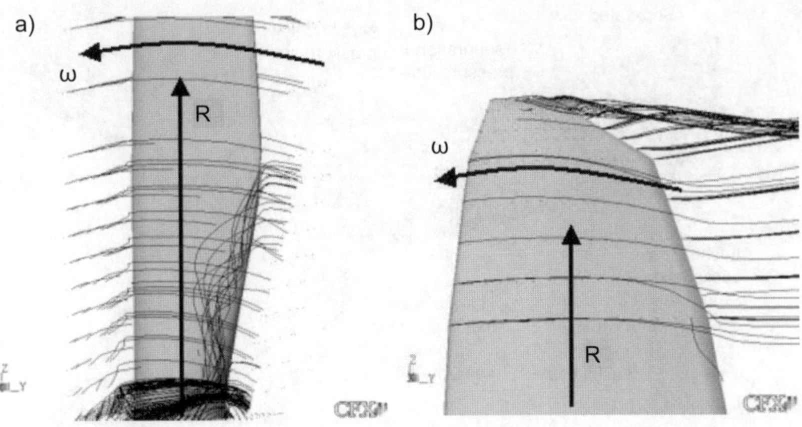

Fig. 6-33 Streamlines (a) at the blade root and (b) at the blade tip of a rotor with 60 m diameter [15]

The quality of the representation of transition determines how close the simulated flow is to the real flow and how accurately the aerodynamic forces at the blade can be determined. Roughness effects may occur when aerodynamic measures are applied (e.g. vortex generators or trip wires on stall-controlled blades) or as a result of dirt.

The influence of the Reynolds number is an issue since it varies greatly at one and the same rotor blade because the relative velocity w rises as the radius increases, while the local chord length c decreases at the same time. Research is also being done on the direct combination of CFD and blade element momentum method (e.g. in the actuator-disk model, [21])

The *flow around the entire wind turbine* is highly complex and has a three-dimensional structure (cf. Fig. 3-19). Besides the vertical wind profile of the inflow in the atmospheric surface boundary layer the transition of the flow around the airfoil from laminar to turbulent occurs, in reality, in the front part of the blade. There are also effects of flow separation at the trailing edge, especially under stall conditions. The vortices at the blade tip and the interaction of rotor, nacelle and tower represent further phenomena. Therefore, the size of the flow region which has to be discretised is an important parameter: If the modelled flow region is too small, e.g. the leeward development of the wake cannot be adequately computed (similar boundary effects occur in the wind tunnel if the model is too large relative to the wind tunnel cross section and length). If the simulated flow region is too large it is impossible to obtain results in an acceptable time frame. The local grid size is important as well. If the grid close to the walls (Fig. 6-28, inside the O-grid) is too coarse and not designed according to the chosen turbulence model, the boundary layer is not modelled accurately. A smooth airfoil surface then may get rough "numerical" steps which will greatly distort the flow.

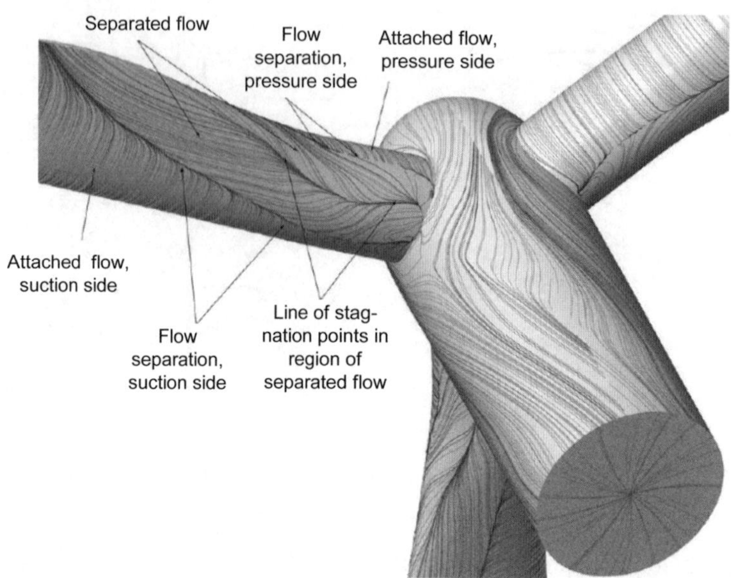

Fig. 6-34 Near-wall streamlines (shear stress lines) from a CFD simulation of rotor and nacelle; wind velocity $v_1 = 10.8$ m/s and $\lambda = 7$, [23]

In the area more distant from the walls, a coarse grid may reduce the computing time significantly without negative impact on the simulation results.

Due to increased computational capacities it is now possible to simulate larger flow regions and finish the computation within an acceptable time. Fig. 6-33 shows the near-wall streamlines on the rotor of a MW wind turbine with a simplified nacelle. A 120° section was computed and copied at the periodic boundaries. The inlet is 3 times the rotor diameter upwind, the outlet 8 times the rotor diameter downwind of the turbine in order to allow the formation of the inflow and wake. In the inner rotor region there are large regions of flow separation beginning at the cylindrical blade root; the interaction of blade and nacelle is also clearly visible.

In the inner region of the rotor some of the air fails to flow over the rotor blade but flows through the rotor in the region of the cylindrical blade root close to hub and nacelle housing without being utilized for energy extraction.

In this context, CFD simulations are very good for visualising the improved flow over the drop-shaped nacelle of the ENERCON wind turbines achieved by the modified aerodynamically optimized, yield increasing inner blade geometry (which is more like the Schmitz blade) [22].

CFD has also been used for the determination of the recommended minimum distances between the turbines for the wind farm siting, in dependence of the wind conditions. In one simulation study for different given mean wind speeds and ambient turbulence intensities the corresponding allowed minimum distances were determined, at which the increase of loads due to induced turbulence intensity (cf. Fig. 4-33) and the yield loss due to the reduced mean wind speed in the wake were still acceptable. Compared to the usually site-independent, fixed values of the recommended minimum distances between the wind turbines, these wind regime dependent minimum distance values allow for a more sophisticated wind farm siting.

Concluding the discussion of the application of numerical flow simulation to wind turbines, it should be mentioned that there is still research needed before CFD becomes a commonly used tool in wind turbine and wind farm design.

References

[1]	Althaus, D.: *Profilpolaren für den Modellflug (Profile characteristics for model flight)*, Neckar Verlag, Villingen Schwenningen, Germany 1985

[2]	Glauert, H.: *The Analysis of Experimental Results in the Windmill Brake and Vortex Ring States of an Airscrew*, Reports and Memoranda, No. 1023, 1926

[3]	Naumann, A.: *Luftschrauben im Bremsbereich (Air screws for braking)*, Yearbook of German aviation research 1940, pp. 1745

[4]	Prandtl, L.; Oertel, H. (Hrsg.): *Prandtl – A guide through fluid mechanics*, F. W. Durand: Aerodynamics, vol. 4, Vieweg, Braunschweig, Germany 1990

[5]	Eggleston, D. M.; Stoddard, F. S.: *Wind Turbine Engineering Design*, Van Nostrand Reinhold, New York, 1987

[6]	Fateev, E. M.: *Windmotors and Windpowerstations*, Moscow 1948

[7]	Ostowari, C., Naik, D.: *Post Stall Studies of Untwisted Varying Aspect Ratio Blades with NACA 44XX Series Airfoil Sections* - Part II, Wind Engineering, Vol 9 No. 3 1985, p. 149ff

[8]	Hoerner, S. F.: *Fluid Dynamic Drag*, self-published, Bricktown, N. J., 1965 und *Fluid Dynamic Lift*, published, Albuquerque, 1985

[9]	Snel, H. et al.: *Sectional Prediction of 3D-Effects for Stalled Flow on Rotating Blades and Comparison with Measurements*, Proceedings ECWEC 1993, Travemünde

[10]	Sörensen, J.N. (Editor): *VISCWIND – Viscous Effects on Wind Turbine Blades*, Department of Energy Engineering, Technical University of Denmark, 1999

[11]	Øye, S.: *Dynamic Stall – simulated as time lag of separation*, IEA 4th symposium on aerodynamics for wind turbines, Rome 1990

[12]	Schepers, J.G., Snel, H.: *Dynamic Inflow, Yawed Conditions and Partial Span Pitch Control*, ECN-C—95-056, Petten 1995

[13]	Ferziger, J.H., Peric, M.: *Computational Methods for Fluid Dynamics*, Springer Verlag, 1997

[14]	Siekmann, H. E., Thamsen, P.U.: *Strömungslehre – Grundlagen (Fluid dynamics - Basics)*, Springer Verlag, 2007

[15]	Hesse, J.: *Numerische Untersuchung der dreidimensionalen Rotorblattumströmung an Windkraftanlagen (Numerical investigation of the three-dimensional flow around rotor blades of wind turbines)*, Diploma thesis, Technical University of Berlin, Germany 2004

[16]	Himmelskamp, H.: *Profile investigation on a rotating airscrew*, Doktorarbeit (Ph. D. thesis), Göttingen, 1945

[17]	Milborrow, D.J., Ross, J.N.: *Airfoil characteristics of rotating blades*, IEA, LS-WECS, 12th expert meeting, Kopenhagen, Denmark 1984

[18]	Sörensen, J.: *Prediction of three-dimensional stall with a wind turbine blade using three level viscous-inviscid-interaction model*, Proc. European Windenergy Association Conference and Exhibition, Rom, 1986

[19]	Corten, G.P.: *Flow Separation on Wind Turbine Blades*, PhD thesis, University of Utrecht, Netherlands 2001

[20]   Timmer, W. A.,, Schaffarczyk, A. P.: *The effect of roughness on the performance of a 30% thick wind turbine airfoil at high Reynolds numbers*, Proc. EWEA Conference: *The science of making torque from wind*, Delft, Netherlands, 2004

[21]   Phillips, D., Schaffarczyk, A. P.: *Blade Element and Actuator Disk Models for a Shrouded Wind Turbine*, Proc. 15th IEA Expert Meeting on Aerodynamics of Wind Turbines, NTUA, Athenes, Greece, 26/27th Nov. 2001

[22]   ENERCON: Data sheet of the wind turbine E33, 2006

[23]   Krämer, T., Rauch, J.: *Dreidimensionale Numerische Simulation der Umströmung eines WKA-Rotors mit besonderem Fokus auf den Nabenbereich (Three- dimensional numerical simulation of the flow around a wind turbine rotor focusing on the near-hub region)*, Project work, Technical University of Berlin, Germany 2006

[24]   Paynter, R., Graham, M.: *Wind turbine blade surface pressure measurement in the field*, Proceedings BWEA 17, Warwick 1995

[25]   Bierbooms, W.A.A.M.: *A comparison between unsteady aerodynamic models*, Proceedings EWEC 1991, Amsterdam

[26]   Spera, D.A.: *Wind Turbine Technology*, ASME Press, New York 1994

[27]   Björck, A.: Dynamic *Stall and Three-dimensional Effects*, FFA TN 1995-31, Stockholm

[28]   Rasmussen, F. et al.: *Response of Stall Regulated Wind Turbines – Stall Induced Vibrations*, Risø-R-691(EN), Risø 1993

7 Scaling wind turbines and rules of similarity

Wind turbines are used in a variety of applications with very different performance requirements. In terms of power supply, a small holiday cottage requires electrical energy of approx.1.5 to 2 kW, a medium-sized restaurant approx. 75 kW with a base load of approx. 15 to 25 kW, and a large farm approx. 50 to 100 kW. In the first case, given a rated wind speed of $v = 9$ m/s, a turbine with a rotor diameter of 3.5 m would be sufficient. For the base load of the restaurant, a rotor diameter of 7 to 8 m would be required, whereas for the 50 to 100 kW, a turbine rotor with a diameter of 15 to 20 m is needed. Large turbines with a diameter of 80 to 100 m, can supply power of up to 3 MW.

It is thus useful to develop a family of wind turbines suitable for such diverse requirements. Cost can be reduced if, for a larger turbine, the experience gained from the smaller turbine can be used, so that development does not have to start again from scratch.

At times, it is also useful to test a scaled-down model in the wind tunnel in order to analyse the power, torque and thrust characteristics of a planned larger system and thus eliminate risks.

For both situations, the theory of similarity allows us to save calculation time and costs.

7.1 Application and limits of the theory of similarity

The wind turbine, which belongs to the group of turbo machines, is characterised by dimensionless curves for the power c_P, the thrust c_T and the torque moment c_M. These were introduced in the chapters on dimensioning and performance characteristics calculation. If these parameters are to have the same value for two differently-sized wind turbines, the flow conditions must be the same. Using the theory of similarity, this is achieved in the following way:

a) Maintaining the tip speed ratio, i.e. the ratio of circumferential speed at the blade tip and wind velocity upstream the rotor:

$$\lambda = \Omega \frac{R}{v_1},$$ (7.1)

b) Maintaining the blade profile, the number of blades and the materials used,

c) Making proportional adjustments to all dimensions (radius, profile chord, spar size).

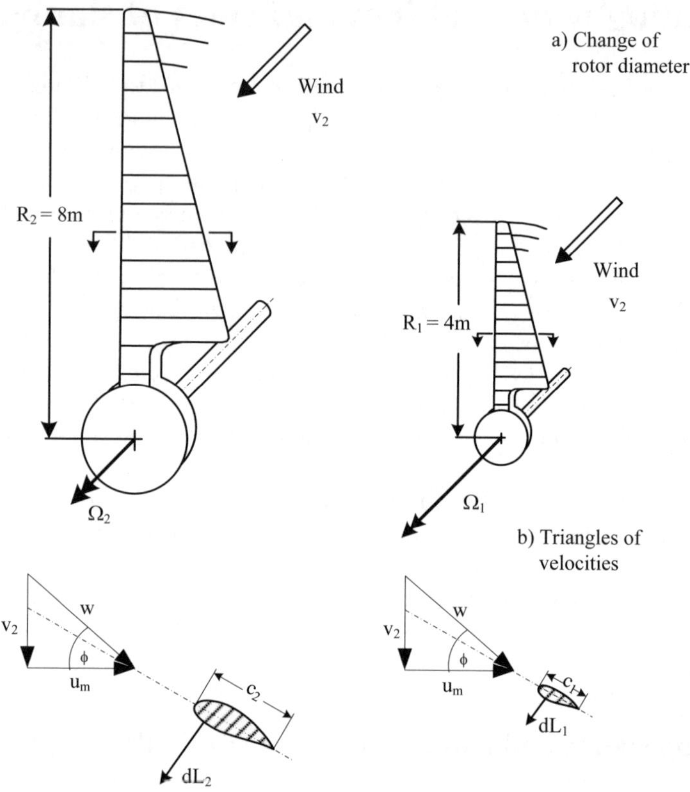

Fig. 7-1 Scaling according to the rules of similarity, flow conditions in the corresponding sections of a small and scaled-up turbine (radius and chord length doubled)

Fig. 7.1a shows two rotors, the second of which has twice the blade length of the first and was scaled up based on the proportions of the first rotor in line with the above requirements. Fig. 7.1b shows the respective flow conditions at two related profile cross sections.

As it has the same tip speed ratio, the rotational (or angular) speed Ω must be halved if the radius is doubled. The circumferential velocities at the related profile sections thus remain constant, as do the triangles of velocity and the relative velocity angles. As the same profile is used, the lift coefficient c_L (α_A) and the drag coefficient c_D (α_A) are preserved, unless the influence of the Reynolds number $Re = w\,c/v$ becomes relevant. Therefore, both rotors operate with the same power coefficient $c_P = c_P$ (λ), thrust coefficient $c_T = c_T$ (λ), and torque moment coefficient ($c_M = c_M$ (λ)).

Based on these considerations we can determine what the effects are of a change in rotor diameter or blade length on the performance characteristics of the rotor, on the forces at the blade, on the stress at the blade root and on those values that affect the dynamics, provided that the rules of similarity are applied.

Table 7.1 lists the outcomes of this process. Before dealing with how this table was derived, the results will be discussed and interpreted.

For a constant tip speed ratio, the *rotational speed* must change in inverse proportion to the change in blade length. Power and thrust increase with the square, the driving torque with the cube of the change in length, as can be seen from the equations.

The *aerodynamic forces* increase with the square. For thrust and tangential force, this can be derived directly from total thrust or driving torque of the rotor. Naturally, it also applies to the lift and drag forces (Fig. 7.1b). Analogous to the aerodynamic forces, the *centrifugal forces* increase with the square of the change in length, whereas the *weight*, due to the volume, increases with the cube.

For the *stresses* at the blade root it follows that the stress from the aerodynamic forces and the centrifugal forces is independent of the change in blade length. The stress from the weight, however, increases proportionally to the blade length. Hence, the limiting factor for further blade extension (up-scaling) is not the aerodynamic and centrifugal forces but the weight.

The dynamic values respond to a change in blade length in a similar way as the aerodynamic forces and the corresponding stress. The *natural frequencies* of the blades decrease in proportion to the increase of the blade length, but the same holds true - as was mentioned above - for the rotational speed. As the excitations e.g. from tower shadow or wind profile, are always proportional to the rotational speed, the *frequency ratio* of excitation frequency and natural frequency remains constant, independent of a change in blade length. A wind turbine operating free of resonance will also remain free of resonance when scaled up in accordance with the rules of similarity.

A tested and functioning turbine can be used as a prototype for developing a series of turbines following the simple rules (a) through (c), provided that:

  - the weight does not become the decisive factor (which it is for large wind turbines) and
  - the Reynolds number does not fall below the critical Reynolds number of approx. $Re_{crit}=200{,}000$, e.g. in battery chargers (cf. section 7.7.).

Table 7.1 Impacts of scaling a wind turbine of radius R_1 to R_2

	Relative	**Absolute**	**Proportional**
Power	$P_2/P_1 = R_2^2/R_1^2$	$P = (\rho/2)c_P(\lambda)v^3 R^2 \pi$	$\sim R^2$
Torque	$M_2/M_1 = R_2^3/R_1^3$	$M = (\rho/2)c_M(\lambda)v^2 R^3 \pi$	$\sim R^3$
Thrust	$T_2/T_1 = R_2^2/R_1^2$	$T = (\rho/2)c_T(\lambda)v^2 R^2 \pi$	$\sim R^2$
Angular speed	$\Omega_1/\Omega_2 = R_2/R_1$		$\sim R^{-1}$
Weight	$W_2/W_1 = R_2^3/R_1^3$		$\sim R^3$
Aerodynamic forces	$F_{a2}/F_{a1} = R_2^2/R_1^2$		$\sim R^2$
Centrifugal forces	$F_{c2}/F_{c1} = R_2^2/R_1^2$		$\sim R^2$

Stress due to	**Relative**	**Proportional**
Weight	$\sigma_{W2}/\sigma_{W1} = R_2/R_1$	$\sim R^1\,!$
Centrifugal forces	$\sigma_{Fc2}/\sigma_{Fc1} = 1$	$\sim R^0$
Aerodynamic forces	$\sigma_{Fa2}/\sigma_{Fa1} = 1$	$\sim R^0$

Dynamics:		
Natural frequencies	$\omega_{R2}/\omega_{R1} = R_1/R_2$	$\sim R^{-1}$
Frequency ratio	Ω/ω_n	$\sim R^0$
Damping ratio D		$\sim R^0$

It must now be proved that proportional relations for stresses at the blade root and for the dynamic stresses stated in the Table 7.1 are valid. For the stresses, the proof will be limited, in the following, to tensile and bending stress. The statements are also valid for the shear stresses.

7.2 Bending stress in the blade root from aerodynamic forces

The bending moments $M_{B,br}$ in the blade root caused by the aerodynamic forces, are produced in wind direction by the thrust T_{ax} and in circumferential direction by the circumferential forces resulting from the driving torque M divided by the number of blades z (Fig. 7-2).

$$T_{ax} = \frac{T}{z} = \frac{1}{z} c_T(\lambda) \frac{\rho}{2} v_1^2 \pi R^2 \tag{7.2}$$

$$M_{B,br} = \frac{M}{z} = \frac{1}{z} c_M(\lambda) \frac{\rho}{2} v_1^2 \pi R^3 \tag{7.3}$$

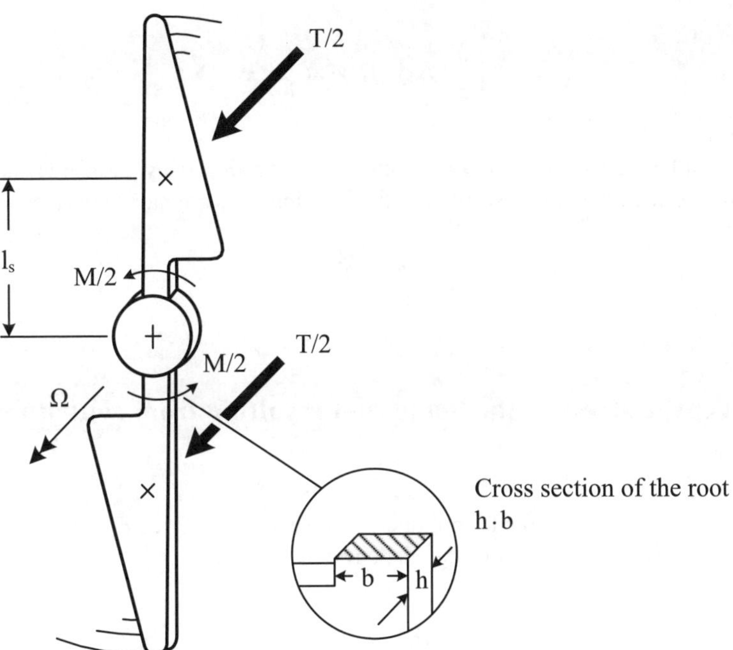

Fig. 7-2 Bending moment at the blade root of a two-bladed rotor due to aerodynamic forces

For a rectangular cross section of the blade root, the bending stress can be obtained by using the section modules Z_{ax} and Z_{circ}:

$$\sigma_{B,ax} = \frac{T_{ax}\, l_S}{Z_{ax}} = \frac{1}{z}\; \frac{c_T(\lambda)\dfrac{\rho}{2}\,\pi R^2\, l_S\, v_1^2}{\dfrac{b h^2}{6}}$$

$$= \left(\frac{3}{z}\; c_T(\lambda)\rho v_1^2\, \pi\right)\frac{R^2\, l_S}{b h^2} \tag{7.4}$$

$$\sigma_{B,circ} = \frac{M_{B,br}}{Z_{circ}} = \frac{1}{z}\; \frac{c_M(\lambda)\dfrac{\rho}{2}\,\pi R^3\, v_1^2}{\dfrac{h b^2}{6}}$$

$$= \left(\frac{3}{z}\; c_M(\lambda)\rho v_1^2\, \pi\right)\frac{R^3}{h b^2} \tag{7.5}$$

Now, if all lengths (R, b, h, l_s) are enlarged or reduced by the same factor, in accordance with the law of similarity, the bending stress remains constant!

$$\sigma_B \sim R^0 \tag{7.6}$$

7.3 Tensile stress in the blade root resulting from centrifugal forces

The tensile stress in the blade root due to the centrifugal force F_c, can be obtained by the cross sectional area A of the blade root:

$$\sigma_{Fc} = \frac{F_c}{A} \tag{7.7}$$

Fig. 7-3 shows the change in centrifugal forces for a blade extension according to the rules of similarity.

For a given distance l between rotor axis and the blade's centre of gravity, the centrifugal forces are:

$$F_{c1} = m_1 \, l_1 \, \Omega_1^2 \tag{7.8}$$

$$F_{c2} = m_2 \, l_2 \, \Omega_2^2 \tag{7.9}$$

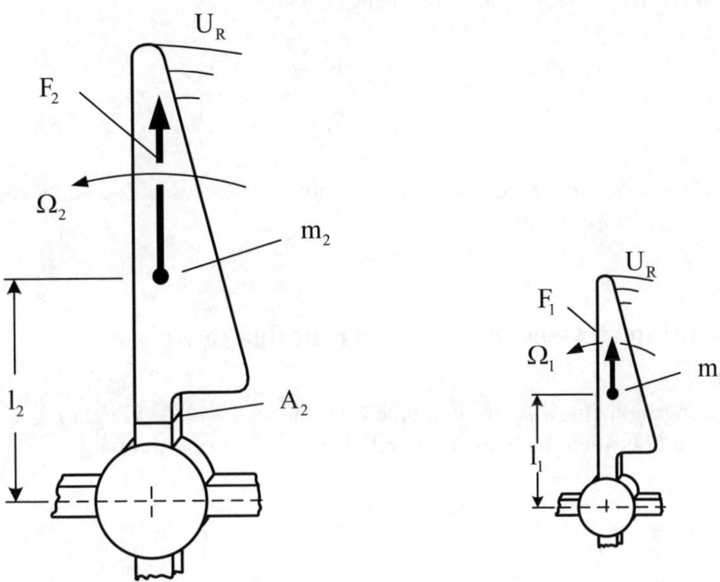

Fig. 7-3 Centrifugal forces at the blade root when changing the rotor diameter

Applying the rules of similarity yields:

$$m_2 = m_1 \left(R_2/R_1\right)^3, \tag{7.10}$$

$$l_2 = l_1 \left(R_2/R_1\right) \text{ and} \tag{7.11}$$

$$\Omega_2^2 = \Omega_1^2 \left(R_1/R_2\right)^2. \tag{7.12}$$

We thus obtain:

$$F_{c2} = m_1 \, l_1 \, \Omega_1^2 \left(R_2/R_1\right)^2 \tag{7.13}$$

The cross sectional area of the blade root is:

$$A_2 = A_1 \left(R_2/R_1\right)^2 \tag{7.14}$$

If this term is inserted into equation (7.7), we obtain the following equation for the tensile stress in the blade root of the enlarged blade:

$$\sigma_{Fc2} = \frac{F_{c2}}{A_2} = \frac{m_1 \cdot l_1 \cdot \Omega_1^2 \cdot \left(R_2/R_1\right)^2}{A_1 \cdot \left(R_2/R_1\right)^2} = \frac{F_{c1}}{A_1} = \sigma_{Fc1} \tag{7.15}$$

The tensile stress in the blade root due to centrifugal forces is then independent of a change in the rotor diameter!

7.4 Bending stresses in the blade root due to weight

The bending moment $M_{B,W}$ in the blade root due to the blade weight mg acting at the distance l between the blade root and blade's centre of gravity is:

$$M_{B,W} = m \ g \cdot l \ . \tag{7.16}$$

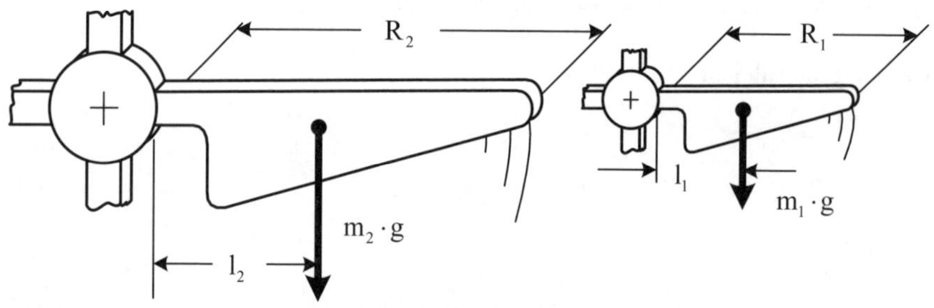

Fig. 7-4 Bending moment at the blade root due to weight influence when changing the rotor diameter

Fig. 7-4 illustrates the impact of weight when comparing two blades of different length. From the rules of similarity we obtain again for the scaling factor (R_2/R_1):

$$m_2 = m_1 \left(R_2/R_1\right)^3 \text{ and} \tag{7.17}$$

$$l_2 = l_1 \left(R_2 / R_1 \right). \tag{7.18}$$

This gives

$$M_{\mathrm{B},2} = M_{\mathrm{B},1} \left(R_2 / R_1 \right)^4 . \tag{7.19}$$

The section modulus Z for a rectangular blade root, according to Fig. 7-2, is:

$$Z_2 = \frac{h_1 b_1^2}{6} \left(R_2 / R_1 \right)^3 = Z_1 \left(R_2 / R_1 \right)^3 \tag{7.20}$$

Thus, the ratio of the bending stresses due to the weight is:

$$\sigma_{B,2} = \frac{M_{\mathrm{B},2}}{Z_2} = \frac{M_{\mathrm{B},1}}{Z_1} \left(R_2 / R_1 \right) = \sigma_{B,1} \left(R_2 / R_1 \right) \tag{7.21}$$

This means that the bending stress in the blade root due to the weight rises linearly with the increase in the rotor diameter. This explains why the wind turbines cannot be enlarged ad infinitum. It also limits the application of the rules of similarity for extremely large turbines, where weight is the limiting factor for the turbine size. Light weight design then becomes extremely important.

7.5 Change in the natural frequencies of the blade and in the frequency ratios

To illustrate how the natural frequencies of the blades change if they are enlarged or reduced in accordance with the rules of similarity, we assume (for the sake of simplicity) an untwisted, homogeneous rectangular blade with a rigid attachment to the hub (Fig. 7-5).

According to [4] the natural frequencies of this vibrating beam are as follows, if we ignore the centrifugal stiffening and take the eigenvalues Λ_n from table 7.2:

$$\omega_n = \frac{\Lambda_n^2}{R^2} \cdot \sqrt{\frac{E \cdot I}{\mu}} \tag{7.22}$$

Table 7.2 Eigenvalues Λ_n of the fixed end beam (cantilever)

n	1	2	3	....	n (large values)
Λ_n	1.8571	4.6941	7.8531	....	$\approx (2 \cdot n - 1) \cdot \pi/2$

For the natural frequencies in direction of the flap deflection (low stiffness), we obtain the nth natural frequency ω_n, using the mass per (blade) length

$$\mu = \rho \, b \, h \tag{7.23}$$

and the geometrical moment of inertia,

$$I = \frac{b h^3}{12} \text{, thus} \tag{7.24}$$

$$\omega_n = \frac{\Lambda_n^2}{R^2} \sqrt{\frac{E b h^3}{\rho 12 b h}} = \frac{\Lambda_n^2 h}{R^2} \sqrt{\frac{E}{12 \rho}} \; . \tag{7.25}$$

For a change in rotor diameter, this results in:

$$\frac{\omega_{n,2}}{\omega_{n,1}} = \frac{h_2 \cdot R_1^2}{h_1 \cdot R_2^2} = \frac{R_1}{R_2} \tag{7.26}$$

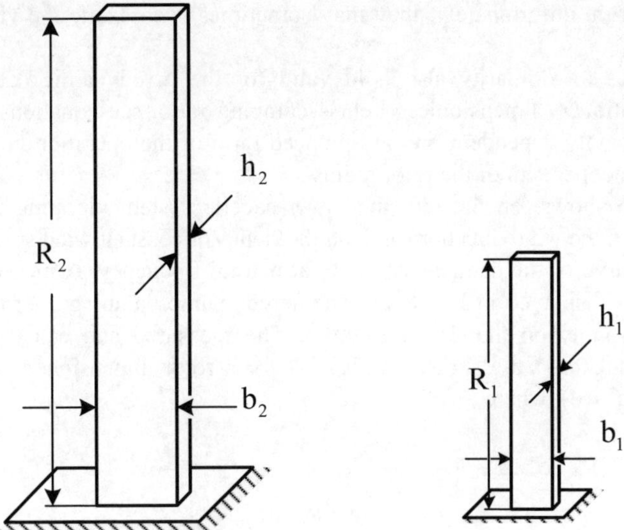

Fig. 7-5 Untwisted, homogeneous rectangular blade; influence of the change of the rotor diameter on the flapwise natural frequency

The natural frequencies in direction of edgewise vibration are obtained in the same way using the geometric moment of inertia for the stiff direction, i.e. b and h are interchanged. It follows that the natural frequencies change in inverse proportion to the change in rotor diameter. This law for the natural frequencies holds true even for tapered and twisted blades.

The excitation frequencies of the blade due to the tower shadow effect, mass unbalance, aerodynamic unbalance or the height-related wind profile are all proportional to the rotational speed Ω. But as Ω, like the natural frequencies, is in inverse proportion to the change in rotor diameter, the frequency ratio η_n between excitation frequencies and natural frequencies remains constant whether the rotor is scaled up or down!

$$\eta_n = \frac{\Omega}{\omega_n} \sim R^0 \tag{7.27}$$

7.6 Aerodynamic damping

Since the structural damping at a wind turbine is small, the damping of vibrations comes, on the one hand, from the generator, which can only be manipulated in the rotational degree of freedom, and on the other, from the aerodynamic forces at the rotor - though, unfortunately, they may sometimes even excite the vibrations [2, 6].

The rules of similarity also hold valid for the aerodynamic damping: The damping ratio D (dimensionless Lehr's damping ratio, see equations (7.30) and (8.8.)) is heavily dependent on the tip speed ratio of the operation points (partial load!) but independent of the rotor radius.

Fig. 7-6 shows, on the left, the tower-nacelle system vibrating in the axial direction (i.e. the wind direction) and, on the right, for constant wind speed v_{Wind}, the decaying curve of the damped vibration at natural frequency. A measured decaying curve is displayed in Fig. 8-11. The aerodynamic damping coefficient $d_{11}(\lambda)$ depends strongly on the tip speed ratio. The measured and calculated dimensionless values $d_{11}*$ are shown in Fig. 7-7 for a rotor dimensioned according to Schmitz ($\lambda_{\mathrm{D}} \approx 6$) where, [6],

$$d_{11}(\lambda) = d_{11}* \ R^2 \ v_{\mathrm{Wind}} \ \frac{\rho}{2} \ . \tag{7.28}$$

The damping force is obtained by multiplying of d_{11} with the vibration velocity $\dot{u}$ which becomes obvious when comparing the dimensions.

The decay factor δ, describing the envelope curve of the decaying natural vibration, is given by [3, 4]

$$\delta = d_{11} / (2 \ m) \ . \tag{7.29}$$

The dimensionless damping ratio D is obtained from the ratio of decay factor δ and the natural frequency ω_0 (cf. chapter 8, dynamics) of the system without damping

$$D(\lambda) = d_{11}(\lambda) / (2 \ m \ \omega_0) \tag{7.30}$$

According to the rules of similarity the following relations are valid

$$
\begin{array}{lll}
\text{mass} & m \sim R^3 \\
\text{natural frequency} & \omega \sim 1/R & \text{and,} \\
\text{as explained above,} & d_{11} \sim R^2.
\end{array}
$$

Therefore, the damping ratio D of the axial tower-nacelle vibration is independent of the radius of the wind turbine.

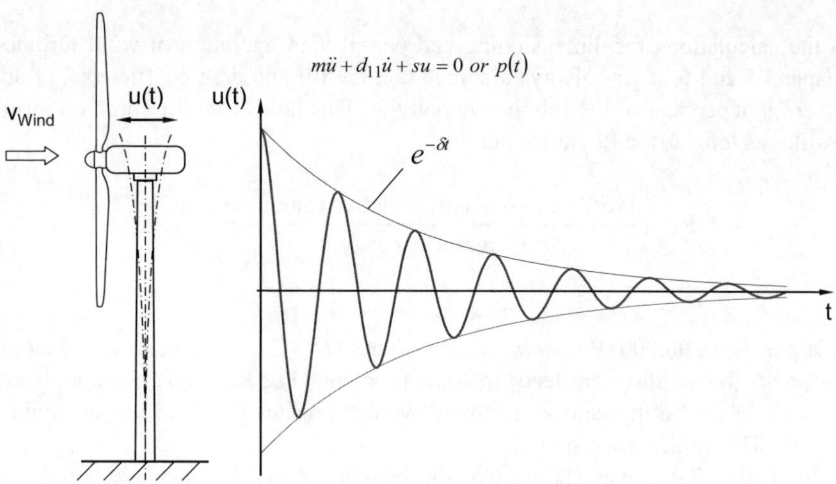

Fig. 7-6 Axial deflection u(t) of the tower-nacelle vibration; fading out due to the decay factor δ of the damping

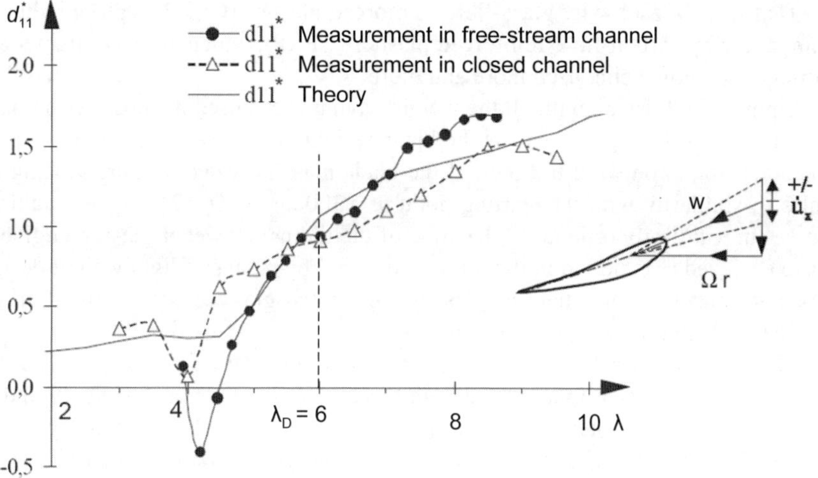

Fig. 7-7 Dimensionless damping coefficient $d_{11}^{\ast}(\lambda)$ for a wind turbine in axial direction; comparison of the theoretical curve and measurements in two different wind tunnels [from [2], chapter 25, lit. 10]

There is a significant loss of damping when the flow separates, i.e. $\lambda < 5$.

7.7 Limitations of up-scaling - how large can wind turbines be?

In the calculations for dimensioning and partial load operation of wind turbines, chapters 5 and 6, it was always assumed that the lift and drag coefficients, c_L and c_D, are independent of the relative velocity w. This holds true for common airfoil profiles as long as the Reynolds number

$$\text{Re} = \frac{\text{Profile chord length } \cdot \text{ relative velocity}}{\text{kinematic viscosity}} = \frac{c\,w}{\upsilon}$$

is larger than 200,000. For *small wind turbi*nes ($D < 5$ m) and in *wind tunnel experiments* this requirement leads to some problems. For Reynolds numbers below Re < 50,000, even thin and sharp low-Reynolds profiles [5, 7] no longer produce any lift. This limits *down* sizing.

In section 7.4 it was shown that the bending stress due to blade weight increases proportionally to the blade radius when scaling up according to the rules of similarity - and thus creates an *upper* limit for wind turbine construction. This limit is higher

-       the higher the fatigue strength against alternating bending, σ_{BW} , and
-       the smaller the density ρ of the material used.

That is the reason why glass-fibre reinforced plastics (GFRP) replaced aluminium, and why carbon-fibre reinforced plastics (CFRP), which are even lighter and stronger, are now being used more and more.

Light-weight design with all the weight-saving tricks used in construction plays an important role. Fig. 7-8 shows the blade weight versus the rotor radius, or rotor diameter, for some wind turbines. If the blade mass is scaled up purely using the rules of similarity with the starting point $m = 300$ kg at $D = 21$ m, we obtain the curve (m $\sim R^3$) proportional to the cube of the radius. However, design engineers have achieved a reduction in the rotor blade mass by means of light weight design: For both material combinations glass/polyester and glass/epoxy the fitted curves for the blade mass are proportional to ($R^{2.2}$).

Moreover, practical reasons prevent the rigid application of the scaling rules. Wind turbines situated inland need high towers ($H/D > 1$) to escape the boundary layer close to the ground. Wind turbines at the coast need only smaller towers ($H/D < 1$). For small wind turbines or battery chargers, the ratio is often $H/D \geq 2$.

The laws of similarity allow quickly to find the shape of a future design. And they reveal where problems are to be expected.

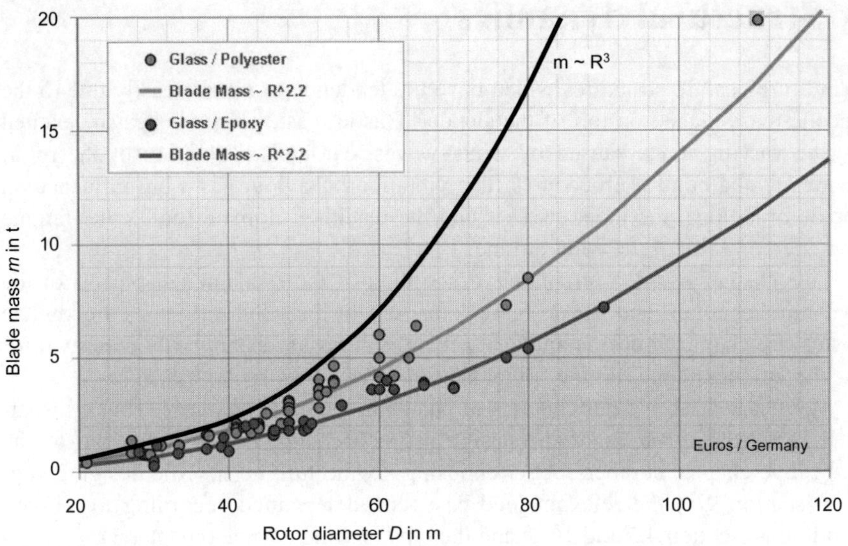

Fig. 7-8 Increase of the blade mass with the diameter; real mass (interpolation $m \sim R^{2.2}$) and theoretical curve ($m \sim R^3$)

References

[1]   Wiedemann, J.: *Leichtbau (Light weight construction)*, Band II, Kapitel 4, Springer Verlag, Berlin 1989

[2]   Nordmann, R, Gasch, R., Pfützner, H.: *Rotordynamik (Rotor dynamics)* (2nd edition), Springer Verlag Berlin, 2002

[3]   Magnus, K., Popp, A.: *Schwingungen (5. Auflage) (Vibrations, 5th edition)*, Teubner Verlag, Stuttgart, 1997

[4]   Gasch, R., Knothe, K.: *Strukturdynamik I und II (Structural dynamics I and II)*, Springer Verlag Berlin, 1987/89

[5]   Althaus, D.: *Profilpolaren für den Modellflug (Profile polar characteristics for aeromodelling)*, Neckar-Verlag, VS-Villingen, 1980

[6]   Kaiser, K.: *Luftkraftverursachte Steifigkeits- und Dämpfungsmatrizen in Windturbinen (Aerodynamic damping and stiffness matrices of wind turbines)*, VDI-Fortschritt-Bericht, Reihe 11 (VDI progress report, series 11) , Nr. 294, Düsseldorf 2000

[7]   Althaus, D.: *Niedriggeschwindigkeitsprofile(Low-Reynolds airfoil profiles)*, Vieweg Verlag, Wiesbaden 1996

8 Structural dynamics

Wind turbines are structures which have the tendency to vibrate easily due to the nacelle mass placed on top of the slender, elastic mast. They are heavily excited by the varying loads caused by wind, waves, earthquake, rotation of the rotor, switching and control procedures. The vibration behaviour has a big influence on the deformations, the inner stresses and the resulting ultimate limit state, fatigue and operating life of the wind turbine.

This chapter offers a qualitative description of the fundamental aspects of the dynamic behaviour and the design calculations of the wind turbine as a complete system. In chapter 9 design and analysis procedures are exemplarily presented for the three typical wind turbine components tower, hub and rotor blade.

Many issues from other chapters of this book are drawn on here. This is visualized in the presentation of the design procedure, Fig. 8-1, by referring to the respective chapter numbers. After choosing a wind turbine specific design guideline (section 9.1) the relevant load cases are determined according to the site conditions (section 4.5 and 16.1) and the type of wind turbine (chapter 3).

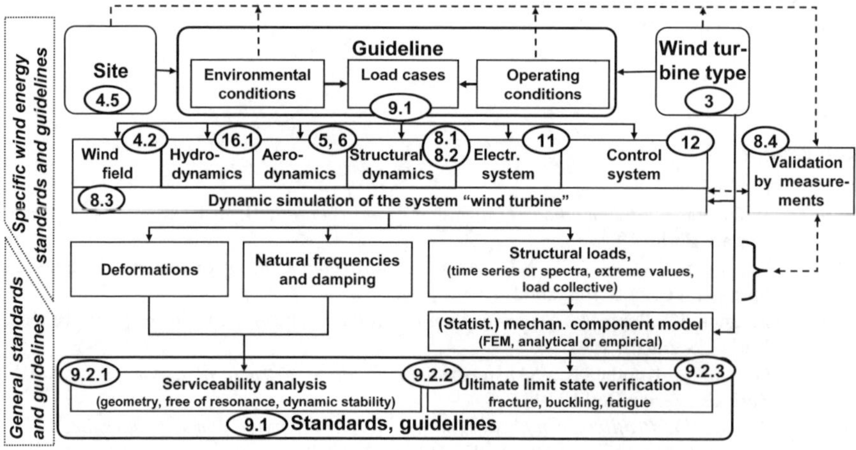

Fig. 8-1 Design procedure for wind turbines; determination of design load cases from site, type of wind turbine and applied guidelines, dynamic simulation and final scope of verification (numbers refer to corresponding chapters of this book)

The internal forces and moments at different component sections as well as the global deformations and natural frequencies are then determined numerically through simulation of the structural dynamics for the complete wind turbine system (section 8.3). Typically, this done by simultaneously calculating the aspects of the wind field, the aerodynamics, hydrodynamics and structural dynamics as well as the electrical system and the supervisory and control system in the time domain in one single model. In a second step, detailed models are used to determine the stresses on the component level according to general technical standards (section 9.1) in order to perform the various analyses (section 9.2). The operating and performance behaviour as well as the modelling of the structural dynamics are validated by measurements on erected wind turbines (section 8.4)

Sections 8.3 and 8.4 of this chapter will discuss the simulation and its validation by measurements. Prior to that, the following two sections will present the basic aspects of dynamic excitations and the phenomenology of the vibrational behaviour of the wind turbine system.

8.1 Dynamic excitations

Types of vibration excitation

For the design and the fatigue analysis of most wind turbine components, the dynamic behaviour in the frequency range of up to approx. 10 Hz is relevant. Phenomena of higher frequencies occur, for example, in the electrical system, the drive train and the gearbox (teeth contact frequencies, shaft and casing elasticity, bearing clearance). The description of the behaviour often relates to two time ranges of different length. Short-term events, so-called transient phenomena, such as switching procedures or impulse effects, are investigated using short time series of up to approximately 1 minute. By contrast, the operating characteristics for stochastic excitation are analysed in time series between 10 minutes and 1 hour in which the statistic environmental conditions, e.g. the mean wind speed and turbulence or the swell, may be considered as constant, cf. Fig. 4-23.

The excitations, i.e. the external loads acting on the wind turbine, may be distinguished according to their time history, cf. table 8.1:
- constant (quasi-steady)
- cyclic (periodic)
- stochastic (random)
- short-time (transient)

Moreover, it is helpful to distinguish the exciting forces according to their physical origin. There are mass forces, inertia forces and gravitational forces due to the operating conditions as well as aerodynamic and hydrodynamic loads from the environmental conditions. In particular, for transient manoeuvres and malfunction there are additional relevant electrical and electromechanical forces, frictional and braking forces, hydraulic or electro-mechanic actuating forces. An overview of the most important exciting forces is given in table 8.1.

Table 8.1 Classification of some exemplary exciting forces according to time history and origin

Origin / Time history	Type of force	Source	Operating condition
Constant (quasi-steady)	Gravity force, Centrifugal force, Mean thrust	Weight, Rotor revolution, Mean wind	Normal operation
Cyclic (periodic)	Mass unbalance, Aerodynamic forces	Unbalance, Tower dam, Oblique flow, Blade passage	Normal operation, Malfunction
Stochastic (random)	Aerodynamic and hydrodynamic forces	Turbulence of the wind, Swell, Earthquake	Normal operation
Short-time (transient)	Frictional and braking forces, Aerodynamic forces	Shut down of the wind turbine, Yawing of the nacelle	Manoeuvre, Malfunction, Extreme conditions

8.1.1 Mass, inertia and gravitational forces

Gravitational forces and excitations due to weight

For the *non-rotating* components of the wind turbine, e.g. nacelle, tower, foundation, there are constant steady loads due to the dead weight. However, the internal forces and moments in the tower sections should be determined for the deformed system using the 2nd order theory (buckling) because the horizontal displacement of the nacelle induces additional bending moments.

The dead weight of rotating components, e.g. the blades, has different effects. Here, the direction of the gravitational force relative to the blade is changing permanently, resulting in strong alternating loads with the frequency of the rotational speed (1Ω).

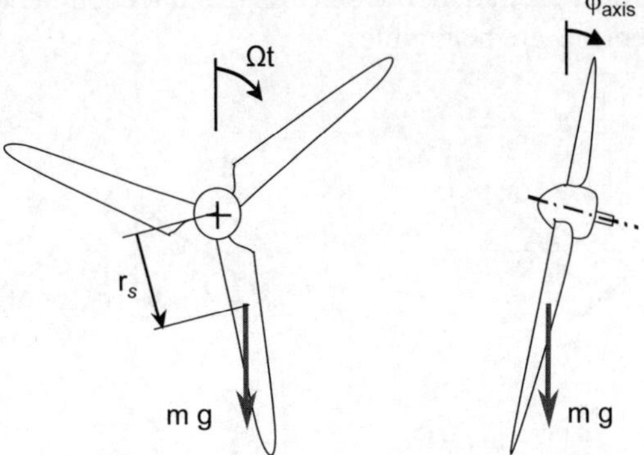

Fig. 8-2 Alternating loads due to blade weight in edgewise and flapwise direction (r_s = center of gravity radius)

The highest alternating bending moment at the blade root due to weight occurs in the edgewise direction, Fig. 8-2

$$M_{\text{edgewise}} \cong m \; g \; r_{\text{S}} \sin \Omega t \quad . \tag{8.1}$$

The inclination of the rotor shaft by the angle φ_{axis} increases the distance between blade tip and tower and is often chosen between 4° and 5°. Thus, the blade weight also causes small 1Ω-excitations of the flapwise bending moment at the blade root

$$M_{\text{flapwise}} = m \; g \; r_{\text{S}} \sin \varphi_{\text{axis}} \sin \Omega t \quad . \tag{8.2}$$

Centrifugal forces and mass unbalance

For constant rotational speed the centrifugal forces F cause only steady tensional loading at the rotating blade . This is also the case if one blade is slightly heavier than the others having an excess mass Δm, Fig. 8-3.

Relative to the rotational axis of the rotor (i.e. in the inertial system) the star of the three centrifugal forces F is basically balanced. Only the excess mass Δm causes a rotating excitation force

$$\Delta F = \Delta m \; r_{\text{s}} \; \Omega^2 \quad . \tag{8.3}$$

Its horizontal component excites lateral vibrations of nacelle and tower,

$$F_{\text{hor}} = \Delta m \; r_{\text{s}} \; \Omega^2 \sin \Omega t \quad . \tag{8.4}$$

There is also a vertical excitation, but since in this direction the tower is very stiff the resulting movements are negligible.

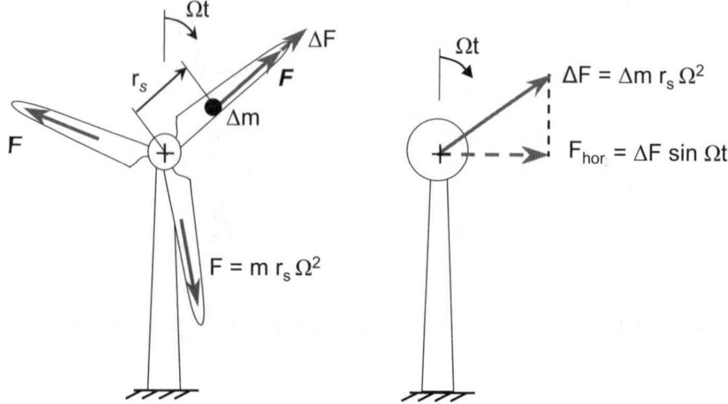

Fig. 8-3 Centrifugal forces F in blades and additional centrifugal force ΔF from mass unbalance $\Delta m\, r_s$, causing lateral vibrations of tower and nacelle

8.1.2 Aerodynamic and hydrodynamic loads

There is a wide variety of aerodynamic excitations of the wind turbine. In the following, we will first describe some periodic loads, which origin from deterministic disturbances of the uniform inflow (deterministic wind field). After that, stochastic wind loads (turbulent wind field) and transient aerodynamic forces are presented. When considering the aerodynamics it must be kept in mind that due to the dependency of the air density on temperature and altitude, the loads may vary between up to 20 and 30% depending on the site conditions and season.

Excitations by the vertical wind profile

In section 4.2 the properties of the surface boundary layer concerning the vertical wind profile, the gustiness and, the turbulence of the wind were already presented. Apart from these two aspects further phenomena, e.g. oblique flow cause strong aerodynamic excitation of the wind turbine. To simplify matters we consider that the *vertical wind profile* in the rotor area increases linear with increasing height, Fig. 8-4. Viewed from the individual rotating blade at the flow (rotating observer), the wind speed in a blade section varies periodically around the mean value v_m between $v_2 = v_m + \Delta v$ at the top position and $v_2 = v_m - \Delta v$ at the bottom position, according to $v_2 = v_m + \Delta v \cos \Omega t$.

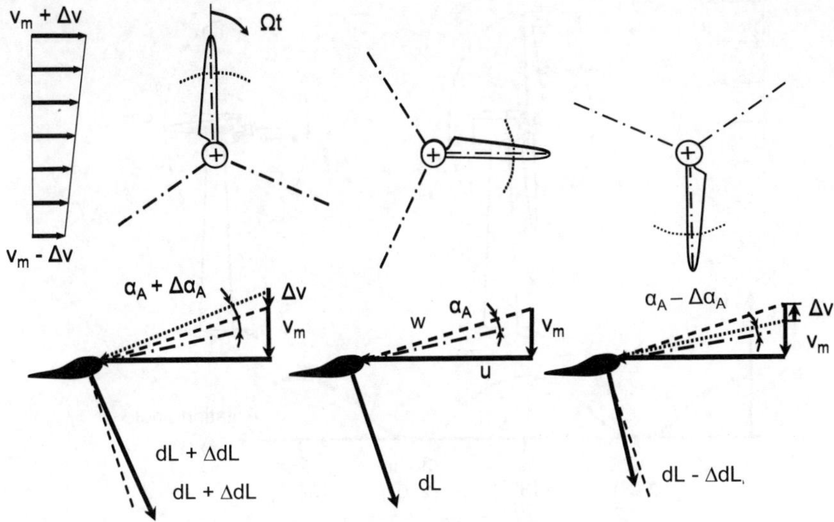

Fig. 8-4 Variation of the flow conditions in the plane of rotation and lift force dL, due to the vertical wind profile and the angular position of the blade

The angle of attack α_A and the relative velocity w oscillate correspondingly. Only the circumferential speed u is independent of the angular blade position Ωt. Therefore, the aerodynamic forces at the rotating blade fluctuate with the angular frequency 1Ω around the mean value.

Higher harmonics of the rotational frequency, which are caused by the non-linear curves of the c_L- and c_D coefficients and the square dependency of lift and drag on the relative velocity w, have weak effects.

For nacelle and tower (inertial system) of a wind turbine with a *three-bladed rotor*, these fluctuations are completely balanced out by the superposition of the values from the three blades arranged at 120° to one another. The rotor acts like a disk where the pressure point lies above the rotation axis – it is shifted upwards a little due to the vertical wind profile. This causes a constant tilting moment around the lateral nacelle axis. By contrast, viewed from the rotating main shaft, this inertially fixed pressure point causes for every revolution alternating bending stress which oscillates with the rotor's angular frequency (1Ω).

The *two-bladed rotor* behaves quite differently. Here the fluctuating blade loads are *not* equalized with respect to the inertial system of nacelle and tower, Fig. 8-5. If the blades are vertical (0°) a strong tilting (and yawing) moment is produced in the inertial system, which disappears at the horizontal blade position (90°) but again reaches its maximum value at 180°. This cycle is repeated once again until at 360° one revolution is finished.

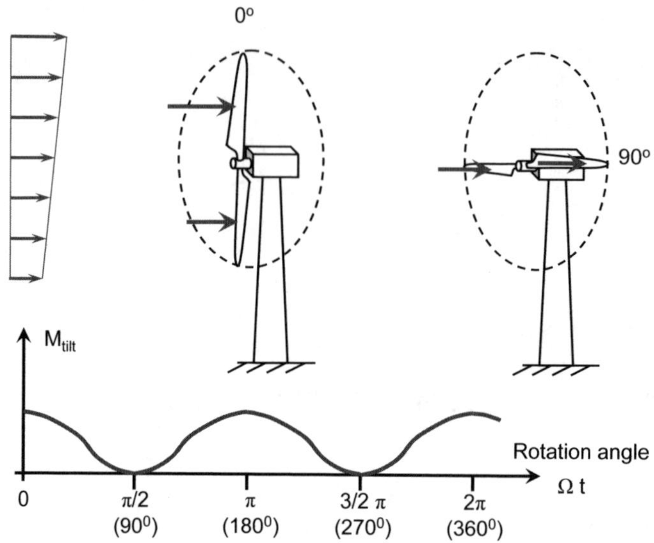

Fig. 8-5 Bending moment (twice per revolution) around the lateral nacelle axis for a two-bladed wind turbine due to the vertical wind profile

The resulting tilting moment, as well as the yawing moment, fluctuates with *twice* the angular frequency which produces high fatigue loads for nacelle frame and tower. In the rotating main shaft there is again alternating bending stress with the rotor's angular frequency (1Ω).

Since in the two-bladed wind turbine the natural frequencies of the tower-nacelle system depend on the angular blade position there are *parameter-excited vibrations* which cause dynamics that are far more problematic than in the three-bladed wind turbine.

Excitations by tower dam, tower wake and Kármán vortex street

The tower dam and tower wake are caused by the tower being an obstacle that retards and deflects the flow, which then separates behind the tower and becomes very turbulent. These separations occur periodically, alternately right and left, and are washed downstream which causes the so-called Kármán vortex street, Fig. 8-6. This may excite lateral vibrations of the tower if the vortices shed with a frequency which is close to the tower's natural bending frequency. In reality, these conditions only occur during erection before the heavy nacelle is mounted on the tower. The tower dam of wind turbines with an upwind rotor (and the even more severe tower wake of the downwind rotor turbines) has strong effects on the vibrations of blade and tower-nacelle system. The short-term collapse of the aerodynamic forces at the *blade* when the blade approaches the tower once per revolution provokes a strong excitation with the 1Ω angular frequency with many additional harmonics (2Ω, 3Ω, 4Ω, etc.); with growing order the intensity decreases only slightly.

In the inertial *tower-nacelle system* there is a loading impulse from each blade passage, Fig. 8-7. For a three-bladed rotor, the tower-nacelle vibrations are excited with 3Ω, three times the angular frequency, and its multiples, 6Ω, 9Ω, 12Ω, etc. Correspondingly, the excitation frequencies for the two-bladed rotor are 2Ω, 4Ω, 6Ω, 8Ω, etc.

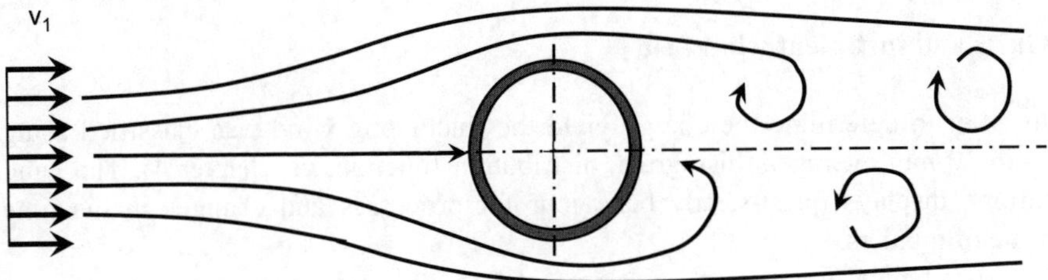

Fig. 8-6 Tower dam, wake and Kármán vortex street

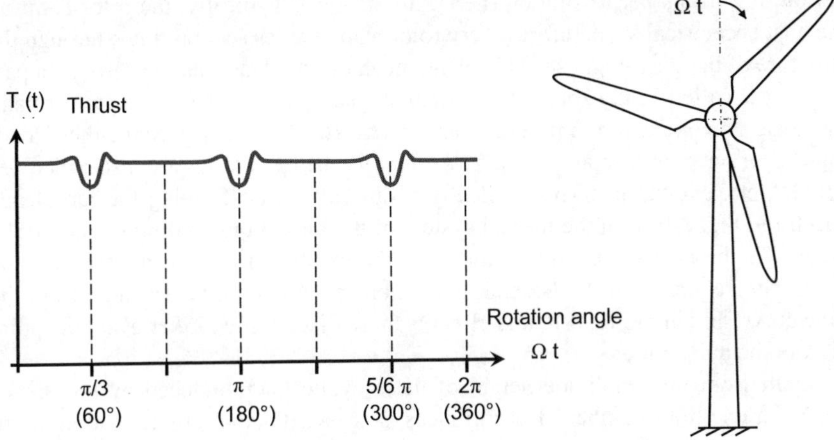

Fig. 8-7 Effects of tower dam: thrust versus rotation angle for a three-bladed wind turbine

Oblique flow and aerodynamic unbalance

There is almost always a slight *oblique inflow* (i.e. inclined inflow) at the rotor, for instance due to effects of a hill, delayed yawing of the nacelle or the inclination of the rotor axis to the horizontal plane. The blade loads vary with the rotor's angular frequency 1Ω, and the tower-nacelle system experiences the blade passing frequency 3Ω. Calculating the aerodynamics of a rotor operating under oblique inflow is very complicated.

Errors in the blade angle adjustment, e.g. during assembly, as well as production tolerances of the outer blade shape, may lead to different aerodynamic forces for the individual blades of a rotor. These aerodynamic unbalances periodically act on the tower-nacelle system in axial and lateral direction with the angular frequency 1Ω, which may excite heavy vibrations.

Gusts and turbulent wind field

In order to determine the energy yield the fluctuating wind was classified using 1- to 10-min averages (histogram, distribution function, cf. chapter 4). The wind turbine displays quasi-steady behaviour for processes and changes in the time scale of minutes.

Gusts typically have a passing time of 3 to 20 s and a spatial extension in the lateral direction of 10 to 100 m. They excite the relevant natural frequencies of the structures in the range between 0.2 and 10 Hz. During operation, the rotor has a period of 2 to 5 s per revolution (i.e. 12 to 30 rpm). Typically, the rotor is hit by the gust eccentrically; therefore, every rotor blade passes several times through the structure of the gust, Fig. 8-8. This phenomena of the blade rotating through a partial gust is called "eddy slicing" or "rotational sampling". As with the tower dam, the gusts provoke a harmonic excitation of 1Ω, 2Ω, 3Ω, etc. [1]. Since these loadings occur once or several *times per revolution,*, the excitations are also called $1P$, $2P$, $3P$, etc. excitations. This is clearly visible after transforming the turbulence spectra in Fig. 8-9 from the inertial system of the hub to the rotating system of the blade. The broad band energy of the turbulent excitation is concentrated near the rotational frequency of 0.5 Hz and its harmonics. As could be assumed from the time curve shown in Fig. 8-8, this effect is more distinct the farther away the point is from the hub centre.

Apart from the amplitude height of these cyclic loads induced by turbulence their extreme high number of occurrences is crucial for the fatigue loading of an operating wind turbine. During the expected lifetime of 20 years, the rotor of a modern wind turbine revolves approx. 100,000,000 times typically causing $5 \cdot 10^8$ to $1 \cdot 10^9$ load cycles. This is an order of magnitude far beyond the load cycle numbers occurring in other technical applications, e.g. automobiles or airplanes, cf. section 9.2.3.

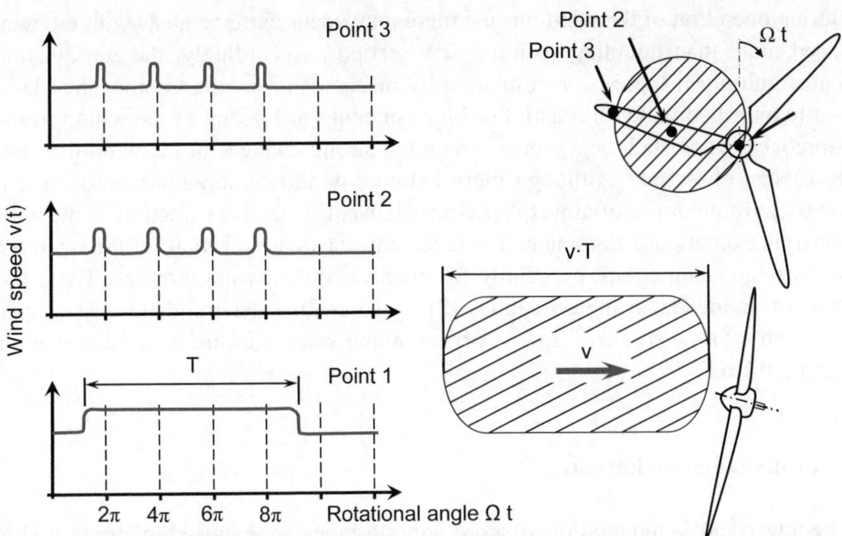

Fig. 8-8 Left: time history of wind speed for three points on the rotor blade during the passage of a partial wind gust: point 1 inertial at hub centre, point 2 rotating in the middle of the blade, point 3 rotating near blade tip; right: spatial structure of the eccentric partial wind gust passing the rotor, front view and side view

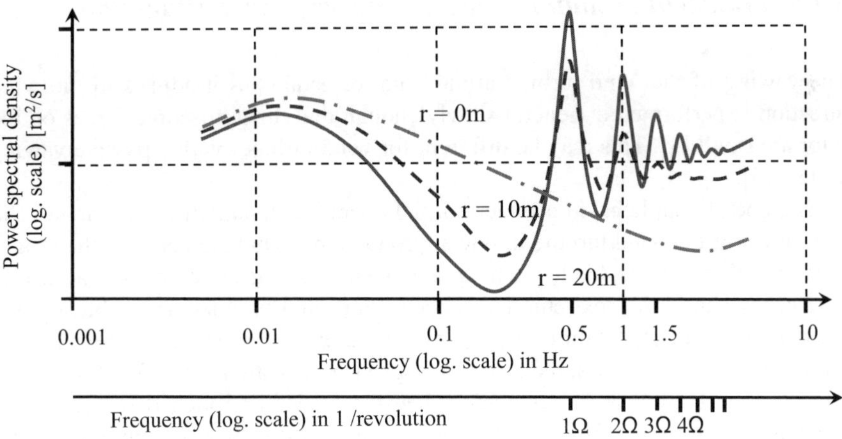

Fig. 8-9 Transformation of the turbulence spectrum from the inertial coordinate system of the hub ($r = 0$ m) into the rotating coordinate system of the rotor blade for 2 different radial positions ($r = 10$ m and $r = 20$ m) (example: rotor diameter 40 m, rotor speed 30 rpm), [1]

Transient aerodynamic loads

During operation of the wind turbine there may occur extreme gusts with the wind speed more than doubling within a few seconds. Accordingly, the aerodynamic loads multiply if there is no limitation by means of flow separation at the blades (stall-controlled wind turbines), fast blade pitching or braking of the wind turbine. Moreover, when such heavy gusts come up, strong changes in the wind direction have been observed. Although these extreme wind conditions are embedded in the background noise of atmospheric turbulence, the transient fraction of the aerodynamic excitation is dominant. These extreme cases are relevant for the wind turbine design calculations, especially for pitch-controlled wind turbines. Therefore, they are modelled in the guidelines [2] by short-time deterministic gust shapes, e.g. with a (1 - cos) shape and a corresponding change in the wind direction, cf. section 9.1.6.

Hydrodynamic excitations

The tower and foundation of offshore wind turbines are exposed to strong hydrodynamic excitations by the waves, but these only have an effect on nacelle and rotor in exceptional cases. Ice floes driven by wind and ocean currents may produce high loads on the foundation. Section 16.1 gives an overview of the problems with hydrodynamic loads [3].

8.1.3 *Transient excitations by manoeuvres and malfunctions*

The yawing of the wind turbine around the tower axis as it adjusts to the wind direction is performed in general slowly enough that the gyroscopic forces on the rotor are small [4]. This may be different for wind turbines with a passive yawing system [5].

The operational behaviour under normal operation conditions, e.g. start-up and shut-down of the wind turbine in power production, switching between the ranges of the rotational speed, blade pitching and yawing, is in general not critical. But in emergency shut-down, extreme loads occur. For wind turbines with a blade-pitch system the load case "failure of pitch system after over-speeding" is relevant for the design. Here, it is assumed that overspeed occurs after a malfunction of the wind turbine control and, moreover, when the blades are pitched to feather for braking the rotor one of the blades fails to move. In the 10 to 15 s until the other blades and the mechanical brake have brought the wind turbine to standstill there are very heavy tilting and yawing loads due to the large rotating aerodynamic unbalance [1].

In an emergency shut-down of the frequency converter or a short-circuit between two phases of the generator high torque peaks may occur in the drive train. In such a case of generator short-circuit there is a load pulsating with the grid frequency and with an amplitude which may reach many times nominal torque.

Loads from earthquakes are relevant for the tower design at special sites. In general they are described by a stochastic excitation of the foundation with transient modulation [2].

8.2 Free and forced vibrations of wind turbines - examples and phenomenology

The numerical simulation of the dynamic behaviour of wind turbines is nowadays performed using extensive multi-degrees-of-freedom models. But a basic understanding, which is important for the interpretation and qualitative verification of numerical results, is at best obtained with suitably simplified sub-models.

8.2.1 Dynamics of the tower-nacelle system

In the following we will discuss how the different exciting forces act on an idealised tower-nacelle system with one degree of freedom. Despite this radical simplification this model approximates quite well the behaviour of a three-bladed wind turbine. If we assume three rigid rotor blades, the rotor may be considered as a massless rotating disk fixed to the rotor and nacelle mass which is concentrated at a single point, Fig. 8-10. The slender tower is especially flexible in the directions of lateral and axial bending. In this simplified model only the horizontal displacement of the tower head is considered. It may be determined analytically or by a simple finite-element model with a few beam elements and a clamping stiffness at the foundation, section 9.3.3.

The enormous mass of the tower is separated into a part fixed to the ground and a part which follows the vibrations, the effective tower mass. It was found that for the latter the value of 25% of the tower mass is a good estimate.

After estimating the active tower head mass ($m = m_{Rotor} + m_{Nacelle} + 0.25\,m_{Tower}$), the translatory stiffness s and the damping coefficient d, the equation of motion of the tower head displacement u(t) in the axial direction is [6, 7]:

$$m\,\ddot{u} + d\,\dot{u} + s\,u = F(t) \quad \text{or} \quad 0. \tag{8.5}$$

$F(t)$ is the excitation force from the thrust forces.

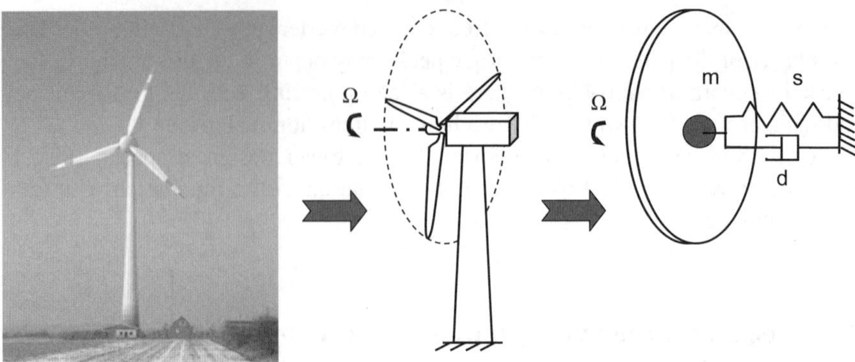

Fig. 8-10 Real wind turbine (Enercon E112) and simplified model of the tower-nacelle dynamics with one degree of freedom

Natural vibrations

If there is no excitation, the right side of the differential equation (8.5) is zero. In this case, vibrations result only from the initial conditions:

$$u(0) = u_0 \qquad and \qquad \dot{u}(0) = \dot{u}_0 \quad . \tag{8.6}$$

This vibration type is called *natural vibrations* and occurs, for example, if the emergency-stop button is pressed on wind turbine in power production. Then the wind turbine is shut down within a few seconds. The rotor thrust, which for bigger wind turbines causes a static displacement u_0 of approx. 0.5 m, collapses abruptly. The wind turbine vibrates with the initial conditions $u_0 = 0.5$ m and $\dot{u}_0 = 0$ m/s back to the rest position and beyond and then oscillates with a settling time of up to a minute or more. The analytic solution of the equation of motion (8.5) is for this case

$$u(t) = e^{-\delta t}\left[u_0 \cos \omega t + \frac{\dot{u}_0 + \delta u_0}{\omega} \sin \omega t\right] \quad , \tag{8.7}$$

cf. [7], where

$\omega_0 = \sqrt{s/m}$ in rad/s is the natural angular frequency,

$f_0 = \omega_0/(2\pi)$ is the natural frequency in Hz of the undamped system,

$\delta = d/(2\,m)$ is the decay factor, which gives

$D = \delta/\omega_0$ the dimensionless damping ratio, when divided by the undamped
          natural angular frequency.

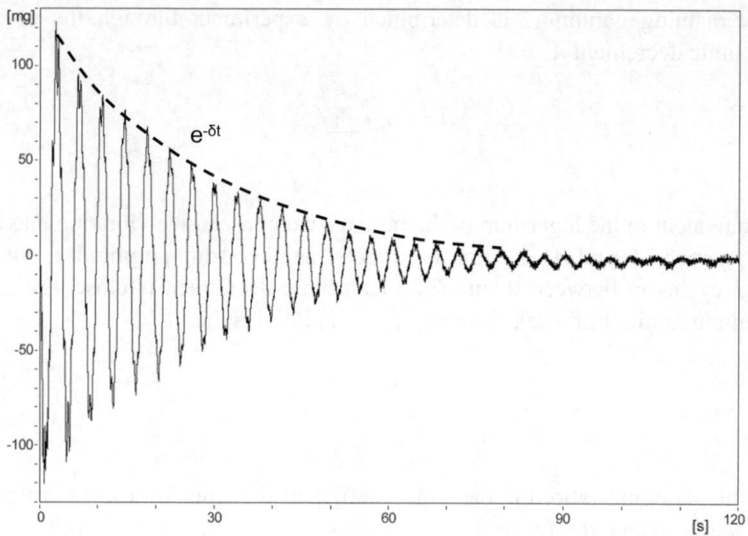

Fig. 8-11 Measured acceleration of the tower-nacelle system of a 2 MW wind turbine after an abrupt stop (1st natural bending frequency 0.26 Hz, damping factor including liquid damper approx. 2%), [8]

The natural angular frequency is slightly reduced by (weak) damping in the system

$$\omega \;=\; \omega_0 \sqrt{1 - D^2} \quad . \tag{8.8}$$

The oscillation is a decaying process for a damping ratio $D < 1$, as in the described case for the wind turbine. If $D > 1$, then there is only creeping into the rest position (supercritical damping $d > 2\sqrt{s\,m}$).

Fig. 8-11 shows a measured decay curve of the tower-nacelle system of a 2MW wind turbine after an abrupt stop. The envelope curve $e^{-\delta t}$ was inserted later. The acceleration (in mg, i.e. a multiple of the gravitational constant g /1000) is measured and shown, and not the displacement. The acceleration can be measured more easily up in the wind turbine nacelle.

The decaying lasts more than one minute because the damping is so small. It comes mainly from the flange connections in the tower, installations in the tower and from the soil (elastic half space). In this wind turbine there is additionally a passive liquid damper installed in the tower head which increases the damping from approx. 0.5% to approx. 2%.

During power production there is additional aerodynamic damping which may be quite large, cf. chapter 7, for which it is possible to calculate an estimate quite easily, in contrast to the "remaining damping" from material and soil. Therefore, this "remaining damping" is determined by experiment through the so-called logarithmic decrement Λ

$$\Lambda = \frac{1}{n-1} \cdot \ln \frac{A_1}{A_n} \qquad . \tag{8.9}$$

It is equivalent to the logarithm of the ratio of two consecutive vibration maxima.

In practice, A_1 and A_n are not the directly neighbouring amplitudes, but there are *n-1* cycles in between them. The logarithmic decrement is closely related to the damping ratio. For weak damping ($D \ll 1$) it holds that

$$D = \frac{\Lambda}{2 \cdot \pi} \qquad . \tag{8.10}$$

With this damping ratio, the damping coefficient for equation (8.5) is calculated "backwards" using $d = D\ \omega_0\ 2\ m$.

Periodic vibrations from mass unbalance

Periodic vibrations of the nacelle due to a mass unbalance of the blades, or also aerodynamic excitation by tower dam or tower wake, play a big role during operation of wind turbines. We want to more closely consider the case of unbalance causing heavy vibrations of the tower-nacelle system in the lateral direction, perpendicular to the wind.

With the degree of freedom $v(t)$ of the lateral displacement (Fig. 8-3) and the exciting unbalance force from equation (8.4) the differential equation for the vibrations is now

$$m\,\ddot{v} + d\,\dot{v} + s\,v = -\Delta m\ r_\mathrm{s}\ \Omega^2 \sin\Omega t = \hat{F} \sin\Omega t \tag{8.11}$$

Due to the assumed "roundness" of the tower-nacelle system the coefficients m, d, s are practically the same as for the axial vibration $u(t)$ in equation (8.5). Choosing a suitable starting time the minus sign on the right side of the equation is eliminated. The vibration response $v(t)$ to the harmonic excitation force $F(t) = \hat{F} \sin\Omega t$ is as follows [7]

$$v(t) = \frac{\hat{F}}{s} \frac{1}{\sqrt{\left(1-\eta^2\right)^2 + (2\,D\eta)^2}} \sin\left(\Omega t + \gamma\right) = \frac{\hat{F}}{s}\ V(\eta, D)\sin\left(\Omega t + \gamma\right) \tag{8.12}$$

where V is the so-called magnification function – a magnification factor of the static displacement $\hat{F}/s$, which is dependent on the frequency ratio $\eta = \Omega/\omega_0$ of the exciting angular frequency Ω to the (undamped) natural angular frequency ω_0, and moreover on the dimensionless damping ratio D. A similar equation describes the phase shift γ [7] which is not discussed here. Close to the resonance frequency $\Omega = \omega_0$, i.e. $\eta = 1$, the vibration amplitudes of the tower may reach very high values. The magnitude ultimately depends only on the damping ratio, which can be easily ascertained

$$V\big|_{\eta=1} = \frac{1}{2 \cdot D} \quad . \tag{8.13}$$

In the case of an unbalance, the amplitude $\hat{F} = \Delta m\, r_s\, \Omega^2$ of the exciting force itself depends through the angular speed Ω on the square of the rotational speed: Therefore, it is functional to rearrange equation (8.12)

$$
\begin{aligned}
v(t) &= r_s\, \frac{\Delta m}{m}\, \frac{\eta^2}{\sqrt{\left(1-\eta^2\right)^2 + \left(2D\eta\right)^2}}\, \sin\left(\Omega t + \gamma\right) \\
&= r_s\, \frac{\Delta m}{m}\, V(\eta, D)\, \eta^2\, \sin\left(\Omega t + \gamma\right) \quad .
\end{aligned}
\tag{8.14}
$$

Fig. 8-12 shows the amplitude of the response to an unbalance as a function of the dimensionless exciting frequency (i.e. the frequency ratio η). The response is limited in the case of resonance only by the damping and approaches a fixed value for high excitation frequencies, i.e. high rotational speeds.

This implies two consequences:

Firstly, the rotor must be balanced as well as possible, since then $r_s\, \Delta m\, /\, m$ is very small.

Secondly, the tower design has to be either

- very stiff so that its natural frequency is significantly higher than the operating range ($\Omega < 0.9\ \omega_0$, stiff, high tuned tower, i.e. subcritical operation), or
- quite flexible in order to ensure that the resonance frequency is below the operating range ($\Omega > 1.1\ \omega_0$, low tuned tower, i.e. supercritical operation).

Both design concepts are applied in practice, cf. Fig. 9-7. In any case, one should avoid operating too close to resonance.

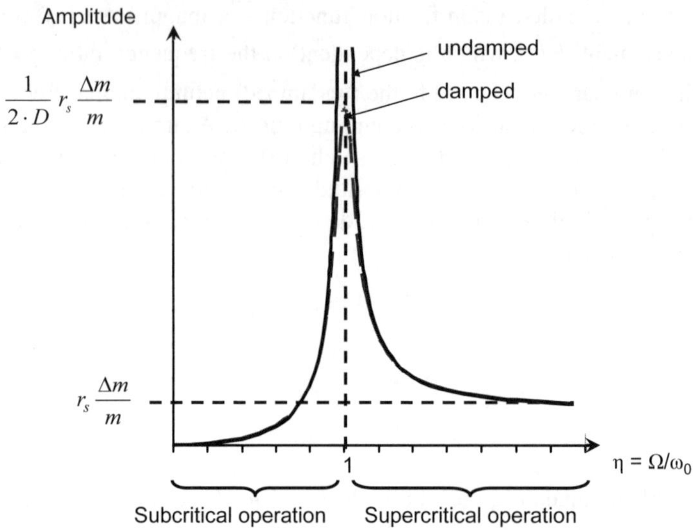

Fig. 8-12 Amplitude of lateral nacelle vibrations due to mass unbalance

Periodic nacelle vibrations due to tower dam

In section 8.1 we saw that due to the tower dam or, in some cases, the tower wake there is a periodic excitation force in the axial direction (but also perpendicular to it, i.e. lateral). However, apart from the dominating blade passage frequency this kind of excitation contains many harmonics with the amplitudes decreasing only slowly with increasing order k. Altogether, the right side of the equation of motion (8.5) of the axial tower vibrations is

$$F\left(t\right) = F_0 + \hat{F}_3 \cos\left(3\Omega t\right) + \hat{F}_6 \cos 2\left(3\Omega t\right) + \hat{F}_9 \cos 3\left(3\Omega t\right) + \ldots$$

$$+ \hat{F}_{3k} \cos k\left(3\Omega t\right) \tag{8.15}$$

where F_0 is the mean thrust which contributes the highest portion. Here as well, as for the case of unbalance, equation (8.14), resonance occurs if the exciting frequency matches the natural frequency ω_0 of the tower-nacelle system, i.e. $\omega_0 = 3$ Ω; 6 Ω; 9 Ω, etc. Fig. 8-13, bottom, shows qualitatively the tower amplitudes for this case.

In essence, the behaviour in the lateral direction, i.e. perpendicular to the rotor axis, is quite similar. The thrust at the blade collapses whenever the blade passes the tower, Fig. 8-7, as does the driving circumferential force.

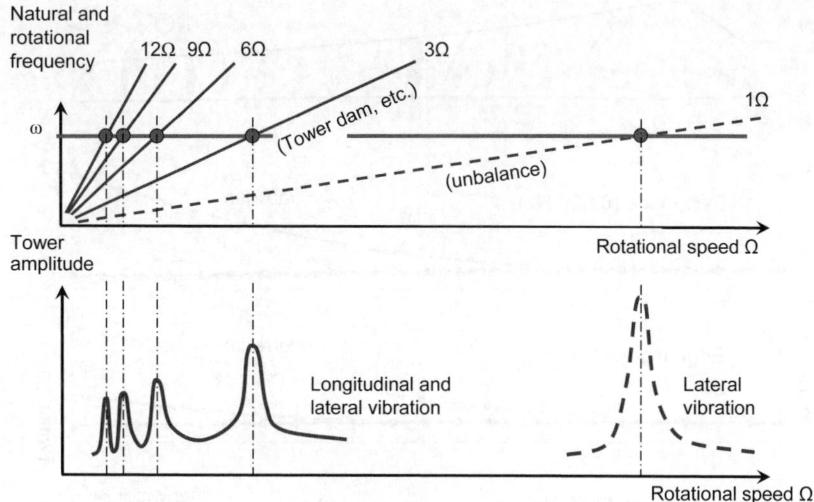

Fig. 8-13 Resonance peaks due to blade passage for a three-bladed wind turbine; top: Exciting frequencies (1Ω, 3Ω,…) versus rotational speed and natural frequency (Campbell diagram), bottom: tower amplitudes (qualitatively)

Therefore, lateral nacelle vibrations are also excited with excitation frequencies of $3\,\Omega$, $6\,\Omega$, $9\,\Omega$ etc. Since the natural frequency of the tower-nacelle system in the lateral direction is nearly identical to the one in the axial direction, the resonance diagram is nearly the same as in Fig. 8-13. But for the lateral vibrations of the tower-nacelle system the influence of the $1\,\Omega$ unbalance excitation is generally dominant, which is of no importance for the axial direction. In Fig. 8-13, it is therefore drawn with a dashed line.

8.2.2 Blade vibrations

Vibrations of the blade are excited, as explained in section 8.1, by the vertical wind profile or by oblique flow with the rotational angular frequency 1Ω and by the tower dam with the rotational frequency and its harmonics, i.e. 1Ω, 2Ω, 3Ω. The blades are elastic for bending and torsion and are modelled by transfer matrices or the finite-element method where it is sufficient to model beam and shell elements. As mentioned in chapter 3, Fig. 3-9, the position of the centre of gravity line, the centre of elasticity line, the radial line (through the flange centre) and the aerodynamic pressure point line (points of action of the aerodynamic forces) have a strong influence on the blade dynamics, above all the fluttering behaviour.

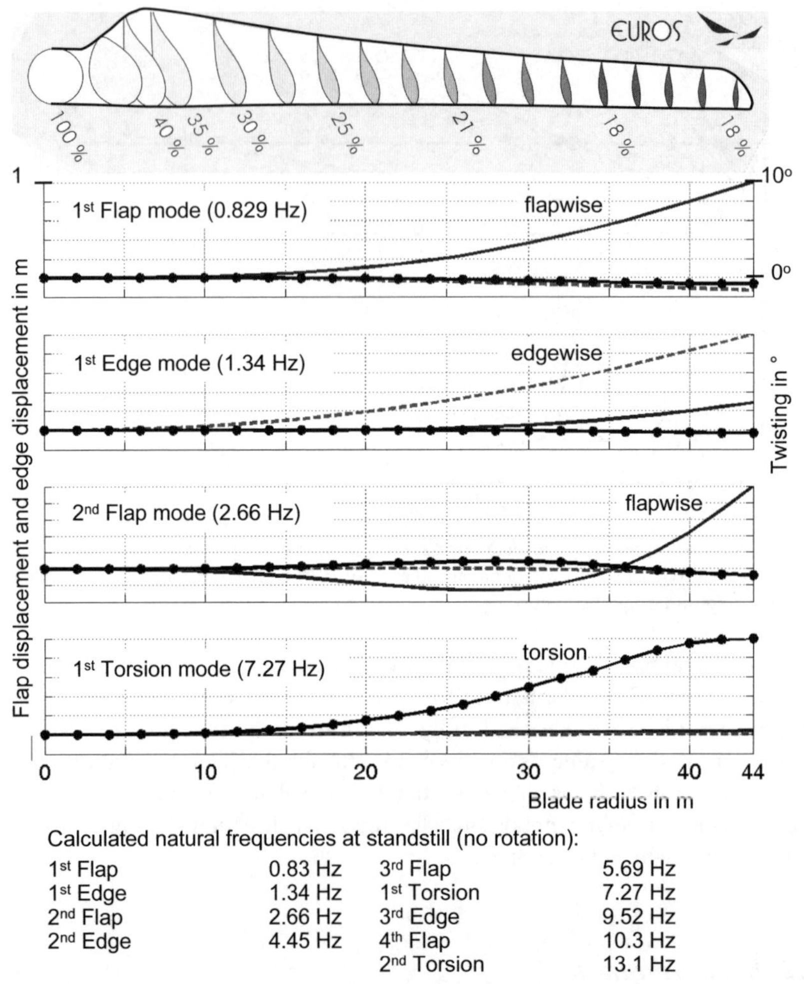

1st Flap	0.83 Hz	3rd Flap	5.69 Hz
1st Edge	1.34 Hz	1st Torsion	7.27 Hz
2nd Flap	2.66 Hz	3rd Edge	9.52 Hz
2nd Edge	4.45 Hz	4th Flap	10.3 Hz
		2nd Torsion	13.1 Hz

Fig. 8-14 Eigenmodes and natural frequencies for a rotor blade of 44 m length (EUROS EU 90.2300-2), [9]

The first natural frequencies and their eigenmodes occur mostly in the following order:

ω_1 1st flapwise bending natural frequency

ω_2 1st edgewise bending natural frequency

ω_3 2nd flapwise bending natural frequency

ω_4 2nd edgewise bending natural frequency

ω_5, ω_6 natural frequencies with strong torsional activity

Fig. 8-14 shows a blade of 44 m length with its first eigenmodes. The corresponding natural frequencies as well as the exciting frequencies 1Ω, 2Ω, 3Ω etc. of the blade are drawn in the resonance diagram (Campbell diagram) of Fig. 8-15.

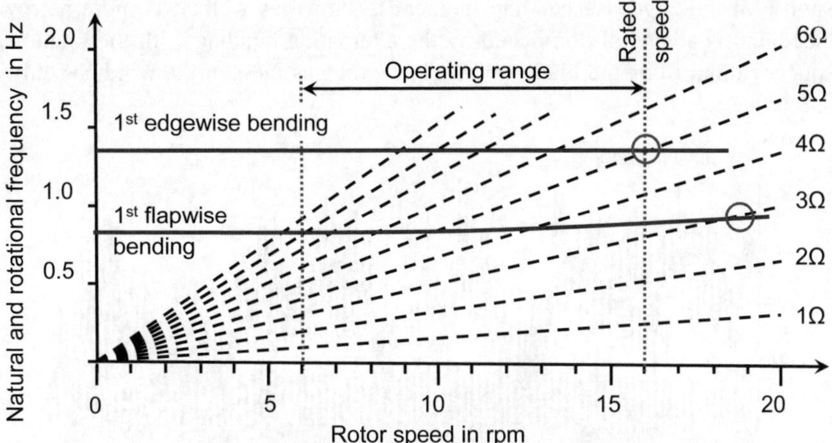

Fig. 8-15 Resonance diagram (Campbell diagram) of the rotor blade in Fig. 8-14

The flapwise natural frequencies increase slightly due to centrifugal stiffening under operation which can be seen for the first flapwise bending frequency in Fig. 8-15. The influence of the rotational speed on the edgewise frequencies is smaller. Potential spots of resonance close to the rated rotational speed are marked with circles in Fig. 8-15.

Around the rated rotational speed of 16 rpm, i.e. 0.27 Hz, the blade is nearly free of resonance problems. A 3 Ω-excitation of the first flapwise natural frequency becomes relevant only at a higher rotational speed of approx. 18 rpm.

The 5 Ω-excitation occurs for the first edgewise natural frequency, but since this excitation only has a small amount of energy the vibration amplitudes of the resonance will be small. In the partial load range where the wind turbine operates with variable rotor speed there are several points of resonance of the fourth and higher harmonics. Due to the permanently fluctuating wind speed and correspondingly varying rotational speed it can be expected that there are only weak resonance effects. If the natural frequencies were excited by lower harmonics of the rotational speed more severe resonance would be likely to occur.

The manufacturers check the expected natural frequencies calculated in the simulations by means of a *vibration test at standstill* (the blade is fixed to a big, rigid foundation) and determine the modal damping of the first eigenmodes as well. The behaviour of the blade under the more or less stochastically fluctuating wind with its strong *turbulence* is far more complicated than the periodic excitations considered up to now.

Fig. 8-16 shows an example of the simulated bending moments at the blade root of a 500 kW wind turbine in flapwise and edgewise direction. The energy-rich gusts with passing times between 10 and 20 s cause a low-frequent variation of the mean loading, whereas the turbulence with its higher frequencies but less energy strongly excites the first flapwise natural frequency of the blade. The response of the edgewise bending moment is shown as well. It is more narrow-banded and is above all dominated by the alternating bending with the rotational frequency induced by the blade weight. Here, the gustiness of the wind has only a small influence.

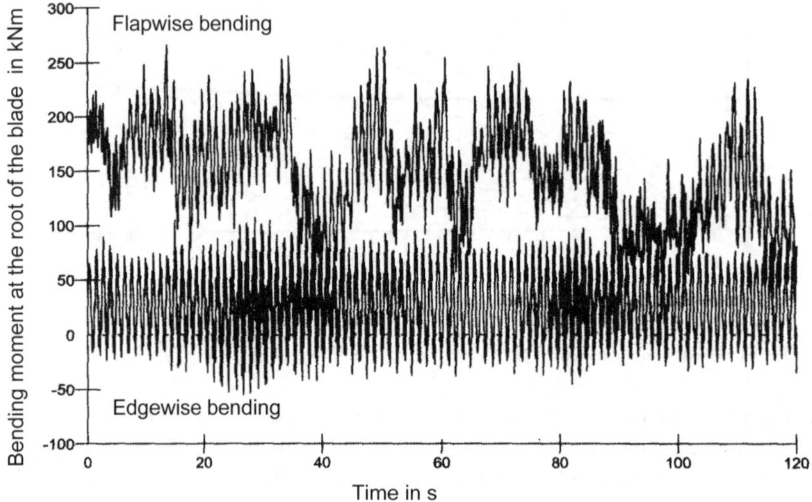

Fig. 8-16 Simulated response of the flapwise and edgewise bending moment at the blade root due to stochastic excitation by the wind, wind turbine of 500 kW

8.2.3 Drive train vibrations

In essence, the excitations of the vibrations in the drive train stem from:
- the blade passage: number of blades times the rotational frequency and its harmonics,
- the turbulent wind field,
- the vertical wind profile for one- and two-bladed wind turbines,
- action of the control, blade pitching, torque moment changes in the generator,
- teeth contact frequencies,
- etc..

The sufficient treatment of the dynamics of torsion of smaller wind turbines ($D < 50$ m) is possible by modelling the drive train as a system with 2 or 3 degrees of freedom (DOF), Fig. 8-17 left and middle. However, magnetic forces act between rotor and stator of the generator which may be modelled for a *synchronous generator* a as load dependent *torsional spring*. The forces for an *asynchronous generator*, which has a slip to the grid frequency, may be described by a *torsional damper*, as shown in Fig. 8-17, where the damping coefficient results from the slope of the torque characteristic at the synchronous frequency, cf. chapter 11. For large wind turbines the first edgewise eigenmode of the blades has to be considered in the dynamics of torsion and the fact that the casing of the gearbox is mounted on the nacelle frame in elastic rubber bushings. Thus, one more degree of freedom comes into play, and the drive train has four degrees of freedom - possibly even five if individual inertia is assigned to the gear wheels in the model, Fig. 8-17, right.

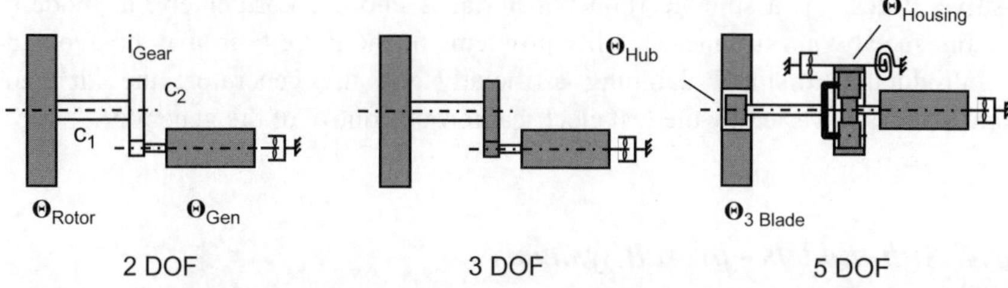

Fig. 8-17 Modelling of the torsional drive train dynamics with different numbers of degrees of freedom (in grey: moment of inertia Θ; torsional stiffnesses c_1, c_2 , etc.; spring and damping systems from the gearbox housing and generator)

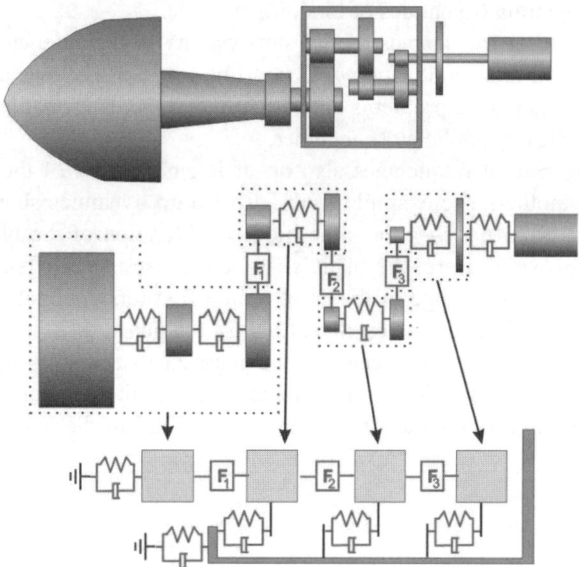

Fig. 8-18 Modelling of the drive train dynamics as a multi-body system [10]

Recent studies on the three-dimensional situation of the loading at the hub and in the entire drive train as well as the damages observed in reality with larger wind turbines show clearly the limitations and the inadequacies of the models described above. As a result, multi-body-system models are being developed. Fig. 8-18 shows a model with torsional and translational degrees of freedom [10]. These investigations become very complex if the gaps between the teeth flanks and, moreover, the elasticities of the support provided by the nacelle frame and the tower are considered.

The drive train is only slightly damped from the mechanical and aerodynamic side. The edgewise movement of the blades merely provides an aerodynamic torsional damping (see below). Only the elastic rubber support of the casing has a positive effect – if a suitable rubber material is chosen. Fortunately, in modern variable-speed wind turbines stability problems of the drive train may be avoided by introducing torsional damping artificially via the generator: the air gap moments can be varied by the fast electrical torque control of the generator.

8.2.4 Sub-models - overall system

For dynamic investigations at an early design stage it is advisable to cut the over-all system of a wind turbine into suitably selected sub-systems, e.g.
- The tower-nacelle system with rigid blades
- The blade rigidly clamped at the root
- The drive train (dynamics of torsion), etc.

Drawing the natural frequencies of the sub-systems in their dependency on the rotational speed in one single *Campbell diagram*, Fig. 8-19, allows us not only to detect potential resonance problems but also to judge the danger of interference of natural frequencies which lie close together.

Changes in the natural frequencies also occur if eigenmodes of the sub-systems influence one another. The example of the drive train dynamics shows that especially for *large wind turbines* the simplified consideration of the blades as rigid bodies is inadmissible. Here, the blade's first edgewise natural frequency is that low that it falls into the range of the drive train's first torsional natural frequency at which the rotor of the turbine oscillates against the generator. Thus a coupling is formed: blade and drive train dynamics are married inseparably together. Some vibration modes are even influenced by the tower dynamics. Therefore, the natural frequencies determined for the sub-systems are shifted in the assembled overall system.

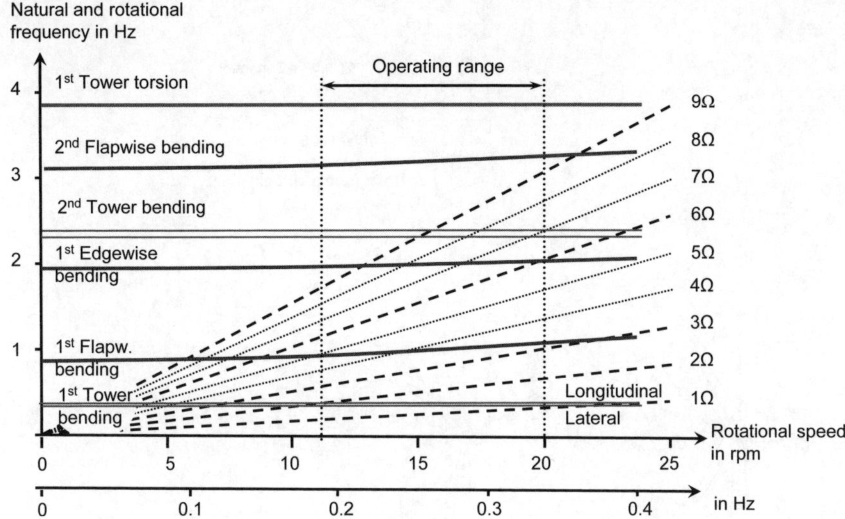

Fig. 8-19 Campbell diagram with natural frequencies of the tower-nacelle system and the coupled blade – drive train system of a typical 1.5 MW wind turbine with variable rotational speed

Fig. 8-20 top shows a simplified drive train model, where the blades are elastic. In eigenmode I the blades and the hub oscillate together *against* the generator. In eigenmode II, which is sometimes called lambda-eigenmode due to its λ-shape, two blades bend together in the edgewise direction and the third blade balances them with doubled amplitude. Eigenmode III is characterised by two blades vibrating against each other while the third stays calm (*Y* or tuning fork eigenmode).

Eigenmodes II and III are not coupled with the drive train torsion. The bending moments at the hub are balanced. The corresponding natural frequencies are (nearly) similar to the natural frequencies obtained for a rigid clamping of the blade. However, with eigenmode III the tower shows a small accompanying motion. In eigenmode IV the blades retain the packet-shaped vibration of Eigenmode I, but now the body of the hub vibrates *in opposition* to the blades. Since in the model we allowed five degrees of freedom (for each blade an individual edgewise degree of freedom, plus one for the hub-gearbox unit and one degree of freedom for the generator) there is one natural frequency and eigenmode missing: it is the "rigid-body degree of freedom", the free unbound rotation of the entire drive train which has the natural frequency of zero.

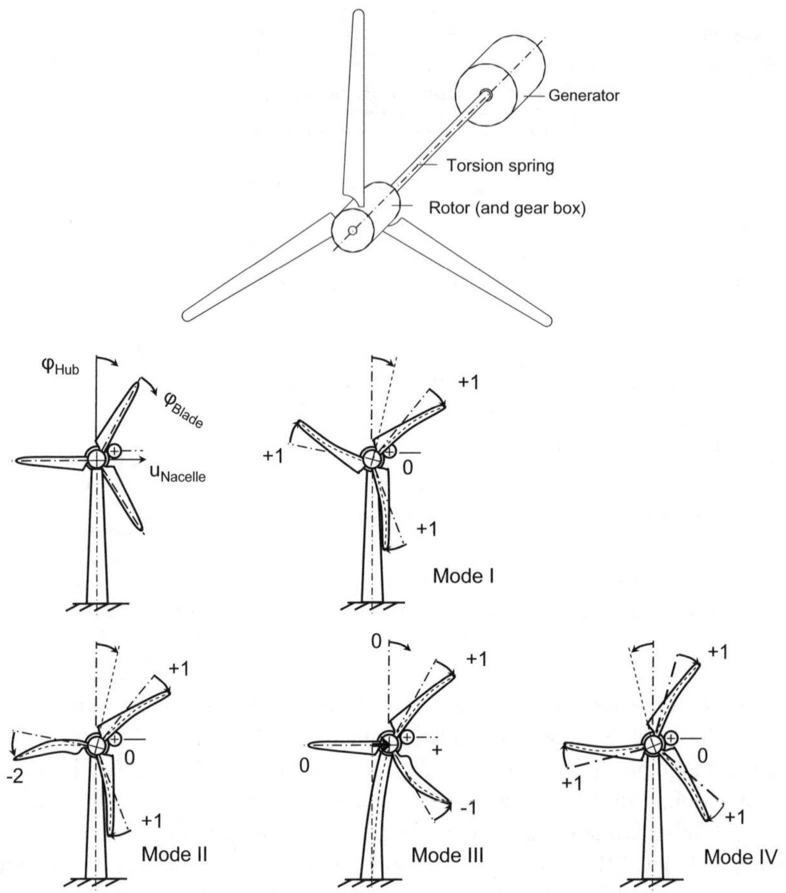

Fig. 8-20 Eigenmodes of coupled blade-drive train system, eigenmodes I, II and IV coupled with the drive train torsion, mode III coupled with the lateral nacelle vibration

1Ω-Shift

Another typical effect in the overall system is the 1Ω-shift. Mass unbalances as well as an aerodynamic unbalance (blade angle or twist angle deviations) excite vibration of the tower head with the frequency 1Ω equal to the rotational frequency in the lateral direction of the nacelle, see section 8.2.1.

These lateral vibrations of the nacelle act in the rotating system as a 2Ω base excitation (with twice the rotational speed) of the rotor blades in the edgewise direction [11]. If the first two natural frequencies of the blades are close to double or triple the maximum rotor speed, resonance problems may be provoked. This 1Ω-shift is familiar to us already from gravitational influences: considered in the inertial system without rotation, 0Ω, the gravity force acts constantly on the blades, whereas in the rotating system it acts periodically with the rotational frequency 1Ω. Therefore, in the resonance diagram of a three-bladed wind turbine, Fig. 8-19, not only are the beams of the 1Ω-, 3Ω-, 6Ω-, etc. excitations shown but also the one for the 2Ω excitation.

Simulation of the overall system dynamics

For larger wind turbines in the early design stages a model for the overall system dynamics is developed, which considers all the couplings between the subsystems. It also comprises the electrical and electro-mechanical (or hydraulic) degrees of freedom of the control and controllers. The solution of these "equations of motion" is then calculated in the time domain by digital simulation. Compared to the consideration in the frequency domain this has the advantage that even nonlinearities and transient processes may easily be considered. Section 8.3 gives an introduction to the common procedures.

8.2.5 *Instabilities and further aeroelastic problems*

Aeroelastic instabilities

In the discussion of the shut-down of a wind turbine we saw how slowly the natural vibrations of the tower-nacelle system decay, Fig. 8-11, since in this case there are only small damping forces from the flange connections of the tower and from the soil.

However, if the machine operates close to the design tip speed ratio, the nacelle vibrations in the axial direction and the blade vibrations in the flapwise direction are well damped by aerodynamic forces. Fig. 8-21 shows in the middle how a (flapwise) movement of the blade against the wind direction changes the triangle of velocity. An additional lift ΔdL is produced which acts against this movement, i.e. damps it. However, it also becomes clear that if the wind turbine operates in the region of separated flow (i.e. stall) which is found to the right of the maximum of the lift curve $c_A\,(\alpha_A)$ then the deviation $c_A{}'$ of the lift coefficient changes its sign. The additional lift ΔdL produced by the flapwise movement becomes negative which means that vibrations are "fuelled" by the reduced damping. If there is insufficient "structural damping" then the natural vibrations increase and instability occurs. Of course, in the stall region the drag forces from c_D have to be considered as well, so this case again becomes a little more complicated [12, 13].

The edgewise movements of the blade provide only very small aerodynamic damping, which may be concluded from Fig. 8-21 (bottom). Since the circumferential speed u is much bigger than the wind speed v_2 in the plane of rotation, the aerodynamic forces – and above all the angle of attack α_A - merely change if edgewise movements occur. The additional edgewise velocity Δu_{edge} is small compared to the circumferential speed u, consequently vibration modes where the edgewise movement dominates are merely damped and therefore imply the danger of instability. Therefore, there are sometimes dampers for the edgewise movements installed in the blade tips of stall-controlled wind turbines (e.g. LM blades).

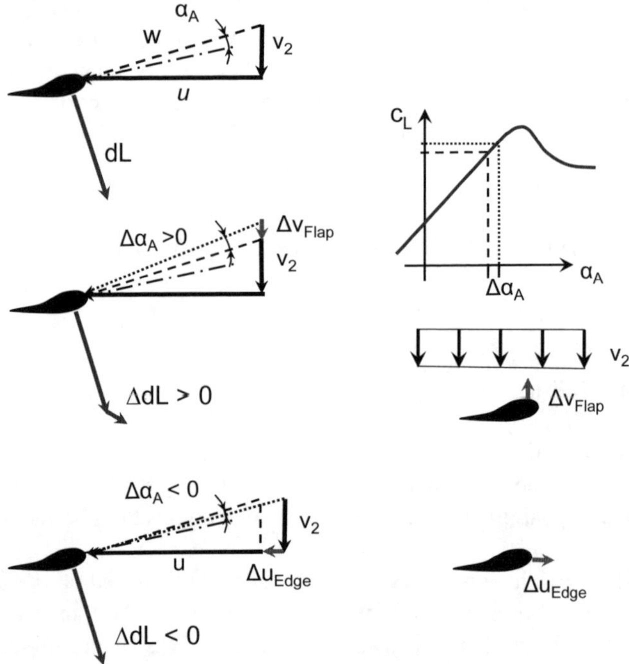

Fig. 8-21 Effect of flapwise bending and edgewise bending on the aerodynamic forces, aerodynamic damping and negative damping. Reference location (top), flapwise movement (middle), edgewise movement (bottom)

Further aeroelastic problems

There are further aeroelastic problems which may occur, e.g.

- Stall-induced edgewise vibrations, provoked by flow separation which may be reduced by dampers close to the blade tip (see above and cf. chapter 3) [14]
- Unstable torsional drive train vibrations with the strong involvement of edgewise blade movement and generator (or generator-converter system)
- Vibrations of the blade pitch controller, where the axial vibrations of the tower-nacelle system (u, $\dot{u}$) suggest incorrect wind speeds $v_2 \pm \dot{u}$, instead of the real wind v_2. The consequent action of the controller reduces the aerodynamic damping or even actively amplifies the tower-nacelle vibrations (see chapter 12)
- The nacelle whirl [11, 14] where the hub centre of the rotor propagates on an elliptic orbit around its rest position
- Unstable lateral tower-nacelle vibrations [13]
- The fluttering known from aviation and the aerodynamics of buildings where combined torsional and flapwise vibrations of the blade have an unfavourable influence on the aerodynamic forces so that vibrations may become unstable.

8.3 Simulation of the overall system dynamics

Since the wind and operating conditions of wind turbines are non periodic or transient, an analytical model may only rarely be applied in a way that makes it possible to determine deformations and stresses directly. Instead, the analysis of the wind turbine behaviour as well as the design of the force-transmitting components or the yield estimation is commonly based on simulations in the time domain [1, 15].

Simply speaking the mechatronic model of the wind turbine is exposed for a manageable time period to a "numerical wind" which is as close to reality as possible. This procedure is enhanced for offshore wind turbines by the simulation of the swell, and in special cases also by the determination of loads during earth quakes. The simulation calculations are used in essence to obtain the following information:

- Internal forces and moments in certain relevant turbine components as input for the component design procedure and for certification
- Determination of the wind turbine behaviour under all occurring operating conditions for design validation and determination of the power curve
- Design and optimization of the control system

8.3.1 Modelling in simulation programs

In general, the applied simulation models are divided into three sub-models, Fig. 8-22:

- Time series model of the environmental conditions (wind field, swell, etc.) as input
- Model of the aerodynamics at the rotor blades
- Model of the structural dynamics of the wind turbine (including electrical system and control behaviour)

The basic consideration in selecting the applied models is

Depth of modelling versus speed of simulation.

Therefore, different kinds of models are applied in parallel, of course depending on the requirements which have to be satisfied. Pre-design studies serving for a first estimation of the power performance and the wind turbine stress may be calculated with models of low complexity which require less computing time. Extended analysis of the controller behaviour requires partly a very complex wind turbine model and consumes correspondingly more computing time. A complete calculation for certification comprises the analysis of up to several hundred simulation runs which have to be performed several times in the iteration loops of the design process. Therefore, the applied models are just detailed enough for a sufficient determination of the internal forces and moments required for certification – a closer consideration of the components is then not possible.

Modelling of the wind field

One of the challenges of the wind turbine simulation is modelling the (unsteady) wind as simulation input. This wind is basically one main reason for the necessity of the simulations. Not long ago, it was thought that in the course of the design calculations it would be sufficient to apply simple wind models consisting of a steady inflow, deterministic gusts and superimposed vertical wind profile or oblique flow. Many loads could then be determined with models according to the ones discussed in the sections 8.1 and 8.2. But the growing influence of the fatigue analysis in the course of the design led to a stronger emphasis on the turbulent fraction of the wind.

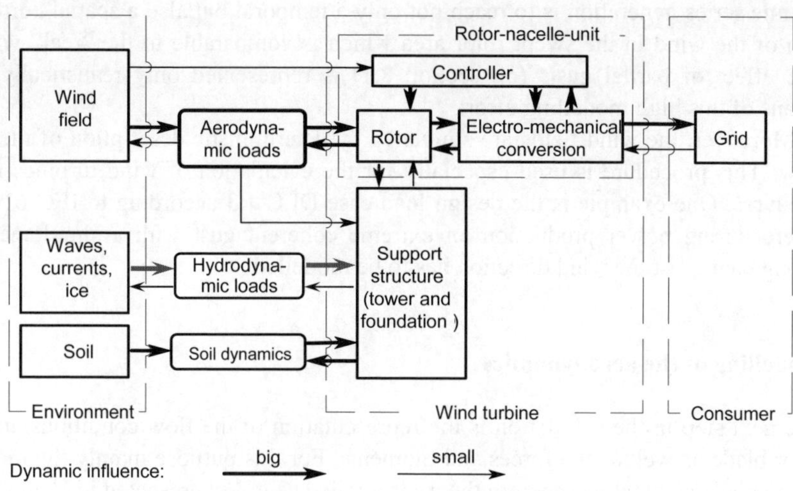

Fig. 8-22 Block diagram of a simulation program with wind field model, rotor aerodynamics, structural dynamics, influence of the generator and control system

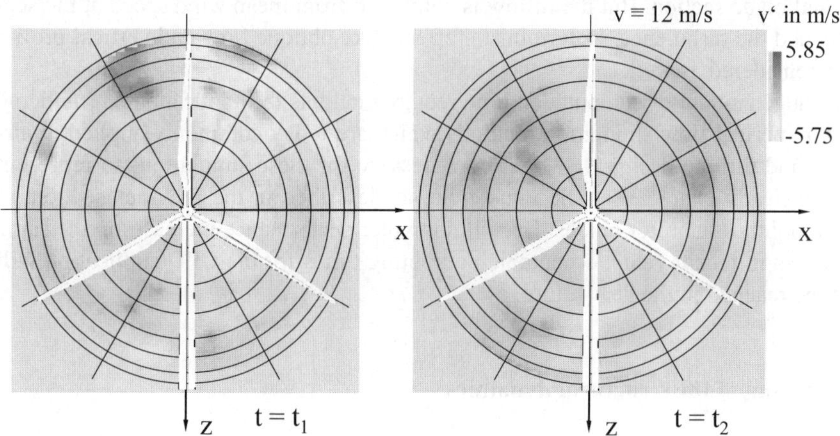

Fig. 8-23 Simulated wind field in the rotor plane of a wind turbine at two consecutive times; mean wind speed of 12m/s is subtracted

For this purpose, times series of the wind are generated which are based on the statistical information of the wind under consideration, in general the mean wind speed, the turbulence intensity, the length scale parameters and the turbulence spectra (cf. section 4.2.5). These time series, which also incorporate wind components perpendicular to the main wind direction, are applied to the swept rotor area at up to several hundred points, Fig. 8-23. An important challenge in this context of time series generation is to reach not only a temporal but also a spatial correlation of the wind in the swept rotor area which is comparable to the "real" wind. The effect of partial gusts (cf. section 8.1) is represented only realistically by means of this high modelling effort.

Moreover, the wind fields are simulated by deterministic description of the inflow. This procedure is used especially for the calculation of wind turbines manoeuvres. One example is the design load case DLC 1.3 according to IEC 61400 where during power production an extreme coherent gust with a simultaneous strong change of the wind direction has to be modelled.

Modelling of the aerodynamics

The next step in the simulation is the representation of the flow conditions at the rotor blade as well as the forces and moments. For this purpose mainly the model of the blade element momentum theory is applied as it was presented in chapters 5 and 6. For every time step a complete iteration for the determination of the balance of forces and flow conditions is calculated. Thereby, the aerodynamic influences of 3D effects (stall delay, etc.) are transferred as well as possible from a 3D to a 2D calculation. Basically, the equations of chapter 6 are applied to each individual blade section. But the inflow is combined from mean wind speed at the section and the turbulence. Deterministic effects like oblique flow and vertical profile are considered as well.

The huge amount of computation time required inhibits the application of "modern" 3D numerical flow simulation methods which are today commonly applied in aircraft and turbine design. It would be necessary for these simulations to represent the unsteady character of the wind field, as described in the previous section, at each point of the rotor blades. Even if computer performance increases in the same order as in recent years, acceptable computing times for the CFD application will not be reached in the near future.

Modelling of the structural dynamics

This part of the simulation is essential for the overall results but is by no means specific to wind turbines. There are as many methods as there are approaches to structural modelling. The particular challenge of the strong deformations and the consecutive non-linearities are no longer a problem for the commonly applied methods. For this issue, in particular, the modelling depth and the applied method strongly influence the simulation speed and therefore the application purpose. At present, approaches based on multi-body simulation are mainly used, which are then linked to the modal analysis for components which may be reduced to an elastic beam model. To speed up the modelling process it is rare to include more than the first two modal degrees of freedom. To increase the depth of modelling for a more detailed investigation of components an FEM approach is later used which means increasing the computation time. The aeroelastics additionally take into account the change of velocity triangles and the orientation of the blade section by deformations and movements, not only of the blade but of the entire structure.

8.3.2 *Application of simulation programs*

As already mentioned in the introductory section of this chapter, the main areas of the application of simulation programs are the load and certification calculations, the initial design and power curve computations as well as wind turbine optimisation. These fields of application are discussed more closely in the following.

The entire determination of the loads and the calculation for certification has to be divided into two completely different branches: fatigue analysis (for operating loads) and ultimate limit state calculation (for extreme loads), which are discussed in chapter 9 in detail.

The basic procedures in the simulation runs and the applied models are identical, but the simulation control and evaluation significantly differ. The calculation of the loads for fatigue analysis is performed through the simulation of different mean wind speed classes for (commonly) 10-min time series. The amplitudes and the load distribution are directly determined by counting the simulated loads, followed by a temporal weighting using a Weibull distribution: the results of each class are extrapolated according to their class frequency to a lifetime of 20 years. For the ultimate limit state verification the manoeuvres of a wind turbine are simulated as closely as possible for different environmental conditions, e.g. the above mentioned design load case 1.3 according to IEC, see also section 9.1.7.

Another field of application is the direct coupling of the controller software of a wind turbine with the simulation ("hardware in the loop"). For this purpose, the modelling of the wind is directly coupled to a "real" hardware controller, which is given the illusion of a real environment by the sensor impulses. The reactions are then fed back into the simulation model of the turbine. The aims are, on the one hand, the verification of the controller and, on the other, gaining knowledge on the loading of the turbine structure during real controller actions. Nowadays, exactly the same software is more and more incorporated in both the real and simulated controllers to facilitate the exchange between real and simulated control. The software only has to be translated for the different platforms.

8.4 Validation by measurement

The application of programs for wind turbine design and simulation is only reasonable if the computed results are confirmed by measurement results from real wind turbines. Different perspectives are relevant from the point of view of the design engineer:
- Verification of important model parameters, e.g. natural frequencies, damping, time constants
- Influence of the wind turbine behaviour and the controller on the power curve and the energy yield
- Validation of the fatigue loads
- Traceability of the loads during operation and manoeuvres
- Additionally: validation of the applied modelling of the wind

The most striking problems are always the reliable measurement of the wind speeds and their direct comparison with the synthetic wind speed in the entire swept rotor area. This would require virtually the exact measurement of the wind speed in the entire swept rotor area – e.g. at all points shown in Fig. 8-23. This is practically impossible for physical and metrological reasons.

Therefore, already the reliable determination of the power curve is a problem which is aggravated by problems of calibration. Thus, useful results are obtained only by the statistics of long-term measuring campaigns. The 10-min mean values are recorded and sorted using the *method of bins*. In addition to classification of the wind speeds into bins further classification is performed according to the turbulence intensity.

Fig. 8-24 shows a typical matrix ("capture matrix") for classifying the measured 10-min values. Each bin (i.e. matrix cell) is defined by the pair of wind speed class of the column and turbulence intensity class of the row and notes how many corresponding 10-min values were measured. According to this classification, the simulation of the 10-min time series is performed. The simulation results are sorted accordingly in order to achieve an easy statistical comparison with the measured forces and moments. They need only to be weighted with an arbitrary temporal distribution (e.g. by the Weibull distribution as required in the IEC 61400 standard) and may then be noted directly as vibration amplitude distribution (and then compared with the calculated result).

The following points are considered for the evaluation:
- bending moments at the blade root (or moments in the entire blade),
- forces and moments at the bearings,
- moments in the drive train as well as
- bending moment at the tower base.

Then, from this data set, a variety of loadings of the specific wind turbine components may be determined or derived.

Operation at rated power Wind speed class per 1 m/s Turbulence class per 2%																		
Time series length Number of 10-min records per bin																		
Wind speed in m/ s Turbulence intensity I in %																		
	4	5	6	7	8	9	10	11	12	13	14	15	16	17	18	19	20	21	23
<3	0	1	0	0	0	0	0	0	0	0	0	0	0	0	0	0	0	0	0
3-5	2	10	2	0	0	0	0	0	0	0	0	0	0	0	0	0	0	0	0
5-7	5	23	9	6	1	0	0	0	0	0	0	0	0	0	0	0	0	0	0
7-9	8	30	28	23	16	5	3	1	3	1	0	0	0	0	1	0	0	0	0
9-11	13	45	62	51	45	45	21	15	10	7	8	1	6	2	6	4	0	0	0
11-13	14	74	121	170	125	88	75	52	48	43	18	18	29	29	25	10	2	2	0
13-15	15	89	116	191	200	146	103	77	86	64	46	49	51	50	23	6	2	0	0
15-17	17	70	101	160	147	108	109	65	51	61	49	57	39	18	11	6	2	0	0
17-19	9	29	53	57	48	49	36	29	22	30	34	31	32	8	4	0	0	0	0
19-21	5	16	21	13	15	16	10	9	2	8	9	6	4	2	0	0	0	0	0
21-23	4	1	1	10	6	3	6	1	0	4	1	0	0	0	0	0	0	0	0
23-25	0	1	4	3	1	0	0	0	0	1	0	0	0	0	0	0	0	0	0
25-27	1	3	0	1	1	0	0	1	0	0	0	0	0	0	0	0	0	0	0
27-29	1	0	1	0	0	0	0	0	0	0	0	0	0	0	0	0	0	0	0
>29	1	1	1	0	0	0	0	0	0	0	0	0	0	0	0	0	0	0	0
Number of turbulence classes with more than 3 series	9	10	9	10	8	8	8	6	6	7	6	5	6	4	5	4	0	0	0
Total number	95	393	520	695	613	468	363	256	228	226	171	167	167	113	75	30	6	2	0

Capture matrix: normal operation, Wind speeds: mean values of the class

Fig. 8-24 "Capture matrix": Classification of measured 10-min time steps by mean wind speed and turbulence intensity

References

[1]   T. Burton, et al. : *Wind Energy Handbook*, Wiley, 2002

[2]   IEC 61400-1 ed.3

[3]   Kühn, M.: *Dynamics and Design Optimisation of Offshore Wind Energy Conversion Systems*, Dissertation, TU Delft, 2001

[4]   Gasch, R., Twele, J.: *Windkraftanlagen (Wind Power Plants)*, 3rd edition, p.188, Teubner, Stuttgart, 1996

[5]   Maurer, J.: *Windturbinen mit Schlaggelenkrotoren – Baugrenzen und dynamisches Verhalten (Wind turbines with flapping hinge rotors - constructional limits and dynamic behaviour)*, VDI-Verlag, series 11, No. 173, Düsseldorf, Germany 1992

[6]   Magnus, K.; Popp, K.: *Schwingungen (Vibrations)*, 5th edition, Teubner, Stuttgart, 1997

[7]   Gasch, R.; Knothe, K.: *Strukturdynamik (Structural dynamics)*, Vol. 1, Springer, Berlin, 1987

[8]   Measurement by Deutsche WindGuard Dynamics GmbH, Berlin

[9]   Data from EUROS Entwicklungsgesellschaft für Windkraftanlagen GmbH, Berlin

[10]   Institut für Maschinenelemente und Maschinenkonstruktion (Institute of Machine Elements and Machine Design), Technical University Dresden, 30th July 2005, http://www.me.tu-dresden.de/forschung/dynamik.shtml.de

[11]   Gasch, R.; Nordmann, R.; Pfützner, H.: *Rotordynamik (Rotordynamics)*, 2nd edition, p. 637, Springer, Berlin, 2002

[12]   Det Norske Veritas (DnV): *Guidelines for Design of Wind Turbines*, 2nd ed., 2002

[13]   Kaiser, K.: *Luftkraftverursachte Steifigkeits- und Dämpfungsmatrizen von Windturbinen und ihr Einfluß auf das Stabilitätsverhalten (Stiffness and damping matrices of*

wind turbines caused by aerodynamic forces, and their influence on stability), VDI-Verlag, Series 11, No. 294, Düsseldorf, 2000

[14]   Hansen, M.H.: *Improved modal dynamics of wind turbines to avoid stall-induced vibrations*, Wind Energy, 6, 179-195, 2003

[15]   Quarton, D.C.: *The Evolution of Wind Turbine Design Analysis – A Twenty Years Progress Report*, Wind Energy, 1, 5-24, 1998

9 Guidelines and analysis procedures

9.1 Certification

In general, *certification* is understood as the verification of entire companies, operating procedures or products for their compliance with certain criteria. Today, the certification of products and manufacturers is a standard in the sector of wind energy. The certification (of conformity) is a measure carried out by independent institutions (or persons) which document that a certain product, procedure or service is in accordance with a certain standard or some other specific normative document (EN 45011, EN 45012, EN 45013).

Fig. 9-1 gives an overview of the fundamental elements of the wind turbine type certificate.

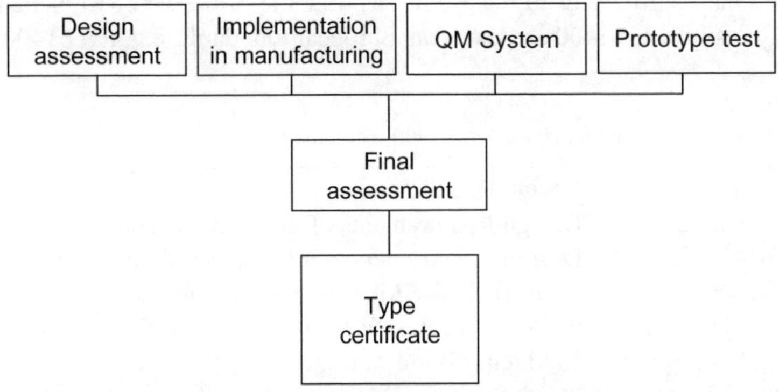

Fig. 9-1 Elements of the wind turbine type certificate [1], modified

Certification is based on the above mentioned standards or normative documents. In this context, distinction should be made between *specific standards for the wind energy sector* and *general standards*. At first an overview of the specific standards for the wind energy sector is given, subsequent sections will make reference to general standards.

Apart from the certification of product and manufacturer there is partly an additional *project specific assessment*, especially if the site conditions are not covered by the general certification scheme.

The so-called *due diligence* is also a site-specific assessment which focuses on the assessment of the recoverability of a project. This comprises design aspects, e.g. if the site conditions are covered by the general certification and if all design assumptions are applicable. Since in general the client of such a due diligence is the future project operator or the investor, this assessment is extended to questions

"

on the accuracy of the yield prognosis as well as the prospective guarantee and service concepts.

9.1.1 Standard for certification: IEC 61400

The International Electrotechnical Commission (IEC) has bundled together under the number 61400 several standards relating to different sectors of the wind energy. The work at these standards is a *continuous process* which led to several revisions of some of these standards in the past years. *The reader should make sure that he is always informed on the current revisions of the standards.*
All these standards have been developed especially for the wind energy sector. The *IEC 61400-1* is of very high importance because it describes the basic - relationships between site data and load assumptions for the simulation of a wind turbine and is therefore the starting point for any design procedure.
It has to be noted that due to their date of issue the guidelines mentioned in the following may partly refer to older editions of some of the IEC 61400 standards. Several parts of IEC 61400 are issued as European standards, e.g. EN 61400-1.

Table 9.1 Overview of the standards IEC 61400 according to [13]

IEC 61400–1	Design Requirements
IEC 61400–2	Design Requirements of Small Wind Turbines
IEC 61400–3	Design Requirements for Offshore Wind Turbines
IEC 61400–11	Acoustic Noise Measurement Techniques
IEC 61400-12-1	Power Performance Measurements of Electricity Producing Wind Turbines
IEC 61400-13	Measurement of Mechanical Loads
IEC 61400-14	Declaration of apparent sound power level and tonality values
IEC 61400-21	Measurement and Assessment of Power Quality Characteristics of Grid Connected Wind Turbines
IEC 61400-22	Conformity testing and certification
IEC 61400-23	Full-Scale Structural Testing of Rotor Blades
IEC 61400-24	Lightning Protection
IEC 61400–25	Communications for Monitoring and Control of Wind Power Plants

9.1.2   Guidelines for the Certification of Wind Turbines by Germanischer Lloyd

This document [1] is not a standard but a *guideline*, a recommendation for a certain procedure. The fundamental elements are based on the IEC 61400 standards, but the main focus is the description of the *design and analysis procedures for the components of a wind turbine*. In this context, the major focus is the product safety. The latest edition in 2010 mentions the national requirements in Germany, Denmark, Netherlands, Canada, India, China and Japan.

9.1.3 Guidelines for Design of Wind Turbines by DNV

Similar to the GL guideline this document [2] by Det Norske Veritas, DNV, describes several aspects of the design procedure of wind turbines. The fundamental points and relationships are treated which are essential to attain certification. Again, the basis for this procedure is the IEC 61400 and further relevant standards.

9.1.4 Regulation for Wind Energy Conversion Systems, Actions and Verification of Structural Integrity for Tower and Foundation by DIBt

In Germany, the tower and foundation of a wind turbine are considered as a building structure and are subject to the relevant permission procedure [3]. The Deutsche Institut für Bautechnik (DIBt, German Institute for Civil Engineering) issued a specific regulation which describes the required analyses for a wind turbine. Although this regulation is not a standard it has a normative character. Deviations from the given approaches are not allowed. This regulation was reissued in 2004 in a second, heavily revised edition with load case definitions based on the 2nd edition of IEC 61400-1. In other parts of the regulation, concepts from civil engineering are applied.

9.1.5 Further standards and guidelines

Apart from the standards and guidelines mentioned above there are some country-specific versions or differing approaches. In the Dutch NVN guideline the IEC standards are adopted, but in some parts there are modifications, e.g. concerning the required safety factors. The Danish standards DS472 (Loads and Safety of Wind Turbine Constructions) and a series of guidelines govern the requirements for wind turbines in Denmark (e.g. http://www.wt-certification.dk/UK/Rules.htm).

9.1.6 *Wind classes and site categories*

One basic concept of the IEC standards, and the guidelines which refer to it, is the classification of the environmental conditions in *wind classes*. In the IEC 61400-1, 2nd edition, there are four different wind classes defined. In each class there are fixed relationships between the fundamental design parameters: annual average (i.e. mean) wind speed V_{ave}, characteristic turbulence intensity T_I and extreme wind speed V_e, shown in table 9.2. The 3rd edition of IEC 61400-1 was issued in 2005, but many national standards and guidelines are still based on the 2nd edition.

Table 9.2 Design wind speeds V and characteristic turbulence intensities T_I of the type classes according to IEC 61400-1, 2nd. ed.

Wind class	I	II	III	IV		
Extreme load					Unit	Comment
$V_{ref} = V_{m50}$	50.0	42.5	37.5	30.0	m/s	Expected extreme 50-years wind, 10-min average
V_{e50}	70.0	59.5	52.5	42.0	m/s	Extreme 50-year wind, 3-sec average
V_{m1}	37.5	31.9	28.1	22.5	m/s	Expected annual extreme wind, 10-min average
V_{e1}	52.5	44.6	39.4	31.5	m/s	Expected annual extreme wind, 3-sec average
Fatigue strength						
V_{ave}	10.0	8.50	7.50	6.00	m/s	Annual average wind speed at hub height
T_{IA}	18.0	18.0	18.0	18.0	%	Characteristic turbulence intensity at V = 15 m/s, Class A
T_{IB}	16.0	16.0	16.0	16.0	%	Characteristic turbulence intensity at V = 15 m/s, Class B

If one follows this concept for the wind turbine design, a wind class must be selected at the *beginning of the design procedure* which corresponds as much as possible to the conditions in the target market and therefore allows a design which is not only economically efficient but also meets the expected loads. A series of load case definitions is derived from the chosen wind class, which form the basis for analysing the wind turbine.

Later, in the *realisation planning* of a certain wind farm project (cf. chapter 15) a site-specific assessment of the chosen wind turbine must be made for checking whether the design wind class covers the real on-site wind conditions including the wind farm effects. Otherwise, new load calculations must be performed in order to show that the wind turbine design includes enough reserves and is therefore still suitable. For sites which do not correspond to the given relationships it is possible to define a "special class" in which the site-specific relationships may defined arbitrarily.

9.1.7 Load case definitions

In the concept of the IEC 61400 *two different categories of load cases* are analyzed: on the one hand, the extreme load cases where the stability of the wind turbine has to be assured (ultimate limit state), and on the other hand, the load cases under operation which are used for the fatigue strength analysis.

Extreme load cases are mainly defined by the occurring *gust wind speeds*. Typical design load cases (DLC) which cause high stress are the 50-year extreme wind speed (DLC 6.1) and the extreme operating gust combined with an extreme change of the wind direction (DLC 1.3). However, there are also operating conditions which result from the failure of the control systems and cause critical load cases, e.g. the failure of a pitch drive followed by an emergency stop at overspeed, cf. section 8.1.3. In order to describe the spatial structure of the gusts, deterministic gust shapes are assumed for some of the load cases, but stochastic time series with high turbulence intensity are applied as well.

The *fatigue strength load cases* are basically calculated from stochastic time series of the three-dimensional wind field. For several mean wind speeds between cut-in wind speed and cut-out wind speed, 10-min time series of the wind are generated and then applied in the simulation program to obtain the stresses. These are then weighted according to the mean wind speed of the selected wind class and extrapolated to the operating lifetime of 20 years, cf. section 8.3. Additionally, the start-up and shut-down cycles are simulated in order to take into account the corresponding loadings as well.

Safety factors are applied to all load cases. For the individual load cases and load components different magnitudes of the factor are defined according to the accuracy of the predictability and the probability of occurrence, as well as the consequences for the wind turbine safety.

9.2 Analysis concepts

This section treats the basics of the analysis of wind turbine components. In the following sections 9.3 to 9.5 peculiarities are discussed for the three typical components: tower, rotor hub and rotor blade.

The analysis of the wind turbine components is based on the assumption that certain states, so-called ultimate limit states, may not be exceeded. The ultimate limit states of the component stability comprise:
- *Exceedance of the maximum strength* (fracture, buckling, fatigue) and
- *Loss of equilibrium of the steady-state condition of a structure*, e.g. rigid body tilting.

9.2.1 *Ultimate limit state and the concept of partial safety factors*

In order to achieve a sufficient reliability the *ultimate limit state* must be analyzed for each component in the course of the design. For this purpose, the *internal forces and moments* (i.e. bending moment, transverse force and torsion) acting on the component are considered. The *partial safety factors γ_F for the internal forces and moments* account for the probability of the particular load occurring (e.g. normal and extreme load, fatigue load), the probable deviation of the load from the characteristic values and the accuracy of the load determination, Fig. 9-2. The characteristic loads F_k are obtained in general by simulation. The design loads F_d to be assumed during the analysis are then $F_d = \gamma_F F_k$.

The *partial safety factors γ_M for the materials* (i.e. material strength of the component) are used to take into account the dependence of the strength on the type of material, the processing, the component shape and the influence of the manufacturing process. The characteristic resistances R_k divided by the partial safety factor then gives the design resistances $R_d = R_k / \gamma_M$.

The *analysis procedure for the component* is then performed in such a way that for the individual ultimate limit state the stress S corresponding to the relevant design load F_d is determined: $S = S(F_d)$. The resulting stresses (i.e. stress, strain, bending) then have to remain below the design resistances: $S < R_d$.

The values of the partial safety factors γ_F and γ_M are given in the standards and guidelines (GL guideline, IEC 61400-1). Since in the individual documents the safety factors may differ, *the analysis of a wind turbine must be performed with a system of consistent standards.*

To some extent national rules and guidelines may apply other systems than the presented system of partial safety factors. In Germany, for example, the foundation is analysed according to DIN1054 using global safety factors, i.e. no difference is made there between the safety factors for load and material.

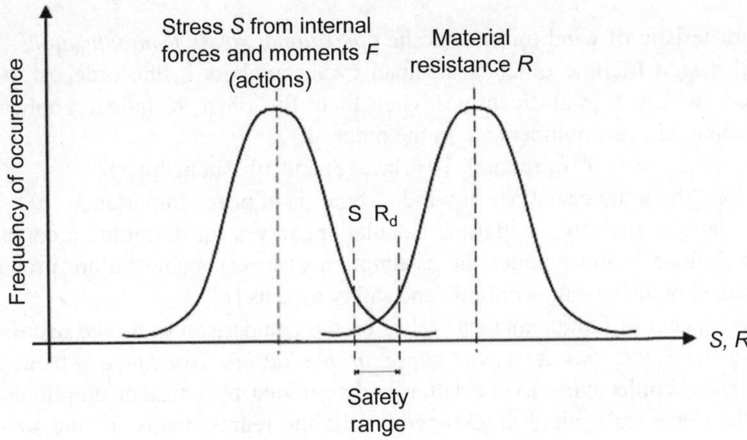

Fig. 9-2 Frequency distribution of the stress S in the section and resistances R

9.2.2 Serviceability analysis

A typical criterion for serviceability is the *distance between the tip of the rotor blade and the tower*. In operation, the blade tip should never touch the tower surface. Since the rotor blade bends strongly due to the thrust forces the bending of the blade and the distance of the rotor blade tips to the tower is analyzed for every load case (extreme loads and fatigue loads, cf. section 9.5.1). At maximum bending this distance may not go below 30% of the static rest position of the rotor blade [1].

With regard to the *vibration behaviour of the entire wind turbine* it has to be avoided that under operation the rotational frequency or its harmonics coincide with natural frequencies of the system, cf. chapter 8. For the *rotor blade,* the typical natural frequencies of the sub-system are the first flapwise and the first edgewise bending natural frequency. Higher order natural frequencies of the rotor blade are in most cases in uncritical ranges. For the *sub-system of nacelle mass, tower and foundation* the first and second bending natural frequency of the tower are considered. In lattice towers which are, compared to steel tube and concrete towers, relatively soft in terms of torsion, at least the first torsional natural frequency must be taken into account. For the analysis, the frequency ratios are put into the so-called Campbell diagram for all relevant rotational speeds in the operating range, cf. sections 8.1 to 8.3.

9.2.3 Basics of fatigue analysis

A characteristic of wind turbines is the *operational stress from vibrations*. In the typical design lifetime of 20 years load cycle numbers in the order of 10^9 are reached, which is significantly higher than for other technical applications. Common load cycle numbers are in the order of

$$5 \cdot 10^6 \text{ (airplane), } 10^7 \text{ (bridge) and } 10^8 \text{ (helicopter).}$$

Therefore, the fatigue analysis of wind turbines is of prime importance.

The fatigue analysis, or lifetime calculation, serves for designing a component with a defined lifetime under the assumption of a reasonable failure probability determined by technical, economic and safety aspects [5].

The concept of fatigue analysis relies on the comparison of a *time series of the stresses* with the *characteristic curve of the design resistance* which is the *S/N curve* (Wöhler curve). The latter is determined by constant amplitude tests until the (technical) initial crack occurs. It is the representation of the stress (or strain) amplitude S versus the number N of stress cycles performed.

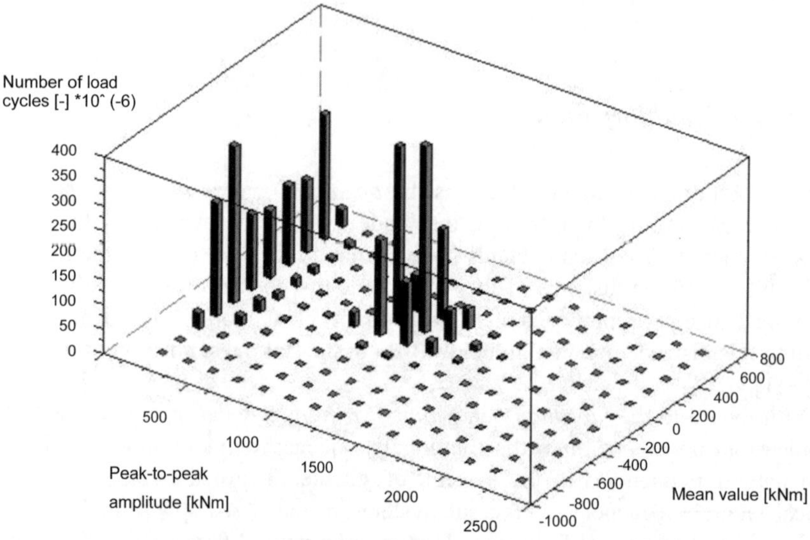

Fig. 9-3 Load cycles versus mean value and amplitude (peak-to-peak) from rainflow count

For the damage accumulation calculation (i.e. the comparison of the stresses with the *S/N* curve) the time series of the stresses is either determined by measurement or simulation. It is divided into the single stress cycles and then compressed into a spectrum. Different counting methods are common, but the damage content is at best represented by the rainflow count method. This is a two-parametric method with which information on the amplitudes and the corresponding mean values is obtained and displayed as a matrix, Fig. 9-3.

For an existing load matrix the relevance for damage accumulation of the load must be determined depending on the component (i.e. its shape). Component failure is accelerated, for example, by notches. This behaviour is expressed by the stress concentration factor and the notch effect number. Depending on the shape and material, the magnitude of the mean stress leads to an accelerated failure. The choice of an *adapted S/N curve* is the key to considering these influences. *S/N* curves for different load cases are given and sorted by categories in [6, ECCS Technical Committee]. The sensitivity to the mean stress is quantified there as well. The principle is shown in Fig. 9-4. $P_{\ddot{U}}$ is the failure probability.

In a next step, the contribution of each stress cycle to the cumulative damage ratio is determined. Typically, the linear damage accumulation hypothesis according to Palmgren-Miner is used where the stress is classified according to mean stress S_{mi} and applied stress amplitude S_{ai}, which is half of the applied peak-to-peak stress range $\Delta S = 2\,S_{ai} = S_{max} - S_{min}$. In the class i, N_i is the number of endured stress cycles according to S-N curve which is tolerable in the design fatigue life, and n_i is the applied number of stress cycles drawn from the fatigue load spectra. The partial damage ratio ΔD_i of class i is then

$$\Delta D_i = n_i\,/\,N_i \qquad \text{with } n_i = n\,(S_{ai},\,S_{mi}) \text{ and } N_i = N\,(S_{ai},\,S_{mi}) , \qquad (9.1)$$

where n_i : Number of applied stress cycles of class i occurring during
              design fatigue life, drawn from fatigue load spectra
         N_i : Number of endured stress cycle of class i according to S-N curve,
         S_{ai} : Applied stress amplitude
         S_{mi} : Mean stress

The cumulative damage ratio is the linear sum of the partial damages:

$$D = \Sigma\,\Delta D_i. \qquad (9.2)$$

A cumulative damage ratio of $D = 1$ corresponds to a calculated component failure exactly at the end of the planned service life, which is in general 20 years. Therefore, the aim of the design is that in all considered sections the cumulative damage ratio D is below one:

$$D < 1 ,$$

meaning that the calculated design fatigue life exceeds the planned service life.

If the considered material has an fatigue strength limit, as it is the case e.g. for structural steel in a mildly corrosive environment, and if the amplitudes of all load cycles are below this fatigue strength limit, then there will be no damage, no matter how many stress cycles occur.

However, it is more often the case, that some of the stress cycles exceed the fatigue strength limit or that the material, e.g. aluminum, has no fatigue strength limit. For this case the fatigue analysis can be performed using the above equations.

For *welding seams* it was found that under a loading, the residual stresses, which are nearly inevitable, have an influence equivalent to the mean stress. Accordingly, the influence of the actual mean stresses is lowered, so the permissible stress cycle number depends in the end only on the stress amplitude. This allows a simplified analysis at which the rainflow matrix is reduced to an equivalent constant-range stress spectrum and the influence of the mean stress is included in the S/N curves. This procedure and the corresponding *S/N* curves are found, for example, in Eurocode 3.

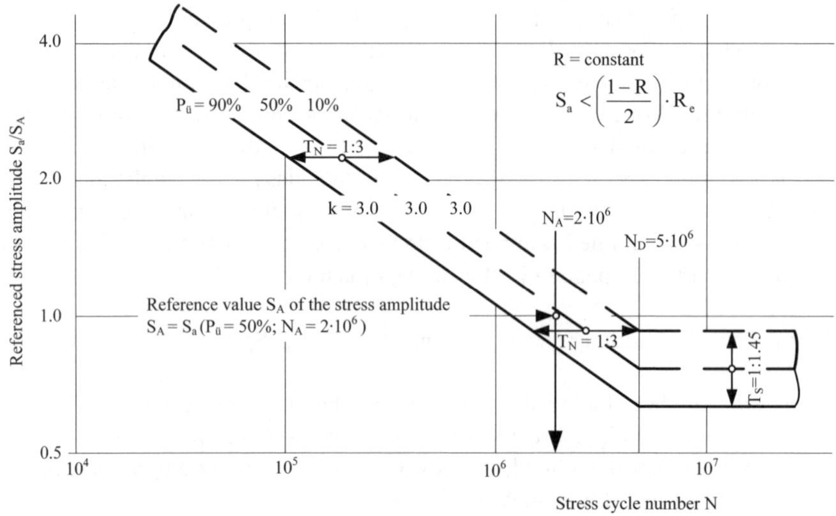

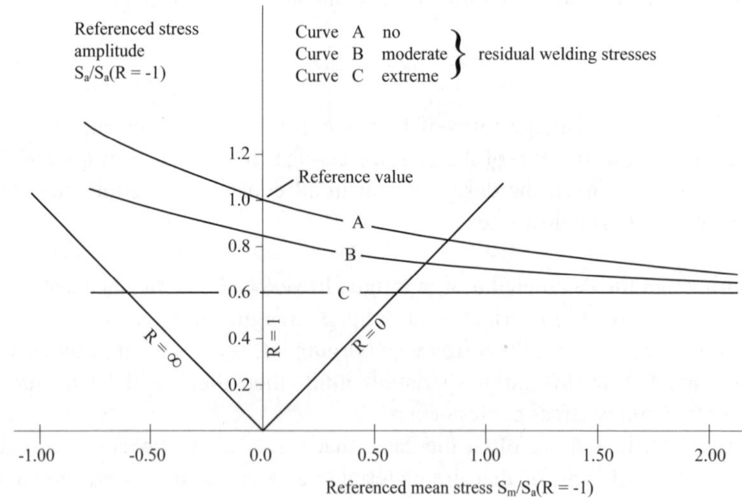

Fig. 9-4 Normalised *S/N* curves (Wöhler curve) for welded structures and dependency on the mean stress [7]

9.3 Example: Tubular steel tower analysis - mono-axial stress state and isotropic material

9.3.1 Ultimate limit state analysis, analysis of extreme loads

Determination of the stress in the wall of the tower

A substantial analysis of the tubular steel tower treats the determination of the stresses occurring in the wall of the tower. Here, mostly global approaches are applied in which the results of the simulation model including the additional loads from 2nd order theory (deformed system) are investigated for several sections. The stresses are then determined section by section using the simple approach: Bending stress equals local moment in the section divided by the local section modulus.

The stresses so obtained are then compared to the admissible values according to DIN 18800 or Eurocode 3 and must be smaller than these.

For edge-stiffened openings, required, at the doors, for example, local FEM analysis is performed for investigating the stress concentration produced by the stiffness changes. These stress concentrations must also be applied in later fatigue analyses. Alternatively, for the stress analysis the complete structure can be modelled using FEM, but this requires a greatly increased computing time.

Analysis of buckling

The tower has a relatively small wall thickness ($t = 10...50$ mm) in relation to its diameter (in the range of meters). Therefore, one possible case of failure is the buckling of the tower shell. In this non-linear failure case there is plastic deformation of the tower shell which leads to a new state of equilibrium, Fig. 9-5. The corresponding analysis is given by DIN 18800-4. For wind turbines towers a
correction for buckling under bending stress must be taken into account since this is the main stress type [8]. At local spots and sections, for example openings, further analysis is required.

Fig. 9-5 Buckling of a tower

Zone 1: Mainly linear progression, clamping force reduction between the flanges with completely compressed contact area

Zone 2: Successively increasing flange gap

Zone 3: Open flange gap with variable slope depending on the geometry. The limiting state for the flange with gap is the load bearing on one edge

Zone 4: Progressing plasticitation of the connection until fracture

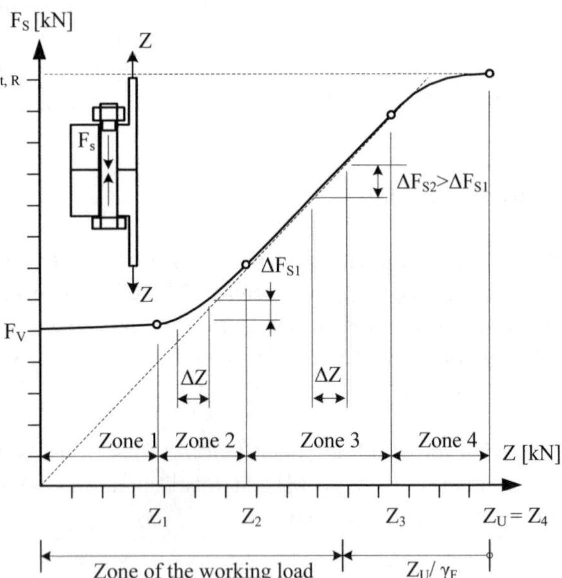

Fig. 9-6 Loading zones of an ideal bolted flange connection depending on the load according to [10]. Preload F_v of the bolt, force Z in the tower segment and additional bolt load F_S

Analysis of the flange connections

Due to production and transportation requirements the towers of wind turbines are manufactured in several sections and bolted together by their ring flange connections during erection on site. For this type of connection there are several analysis procedures [9]. In the following, only the basic relationships are presented.

Depending on the geometry of flange and the bolt as well as the pre-stressing conditions, there is a certain ratio of the force Z in the tower's wall segment and in the bolt F_S. The diagram in Fig. 9-6 shows the different zones for an ideal bolted flange connection.

In order to use the preloading of the flange-bolt system according to Fig. 9-6, the specifications have to be respected. In particular, one must ensure that the flanges are even and parallel since a gap in the flange connection leads to a strong increase in loading on the bolts.

9.3.2 Fatigue strength analysis

The fatigue strength analysis of steel towers follows in general the *nominal stress approach*. The loadings are provided by the simulation programs through the internal forces and moments at the position of the analyzed component and transferred to the local stress level. Then, the damage sum is calculated using the load spectra and applicable class of geometric discontinuity from DIN 4131 or Eurocode 3.

For more complex components, e.g. the flanges and openings, FEM models are necessary to determine the stress level. One must bear in mind that the stress concentrations calculated by FEM analysis may not be combined with the reduced design *S/N* curves for the corresponding component since these curves already take into account the stress concentration.

9.3.3 Serviceability analysis, natural frequencies analysis

In order to determine the natural frequencies of the tower, FEM models created for the global stress calculations are for the most part applied. Alternatively, there are approximation formulae, e.g. according to Morleigh in [9].

The *natural frequency of the tower* should have a distance of more than 10% to the exciting frequencies of the wind turbine, 1Ω and 3Ω. For a large wind turbine the admissible and inadmissible frequency ranges are shown in Fig. 9-7.

When calculating the natural frequencies, correct and reasonable clamping conditions have to be considered. If the soil is too soft (e.g. around the monopiles for offshore wind turbines) this may unduly lower the natural frequencies of the overall system. It is usual to define a required minimum spring stiffness for the tower design, which then has to be verified in the planning procedure by a geological survey. Fig. 9-8 shows the relations between soil spring and natural frequency.

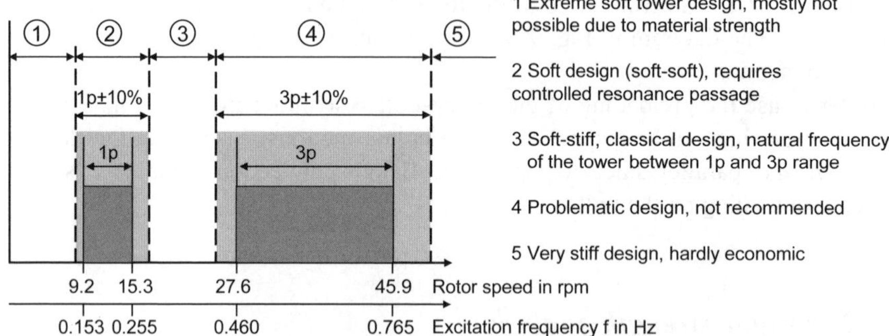

Fig. 9-7 Frequency ranges for the tower design [10]

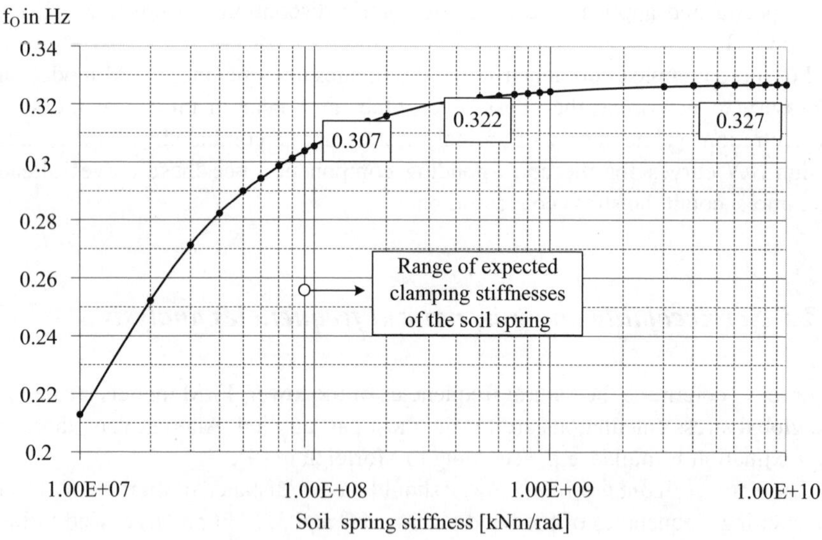

Fig. 9-8 Natural frequency f_O of the tower depending on the soil spring stiffness [9]

9.4 Example: Rotor hub analysis - multi-axial stress state and isotropic material

Rotor hubs of modern wind turbines are usually components cast from EN-GJS-400-18-LT (formerly known as GGG-40.3), cf. chapter 3. This material allows the production of complex geometries with good material strength and relatively low weight.

9.4.1 Geometric design

The hub links the blades to the rotor shaft flange; therefore its shape and size is determined by these components, especially by the diameter of the blade flange connection and the blade.

If the *connection dimensions* are known, the *actual shape of the hub* has to be determined. Due to the complex geometry FEM programs are required for this purpose. Several years ago, it was common to choose a sphere or star shape, Fig. 9-9, where the latter offers the possibility of increasing the rotor diameter using extenders at the cylindrical parts of the hub. Nowadays, the rotor hub is topologically optimised: an optimisation algorithm in the FEM program varies the wall thickness according to the local stress and searches for a solution with minimal use of material. These hubs then have the shape of an apple with additional openings in zones where no material is needed.

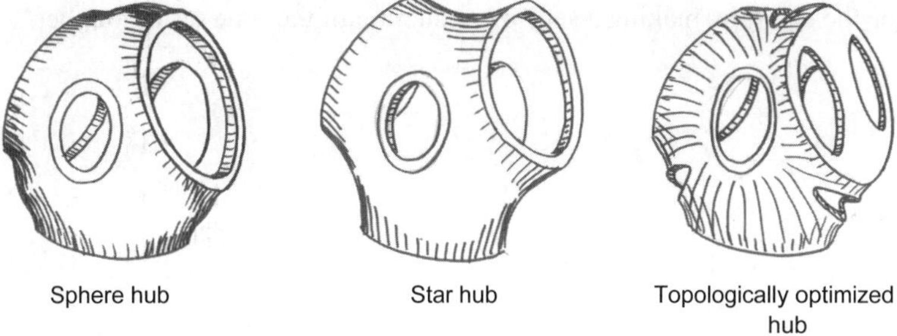

Fig. 9-9 Typical shapes of rotor hubs

9.4.2 *Ultimate limit state analysis - critical section plane method*

In contrast to the analysis of the wind turbine tower presented in section 9.3, there are significant spatial and temporal changes of the principal stresses in the rotor hub, due to its geometry and the temporal superposition of the loadings from the rotor blades. This leads to a greatly increased analysis effort, especially for the lifetime calculation.

A common approach for the periodically changing principal stress is the *method of the critical section plane*. Fig. 9-10 shows the algorithm of this procedure schematically; this is described briefly in the following:
-    For each load acting on the hub a unit load case is calculated.
-    Using these unit load cases the so-called "hot spots" are determined, which are the zones of highest stress in the structure.
-    For these zones of the hub shear and normal stress is then calculated for each time step of the load time series. From these stresses an equivalent stress is determined using common approaches.
-    For the ultimate limit state analysis the result is now given since this load case is only a "special case" of the fatigue strength analysis: it is one time series with only one time step.
-    The actual fatigue strength analysis is carried out, using a modified design S/N curve (cf. section 9.4.3), by a rainflow classification followed by the damage accumulation. This has to be performed individually for each individual hot spot, each section plane and each time step.
-    The following damage accumulation calculation of the partial damages finally gives the plane of maximum damage.

Naturally, this procedure needs excessive computing time, so attempts are made to reduce the effort by making a sensible reduction of the time series number.

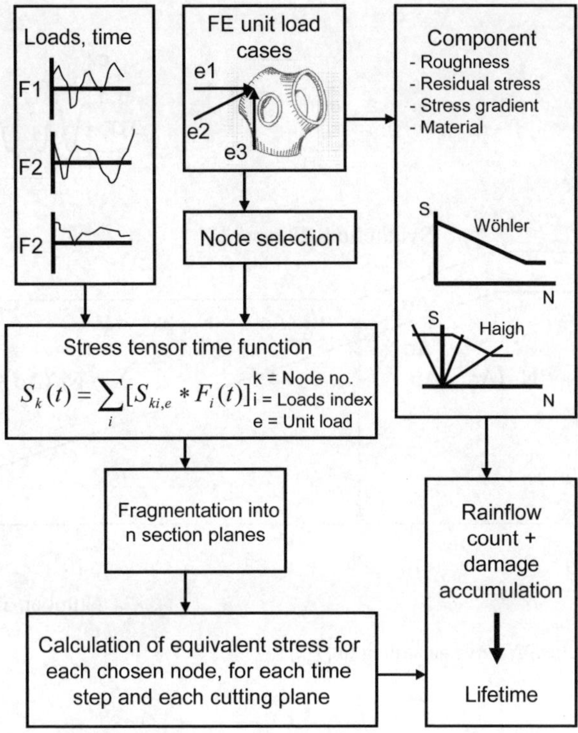

$$S_k(t) = \sum_i [S_{ki,e} * F_i(t)]$$

Fig. 9-10 Flow chart of the lifetime fatigue strength analysis of rotor hubs [11]

9.4.3 Fatigue strength analysis - procedure-dependent S/N curves

While in section 9.3 the fatigue strength analysis was performed using the nominal stress concept, the *determined equivalent stresses of cast components* are "local" stresses which means that they already include the influence of the component shape, local notches and supporting effects. Therefore, a different approach must be selected for the applied *S/N* curves which takes into account in particular the influences not considered up to now, for example the surface quality and the porosity of the material.

For this purpose, synthetic *S/N* curves are generated which describe the influence of exactly these boundary conditions. The procedure relevant for the certification of wind turbines is described in [1]. The following Fig. 9-11 shows the construction of a synthetic *S/N* curve. In this diagram the reduction factors $S_{P\ddot{U}}$ and S_d consider the survival probability and casting flaws including the inspection method used.

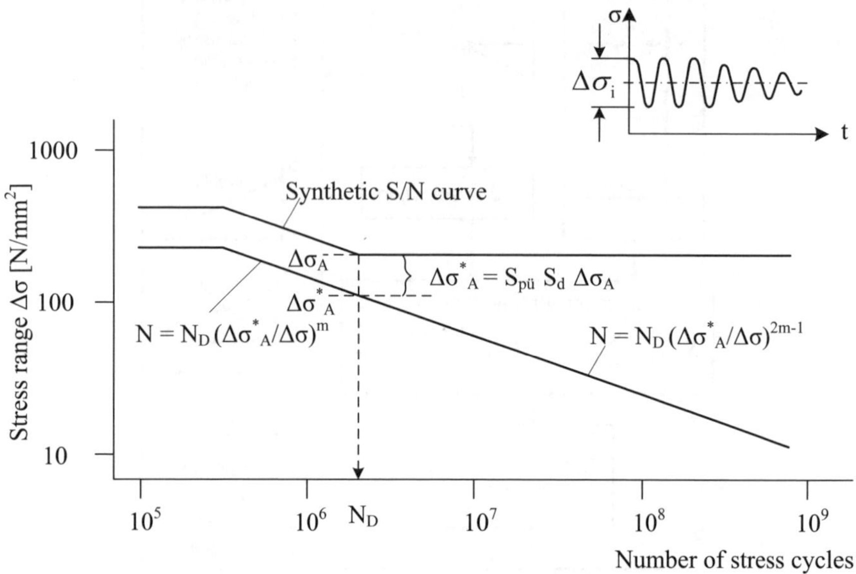

Fig. 9-11 Synthetic *S/N* curve according to [1]

9.5 Example: Rotor blades analysis - mono-axial stress state and orthotropic material

In contrast to the components treated up to now, the particular difficulty in the *design of the rotor blades* is the elementary relationship between geometry and stresses.

If the shape of the rotor blade changes, there are fundamental consequences for the load cases to be analyzed, for the stresses and last but not least for the expected energy yield as well. Therefore, the *blade design is an iterative process* and it is quite often necessary to run through it several times before the optimum compromise between aerodynamic efficiency, weight and costs of the rotor blade is found.

For design and analysis the rotor blade structure is divided into its functional elements, cf. Fig. 3-9, 3-11, 3-12 and 9-12:

- Supporting structure: Chords, webs and other unidirectional reinforcing elements
- Shells
- Connecting elements: blade bolts and bonding

The material strength design concentrates on the dimensioning of the chords, spars, the unidirectional layers and the connecting elements. For the structural mechanics the rotor blade is reduced to a *double-T beam*. For the thin-walled shell the key issue is its stability while, in the first approach, its contribution to the supporting structure is ignored.

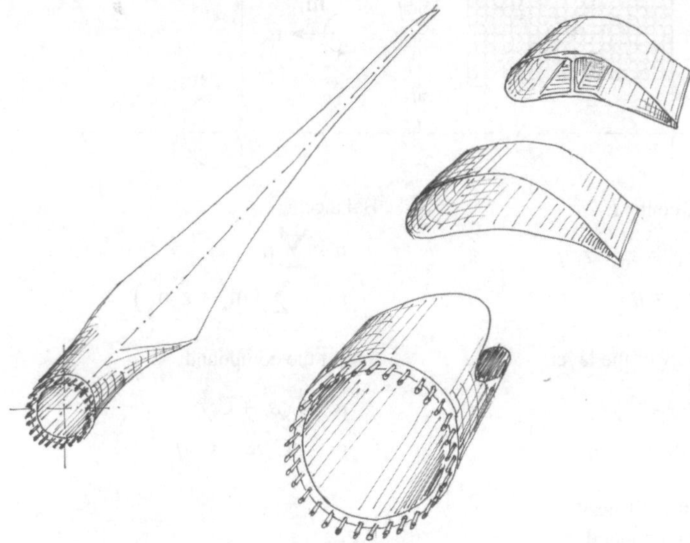

Fig. 9-12 Functional elements of a rotor blade (web, shell, flange)

9.5.1 Concept of admissible strain for analysis of the chords

In a first run a design of the rotor blade is created with the main dimensions and aerodynamic profiles already determined and selected. The loads are then calculated for this design. Section by section these loads are used for the following optimisation of the *chord construction*. For components made out of fibre reinforced plastics the maximum strain occurring in the laminate serves as the design criterion. For given internal sectional loads the strain behaviour may be influenced, on the one hand, by changing the laminate construction, and on the other hand, also by changing the geometry of the individual section, e.g. by modifying the thickness ratio. Due to the anisotropic structure of the laminate the determination of the strain has to be performed for each individual layer of the chord in order to take into account the influence of the fibre direction.

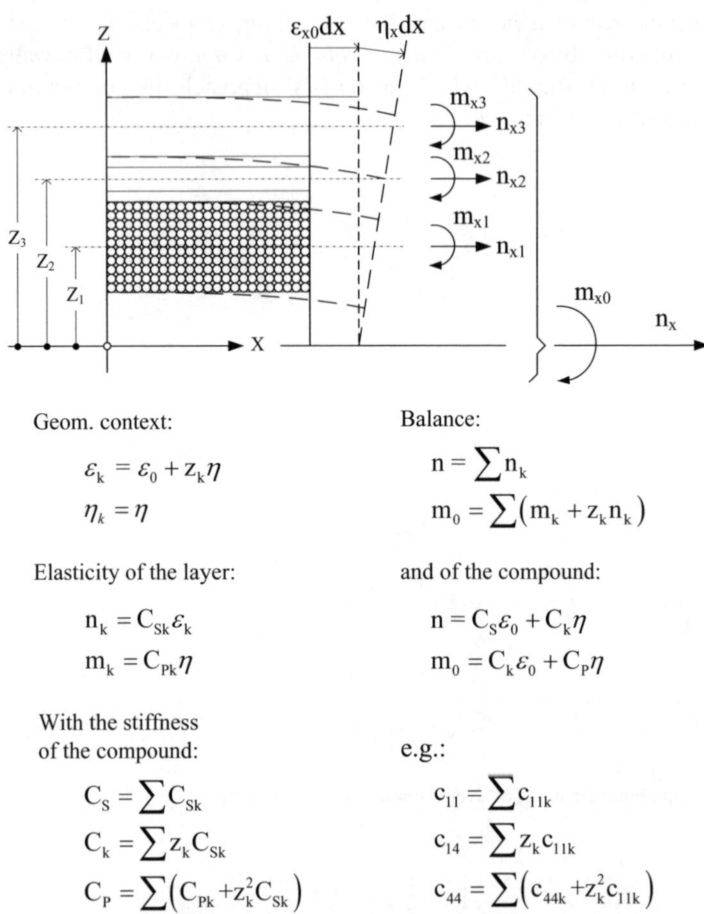

Geom. context:

$$\varepsilon_k = \varepsilon_0 + z_k \eta$$
$$\eta_k = \eta$$

Balance:

$$n = \sum n_k$$
$$m_0 = \sum \left(m_k + z_k n_k \right)$$

Elasticity of the layer:

$$n_k = C_{Sk}\varepsilon_k$$
$$m_k = C_{Pk}\eta$$

and of the compound:

$$n = C_S \varepsilon_0 + C_k \eta$$
$$m_0 = C_k \varepsilon_0 + C_P \eta$$

With the stiffness
of the compound:

$$C_S = \sum C_{Sk}$$
$$C_k = \sum z_k C_{Sk}$$
$$C_P = \sum \left(C_{Pk} + z_k^2 C_{Sk} \right)$$

e.g.:

$$c_{11} = \sum c_{11k}$$
$$c_{14} = \sum z_k c_{11k}$$
$$c_{44} = \sum \left(c_{44k} + z_k^2 c_{11k} \right)$$

Fig. 9-13 Laminate matrix, fibre direction and transformation [12]

If the requirement is fulfilled that the admissible level of strain is not exceeded, the fatigue strength analysis is performed as a last step. The zones of unidirectional layers are critical. Depending on the material, the material values are either found in the literature or, in most cases, material tests for this special blade need to be performed. The strain and the characteristic material values give the total bending of the blade which may not exceed the limiting value given for the corresponding extreme load cases: the blade must neither come into contact with the tower nor fail. If the structure of the rotor blade fulfils the given requirements another load calculation is performed for the geometry and mass distribution obtained. Often, there are now deviations from the original assumptions, so one more loop of the iteration is necessary.

9.5.2 *Local component failure*

In the next step following the design of the supporting structure (chord-web system), single elements of the blades are investigated for local failure. Failure from buckling may not occur, neither in the shell of the rotor blade nor at the webs or chords. Critical zones are found in particular where the distance between supports is large or where the compressive stress is at a high level. FEM models are applied for the buckling analysis.

The bonded joints of the rotor blade must be investigated for shear and peeling. The individual limiting values result from the properties of the chosen bonding material.

The bolting zone of the blade root is investigated using FEM models as well (cf. Fig. 9-12 mad 3-12). Here, depending on the chosen type of connection, the forces that occur must be fed into the material by the bonding of the inserts, or by the surface pressure of the transverse bolts. The blade bolts are analyzed by the common procedure according to VDI 2230.

9.5.3 *Choice of materials and production methods*

The rotor blades for the first series of wind turbines in the 1980s were already manufactured out of glass fibre fabrics in a polyester resin matrix. This material combination is still widely used today because of the following advantages: the purchase price of the basic materials is relatively low, the processability is good and the endurance is relatively high. In parallel, epoxy resins were applied for the matrix because of the better fatigue strength properties of these resin systems (cf. Fig. 3-10.) and less shrinking during the hardening process. Moreover, alternatives to the glass fibre structure are investigated time and again: the stiffness is a crucial design parameter especially for large wind turbines. Therefore, carbon fibre structures are being increasingly used for rotor blades. However, the most relevant obstacles to the application of carbon fibre are the high price, higher production requirements and the problem of lightning protection since the blade structure is then electrically conductive.

Apart from carbon fibre, wood is also applied in combination with epoxy resin as rotor blade material. In this case, the stiffness is not such an issue, but the good damping properties of wood are of importance. These rotor blades are often applied for stall-controlled wind turbines where additional structural damping improves operating behaviour.

Until the end of the 1990s rotor blades were manufactured manually, Fig. 9-14. The laminate structure was built up layer by layer in a mould and then soaked with resin.

With growing piece numbers and component sizes the production is becoming ever more automated. Two different approaches have been established:

- The pre-preg method (also common in the aircraft industry) using already soaked glass or fibre fabrics which are hardened in a heatable mould or
- The vacuum infusion method where the glass or fibre structure is covered with a plastic film to seal the mould, and the resin is soaked in automatically under evacuation, cf. Fig. 3-11.

Both methods improve the product quality (e.g. less air inclusions which reduce material strength), allow control of the amount of resin applied (material savings) and reduce the holding times in the moulds (shorter production time).

Fig. 9-14 Manufacturing of rotor blades at LM, DK; laying of the glass fiber fabrics

References

[1]   Germanischer Lloyd Industrial Services GmbH: *Richtlinie für die Zertifizierung von Windenergieanlagen (Guideline for the Certification of Wind Turbines)*, Hamburg 2003, 2010

[2]   Risoe/DNV: Guidelines for Design of Wind Turbines, 2nd ed. 2002

[3]   DIBt (German Institute for Civil Engineering): *Richtlinie für Wind turbinen Einwirkungen und Standsicherheitsnachweise für Turm und Gründung (Regulation for Wind Energy Conversion Systems, Actions and Verification of Structural Integrity for Tower and Foundation)*, series B, vol. 8 DIBt Berlin, 2004

[4]   Schneider: *Bautabellen für Ingenieure (Construction tables for engineers)*, 16th ed., Werner Verlag Neuwied

[5]   Gudehus, H.; Zenner, H.: *Leitfaden für eine Betriebsfestigkeitsrechnung, Empfehlung zur Lebensdauerabschätzung von Maschinenbauteilen (Guideline for fatigue strength analysis, recommendations for the lifetime estimation of machinery components)* , 3rd ed., Verlag Stahleisen mbH, Düsseldorf 1995

[6]   European Convention for Constructional Steelwork: *Recommendations for the Fatigue Design of Steel Structures*, ECCS Publication 43, 1985

[7]   Haibach, E.: Betriebsfestigkeit: *Verfahren und Daten zur Bauteilberechnung (Methods and data for component design)*, Düsseldorf, VDI-Verlag 1989

[8]   Schaumann P., Seidel M., Messenburg L.: *Weiterentwicklung des Beulnachweises von WEA-Türmen mit elasto-plastischen FEM-Analysen (Improvement of the buckling analysis of wind turbine towers using elasto-plastic FEM analysis)*, DEWEK 1998

[9]   Petersen C.: *Stahlbau - Grundlagen der Berechnung und baulichen Ausbildung von Stahlbauten (Steel construction – Basics of the dimensioning and the constructional formation of steel constructions)*, 3rd revised and enhanced edition. Braunschweig/ Wiesbaden: Vieweg 1993

[10]   Seidel, M.: *Zur Bemessung geschraubter Ringflanschverbindungen von Windenergieanlagen (On the dimensioning of bolted ring flange connections of wind turbines)*, PhD thesis at the Institute of Steel Construction, University of Hannover 2001

[11]   Willmerding, G., Häckh, J., Schnödewind K.: *Fatigue Calculation using winLIFE*; NAFEMS-Tagung Wiesbaden 2000

[12]   Wiedemann J.: *Leichtbau (Lightweight construction)*, Vol. 1: Elements, Vol. 2: Construction, Springer-Verlag, Berlin, 1996

[13]   International Electrotechnical Commission: IEC 61400, preview documents on WWW (Oct. 2010): http://webstore.iec.ch/

10 Wind pump systems

The application of wind mills for water pumping is of lesser importance today than in the past centuries. However, for several reasons, it is useful to discuss this type of wind energy application in a wind energy book targeted at development and planning engineers as well as students.
A wind pump system for water supply is an energy system which serves to illustrate the interaction of the system components: wind turbine, water source and the other system components for the water utilization. An optimum configuration of the complete system is only possible when considering the interaction of the system components with their individual operating behaviour and characteristics.

Moreover, it is likely that the significance of wind pump systems will grow again with the increasing acceptance of renewable energies, particularly in the developing countries. They may contribute significantly to rural development [1]. Today, there are ideas for new fields of application for wind pump systems, not only for water pumping but also for oil pumping in remote and windy areas of the world, e.g. sites in the southern parts of Chile and Argentina.

10.1 Characteristic applications

In Europe, *utilizing wind energy for water pumping* has a long tradition reaching back to the 1500s. Historical examples of applications which were of great economic importance are the drainage of coastal areas in the Netherlands lying below sea level, the irrigation of potato plantations in the Cretan plateau of Lassithi, and the wind pumps on cattle farms in the American Midwest. Today, industrialized countries seldom use wind energy for water pumping. However, in developing countries, where many regions are not connected to a local infrastructure with electrical grid and public water supply, the utilization of wind energy constitutes an economical and environmentally friendly option for improving the water supply.

A schematic diagram of the conversion of wind energy into hydraulic energy by the wind pump system is shown in Fig. 10-1. In the following sections, distinction is made between the *wind pump* consisting of the wind turbine, the gearbox and the pump, on one hand, and the *wind pump system* combining the wind pump and the hydraulic system, on the other.

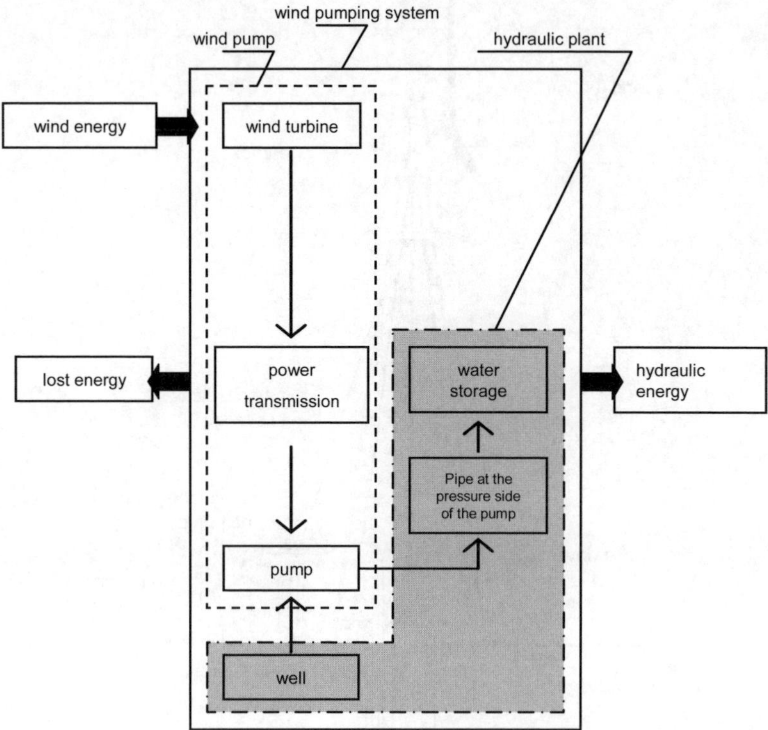

Fig. 10-1 Schematic diagram of the conversion of wind energy into hydraulic energy by the wind pump system

Wind energy is a stochastic input. Losses occur in the course of its conversion into hydraulic energy. Most wind pump systems with a mechanical coupling operate without a control mechanism with the exception of an overspeed control. Therefore, any variation in wind speed directly leads to a change in the hydraulic operating point, particularly in the flow rate. Moreover, the energy conversion is also influenced by varying pumping conditions, e.g. the varying height of the water level in the well or increasing friction losses due to corrosion of the piping system.

Fig. 10-2 shows the *general design of a wind pump system* with a mechanical coupling of wind turbine and pump by gearbox, shaft and belt drive.

The characteristic applications of wind pumps are mainly found in developing countries. There are four *types of applications*:

- Drinking water supply,
- Livestock watering,
- Irrigation and
- Drainage.

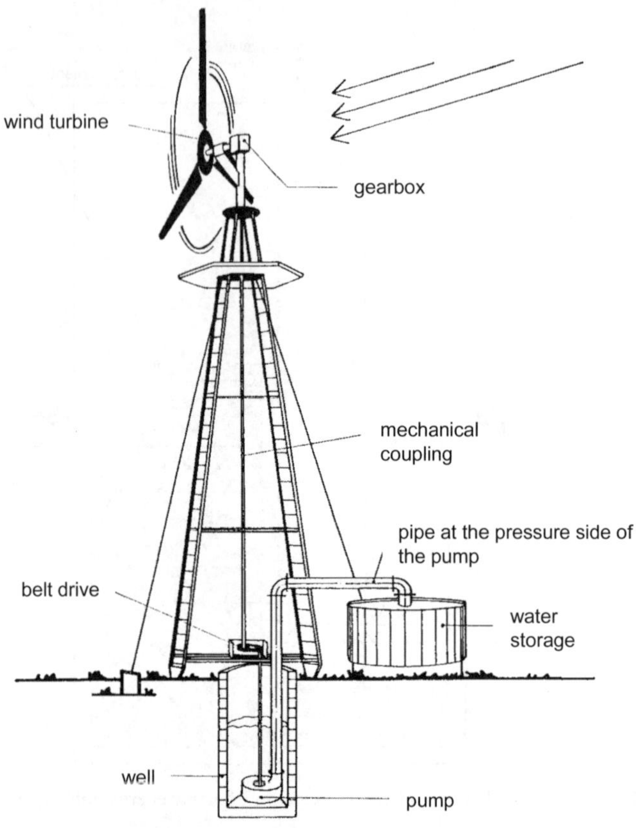

Fig. 10-2 General design of a wind pump system with mechanical coupling of wind turbine and pump

In the developing countries, the majority of operating wind pumps are currently applied for drinking water supply and livestock watering. Recent approaches to the use of wind pumps for irrigation have often failed due to the complexity of this application and not taking the socio-economic conditions into account. However, the use of wind pumps for irrigation and drainage has a great potential in the agriculture of developing countries [2].

The location of the water resource is of great importance when it comes to converting wind energy into hydraulic energy. Depending on the type of application, either underground water or surface water may be used. The *hydraulic power* P_{hydr} is given by the density of water ρ_{w}, the local acceleration due to gravity g, the flow rate Q and the total head H

$$P_{\mathrm{hydr}} = \rho_{\mathrm{w}}\, g\, Q\, H\,. \tag{10.1}$$

According to equation (10.1), the same amount of hydraulic power results either from the combination of a high flow rate with a low head or a low flow rate with a high head. Consequently, three general *pumping conditions* are distinguished assuming that the head is dominated by the geodetic head (i.e. level difference between the water level of the well and the discharge level):

- Low flow rates and a head of more than 20m,
- Medium flow rates and a head between 5 and 20m, and
- High flow rates and a head of less than 5m.

Fig. 10-3 shows how these pumping conditions are related to the four types of application, according to [1].

In order to illustrate the quantitative dimension of this general classification of application types and pumping conditions, Fig. 10-4 shows the calculated output for a common-size wind pump system with a diameter d_{WT} = 5 m of the wind turbine and also the corresponding ranges of the types of applications. Assuming there are eight hours of pump operation per day, the daily discharge volume V_d varies for different heads H along the curves of constant wind speed v. For instance, at a wind speed v = 4 m/s and a head of H = 3 m, a field of one hectare (10,000 m²) can be irrigated daily with approx. 70 m³ [1].

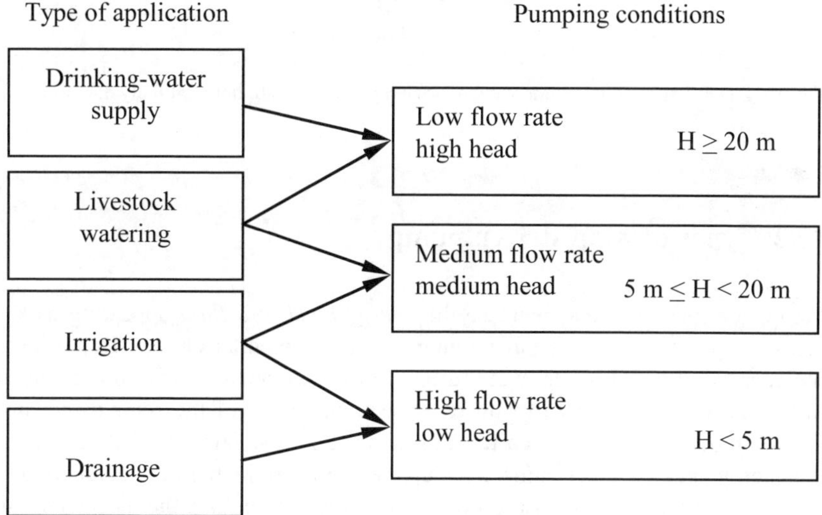

Fig. 10-3 Types of application and the corresponding pumping conditions [1]

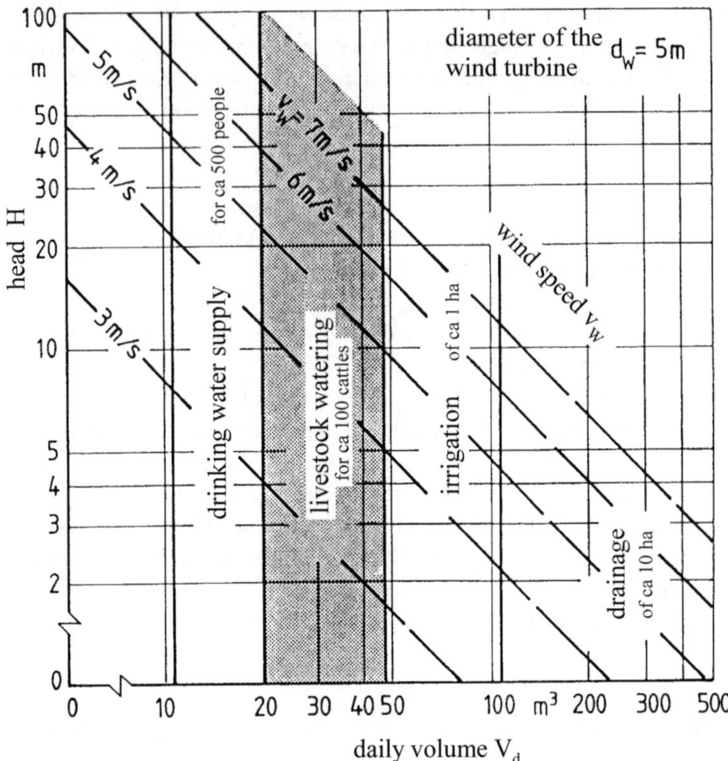

Fig. 10-4 Daily discharge volume yield of a wind pump for eight hours of daily operation [1]

10.2 Types of wind-driven pumps

Pumps are machines that increase the *energy level of a fluid* appearing as an increase in pressure energy, kinetic energy and/or potential energy. In section 5.1, the energy extracted by the wind turbine rotor was calculated from the velocities in the upstream and downstream cross-sectional areas of the rotor using the assumption of the virtual stream tube. Now, for a real stream tube with solid walls, the energy balance at the pump sums up the energy at the pump inlet (suction side, index '*s*'), at its outlet (pressure side or discharge, index '*d*'), the mechanical shaft power and power losses. The energy added to the fluid per second is the hydraulic power

$$P_{hydr} = \dot{m}\, Y = \rho_W\, Q\, Y = \rho_W\, g\, Q\, H. \qquad (10.2)$$

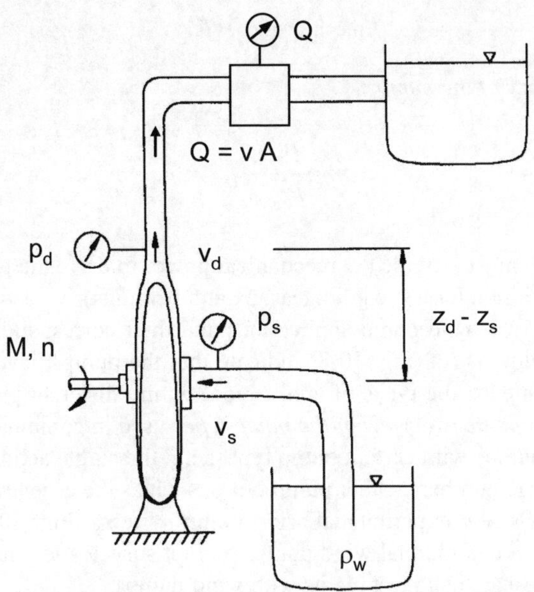

Fig. 10-5 Schematic view of a pump

It is the product of mass flow rate $\dot{m}$ and the *specific work Y* of the pump (i.e. the specific energy in J/kg)

$$Y = g\ H. \tag{10.3}$$

Fig. 10-5 shows schematically a set-up to determine the specific work and the flow rate of a pump. According to *Bernoulli*, the specific work Y results from the energy balance between the pressure (d) and suction (s) side of the pump

$$Y = g \cdot (z_d - z_s) + \frac{p_d - p_s}{\rho_w} + \frac{1}{2} \cdot \left(v_d^2 - v_s^2\right)\ . \tag{10.4}$$

The heights z_d and z_s are measured. The corresponding manometers give the local static pressures p_s and p_d. Using the measured flow rate Q and the cross-sectional areas A_s and A_d, the velocities are determined by the *equation of continuity*

$$\dot{m} = \rho_w\ Q = \text{const} = \rho_w\ v_d\ A_d = \rho_w\ v_s\ A_s \tag{10.5}$$

Often, the measured increase in pressure energy dominates the change in kinetic and potential energy.

The ratio of the 'useful' hydraulic power to the required *mechanical power*

$$P_{\text{mech}} = 2\,\pi\,M\,n \tag{10.6}$$

gives the *efficiency of the pump*

$$\eta = \frac{P_{\text{hydr}}}{P_{\text{mech}}} \, . \tag{10.7}$$

The wind turbine must provide the mechanical power to drive the pump and cover all power transmission losses (e.g. in gearbox and bearings).

The characteristic wind pump applications and their corresponding wide range of pumping conditions (cf. Fig. 10-3) indicate that the pumps have to fulfill very different requirements: the types of application require different pump types. Fig. 10-6 gives a *systematic overview of the pump types* used in combination with wind turbines. Wind pumps with certain pump types, e.g. the single acting piston pump, are applied in large numbers. Other pump types, such as the eccentric screw pump, are used only in a few experimental wind pump systems. Fig. 10-6 is based on data collected by a commercial wind pumps market survey and a literature review of documented research and test plants with wind pumps.

Pump types are principally distinguished by their *working principle*. Moreover, the systematic overview in Fig. 10-6 also shows the characteristics and parameters which are essential for operation in wind pump systems:

- The *Q-H characteristics* of total head H versus flow rate Q describe the operating behaviour of the pump. The rotational speed n is the variable curve parameter for the wind pump operation.
- The data on total head H, specific speed n_q, and the achieved maximum efficiency $\eta_{\text{opt.max}}$ refers to pumps that are suitable for application with wind turbines of a rotor diameter below 10 m. The multi-stage centrifugal pump with a submersible motor is an exception, as it is also used with larger wind turbines [7].
- The *specific speed n_q* of the pump is

$$n_q = n_r \frac{Q_r^{1/2}\, H_q^{3/4}}{H_r^{3/4}\, Q_q^{1/2}} \tag{10.8}$$

with n_q in rpm,

           n_r rated rotational speed of the pump in rpm,

           Q_r rated flow rate of the pump in m^3/s,

           H_r rated total head of the pump in m,

           $H_q =$ 1 m, total head of the reference pump and

           $Q_q =$ 1 m^3/s, flow rate of the reference pump.

The specific speed scales the best point data to that of a reference pump. This normalization allows the description of a the pump independent of its actual size and furthermore the choice of the pump type which is likely to have the best efficiency at the intended rated operating point with H_r and Q_r.

- The *required torque M* depending on the rotational speed n of the pump, i.e. the torque-speed characteristic, is very significant because it determines the operating points of the considered combination of pump and wind turbine.

The *piston pump* (also called reciprocating pump), denomination *A* in Fig. 10-6, is applied in thousands of wind pumps and the most important type of displacement pumps. Due to its slender design with one valve integrated in the piston it is suitable for installation in boreholes. Fig. 10-7 shows a schematic view of the piston pump. Manufactureres offer single-acting piston pumps which are connected to the wind turbine by a crank mechanism or an excentric. For smaller turbines ($d_{WT} < 5m$), the rotational speed of the wind turbine is reduced to the slower speed required by the piston pump by a gear connected to the crank mechanism. The pump is adapted to different wind turbine sizes and to the required total head by selection of a corresponding piston diameter and piston stroke. For applications with a low head, the piston pump can be also installed as a suction pump as long as there are no cavitation problems.

The *H-Q* characteristics of the piston pump show that the total head is independent of the flow rate, unless the leakage losses become relevant in the range of high total heads. For the use in wind pump systems, this brings the advantage that already at low rotational speeds a high total head is delivered and there is a discharge. Disadvantages are the high wear when pumping dirty water (e.g. abrason due to sand or blocking of valves), high dynamic forces due to the oscillating piston, and the high starting torque required. However, the starting behaviour can be substantially improved by design measures: e.g. compensating boreholes in the piston, i.e. a by-pass between the cylinder's pressure and suction sides, a resonance pipe at the cylinder's suction side parallel to the rising pipe, special pressure valves or pre-stressed springs in order to relieve the bottom dead centre [8].

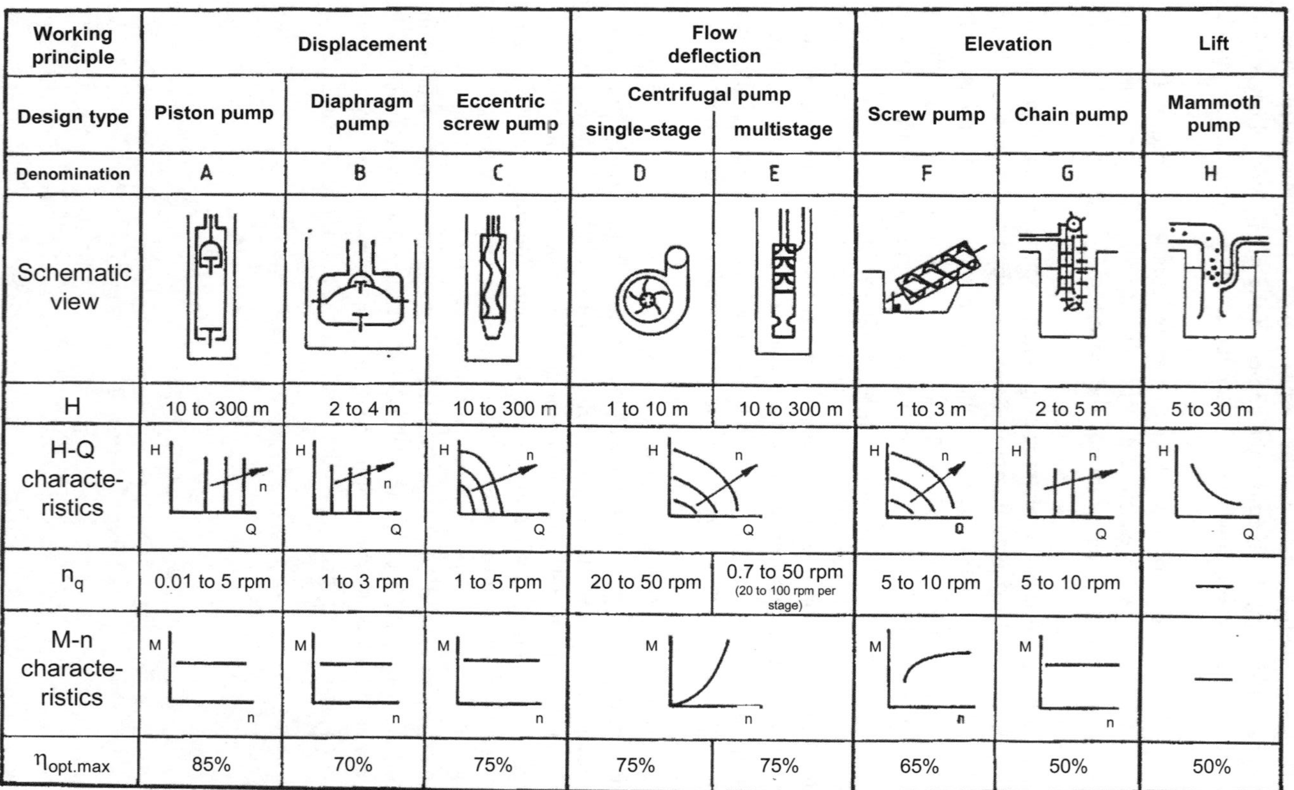

Working principle	Displacement			Flow deflection		Elevation		Lift
Design type	Piston pump	Diaphragm pump	Eccentric screw pump	Centrifugal pump single-stage	Centrifugal pump multistage	Screw pump	Chain pump	Mammoth pump
Denomination	A	B	C	D	E	F	G	H
Schematic view								
H	10 to 300 m	2 to 4 m	10 to 300 m	1 to 10 m	10 to 300 m	1 to 3 m	2 to 5 m	5 to 30 m
H-Q characteristics								
n_q	0.01 to 5 rpm	1 to 3 rpm	1 to 5 rpm	20 to 50 rpm	0.7 to 50 rpm (20 to 100 rpm per stage)	5 to 10 rpm	5 to 10 rpm	—
M-n characteristics								—
$\eta_{opt.max}$	85%	70%	75%	75%	75%	65%	50%	50%

Fig. 10-6 Systematic overview of pumps types suitable for combination with wind turbines [3]

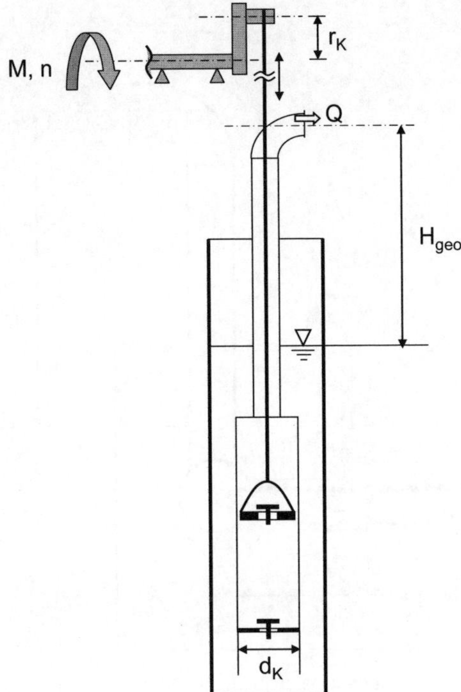

Fig. 10-7 Schematic view of piston pump

The *diaphragm pump*, *B* in Fig. 10-6, is used as a suction pump for low total heads. Its design requires larger diameters than the piston pump, therefore, it cannot be used in boreholes. Usually, the *H-Q* characteristics and the torque characteristics (*M-n* curve) of the diaphragm pump correspond to those of a piston pump, but it achieves higher flow rates. The best efficiency is substantially lower than that of a piston pump. The short service life of the diaphragm prevents a wider application of diaphragm pumps in wind pump systems.

The *eccentric screw pump*, C in Fig. 10-6, is similar to the piston pump in terms of size and total head range. Therefore, it is also primarily used in boreholes. However, the H-Q characteristics show the dependence of the total head on the flow rate. The rotational speed has to be stepped up when driven by a wind turbine. In comparison to other positive displacement pumps, the eccentric screw pump has the advantage that no oscillating forces are introduced into the wind pump system. Similar to other positive displacement pumps, the starting behaviour is disadvantageous in that it has a high starting torque resulting from the static friction between eccentric screw and rubber stator. It remains to be seen if this pump type is suitable for long-term operation in wind pump systems. During water pumping the lubrication is poor and the water is often dirty, containing abrasives like sand. Thus, the service life of the stator material, often made from rubber, must be tested.

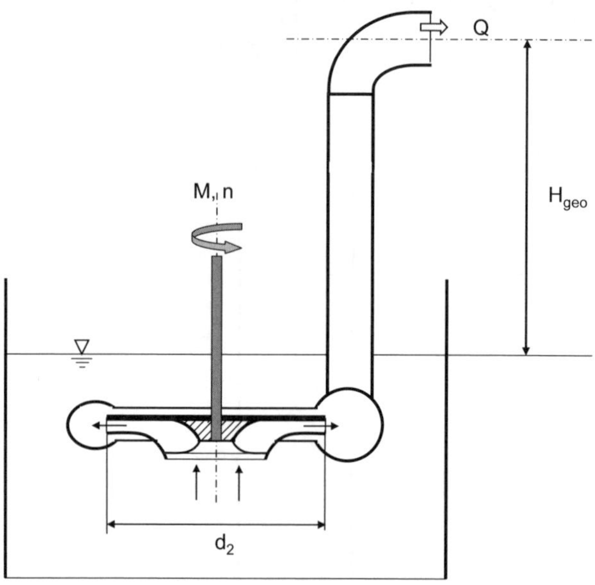

Fig. 10-8 Schematic view of a single-stage centrifugal pump

The *single-stage centrifugal pump*, D in Fig. 10-6, is applied in wind pump systems either as radial or mixed flow centrifugal pump type. They are rotodynamic machines (like wind turbines) using the working principle of flow deflection by the rotating impeller.

The total head of mechanically coupled single-stage centrifugal pumps is limited to 10 m due to the maximum impeller diameter (difficult manufacture of impeller diameters $d_2 > 400$ mm) and the gearbox transmission ratio [2]. Therefore, providing underground water from dug wells or surface water from an intake construction is the main application.

Apart from a mechanical coupling, an electrical coupling of wind turbine and single-stage centrifugal pump is also possible. However, this has been hardly used up to now.

In general, the advantages of the centrifugal pump are that it is not sensible to dirty water (sand) and that it has a good starting behaviour: the required starting torque is low and the torque characteristics of wind turbine and centrifugal pump are compatible because they are both rotodynamic machines, see section 10.3.2.

Todate, *multistage centrifugal pumps*, E in Fig. 10-6, have only been applied in wind pump systems with an electric coupling and a submersible motor. For multistage centrifugal pumps, the maximum total head achieved increases with the number of stages.

Centrifugal pumps of this type suitable for wind pump systems are offered with submersible motors for applications in boreholes in a large variety and in a wide power range. The electrical power transmission has the advantage that the wind turbine site can be selected independent of a well, and that moreover, the electrical energy is also available for other applications than pumping. The disadvantages of electrical power transmission are the additional electrical losses, in comparison to a mechanical coupling. Further advantages and disadvantages of the multistage centrifugal pump are the same as for the single-stage centrifugal pumps.

The application of the *screw pump* (also called 'Archimedean screw'), *F* in Fig. 10-6, with windmills has had a long tradition in the Netherlands, cf. Fig. 2-6. It is still commonly used in China and Thailand. Due to its inclined arrangement, the screw pump can only be used for pumping surface water. The total head is limited to 3 m for mechanical reasons (the distance between the screw bearings). The characteristics show a dependence of total head H on flow rate Q. The major advantage of the screw pump is its easy manual crafting. The large quantities of material required are a disadvantage.

The *chain pump*, *G* in Fig. 10-6, has a range of application similar to the screw pump. It was also commonly used in China and Thailand. The complex design limits its total head; moreover, the well has to provide sufficient space for the installation of the chain pump. Therefore, only dug wells or intake constructions for surface water are suitable. Its advantages include the low requirements of the chain pump regarding the procuction accuracy. This contrasts with its low pump efficiency.

The *mammoth pump*, *H* in Fig. 10-6, transmits the power from the wind turbine to a compressor. The compressed air is blown into the rising pipe where the ascending air bubbles produce the rising flow. The combination of wind turbine and mammoth pump is used when the pumping medium is very aggressive or heavily dirty. When using a mammoth pump, only the rising pipe and the compressed air piping have to be installed in the well, and there are no movable parts in the pump. Therefore, the mammoth pump has a very long lifetime. In addition, the pneumatic coupling to the wind turbine offers nearly the same advantages as an electric coupling. Apart from pumping water, a combination of wind turbine and mammoth pump might be used for water treatment. As mammoth pumps do not have any movable parts, there is no specific rotational speed and *M-n* curve. One reason for not using a mammoth pump is its low efficiency, which decreases even further in the upper total head range of 25 to 30 m. For comparison with the other pump types, the efficiency of the mammoth pump is calculated as the ratio of the hydraulic power P_{hydr}, see equation (10.1), to the power of the air at the injection point. Therefore, the losses of the pneumatic coupling by compressor and air piping are not included in the efficiency given in Fig. 10-6.

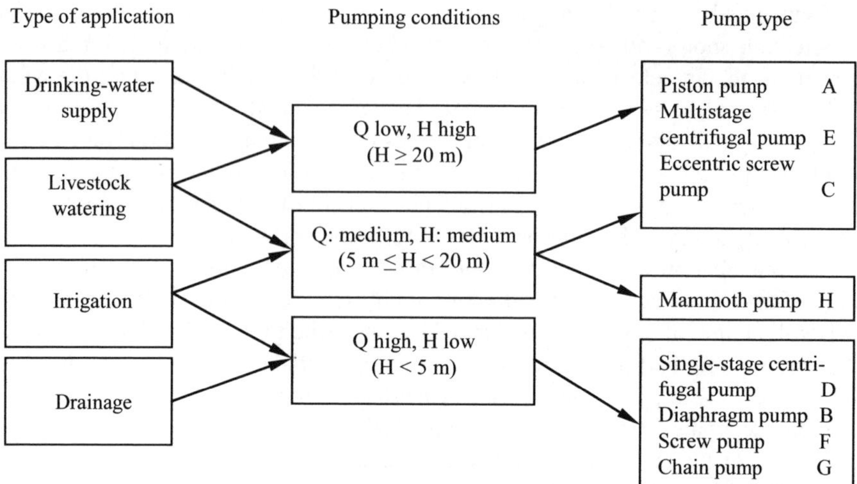

Fig. 10-9 Types of application, pumping conditions and pump types

The selection of a pump design for a particular application is influenced by a number of criteria which have to be assessed individually. These can be physical, technical and economic criteria. Considering the types of application and the related pumping conditions, cf. Fig. 10-3, the correspondingly suitable pump types can be identified, Fig. 10-9.

10.3 Operation behaviour of wind pumps

10.3.1 Suitable combinations of wind turbines and pumps

The design of the wind pump is determined, on the one hand, by the pumping conditions of the particular type of application which is the main selection criterion for the pump type choice, Fig. 10-9. On the other hand, the wind turbine, the transmission ratio, and the pump have to be adapted to each other, taking the local wind conditions of the intended site into account.

To find a suitable *combination of wind turbine and pump*, the torque-speed characteristics of the wind turbine and the pump have to be compared, as it is common practice for the coupling of an engine and a machine. Fig. 10-6 shows the M-n characteristics of the different pump types. Two issues are important for selecting a combination of wind turbine type and pump type. The rotational speeds of the wind turbine and pump under operation have to be adapted to each other by a transmission component, e.g. gears or belt drive. Moreover, the starting torque has to be taken into account in order to achieve a favourable starting behaviour of the wind pump already at low wind speed.

Fig. 10-10 shows the typical *ranges of rotational speeds* and *total heads* for different pump types. The figure also contains the ranges of rotational speeds of the various wind turbine types considering a reference diameter d_{WT} = 5 m of the wind turbines[*]. It becomes clear that in many cases the rotational speeds have to be adapted by a gearbox, as the wide range of speeds of the likely suitable pump type is not covered by the much narrower range of rotational speeds of likely suitable wind turbines. For the pump types on the left of Fig. 10-10, gearing-down is usually required. For the pump types on the right, gearing-up is required. In some cases, the rotational speeds match without gears. Fig. 10-10 illustrates, that on one hand, wind turbines with a low tip speed ratio are suitable for pump types *A*, *B*, *F*, and *G*. On the other, wind turbines with a high tip speed are more suitable for operation with pump types *C*, *D*, and *E*.

The *transmission ratio i* is selected so that the highest system efficiency possible over the entire operating range of the wind pump is achieved for a given wind turbine and pump. Usually, to determine of the transmission gear ratio, a rated wind speed where both wind turbine and pump reach their maximum efficiency, is chosen from the given wind conditions, cf. section 10.4.2.

Apart from matching the rotational speeds, any combination of wind turbine and pump type has to take into consideration whether the torque of the wind turbine is sufficiently large in order to start the pump. Positive displacement pumps (A,B and C), such as piston pumps, diaphragm pumps and eccentric screw pumps require a high starting torque. In contrast, centrifugal pumps (D,E) require only a low torque, cf. Fig. 10-6.

A comparison of the starting torque required by the pumps with the starting torque provided by the wind turbines (see chapter 6, Fig. 6-10) gives the following suitable combinations: Wind turbines with a low tip speed ratio match with positive displacement pumps and wind turbines with a high tip speed ratio match with centrifugal pumps. A suitable compromise has to be found for eccentric screw pumps (*C*) between favourable start-up behaviour and sufficiently high rotational speed under operation, as the eccentric screw pump has both a high starting torque and a relatively high operating speed. A solution for this can be the use of a centrifugal clutch and of a gearbox with a variable transmission ratio, for instance.

---

[*] The rotor diameter d_{WT} = 5.0 m is representative for the wind turbine capacities which have a very high application potential in rural areas (cf. Fig. 10-4).

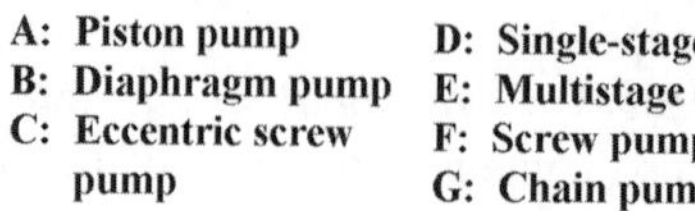

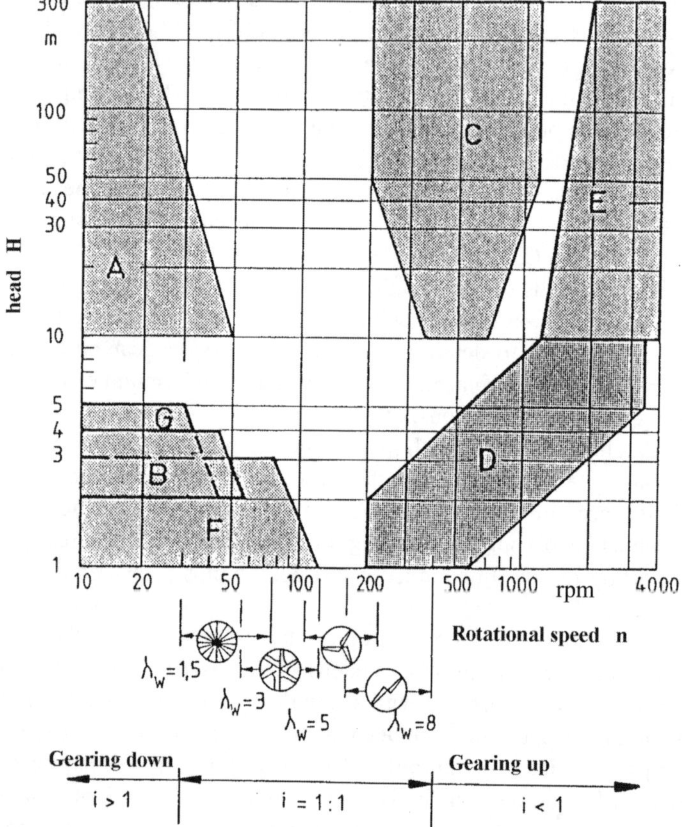

Fig. 10-10 Operating ranges of different types of wind turbines (diameter $d_{WT} = 5$ m) and pumps [5]

Fig. 10-11 shows schematically the suitable combinations of wind turbine and pump types which result from the discussion so far, and their attainable total head H. Every wind pump shown has a limited application range where it reaches a good efficiency. The application range is mainly determined by the operating characteristics of the combined components. There is no such thing as a universally applicable wind pump. If the wind pump is operated outside this application range, the yield will be reduced since the wind energy available is converted into hydraulic energy less efficiently.

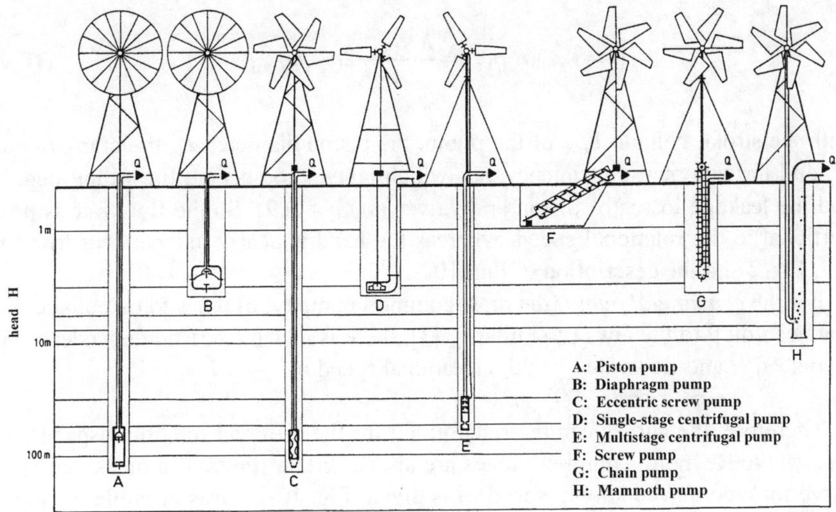

Fig. 10-11 Suitable combinations of wind turbines and pumps [3]

10.3.2 Qualitative comparison of wind pump systems with piston pump and centrifugal pump

In the following, piston and centrifugal pumps are taken as examples, since they are the most important pump types. The *H-Q* diagrams in Fig. 10-12 show the different characteristic behaviour of these two pump types. The pump's rotational speed *n* is the parameter used to clarify the effect of varying rotational speeds on the performance characteristics.

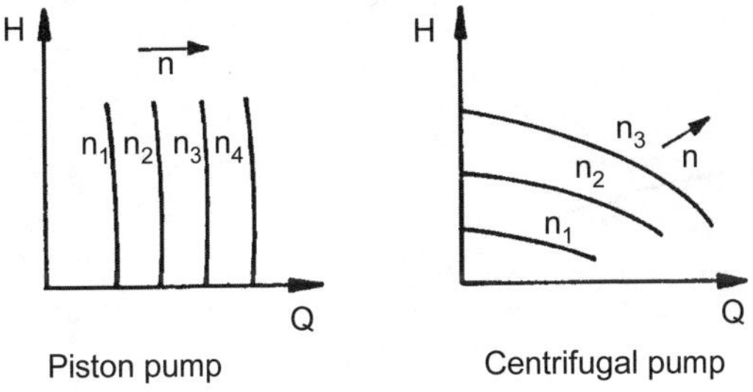

Fig. 10-12 *H-Q* diagram of a piston pump and a centrifugal pump

The flow rate Q of the *piston pump* (Fig. 10-7) is

$$Q = V_{\text{Hub}} n\, \eta_{\text{vol}} = \frac{\pi \cdot d_K^2}{4} 2 r_K\, n\, \eta_{\text{vol}} \tag{10.9}$$

with the stroke volume V_{hub} of the pump, the piston diameter d_K, the crank radius r_K, and the volumetric efficiency η_{vol} which takes into account the filling degree and the leakage losses of piston and valves ($\eta_{\text{vol}} > 0.9$). So the flow rate is proportional to the rotational speed, whereas the head is nearly independent from it, Fig. 10-12 and the description of Fig. 10-7.

For the *centrifugal pump* (the most common pump type) it has to be pointed out that according to the laws of similarity [3], there is a characteristic dependency of total head H and flow rate Q on the rotational speed n:

$$H \sim n^2 \quad \text{and} \quad Q \sim n.$$

Therefore, the curves can be transformed for the different rotational speeds, as long as the Reynolds' number values are above certain limits. If a measured H-Q curve for a certain rotational speed n_1 is given, Fig. 10-13, it is possible to calculate for another rotational speed n_2 the corresponding H-Q curve using these laws of similarity. Moreover, it follows from equation (10.1) that the hydraulic power is proportional to the cube of the rotational speed.

$$\frac{H_2}{H_1} = \left(\frac{n_2}{n_1}\right)^2 , \quad \frac{Q_2}{Q_1} = \left(\frac{n_2}{n_1}\right) , \quad \frac{P_{\text{hydr.2}}}{P_{\text{hydr.1}}} = \left(\frac{n_2}{n_1}\right)^3 \tag{10.10}$$

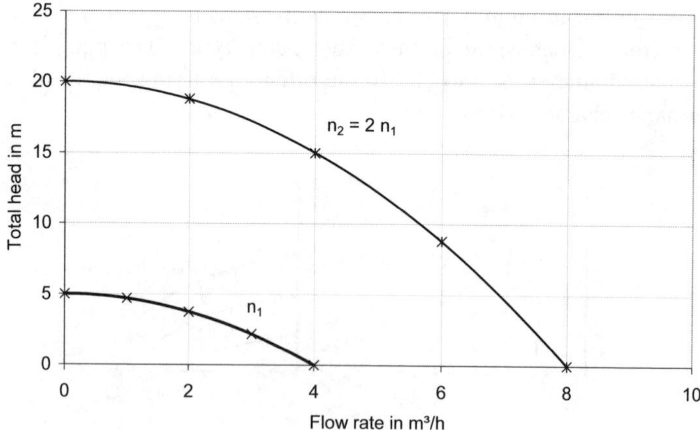

Fig. 10-13 *H-Q* diagram of a centrifugal pump with measured characteristic curve for rotational speeds n_1 and calculated similar curve for n_2

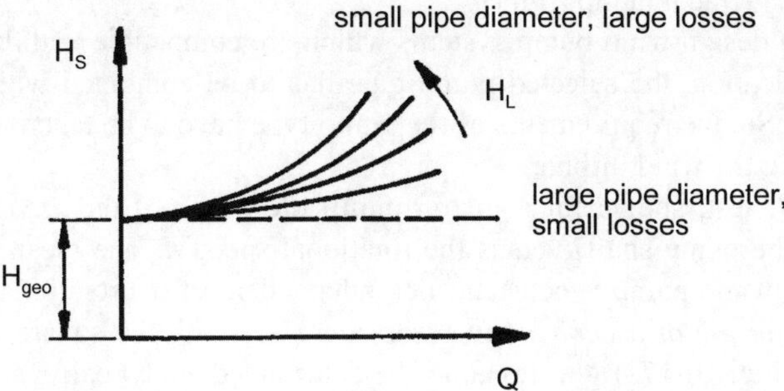

Fig. 10-14 H_S-Q diagram with characteristic curves of the hydraulic system for different head losses

Usually the pump is integrated in a *piping system* which conducts the flow. The pipes that are attached at the pump's suction and pressure side (delivery side) are called the *hydraulic system*. The head H_S of the hydraulic system is calculated on the basis of the geometry of the well and the pipes. In most cases, it is the sum of the *geodetic head* H_{geo} which is the height difference that has to be overcome between system entry and exit, and the head loss H_L:

$$H_S = H_{geo} + H_L \qquad\qquad (10.11)$$

The *head loss H_L* results from the pressure losses (mainly friction) in the pipes, depending on their geometry and surface. It increases with the square of the flow velocity in the pipe, and is therefore proportional to the square of the flow rate Q. Other components (e.g. bends, valves, fittings) also cause pressure losses. The characteristics of the hydraulic system show the system head H_S versus the flow rate Q, Fig. 10-14. Thus, the characteristics of pump and hydraulic system may be drawn in one diagram, Fig. 10-15, to determine the *operating points*. These are given by the intersections of pump and hydraulic system curves.

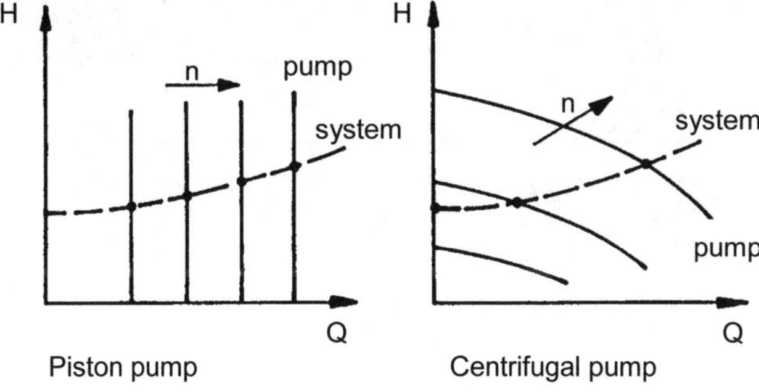

Fig. 10-15 Operating points of a hydraulic system and the pump

In Fig. 10-15 one and the same hydraulic system operates with a piston pump (left) and a centrifugal pump (right).

In order to design wind pump systems which are compatible with the different types of application, the selected pump type has to be combined with a suitable wind turbine. So, the requirements of the pump type have to be taken into account when choosing the wind turbine.

Fig. 10-17, left, shows for a piston pump the curve of the *mean torque M* (required at the pump shaft) versus the rotational speed n. The mean torque of a single acting piston pump is constant, i.e. independent of the rotational speed. By contrast, the *torque of a centrifugal pump* increases with the square of the rotational speed, Fig. 10-17, right. It cannot be determined analytically, and therefore has to be taken from the measured characteristic curve. This shows that for the choice of a suitable wind turbine, these two pump types, piston pump and the centrifugal pump, have completely different requirements.

The *torque of a piston pump* of one revolution is a half-sine wave curve, Fig. 10-16. The mean torque $\overline{M}$ of the piston pump results from integrating this half-sine wave over one revolution of the crank drive:

$$\overline{M} = \frac{1}{\pi}\, M_{\mathrm{max}} \qquad\qquad (10.12)$$

However, the calculation of the starting process cannot be based on the mean torque of the piston pump, because in order to start up the pump, the wind turbine has to provide the maximum torque M_{max}. During operation, the system has kinetic energy, and the oscillating torque curve can be replaced by the mean pump torque $\overline{M}$.

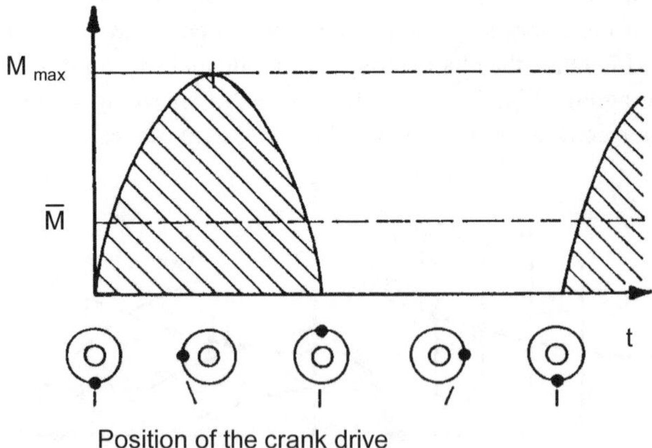

Fig. 10-16 Torque curve of the piston pump for one revolution

Equating mechanical and hydraulic power

$$\overline{M} \, 2 \, \pi \, n \, \eta_\mathrm{m} = \rho_\mathrm{w} \, g \, H \, V_\mathrm{Hub} \, n \qquad (10.13)$$

gives the mean torque of the piston pump under operation

$$\overline{M} = \frac{\rho_\mathrm{w} \, g \, H \, d_\mathrm{K}^{\,2} \, r_\mathrm{K}}{4 \, \eta_\mathrm{m}} \, . \qquad (10.14)$$

The mechanical efficiency η_m takes into account the mechanical losses of the piston pump. The torque of the piston pump is directly proportional to the head H, the crank radius r_K, and the square of the piston diameter d_K. It is independent of the rotational speed n.

The *operation behaviour of the wind pump* is derived from the superposition of torque and power curves of both wind turbine and pump. Fig. 10-17 shows power and torque characteristics of a wind turbine for different wind speeds and the load curves of the piston pump (left), and the centrifugal pump (right) respectively. The intersection of wind turbine torque curve for the actual wind speed and the pump torque curve, respectively of wind turbine power curve and pump power curve, gives the actual operating point of the wind pump.

In the consideration, a constant total head H is assumed, i.e. in the hydraulic system the head loss H_S is negligible compared to the geodetic head H_geo (short pipe length, large pipe diameter). It is evident that the characteristic power curve of the centrifugal pump matches much better with the power characteristics of the wind turbine than that of the piston pump. The piston pump achieves an optimum loading of the wind turbine only at one single operating point, and at higher wind speeds it does not utilise the maximum power of the turbine.

Being a rotodynamic machine (like the wind turbine), the centrifugal pump shows an optimum matching with the wind turbine characteristics at almost all wind speeds if a suitable rotational speed adaption (transmission gear ratio) is used, cf. section 10.4.4.

The wind speeds v and the rotational speeds n of the operating points can now be taken from the diagram. For the piston pump, the corresponding flow rate Q is according to equation (10.9)

$$Q = V_\mathrm{Hub} \, n \, \eta_\mathrm{vol} \, . \qquad (10.15)$$

On the contrary, an analytical determination is not possible for the centrifugal pump. The flow rates Q of the operating points have to be taken from the H-Q diagram (see Fig. 10-15, right) at the intersections of the pump curves with the H_S-Q curve of the hydraulic system.

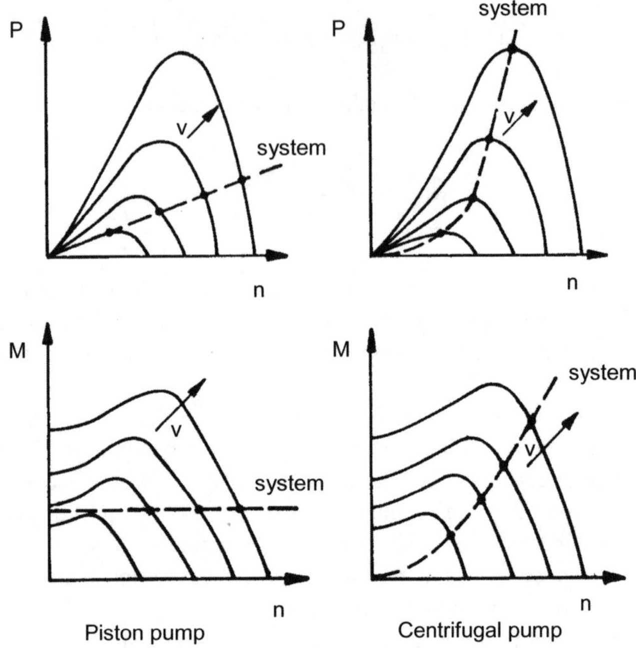

Fig. 10-17 Operating points of a wind pump with a piston pump and a centrifugal pump

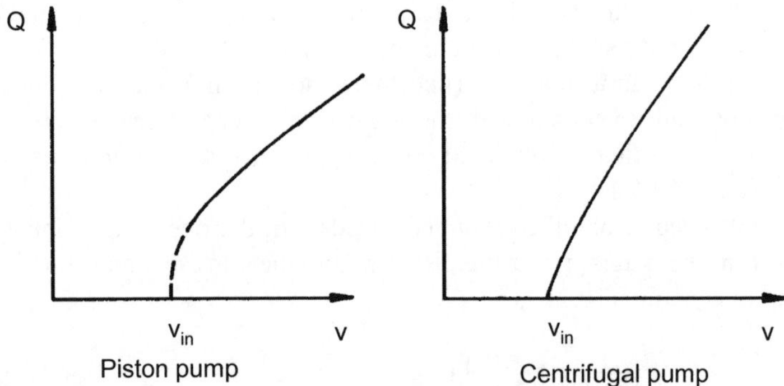

Fig. 10-18 Flow rate Q versus wind speed v of a wind pump with a piston pump and a centrifugal pump (discharge curves)

If the flow rates are plotted versus the corresponding wind speeds, the result is the '*discharge curve*' $Q = Q(v)$ of the wind pump (Fig. 10-18). This diagram relates the characteristic value of the wind energy, the wind speed v, to a value representing the hydraulic energy, the flow rate Q. The wind pump begins to discharge at the start-up wind speed v_{in} .

For a constant total head H, the curve of the total *efficiency* η_{WP} of the wind pump versus the wind speed v (Fig. 10-19) can be determined immediately, as the ratio of hydraulic power and the theoretical power in the wind

$$P_{Wind} = (\rho/2)\, \pi\,(d_{WT}^{\,2}/4)\ v^3 :$$

$$\eta_{WP} = \frac{8\,\rho_{w}\,g\,Q\,H}{\rho\,\pi\,d_{WT}^{\,2}\,v^3} \tag{10.16}$$

In Fig. 10-19 the advantage of the better matching of the wind turbine and centrifugal pump is obvious, as this kind of wind pump has a wider range of high efficiencies in its efficiency curve, in comparison to the one with a piston pump. The latter has a more distinct efficiency peak, and the efficiency is then reduced rapidly with growing wind speeds.

So far, the total head H has been assumed to be constant, but in reality it increases with higher wind speeds because the correspondingly growing flow rate (higher velocity in the pipe) leads to higher pipe friction losses. Additionally, the water level in the well often decreases during continuous pumping at high winds, so the geodetic head increases, too. A qualitative illustration of how a change in the head influences the discharge curve $Q = Q(v)$, valid for both piston and centrifugal pumps, is given in Fig. 10-20.

The discharge curve $Q = Q(v)$ describes the technical system wind pump regarding the conversion of wind energy into hydraulic energy. The calculation of *the yield of the wind pump* requires a description of the wind conditions. The yield is defined as the discharge volume during the period under consideration, e.g. the daily discharge volume V_d. The wind conditions are measured using a wind measuring system with a classification function which provides a wind speed histogram (cf. section 4.3).

The *discharge volume V* of the wind pump is determined by wind speed histogram and discharge curve. The total time T of the period under consideration is multiplied by the summation of the frequencies h_i of the individual wind class v_i and the corresponding flow rate $Q_i(v_i)$:

$$V = T \sum_{i} Q_i\, h_i \ . \tag{10.17}$$

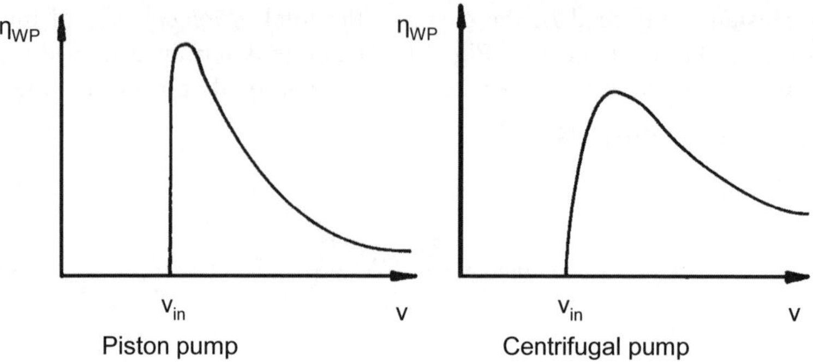

Fig. 10-19 Total efficiency of a wind pump with a piston pump and a centrifugal pump

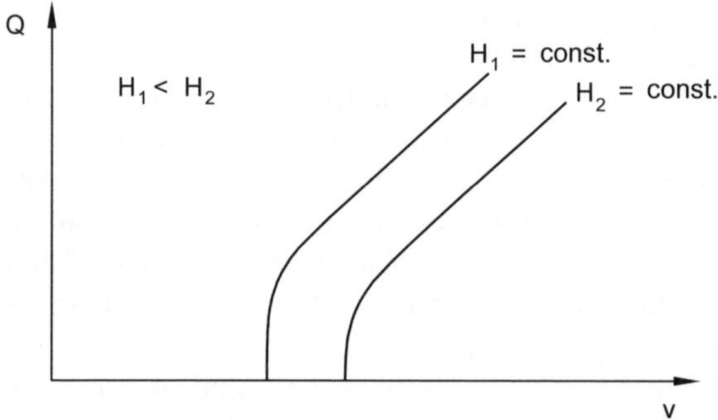

Fig. 10-20 Flow rate versus wind speed of a wind pump for two different constant geodetic heads (discharge curves)

10.4 Design of wind pump systems

10.4.1 Design target

The design calculations of wind pump systems aim for the most efficient conversion of wind energy into potential energy of the water pumped into the water storage, cf. Fig. 10-1. The *quality factor* γ characterizes the efficiency of this energy conversion:

$$\gamma = \frac{g\,\rho_w\,H_{geo}\,V}{E_{Wind}} = \frac{g\,\rho_w\,H_{geo}\,T\,\Sigma\,Q_i\,h_i}{\dfrac{1}{2}\,\rho\,A_{Rotor}\,T\,\Sigma\,v_i^{\,3}\,h_i} \quad . \tag{10.18}$$

It is the ratio of the available potential energy of the discharge volume V to the energy contained in the wind in the considered period T:

$$E_{Wind} = \frac{1}{2}\,\rho\,A_{Rotor}\,T\,\Sigma\,v_i^{\,3}\,h_i \tag{10.19}$$

In contrast to the efficiency of the wind pump according to equation (10-16), the quality factor also considers the losses due to friction in the pipes, and it averages over the considered period. Wind pump systems with a maximized quality factor have the highest discharge volume possible in the period under consideration. This period may be a season or a certain period when water supply is critical. In practice, the design is not based on a maximum quality factor, but the following simplifications are used:

- the rated wind speed v_r is chosen wisely, and
- the requirement that the wind pump reaches its maximum total efficiency η_{WP} at this rated wind speed.

10.4.2 Selection of the rated wind speed for the wind pump design

For typical European wind conditions, it is common practice to choose the *rated wind speed* v_r in relation to the mean wind speed $\bar{v}$ using a factor $v_r / \bar{v} \approx 1.4...1.6$ [2, 4, 7]. This has the positive effect that at wind speeds of a high energy density the energy E_i is converted with high total efficiencies η_{WP}, Fig. 10-21. If a control mechanism or storm control of the wind turbine is already activated at wind speeds below triple the mean wind speed ($3\,\bar{v}$), the factor $v_r/\bar{v}$ reduces to a smaller value. A study on the influence of the wind speed of control start on the determination of the rated wind speed with optimum yield is given in [7].

In addition considering that the *efficiency of the gearbox* is not constant, but decreases in the partial load range below the rated efficiency, the factor $vr /\bar{v}$ moves to higher values $v_r / \bar{v} \approx 1.5...1.95$ [11]. These values are valid for a wind pump system with a centrifugal pump taking into consideration the control of the wind turbine.

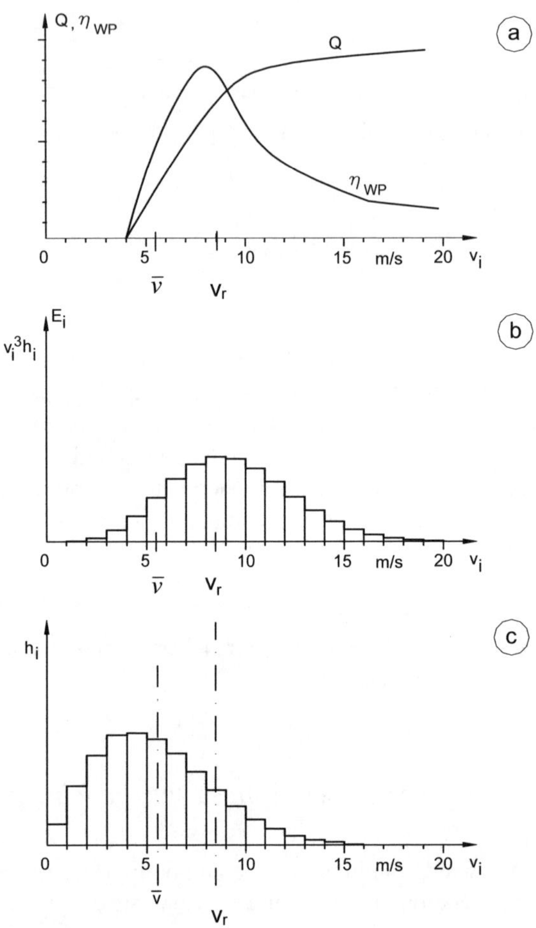

Fig. 10-21 Selection of the rated wind speed $v_r \approx 1{,}5\ \overline{v}$ for a wind pump system, a) Discharge curve and efficiency of the wind pump, b) Energy content of the wind classes, c) Wind histogram (Frequency distribution of the wind), $\overline{v}$ mean wind speed

The wind speed v_{in} at which the discharge starts influences the discharged volume, too. However, for a given hydraulic system it has a fixed relation to v_N. Therefore, it is usually sufficient to dimension the wind pump for the rated wind speed. The rated total head H_r is mainly determined by the geodetic head H_{geo}, but the pipe losses have to be considered with an allowance. For low heads in the range of up to 10m, the rated total head is approximately $H_r = (1.2...1.4)\ H_{geo}$. For a higher rated total head, a common estimation is $H_r \leq 1.2\ H_{geo}$ [3].

The following sections will treat the equations that can be used to determine the parameters of turbine and pump, and especially the required transmission gear ratio. Thus, the focused aim cosists in the wind pump system reaching the maximum total efficiency η_{WP} at the rated wind speed v_r.

10.4.3 Design of a wind pump system with a piston pump

By equating wind turbine power and (mechanical) piston pump power at the rated operating point (Fig. 10-22)

$$\frac{\rho}{2}\, v_r^3\, c_{P.opt}\, A_{Rotor} = \frac{\rho_w\, g\, V_{Hub}\, n_P\, H_r}{\eta_m\, \eta_{vol}} \tag{10.20}$$

a relation is obtained that is useful to determine the *required transmission gear ratio i* of wind turbine's rotational speed n_{WT} to pump's rotational speed n_P

$$i = \frac{n_{WT}}{n_P}. \tag{10.21}$$

Additionally, the optimum tip speed ratio λ_{opt} at which the wind turbine reaches its optimum power has to be considered

$$\lambda_{opt} = \frac{\pi\, n_{WT}\, d_{WT}}{v_r}. \tag{10.22}$$

This gives the transmission gear ratio

$$i = V_{Hub}\, \frac{\rho_w\, g\, H_r}{\eta_m\, \eta_{vol}} \cdot \frac{8\, \lambda_{opt}}{\pi^2\, \rho\, c_{P.opt}\, d_{WT}^3}\, \frac{1}{v_r^2}. \tag{10.23}$$

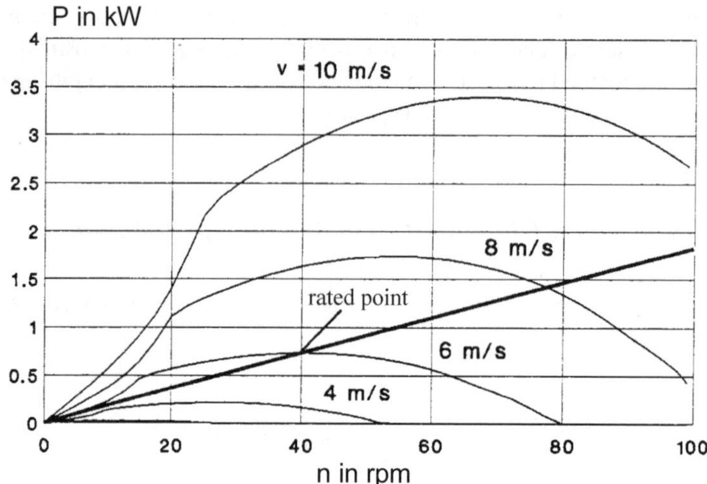

Fig. 10-22 Wind pump system with a piston pump designed for a rated wind speed of $v_r = 6$ m/s

Basically, the goal is to do without a gearbox, i.e. $i = 1$. For larger Western mills ($\lambda_{opt} \approx 1$, $d_{WT} \geq 5$ m), this is achieved by selecting a suitable stroke volume $V_{Hub} = \pi/4 \, d_K^2 \, 2 \, r_K$ or a suitable crank radius. For smaller systems of this type, down-gearing is necessary.

Wind pump systems with a piston pump show a particular operating behaviour during start-up, Fig. 10-23. The wind turbine only starts to rotate when, due to a sufficiently high wind speed, the rotor torque provides, via the eccentric lever, a piston force that is larger than the force from the weight of the water column above the piston area. The static friction of the system is neglected in this consideration. So, for starting the maximum torque M_{max} of the oscillating curve of the pump torque (cf. Fig. 10-16) has to be overcome. In Fig. 10-23, the wind pump starts to operate at the start-up wind speed $v_{in} = 5$ m/s. Moreover, the simplifying assumption is made that the wind speed remains constant during start-up until the mean pump torque $\overline{M}$ is reached.

During operation, the rotor acts as a flywheel. According to the relations discussed in section 10.3.2, the operating behaviour can be described by the mean torque $\overline{M}$ as the rotational speed of the wind turbine does not follow the oscillating torque of the piston pump, in a first approximation. If the wind speed falls below v_{in}, the rotor will keep turning due to its inertia, and the discharge will continue. Only if the wind speed falls below the minimum discharge wind speed v_{min} where the rotor torque is smaller than the mean torque $\overline{M}$, does the system stop. In Fig. 10-23, this happens at $v_{min} = 2.8$ m/s.

The wind pump system with a piston pump has therefore a *characterisic hysteresis in the discharge curve*, Fig. 10-24, in the range between start-up wind speed v_{in} and the minimum discharge wind speed v_{min}. This means that within the hysteresis range, a discharge is only possible if v_{in} is exceeded often enough, i.e.

during a day with relatively low wind speeds, there are often enough gusts to start up the wind pump system.

Equating the individual operating point the mean torque $\overline{M}$ of the piston pump, from equation (10.12) and (10.14), and the torque of the wind turbine:

$$M_{\mathrm{WT}} = \frac{\rho}{2} \; \frac{\pi d_{\mathrm{WT}}^{2}}{4} \; \frac{d_{\mathrm{WT}}}{2} \; v^{2} c_{\mathrm{M}}\left(\lambda\right) \tag{10.24}$$

gives the characteristic wind speeds of the hysteresis range.

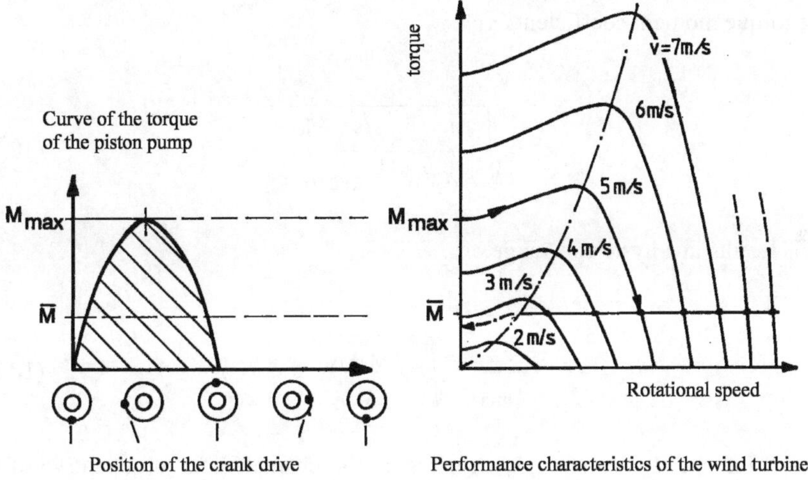

Fig. 10-23 Start-up and shut-down behaviour (hysteresis) in the torque diagram of a wind pump system with a piston pump

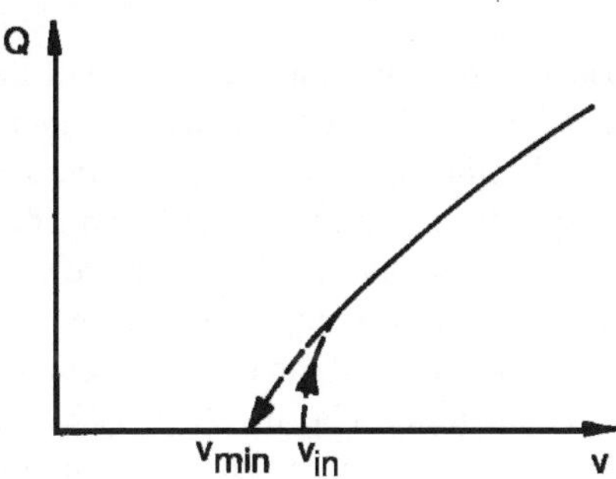

Fig. 10-24 Flow rate versus wind speed of a wind pump system with piston pump, showing the hysteresis in the starting range of the discharge curve

The minimum discharge wind speed v_{min}, is calculated from the torque equilibrium of the mean pump torque $\overline{M}$ and the torque of the wind turbine for the point of maximum torque moment coefficient $c_{\mathrm{M.max}}$:

$$v_{\mathrm{min}} = \sqrt{\frac{4\,\rho_{\mathrm{w}}\,g\,H\,d_{\mathrm{K}}^{\,2}\,r_{\mathrm{K}}}{\rho\cdot\pi\,d_{\mathrm{WT}}^{\,3}\,c_{\mathrm{M.max}}\,\eta_{\mathrm{m}}}}\qquad\qquad(10.25)$$

The start-up wind speed v_{in} results from the torque equilibrium of the required maximum pump torque M_{max} and the wind turbine torque at standstill ($\lambda = 0$) with the torque moment coefficient $c_{\mathrm{M.0}}$

$$v_{in} = \sqrt{\frac{\pi\,4\,\rho_{\mathrm{w}}\,g\,H\,d_{\mathrm{K}}^{\,2}\,r_{\mathrm{K}}}{\rho\,\pi\,d_{\mathrm{WT}}^{\,3}\,c_{\mathrm{M.0}}\,\eta_{\mathrm{m}}}}\,.\qquad\qquad(10.26)$$

This results in a hysteresis range of

$$\frac{v_{\mathrm{in}}}{v_{\mathrm{min}}} = \sqrt{\pi\,\frac{c_{\mathrm{M.max}}}{c_{\mathrm{M.0}}}}\,.\qquad\qquad(10.27)$$

This wind speed ratio strongly depends on the characteristic torque curve of the wind turbine. For a wind turbine with an extremely low tip speed ratio the two torque moment coefficients are nearly identical, $c_{\mathrm{M.0}} \approx c_{\mathrm{M.max}}$, therefore

$$v_{\mathrm{in}} \approx \sqrt{\pi}\, v_{\mathrm{min}}\,.\qquad\qquad(10.28)$$

In order to achieve a load-free or load-reduced start-up of the piston pump, load-relieving measures have to be included for smoothing the torque curve, cf. section 10.2, and [4, 8]. These load-relieving measures allow the wind pump to start the discharge at lower wind speeds or even the installation of a bigger pump with a higher hydraulic power which may increase the discharge volume significantly.

For the yield calculation of a wind pump system with a piston pump, the starting behaviour and the hysteresis range of the discharge curve have to be taken into account. This procedure is suggested e.g. in [9]. It is based on a corrected discharge curve using the probabilities of wind speeds in the hysteresis range which can be determined by the wind speed distribution on site. This curve is then used to determine the yield according to section 10.3.2, equation (10.17).

10.4.4 Design of a wind pump system with a centrifugal pump

For the rated operating point, the power balance of wind turbine and centrifugal
pump is calculated assuming that both wind turbine and centrifugal pump operate
at their maximum efficiency point which is illustrated by Fig. 10-25. For a
mechanical coupling of wind turbine and centrifugal pump with a gearbox
(mechanical efficiency η_G) the power balance is then

$$c_{P,\mathrm{opt}}\,\eta_G\,\frac{\rho}{2}\,\frac{\pi\,d_{WT}^2}{4}\,v_r^3 = \frac{\rho_W\,g\,Q_r\,H_r}{\eta_{\mathrm{opt}}}\,. \tag{10.29}$$

In the following, the rated flow rate Q_r in equation (10.29) is substituted by specific values of the pump, in order to illustrate the relation between the main characteristic values of the centrifugal pump and other system parameters. Similar to
the description of types of wind-driven pumps in section 10.2, the specific speed
n_q (equation 10.8) is used since it is characteristic of the corresponding type of
centrifugal pump.

Using the model laws based on the laws of similarity, the head of the centrifugal
pump can be expressed by the dimensionless *pressure number* ψ [4]:

$$\Psi = \frac{2\,g\,H}{\left(\pi\,n\,d_2\right)^2}\,. \tag{10.30}$$

The impeller diameter of the centrifugal pump is d_2, cf. Fig. 10-8. For the best-
efficiency point the rated rotational speed of the pump can be calculated with the
corresponding pressure number ψ_{opt}:

$$n_r = \sqrt{\frac{2\,g\,H_r}{\Psi_{opt}}}\;\frac{1}{d_2\pi}\,. \tag{10.31}$$

In the power balance of equation (10.29), the rated flow rate Q_r is now substituted
by the specific speed n_q of equation (10.8). Moreover, equation (10.31) is taken
into account, giving

$$\pi^2\,n_q^2\,d_2^2\,\frac{\Psi_{opt}}{\eta_{\mathrm{opt}}}\,\rho_W\,\frac{Q_q}{H_q^{3/2}} = \rho\,\frac{\pi\,d_W^2}{4}\,c_{P.\mathrm{opt}}\,\eta_G\,\frac{v_r^3}{H_{S.r}^{3/2}}\,. \tag{10.32}$$

At the rated operating point, the rated total head $H_{S.r}$ of the hydraulic system is
equal to the rated total head H_r of the pump, cf. section 10.3.2.

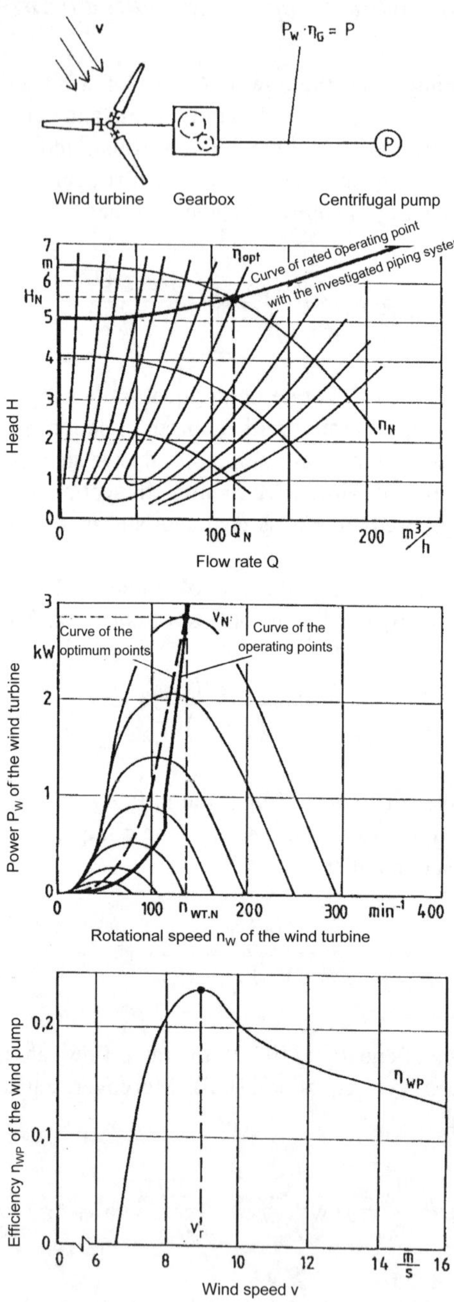

Fig. 10-25 Rated operating point of a wind pump system with a mechanically coupled centrifugal pump

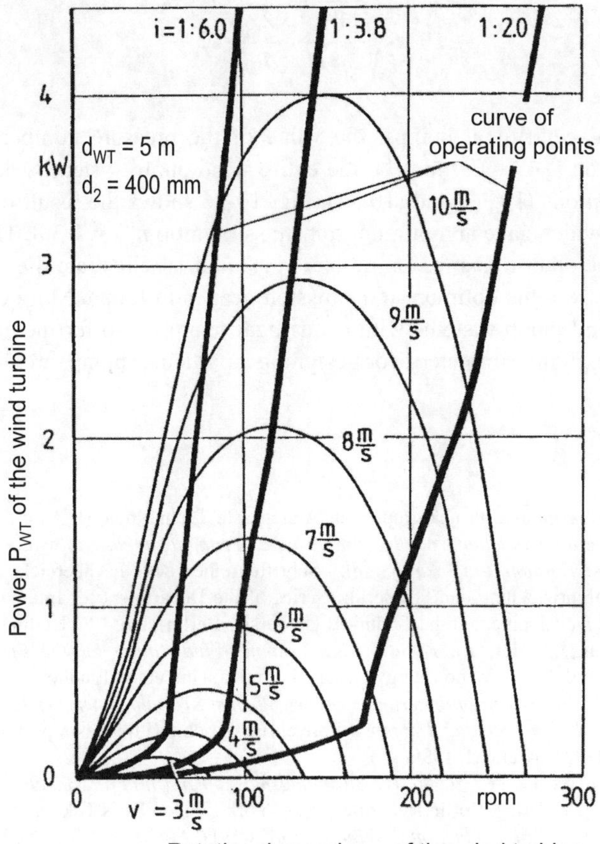

Fig. 10-26 Influence of the transmission gear ratio on the curve of operating points in the power characteristics of a wind turbine, wind turbine diameter d_{WT} = 5 m, tip speed ratio λ_{opt} = 4, impeller diameter of the centrifugal pump d_2 = 400 mm, rated wind speed v_r = 8 m/s

It is considered that a centrifugal pump, suitable for given site conditions and a given wind turbine, is chosen from a product line of similar pumps, scaled according to the model laws. For all these pumps the specific speed n_q, the optimum pressure number ψ_{opt}, and the efficiency η_{opt} in the best point are the same. Therefore, only the impeller radius d_2 has to be determined from equation (10.32):

$$d_2 = \frac{1}{\pi n_q} \sqrt{\frac{1}{\rho_w} \frac{\eta_{opt}}{\psi_{opt}} \frac{H_q^{3/2}}{Q_q} \rho \frac{\pi d_{WT}^2}{4} c_{P.opt} \eta_G \frac{v_r^3}{H_{S.r}^{3/2}}} \ . \qquad (10.33)$$

Considering the design tip speed ratio λ_{opt} of the wind turbine and equation (10.31) gives the required transmission gear ratio i for the rated rotational speed n_r

$$i = d_2 \sqrt{\frac{\Psi_{\mathrm{opt}}}{2g\,H_{\mathrm{r}}}} \; \frac{\lambda_{\mathrm{opt}}}{d_{\mathrm{WT}}} \, v_{\mathrm{r}} \quad .$$
(10.34)

As for radial centrifugal pumps, the value of the pressure number only shows slight variation ($\psi_{\mathrm{opt}} = 0.9$ to 1.1), the entire wind pump system is determined by the two equations (10.33) and (10.34). Fig. 10-26 shows the result of such a system design, which gave an optimum transmission ratio $i_{\mathrm{opt}} = 1{:}3.8$. The additional curves for the other transmission ratios $i = 1{:}2$ and $1{:}6$ illustrate the consequences of deviating from the optimum transmission gear ratio found. More details on the design of wind pump systems with centrifugal pumps, also for the case with free choice of the pump parameters, for designing a particular pump, are found in [3].

References

[1]   Interdisziplinäre Projektgruppe für Angepasste Technologie (IPAT): *Der Einsatz von Windpumpsystemen zur Be- und Entwässerung (The use of wind pump systems for irrigation and drainage)*, Schriftenreihe des Fachbereichs Internationale Agrarentwicklung der TU Berlin (Series of the Department for International Agricultural Development at the Technical University Berlin), No. 120, Berlin, 1989

[2]   Jongh, J.A. de: *Low Head / High Volume Wind Pumps for the Fleuve Region in Senegal*, CWD, Wind Energy Group, Technical University Eindhoven, 1988

[3]   Twele, J.: *Ertragsoptimierung windgetriebener Kreiselpumpen (Yield optimization of wind-driven centrifugal pumps)*, Fortschritt-Berichte (Progress reports) VDI, series 7, No 181, Düsseldorf, 1990

[4]   Pfleiderer, C.; Petermann, H.: *Strömungsmaschinen (fluid flow machines)*, 5th edition, Springer-Verlag, Berlin Heidelberg New York London Paris Tokyo, 1986

[5]   Lysen, E.H.: *Introduction to Wind Energy*, CWD, Amersfoort, The Netherlands, 1982

[6]   Dijk, H. van: *The Volume of Storage Tanks in Water Supply Systems with Windmill Driven Pumps in Cap Verde*, International Institute for Reclamation and Improvement, Wageningen, The Netherlands, 1984

[7]   Staassen, A. J.: *A Model of a Centrifugal Pump Coupled to a Windrotor*, Wind Energy Group, University of Technology Eindhoven, 1988

[8]   Cleijne, H., u.a.: Pump Research by CWD: *The influence of starting torque of single acting piston pumps on water pumping windmills*, European Wind Energy Association, Conference and Exhibition, 7.10.-9.10.1986, Rome, Section E10, p. 163-167

[9]   Meel, J. van; u.a.: *Field Testing of Water Pumping Windmills by CWD*, European Wind Energy Association Conference and Exhibition, 7.10.-9.10.1986, Rome, Section F15, p. 423-430

[10]   Mier, M.; Siekmann, H.; Twele, J.: *Optimization of Winddriven Centrifugal Pumps*, 1st International Congress on Fluid Handling Systems 10.9.-12.09.1990, Essen

[11]   Kortenkamp, R.: *Die Optimierung von Windpumpsystemen mit Kreiselpumpe unter Berücksichtigung des instationären Betriebsverhaltens (The Optimisation of wind pump systems with centrifugal pumps, taking into account unsteady operating behaviour)*, Fortschritt-Berichte (Progress reports), VDI, series 7, No 235, Düsseldorf, 1993

11 Wind turbines for electricity generation - basics

Presently wind turbines are used primarily for electrical power generation. Three-phase alternators (AC generators) are used almost exclusively. Even for applications that require DC the lower-cost alternator/rectifier configuration has superceded the DC-generator. When a three-phase generator feeds directly into a grid, that operates at a fixed frequency (e.g. 50 Hz in Europe, 60 Hz in the U.S.A.) the angular velocity of the generator is fixed - or almost fixed. In this situation the power generation capability of the wind turbine will be fully utilized for just one value of wind speed (approximately 8 m/s in Fig. 11-1a). Thanks to the highly developed converter technology of today it is now possible to operate with variable generator speeds even for grid-connected generators (see Fig. 6-18). This yields better utilization of the wind turbine; during gusty winds it also substantially decreases the mechanical stresses in the blades and the shaft between turbine and generator.

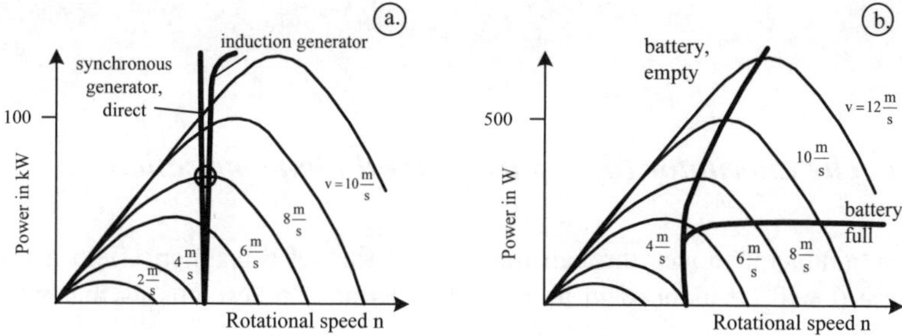

Fig. 11-1 Operating points of wind turbine and generator: a) directly grid-connected operation; b) stand-alone operation as a battery charger

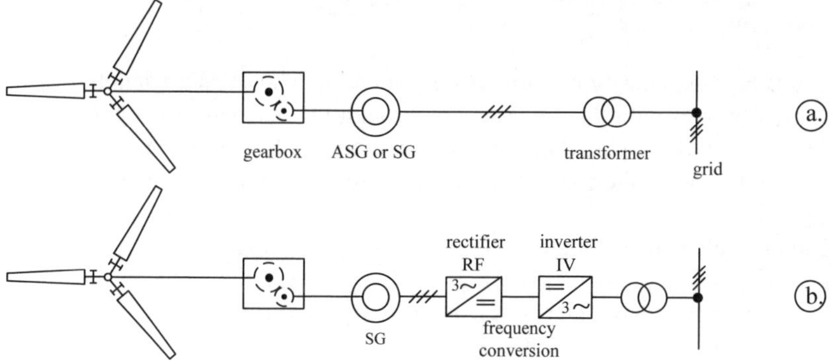

Fig. 11-2 Grid connection of a wind turbine: a) Direct grid connection of asynchronous (ASG) or synchronous generator (SG); b) synchronous generator feeding into the grid via a frequency converter with intermediate DC circuit

We have already considered extensively the aerodynamic properties of rotors in chapters 5 and 6. In the present chapter we will focus on electrical generators and AC-DC-AC converters. This will lead us to an understanding of the overall system. Concepts of control and of plant layout will then be introduced in chapters 12 and 13. Chapter 14 will be devoted to issues of grid connection.

Table 11.1 presents the symbols for the circuit elements that we will use. Note that the table contains elements from semiconductor electronics in addition to the classical electrical components Resistor R, Capacitor C, and Coil L. The dynamic behaviour of the elements is summarised in the second column of the table. For the electronic components the input/output transfer with sinusoidal inputs is shown. These semiconductor components, which are increasing in importance, and their uses will be the topic of section 11.3.

11.1 The alternator - single-phase AC machine

11.1.1 The alternator (dynamo) in stand-alone operation

When a conductor moves through the lines of flux of a magnetic field a source voltage $e(t)$ will be induced in it, Fig. 11-3. Faraday's law tells us that we will have

$$e(t) = B \; l \; v(t). \tag{11.1}$$

Here B is the flux density of a uniform magnetic field and l is the length of a conductor that moves with the velocity $v(t)$ through the magnetic field perpendicular to the lines of flux. The flux density B is commonly expressed in Tesla units ($1 \; T = V\mathrm{s/m}^2$). In an alternator, this effect is used to generate a single-phase alternating voltage.

The most elementary implementation consists of a single one-turn coil rotating in a field of a permanent magnet, Fig. 11-4. The effective conductor length is $2 \; l$ in that case. The voltage $e(t)$ is delivered by two slip rings. Given that the angle $\psi = \Omega t$ is relative to the horizontal axis the voltage $e(t)$ turns out to be:

$$e(t) = B \, 2 \, l \, r \, \Omega \sin \Omega t \; = \; E_S \sin \Omega t. \tag{11.2}$$

Table 11.1 Electrical components and their dynamic behaviour

Electrical components and their dynamic behaviour	
Resistor Resistance R	$u = R \cdot i$ Voltage = resistance x current
Capacitor Capacitance C	$u = \dfrac{1}{C} \int i \, dt$
Inductor Inductance L	$u = L \dfrac{di}{dt}$
Diode	
Thyristor	
Transistor	

Fig. 11-4 shows that the velocity $r\,\Omega\sin(\Omega t)$ is the component of the circumferential velocity $r\,\Omega$ which intersects the lines of the magnetic field at right angles, i.e. it corresponds to $v(t)$ in Fig. 11-3. The amplitude E_S of the source voltage $e(t)$ is proportional to the circumferential velocity $r\,\Omega$, and therefore, it is proportional to the angular velocity Ω and the rotational speed n, $\Omega = 2\,\pi n$. It is also proportional to the magnetic flux density B, and to the number of windings of the coil which is one in Fig. 11-4. The highest voltage is always induced when the change of magnetic flux $\Phi = 2\,B\,(l\,r)\cos\Omega t$ within the one-turn coil is largest, see Fig. 11-5.

For real electrical machines, it is often too complex to determine the operating behaviour based on the exact physical relations. Therefore, equivalent circuit diagrams are used in that case. They only represent the most important physical characteristics.

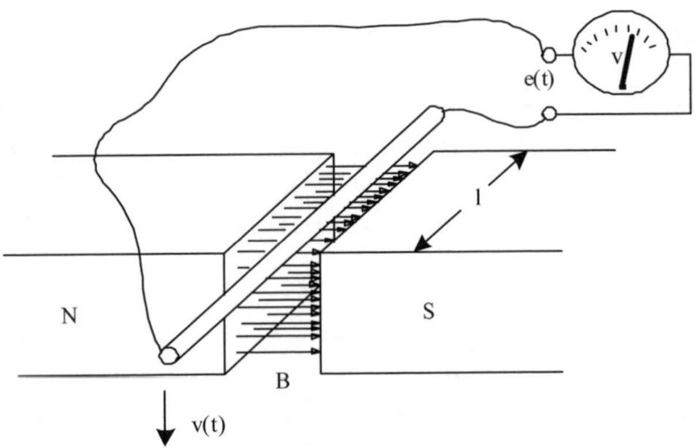

Fig. 11-3 Moving a conductor in a uniform magnetic field. Induction of the voltage $e(t)$

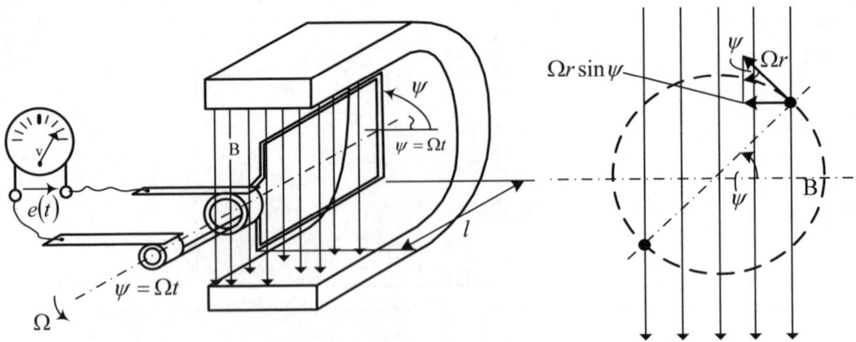

Fig. 11-4 Left: alternator (dynamo); right: active velocity component

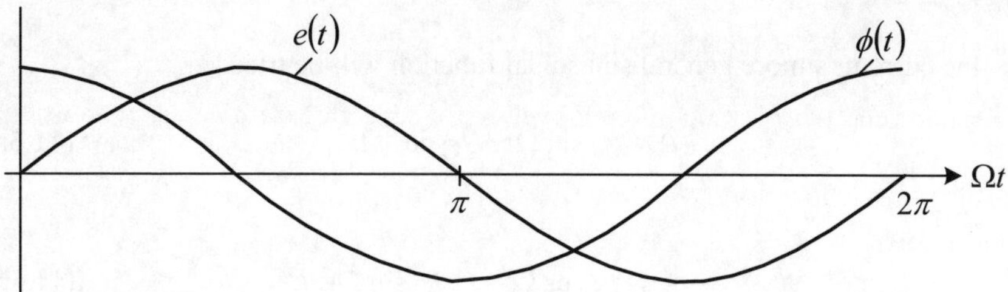

Fig. 11-5 Voltage $e(t)$ and magnetic flux $\Phi(t)$ in a dynamo

The equivalent circuit diagram of an alternator connected to a load resistance R_L, is shown in Fig. 11-6 where $e(t)$ is the source voltage; R_i and L_i are the internal rotor resistance and the rotor inductance, respectively. If the machine is at standstill, the ohmic resistance R_i can directly be measured. The source voltage is obtained by measuring the terminal voltage at the open circuit terminals.

The inductance is obtained from a short circuit experiment (with low voltage level). All electrical machines can be represented by such equivalent circuit diagrams. The required parameters can be usually determined by simple experiments [1, 10, 13, 15].

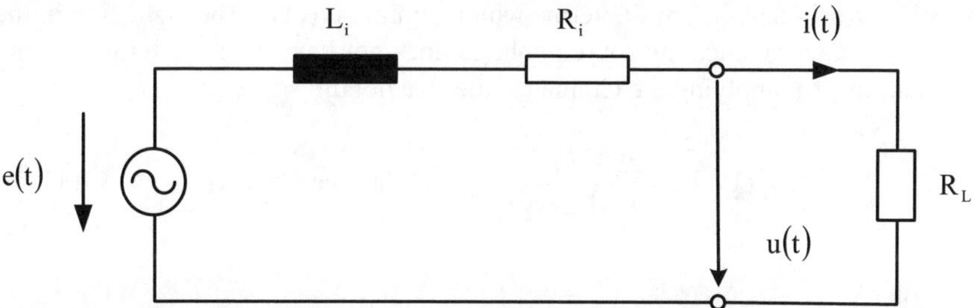

Fig. 11-6 Equivalent circuit diagram of a dynamo with the load resistance R_L, the internal rotor resistance R_i, and the rotor inductance L_i

Now, we will analyze how an alternator behaves if it is connected to a resistive load R_L, e.g. an electrical heater.

According to Kirchhoff's voltage law the induced source voltage $e(t)$ is absorbed by the inductance L_i, and the resistances R_i and R_L, as shown in Fig. 11-6. The differential equation

$$R_\mathrm{i}\, i + R_\mathrm{L}\, i + L_\mathrm{i}\,\frac{di}{dt}=e(t)\,, \tag{11.3}$$

is valid; as the voltage drop in the coil is not proportional to the current, but to the current change versus time di/dt.

As the relation is analyzed for a steady rotational speed Ω, the source voltage is

$$e(t) = E_S \sin \Omega t. \tag{11.4}$$

For the current, a more general sinusoidal function will be tried:

$$i(t) = I_S \sin \Omega t + I_C \cos \Omega t \tag{11.5a}$$

$$\frac{di}{dt} = \Omega I_S \cos \Omega t - \Omega I_C \sin \Omega t. \tag{11.5b}$$

A pure sine approach would not be sufficient, as will be seen soon. Inserting (11.4) and (11.5) into the differential equation (11.3), and assorting according to sine and cosine terms, which each have to be balanced, the following system of equations for the current amplitudes results:

$$\begin{pmatrix} R_i + R_L & -\Omega L_i \\ +\Omega L_i & R_i + R_L \end{pmatrix} \cdot \begin{pmatrix} I_S \\ I_C \end{pmatrix} = \begin{pmatrix} E_S \\ 0 \end{pmatrix}. \tag{11.6}$$

The top row of (11.6) is a collection of terms that had contained $\sin \Omega t$. The bottom row originates from the terms which contain $\cos \Omega t$. The right side in the bottom row is zero since the source voltage does not have a cosine term. Solving the equation (e.g. applying the Cramer's rule) we obtain:

$$I_S = E_S \frac{R_i + R_L}{\left(R_i + R_L\right)^2 + \left(\Omega L_i\right)^2} \qquad \text{(active current)} \tag{11.7a}$$

$$I_C = E_S \frac{-\Omega L_i}{\left(R_i + R_L\right)^2 + \left(\Omega L_i\right)^2} \qquad \text{(reactive current).} \tag{11.7b}$$

Both current amplitudes are proportional to the source voltage amplitude E_S. The sine amplitude I_S, which is in phase with the source voltage, disappears if there is no ohmic resistance ($R_i + R_L = 0$) in the circuit. The cosine amplitude disappears if no inductance is present ($L_i = 0$).

The cosine amplitude is negative, equation (11.7b). This will be discussed later. For the time history of the current we obtain

$$i(t) = E_S \left(\frac{R_i + R_L}{(R_i + R_L)^2 + (\Omega L_i)^2} \sin \Omega t - \frac{\Omega L_i}{(R_i + R_L)^2 + (\Omega L_i)^2} \cos \Omega t \right). \tag{11.8}$$

The sine term is called *active* current, and the cosine term *reactive* current. The reasons become immediately clear if we calculate the power, which is the product of voltage and current

$$P(t) = i(t)\, e(t)$$

or

$$P(t) = \underbrace{E_S\, I_S\, \sin^2\Omega t}_{} + \underbrace{E_S\, I_C\, \sin\Omega t\, \cos\Omega t}_{} \quad . \tag{11.9}$$

        Active power   Reactive power

The first term, the active power, is always positive. However, due to

$$\sin^2\Omega t = \frac{1-\cos 2\Omega t}{2} \tag{11.10}$$

it oscillates around the mean value, cf. Fig. 11-7,

$$P_\mathrm{m} = \frac{1}{2}\, E_S I_S = \frac{1}{2}\, (2\,Blr)^2\, \Omega^2\, \frac{R_\mathrm{i} + R_\mathrm{L}}{(R_\mathrm{i} + R_\mathrm{L})^2 + (\Omega L_\mathrm{i})^2}\;. \tag{11.11}$$

The oscillation occurs at twice the frequency Ω. The second term, the reactive power, oscillates at twice the frequency, too. However, its mean value is zero, therefore it is also called wattless power.

It is common when discussing AC currents to combine the sine and cosine components of equation (11.5a):

$$i(t) = I_S\, \sin\Omega t + I_C\, \cos\Omega t$$

$$= \hat{\imath}\, \sin(\Omega t + \varphi). \tag{11.12}$$

$\hat{\imath}$ is the amplitude of the current $i(t)$, and the phase angle φ gives the shift of the current relative to the source voltage $e(t) = E_S \sin\Omega t$. These two values are determined from I_S and I_C by

$$\hat{\imath} = \sqrt{I_S^2 + I_C^2} \quad and \quad \tan\varphi = \frac{I_C}{I_S}\;. \tag{11.13}$$

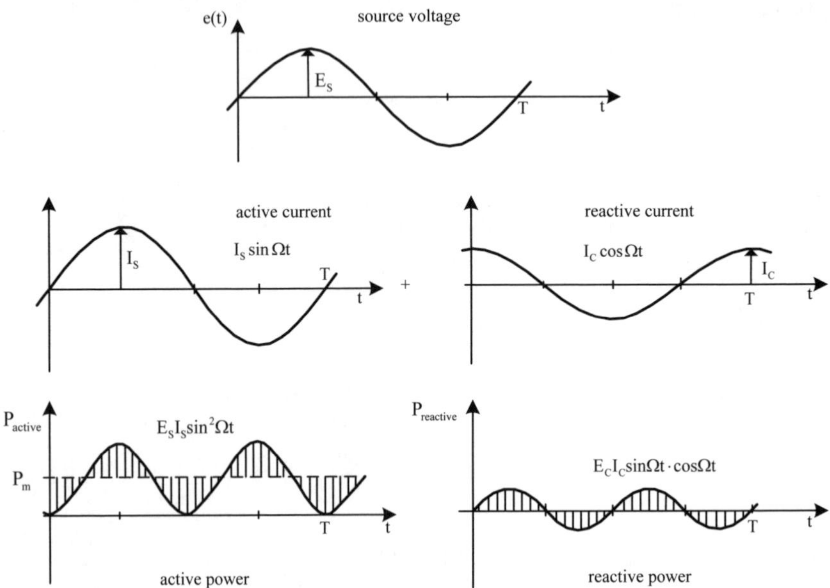

Fig. 11-7 Active and reactive power from multiplication of voltage and current

The derivation shall be mentioned only briefly. Due to

$$\hat{\imath} \sin (\Omega t + \varphi) = \hat{\imath} (\sin \Omega t \cos \varphi + \cos \Omega t \sin \varphi)$$

the comparison of the coefficients with (11.12) gives $I_C = \hat{\imath} \sin \varphi$; $I_S = \hat{\imath} \cos \varphi$. From this follow the two equations (11.3). The advantage of the representation by amplitude and phase angle is obvious, Fig. 11-8: both these values are easily measured – e.g. using an oscilloscope.

The fact that the cosine amplitude of the alternator is negative in equation (11.7b) is considered in Fig. 11-8. Here, it can be seen as well that the current maximum follows the voltage maximum which is typical for a system of inductors and resistors: current follows voltage. The time curves of voltage and current may be considered as projections of the vectors in the diagrams to the left in Fig. 11-8.

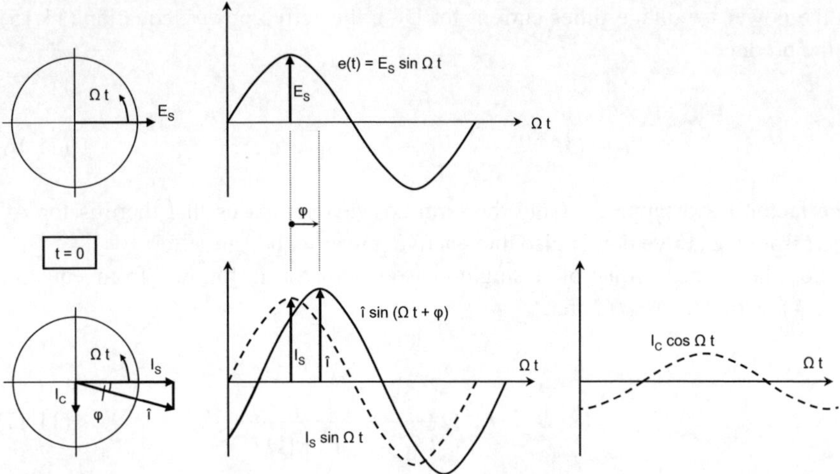

Fig. 11-8 Vector diagram of current and power at the time $t = 0$; current amplitude $\hat{\imath}$; phase angle φ; (left); time curves of current and power versus Ωt (right)

With this representation of the current time curves we obtain the momentary power, equation (11.9), in the form which was already discussed in connection with Fig. 11-7:

$$P(t) = i(t)\,e(t)$$

$$= E_S\,\hat{\imath}\,\cos\varphi\,[\sin^2\Omega t] + E_S\,\hat{\imath}\,\sin\varphi\,[\cos\Omega t\,\sin\Omega t] \ . \qquad (11.14)$$

$$\underbrace{\hphantom{= E_S\,\hat{\imath}\,\cos\varphi\,[\sin^2\Omega t]}}_{\text{Active power}} \quad \underbrace{\hphantom{E_S\,\hat{\imath}\,\sin\varphi\,[\cos\Omega t\,\sin\Omega t]}}_{\text{Reactive power}}$$

The amplitudes of active and reactive current are $\hat{\imath}\cos\varphi = I_S$ and $\hat{\imath}\sin\varphi = I_C$, cf. Fig. 11-8.

The integration of the *reactive power* (i.e. wattles power) over the period T yields zero (cf. Fig. 11-7 right) and for the *active power*, as given in equation (11.11) we obtain the mean value of

$$P_m = (1/2)\,E_S\,\hat{\imath}\,\cos\varphi = (1/2)\,E_S\,I_S. \qquad (11.15)$$

The amplitude of the reactive power is $E_S\,\hat{\imath}\,\sin\varphi = E_S\,I_C$, but it oscillates with twice the frequency around zero.

For historical reasons, in the daily work of electrical engineering, not the amplitudes of the sinusoidal voltages and currents (E, $\hat{\imath}$) are used but the so-called effective values. The effective value corresponds to about 70% of the amplitude value, $I_{\text{eff}} = I/\sqrt{2}$ and $E_{\text{eff}} = E/\sqrt{2}$, respectively. Using this convention (analogous to the power = voltage times current for DC), the active power, equation (11.15), is the product

$$P_{\mathrm{m}} = (1/2)\, E_{\mathrm{S}}\,\hat{\imath}\,\cos\varphi = E_{\mathrm{eff}}\,I_{\mathrm{eff}}\cos\varphi\,. \tag{11.16}$$

The factor 1/2 disappears. Only the term $\cos\varphi$ reminds us that there is for AC apart from the active power also the reactive power with a $\sin\varphi$ term.

For the mean torque of a single-phase alternator it follows from equation (11.11), with $M_{\mathrm{m}} = P_{\mathrm{m}}/\Omega$, that

$$M_{\mathrm{m}} = \frac{1}{2}(2\cdot B\cdot l\cdot r)^2\Omega\,\frac{R_{\mathrm{i}}+R_{\mathrm{L}}}{(R_{\mathrm{i}}+R_{\mathrm{L}})^2+(\Omega L_{\mathrm{i}})^2}\,, \tag{11.17}$$

if small mechanical losses due to friction etc. are neglected.

The expression in parenthesis ($2\,B\,l\,r$) is the machine constant for one winding that has been analyzed so far. With increasing number N of windings the constant is correspondingly larger. A closer look at the mean active power in equation (11.11) reveals that it consists of useful output power at the load resistor R_{L}, and a power loss which is proportional to the internal resistance R_{i} of the machine. The following expressions can be used:

$$P_{\mathrm{m}} = P_{\mathrm{use}} + P_{\mathrm{loss}}$$

$$P_{\mathrm{loss}} = \frac{1}{2}(2\,Bl\,r)^2\Omega^2\,\frac{R_{\mathrm{i}}}{(R_{\mathrm{i}}+R_{\mathrm{L}})^2+(\Omega L_{\mathrm{i}})^2}$$

$$P_{\mathrm{use}} = \frac{1}{2}(2\,B\,l\,r)^2\Omega^2\,\frac{R_{\mathrm{L}}}{(R_{\mathrm{i}}+R_{\mathrm{L}})^2+(\Omega L_{\mathrm{i}})^2}$$

$$\eta = \frac{P_{\mathrm{use}}}{P_{\mathrm{m}}} = \frac{R_{\mathrm{L}}}{R_{\mathrm{i}}+R_{\mathrm{L}}} = \frac{1}{1+\dfrac{R_{\mathrm{i}}}{R_{\mathrm{L}}}}\,. \tag{11.18}$$

The power curve $P_{\mathrm{m}}(\Omega)$ and the torque curve $M_{\mathrm{m}}(\Omega)$ for different speeds are of particular interest for the operation of a generator with a wind turbine. They result from equations (11.11) and (11.18). It becomes clear that the alternator has a very pronounced maximum torque, and that the power tends towards a finite maximum value for high rotational speeds, Fig. 11-9.

For the electrical design of systems the behaviour of current and voltage are important. According to Fig. 11-9, the output voltage of the machine, and the current tend towards a final value for high frequencies, similar to the power. The source voltage, however, increases linearly with the rotational speed.

Changing the load resistance R_L or the excitation (magnetic field B) shifts the load curve, as illustrated in Fig. 11-10. This provides a means to match, within certain limits, the load curve to a given rotor and generator, cf. also Figs. 11-1a and 11-1b.

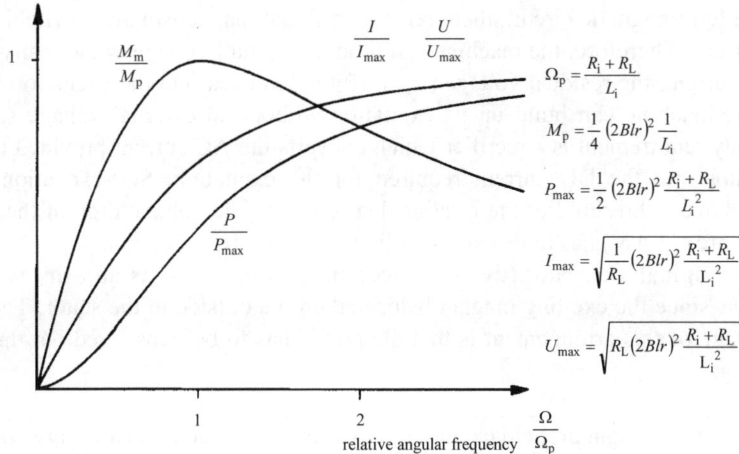

Fig. 11-9 Dimensionless power, torque, current and source voltage versus the relative angular frequency for an alternator with permanent magnets and a load R_L; reference values given at the right margin of the figure

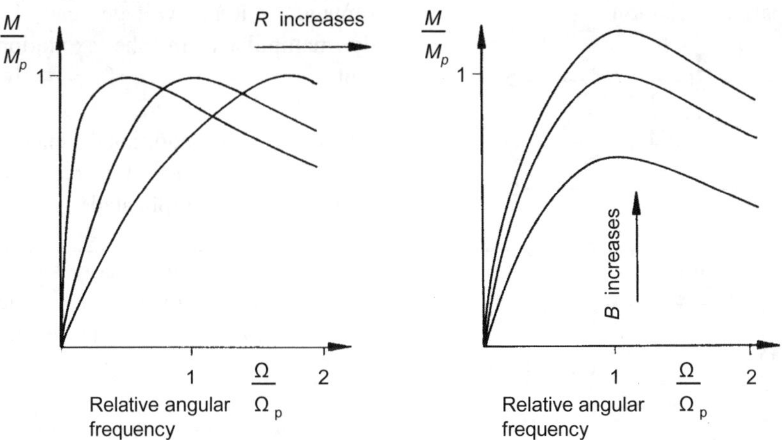

Fig. 11-10 Shifting of the torque curve by changing the load resistance or by changing the exciting magnetic field

11.1.2 Types of excitation, internal and external pole machine

In the alternator described so far, the "*permanent-excited* dynamo", the required magnetic field is created by a permanent magnet. The result is that the output voltage of the machine can only be manipulated by the rotational speed. If the permanent magnet is replaced by an electromagnet, the flux density B can be influenced through the excitation current. The source voltage can now be manipulated for a fixed rotational speed. This is called a *separately-excited* machine, Fig. 11-11b.

In a ferromagnetic circuit, there remains residual magnetism after switching off the current. Therefore, the machine can produce a small voltage without any excitation current, the residual voltage (U_{res}). If it is fed back into the excitation winding, the machine can build up its excitation without an external voltage source. The only requirement is a rectifier which converts the AC current provided by the generator into the DC current required for the excitation. Self-excitation only starts above a threshold of the rotational speed since the voltage drop of the rectifier (approx. 1.4 V) has to be exceeded first, Fig. 11-11c.

The original alternator (dynamo) according to Fig. 11-4 is an *external pole machine* since the exciting magnet is located on the outside in the stator. The disadvantage of this arrangement is that the power has to be transferred out through slip rings.

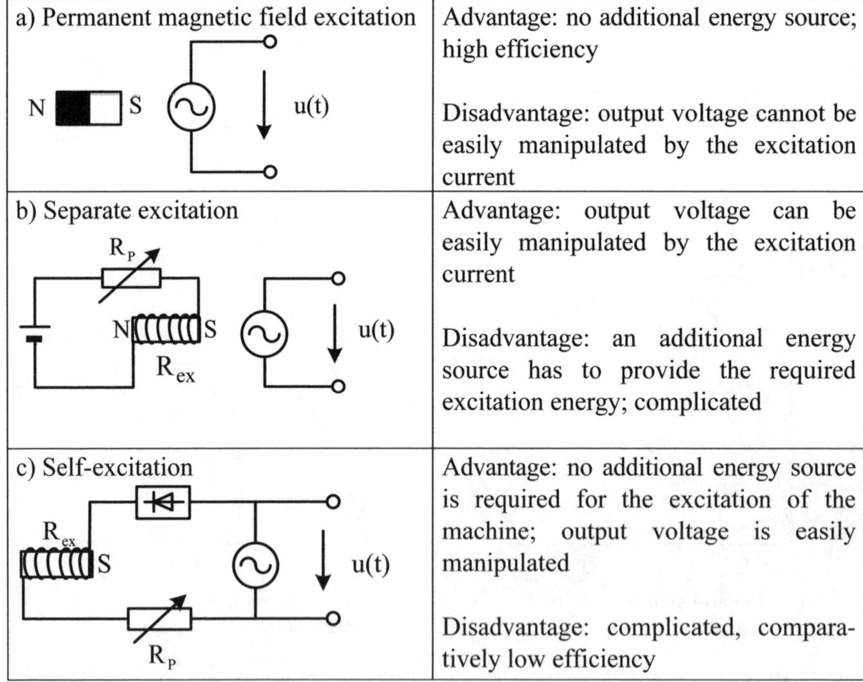

a) Permanent magnetic field excitation	Advantage: no additional energy source; high efficiency Disadvantage: output voltage cannot be easily manipulated by the excitation current
b) Separate excitation	Advantage: output voltage can be easily manipulated by the excitation current Disadvantage: an additional energy source has to provide the required excitation energy; complicated
c) Self-excitation	Advantage: no additional energy source is required for the excitation of the machine; output voltage is easily manipulated Disadvantage: complicated, comparatively low efficiency

Fig. 11-11 Types of excitation

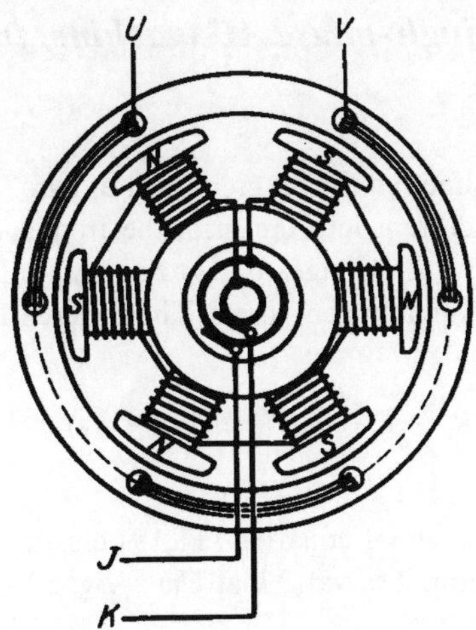

Fig. 11-12 A.C. machine, excitation on the rotor, internal pole arrangement [7]

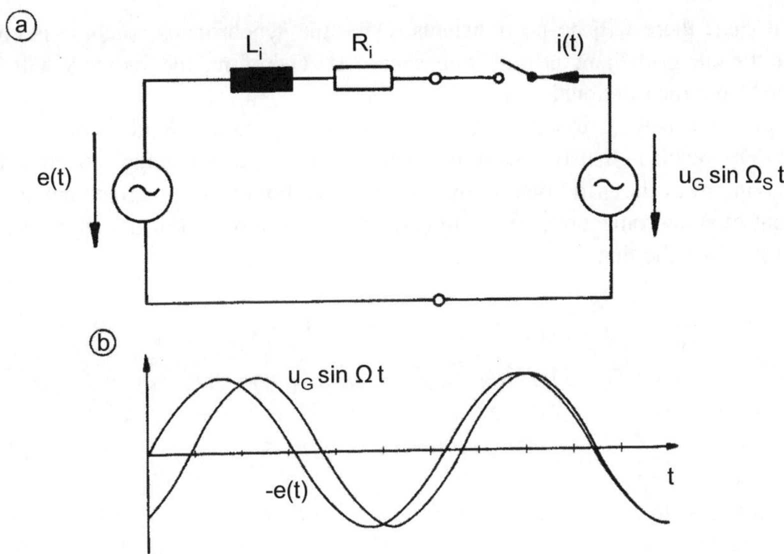

Fig. 11-13 a) Equivalent circuit diagram of a single-phase synchronous machine; (b) process of synchronization

For a large power, this is expensive and susceptible to wear. Therefore, usually, the magnet revolves, the so-called revolving field, and the coil is in the stator, cf. Fig. 11-12. Separately excited *internal pole generators* still require slip rings for the excitation current, but now the power to be transmitted is small in comparison to the rated power (approx. 2 to 10%).

11.1.3 *Alternator (single-phase AC machine) in grid-connected operation*

Bringing a generator on-line requires a great deal of care. At the very moment the generator starts feeding current into the grid, the frequencies (i.e. speed), amplitudes and phase angles of the voltages have to be equal. If these three conditions are fulfilled, there will be no transient terms. The differential equation reads

$$R_i\, i + L_i\, \frac{di}{dt} = U_G \sin \Omega_s t - e(t) \ . \tag{11.19}$$

The right side of the differential equation (11.19) disappears completely if in the very moment of connection, it is valid that $U_G = E_S$, $\Omega = \Omega_S$ und $\alpha_0 = 0$ for the source voltage $e(t)$

$$e(t) = E_S \sin (\Omega t + \alpha_0) \ . \tag{11.20}$$

In that case, there will be no transients. Once the synchronous machine is connected to the grid there is only one speed Ω. Therefore, the index S will be dropped from the rotational speed.

Applying a driving torque to the shaft, causes a leading load angle ϑ, Fig. 11-14. The result is that the maximum value of the source voltage $e(t)$ precedes the maximum of the grid voltage (by the angle ϑ): both were in agreement at the moment of synchronization, with (almost) no driving torque. For the source voltage, it is now valid that

$$e(t) = E \sin(\Omega t + \vartheta)$$

$$= E (\sin \vartheta \, \cos \Omega t + \cos \vartheta \, \sin \Omega t)$$

$$= E_C \cos \Omega t + E_S \sin \Omega t \ . \tag{11.21}$$

By means of a stronger excitation in the revolving field, the source voltage amplitude E can now be increased $E > U_G$, or alternatively decreased. At this point a general sinusoid will be assumed for the current (where I_S is in phase with E_S), cf. equation (11.5):

$$i(t) = I_S \sin \Omega t + I_C \cos \Omega t \tag{11.22a}$$

$$\frac{di}{dt} = \Omega \ I_S \cos \Omega t - \Omega \ I_C \sin \Omega t \ . \tag{11.22b}$$

This is inserted into the differential equation (11.19). We use the simplifying assumption that the internal resistance is very small, $R_i \approx 0$. This is valid for medium-sized and large synchronous generators.

With $R_i \ll \Omega L_i$ and equations (11.21) and (11.22), the differential equation (11.19) becomes

$$\Omega L_i \, (I_S \cos \Omega t - I_C \sin \Omega t) = (U_G - E_S) \sin \Omega t - E_C \cos \Omega t \qquad (11.23)$$

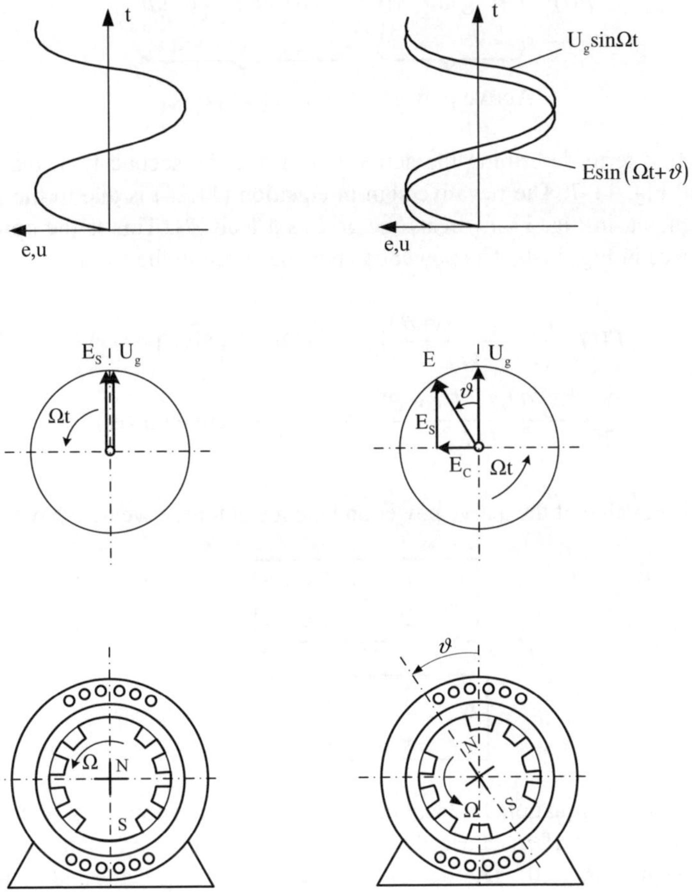

Fig. 11-14 Left: Synchronous machine at the moment of synchronisation, vector diagram and voltages $e(t)$, $u(t)$; right: rotor load angle ϑ due to the driving torque, vector diagram and voltages

Comparing the coefficients of the sine terms and cosine terms, respectively, we obtain the current amplitudes:

$$I_S = -\frac{E_C}{\Omega L} = -\frac{E \sin \vartheta}{\Omega L} \qquad \text{(active current)}, \qquad (11.24a)$$

$$I_C = -\frac{U_G - E_S}{\Omega L} = -\frac{U_G - E \cos \vartheta}{\Omega L} \quad \text{(reactive current)}. \qquad (11.24b)$$

The power results from $P(t) = u(t)\, i(t)$, and is

$$P(t) = U_G\, I_S \sin^2 \Omega t + U_G\, I_C \sin \Omega t \cos \Omega t, \qquad (11.25)$$

$$\underbrace{\hspace{3cm}}_{\text{Active power}} \quad \underbrace{\hspace{4cm}}_{\text{Reactive power}}$$

with the first term describing the active power and the second term the reactive power, cf. Fig. 11-7. The negative sign in equation (11.24) is due to the fact that the referencing in Fig. 11-13 treats the grid as a load [2]. This is the opposite of the reference in Fig. 11-6. The power can now be stated in the form

$$-P(t) = \frac{1}{2} \cdot \frac{U_G\, E \sin \vartheta}{\Omega L} (1 - \cos 2\Omega t) \quad \text{(Active power)}, \qquad (11.26a)$$

$$+\frac{1}{2} \cdot \frac{U_G\, (U_G - E \cos \vartheta)}{\Omega L} \sin 2\Omega t \quad \text{(Reactive power)}. \qquad (11.26b)$$

For the mean value of the active power and the mean torque we now have

$$\boxed{P_m = \frac{1}{2} \cdot \frac{U_G\, E \sin \vartheta}{\Omega L}} \qquad (11.27)$$

$$\boxed{M_m = \frac{1}{2} \cdot \frac{U_G\, E \sin \vartheta}{\Omega^2 L}}. \qquad (11.28)$$

Active current I_S, equation 11.24a, and active power, equation (11.27), are caused by the load angle ϑ. If $\vartheta > 0$ (leading), the machine acts as a generator, power is fed into the grid ($P_m < 0$). The sign is negative, but that was neglected in the equations (11.27) and (11.28). If a torque is required at the machine shaft (motoring mode), the load angle becomes negative, $\vartheta < 0$. Therefore, the sign of $\sin \vartheta$ changes, i.e. power is consumed from the grid ($P_m > 0$). For rated power, the load angle is usually 20° to 30°. If due to overloading, the load angle is forced to become larger than 90°, the machine slips out of step. During motoring, the speed will then drop to a speed below synchronous speed, depending on machine parameters and load moment. In the generating mode, the speed increases steeply. Both would lead to the machine's destruction due to the speed oscillations and the currents. Therefore this state has to be avoided.

The sign of reactive current, equation (11.24b), and reactive power can be manipulated by the value of $E_S = E \cos\vartheta$, i.e. by the excitation current of the revolving field which causes the source voltage amplitude E.

If $\vartheta = 0°$ and for the case of
    overexcitation, $E_S > U_G$, this equation shows that reactive power is fed into the grid (capacitive mode).
In the case of
    underexcitation, $E_S < U_G$, reactive power is taken from the grid (inductive mode).

Often, grid-connected synchronous generators operate with overexcitation in order to compensate for the reactive current consumption of the numerous induction motors, reactance coils and transformers, etc. which are always present in the grid.

Fig. 10.15 shows the rotating vectors of voltages E and U_G, and current I which can be projected to obtain the time curves, cf. Figs. 11-8 und 11-14.
Comparing the active current component $I_S = I \cos\varphi$ in Fig. 11-15 to the component calculated according to equation (11.24a), we obtain the relationship between the current phase angle φ, and the load angle ϑ

$$I \cos\varphi = - \frac{E \sin\vartheta}{\Omega L}. \tag{11.29}$$

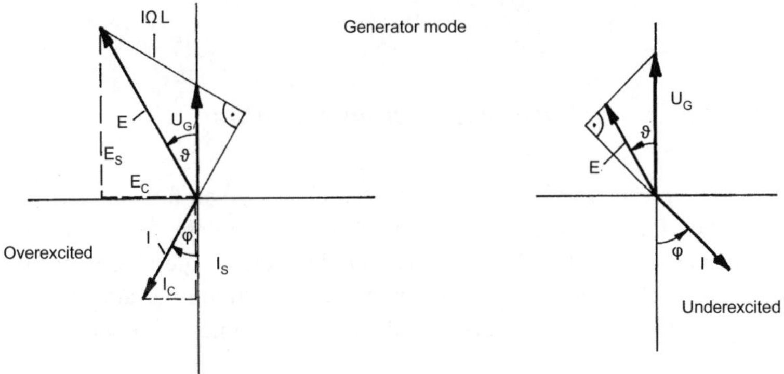

Fig. 11-15 Vector diagram of voltages U_G, E, and current I during electricity generation; $E_s > U_G$ overexcitation, $E_s < U_G$ underexcitation

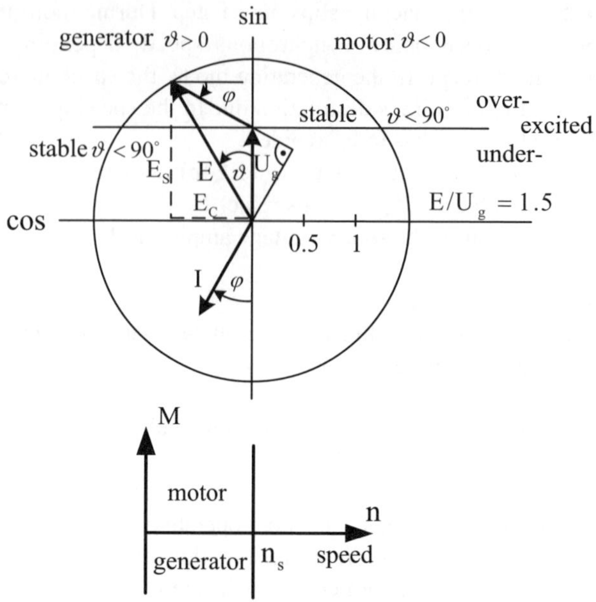

Fig. 11-16 Grid-connected synchronous machine: a) Heyland's vector diagram at $E = 1{,}5\ U_N$; b) torque-speed diagram

This is also the instruction for scaling the length of the current vector in Fig. 11-15 and 11-16. The latter figure gives an overview of the total behaviour of the grid-connected synchronous machine.

11.2 Three-phase machines

11.2.1 The three-phase synchronous machine

The *single-phase* alternator - which has been discussed so far - has the disadvantage that the electrical and the mechanical power oscillates between zero and a maximum value, cf. Fig. 11-7. To prevent this, the single-phase machine was extended into a *three-phase* machine with the three windings arranged in the stator spatially offset by 120°, Fig. 11-17 and 11-18. Machines with such a stator are called three-phase machines.

The *three-phase synchronous machine* works just like the single-phase alternator. The single-phase equivalent circuit diagram is therefore applicable to the three-phase machine. The calculations necessary for determining its operating behaviour can be derived from those for the alternator (cf. section 11.1). The voltage curves correspond to Fig. 11-17. They are those of the alternator supplemented by a phase offset at 120° and one at 240°.

The same applies to phase currents and phase powers. The mean total power of the machine is tripled. The individual phase powers complement one another in such a way that the power output is constant without any oscillation. This becomes clear if the active power curve of Fig. 11-7, bottom,

$$E_S\,I_S\,\sin^2\Omega t = (1/2)\,(1 - \cos 2\Omega t)\,E_S\,I_S$$

is drawn three times with the respective 120° phase shifts.

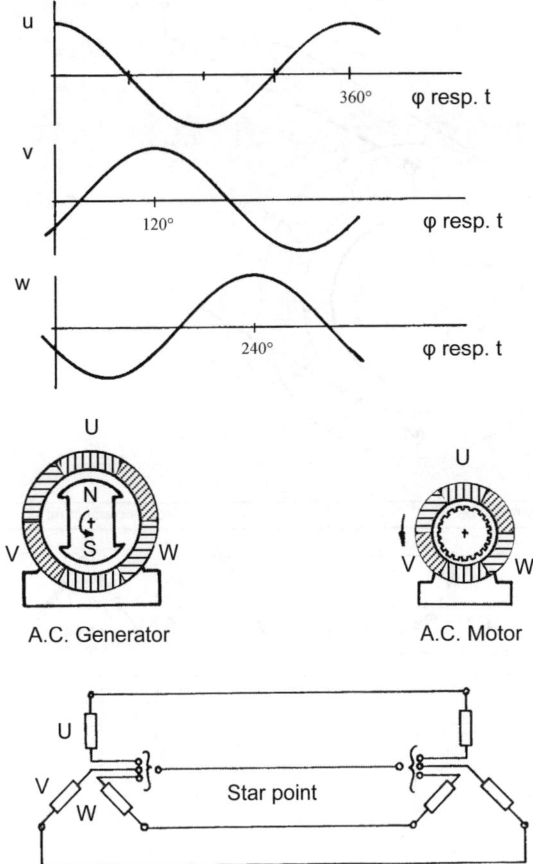

Fig. 11-17 Three-phase (A.C.) generator and motor: a) voltages in the three phases, b) application with a star connection

Adding graphically only the sum of the mean values remains

$$3 \cdot (1/2) \; E_S \, I_S = 3 \, E_{\text{eff}} \, I_{\text{eff}} \cos \varphi \; .$$

The troublesome oscillations are eliminated. The supply of power and torque is constant. Synchronous machines are mainly used as generators today.

Instead of using a three-phase system with six wires to carry the current between generator and motor, it is possible to use only four or three wires, due to the symmetry of the voltages. The result is the star connection of the windings shown in Fig. 11-17. It is mainly used for generators which often operate in an asymmetric load regime, i.e. the phases are not loaded evenly. For larger three-phase loads, e.g. motors, three-phase heaters, etc., the delta connection is used. It does not require a star point, Fig. 11-18.

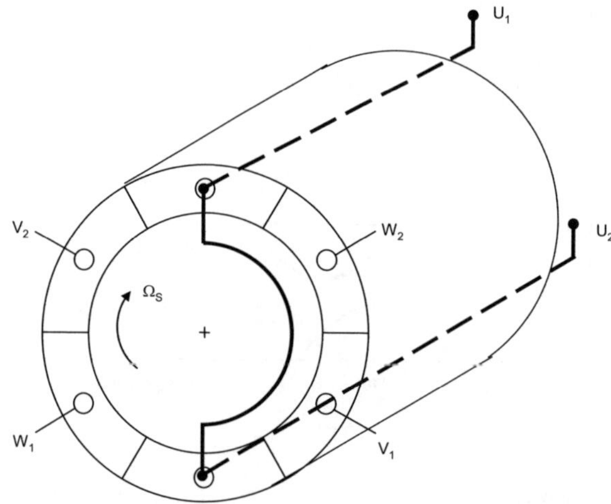

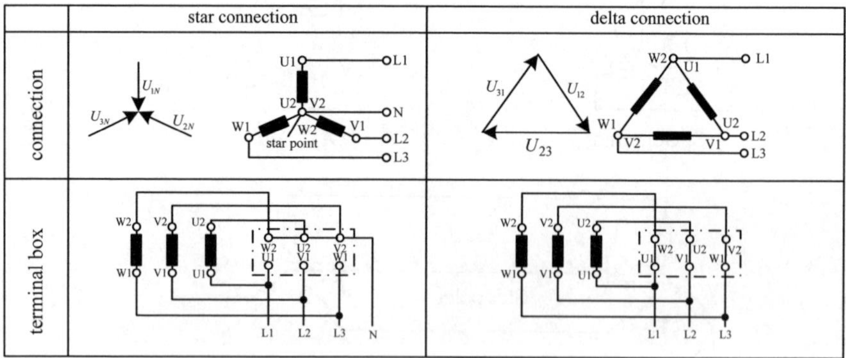

Fig. 11-18 Winding arrangement in the stator of a three-phase machine; possibilities of connecting AC windings; labelling of terminals in the terminal box [8]

This is possible, as the load is usually symmetric in these cases, therefore the phase currents add up to zero.

Using the delta connection, the power of generators and motors is increased by the factor $\sqrt{3} \cdot \sqrt{3} = 3$ because the voltage at the coil endings, e.g. between $L1$ and $L2$, is larger by the factor $\sqrt{3}$ than for the star connection between $L1$ and N, Fig. 11-18. the currents are increased by that factor as well. So it follows that

$$P_{\text{delta}} = 3\, P_{\text{star}}.$$

The stator construction of the three-phase *synchronous machine* which we just discussed, is similar to that of the three-phase induction machine (i.e. asynchronous machine). In contrast to that, the rotor construction of these two classical types of the three-phase machines differs significantly, Fig. 11-19.

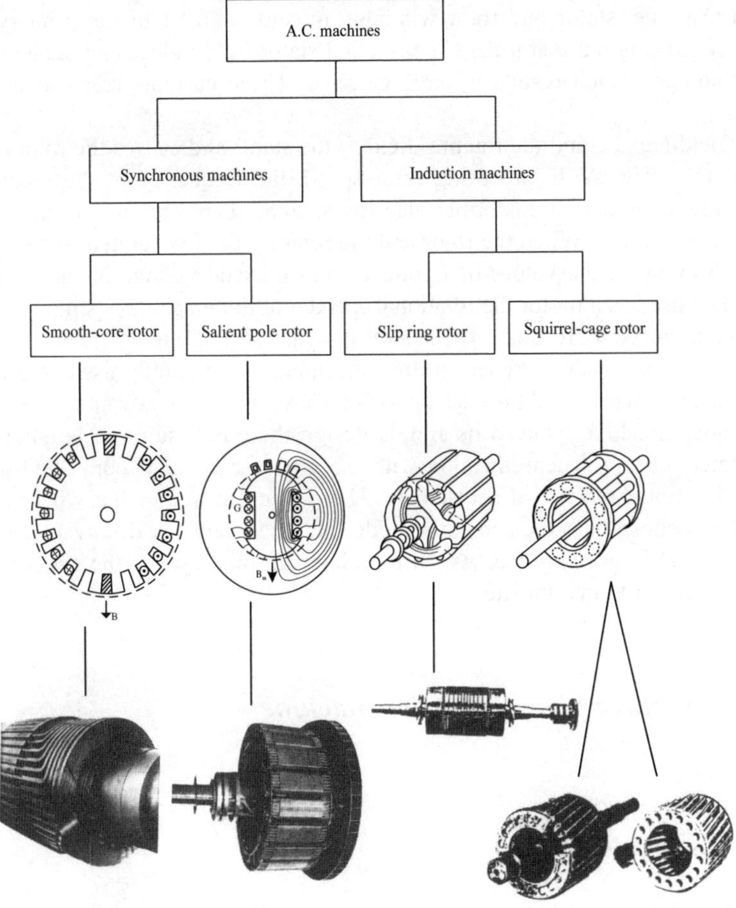

Fig. 11-19 AC machines and their rotor types [2, 4]

The rotor of the three-phase *synchronous machine* carries a pole system that is either a *salient pole rotor* or a *smooth-core rotor*. The excitation is caused by permanent magnets or electromagnets where the excitation current is supplied via slip rings. Some machines are completely without slip rings. In that case, an auxiliary machine is used. It is mounted on the same shaft and supplies the necessary energy (brushless excitation).

For motors, mainly three-phase *induction machines* (asynchronous machines) are used today. Their rotor is equipped with a three-phase winding, too. If the windings are short-circuited directly in the rotor, it is called a *squirrel-cage rotor*. This arrangement allows for a very robust and low cost construction of the machine. If a *slip ring rotor* is used, the rotor circuit can be manipulated. This type of machine carries a three-phase winding on the rotor. The ends of the windings are accessible through slip rings.

If a rotating field (star or delta connection) is generated in the stator of an induction machine, stator and rotor windings function similar to the primary and secondary sides of a transformer. The rotating stator field induces a voltage in the rotor windings, which results in large currents. These currents create their own field.

This field lags behind the rotating field of the stator and causes the rotor of the machine to rotate. With increasing rotor speed, the frequency of the secondary side of the imaginary transformer decreases, and, therefore, the value of the induced voltage, too. When the rotor and the rotating field generated by the stator run synchronously, the values of frequency and induced voltage in the rotor are zero. When used as a motor the rotational speed n of the machine is slightly below synchronous speed n_s. It runs *asynchronously* with a slip of a few percent.

The induction machine (asynchronous machine) is commonly used as *generator* for medium-sized wind turbines up to 1000 kW. It has two advantages over the synchronous machine. Due to its simple design the purchase price is quite low. And, moreover, the synchronization with the grid is easy. It is connected to the grid in the motor mode and accelerates. The driving torque of the wind turbine rotor then pushes it into the generator mode "automatically". A disadvantage is its *reactive power consumption* as we will see later. In this respect, the synchronous machine is easier to manipulate.

11.2.2 The three-phase induction machine

Stator construction

As mentioned above, the stator construction of the three-phase induction machine (i.e. asynchronous machine) is similar to the one of the synchronous machine,

Fig. 11-17 and 11-18. Three windings ($p = 1$) produce a rotating field with the grid frequency f_G, resp. $\Omega_S = 2\,\pi f_G$.

If the stator has a suitably connected system of 6 windings ($p = 2$), then the synchronous angular frequency of the rotating field is halfed to $\Omega_S = 2\,\pi f_G / p$. It rotates with 25 Hz (1500 rpm) instead of 50 Hz (3000 rpm). In pole switchable machines this is used to vary the synchronous frequency of the rotating field and consequently the idling speed of the induction machine. In wind turbines with power limitation by the stall effect pole switching of the induction machine is often used: most often it is switched from $p = 2$ to $p = 3$, i.e. from a synchronous speed of 1500 rpm (in a 50 Hz grid) to 1000 rpm.

Principle of operation

The field which rotates at grid frequency in the stator of the induction motor induces a voltage in the short-circuited rotor winding. The frequency of this voltage depends on the rotor speed. If the rotor does not move, this frequency is the same as the grid frequency (synchronous frequency) $\Omega_S = 2\,\pi n_S$. The system works like a transformer with a short-circuited secondary side. Rotor current and air-gap field, cause a tangential force on the rotor which moves the rotor in the direction of the rotating field.

If a load-free rotor is starting up ($M = 0$, load-free idling), Fig. 11-20, it will accelerate up to the synchronous speed n_S. At synchronous speed, the rotating magnetic field of the stator does not intersect the windings (slip ring rotor) or the conductor bars (squirrel-cage rotor) any longer, since the rotor moves at the same speed as the stator field. Therefore, no rotor current is induced, and there is no torque, either.

If a load is applied to the rotor ($M_L > 0$), its speed decreases slightly relative to the synchronous speed, $n < n_S$. The angular frequency Ω_2 of the rotor voltage is then equal to the difference between the rotating field speed n_S, and the mechanical speed n multiplied by $2\,\pi$

$$\Omega_2 = (n_S - n)\,2\,\pi = s\,\Omega_S . \qquad (11.30)$$

Here s is the slip, i.e. the difference between synchronous speed n_S and real speed n divided by the synchronous speed

$$s = \frac{n_S - n}{n_S} . \qquad (11.31)$$

The currents now induced in the rotor cause torque. During normal operation, the slip is not very large, s < 10 %, i.e. the induction motor normally runs almost at synchronous speed.

If instead of a load a driving torque ($M < 0$) is forced upon the rotor, it accelerates to a speed that is larger than the synchronous speed, $n > n_S$: now the machine acts as a generator, Fig. 11-20.

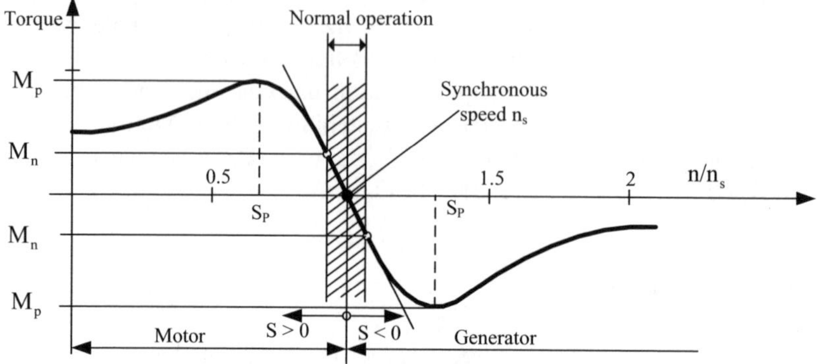

Fig. 11-20 Torque-speed characteristic of an induction machine; pull-out torque M_p, slip at pull-out torque s_P

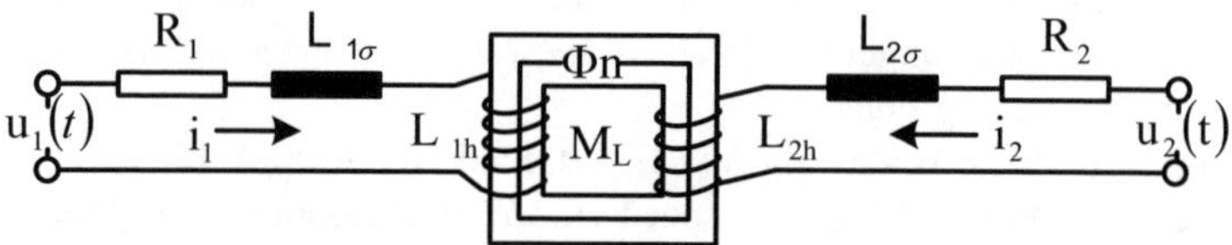

Fig. 11-21 Equivalent circuit diagram of a transformer

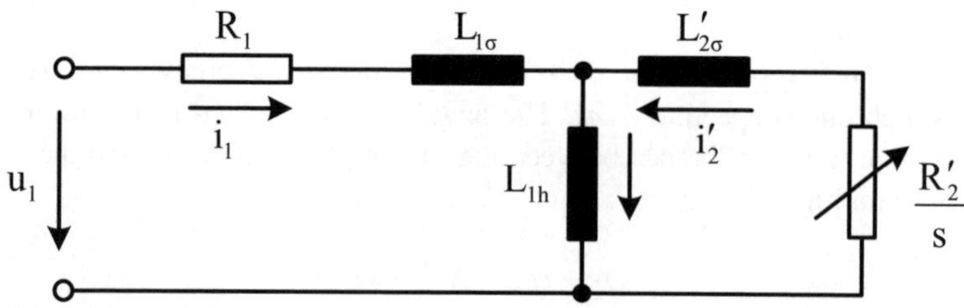

Fig. 11-22 Equivalent circuit diagram of an induction machine, single phase model

In the generator mode the driving torque must never be allowed to exceed the pull-out torque M_p which is 2 to 3 times the rated torque M_n, otherwise the rotor would reach excessively high speeds almost immediately. In the motor mode, the rotor would stop abruptly and overheat if the load torque was higher than the pull-out torque.

The similarities between a transformer and an induction machine were mentioned before. The transformer circuit diagram in Fig. 11-21 can be used to derive the voltage equations [10, 13].

They are as follows:

$$u_1(t) = R_1\, i_1 + L_{1\sigma}\frac{di_1}{dt} + L_{1h}\frac{di_1}{dt} + M_L\frac{di_2}{dt} \tag{11.32a}$$

$$u_2(t) = R_2\, i_2 + L_{2\sigma}\frac{di_2}{dt} + L_{2h}\frac{di_2}{dt} + M_L\frac{di_\Gamma}{dt} \tag{11.32b}$$

with the main inductances $L_{1h} = \Lambda_h w_1^2$ and $L_{2h} = \Lambda_h w_2^2$ in the circuit diagram on the right and the left. M_L is the mutual inductance $M_L = \Lambda_h w_1 w_2$ which couples the two circuits via the windings w_1 and w_2. Λ_h is the magnetic permeance of the main circuit. R_1 and R_2 are the ohmic resistances of the coils, and $L_{1\sigma}$ and $L_{2\sigma}$ are the leakage inductances. They represent magnetic fluxes that are not contained in the iron core, and, therefore, do not link both coils.

For the induction motor with its short-circuited rotor winding it is valid that $u_2 = 0$. The frequency in circuit two is determined by the slip $n_2 = n_S\, s$, i.e. $\Omega_2 = \Omega_S\, s$. This differs from a transformer.

The equivalent circuit diagram of Fig. 11-22 is a fairly good description of an induction machine. For more details, see [3, 10, 13, 15]. The apostrophe (') indicates values which have been transferred to the primary side using the turns ratio.

The current i_2' is calculated from the current of the secondary circuit i_2, according to Fig. 11-22 by multiplying with the reciprocal value of the turns ratio w_1/w_2

$$i_2' = i_2\,\frac{w_2}{w_1}\ . \tag{11.33}$$

For the other quantities, the following is valid:

$$L_{1h} = M_L\,\frac{w_1}{w_2} = L_{2h}\left(\frac{w_1}{w_2}\right)^2 \tag{11.33a}$$

$$R_2' = R_2\left(\frac{w_1}{w_2}\right)^2 \tag{11.33b}$$

$$L_{2\sigma}' = L_{2\sigma}\left(\frac{w_1}{w_2}\right)^2 . \tag{11.33c}$$

The fact that circuit two has the angular frequency Ω_2 is apparent from the equivalent resistance of the rotor, which depends on the slip. A plausibility check of this equivalent circuit diagram considers $s = 0$, i.e. the rotor moves at synchronous speed. As initially discussed, no voltage or current is induced in the rotor circuit, for $s = 0$. In the circuit diagram, this is due to the resistance R_2' / s which becomes infinite then.

For large induction machines the stator resistance is small compared to the main reactance ΩL_{1h}, therefore the equivalent circuit diagram of Fig. 1-22 can be simplified further, Fig. 11-23 top.

The calculation of the currents in an induction machine is beyond the scope of this book. Instead, the Heyland diagram, or Ossanna's circle, will be used. It is based on the short-circuit and idling tests. This diagram shows the amplitude and the phase angle φ of the stator current I_1 relative to the grid voltage U_1, depending on the slip, Fig. 11-23 bottom.

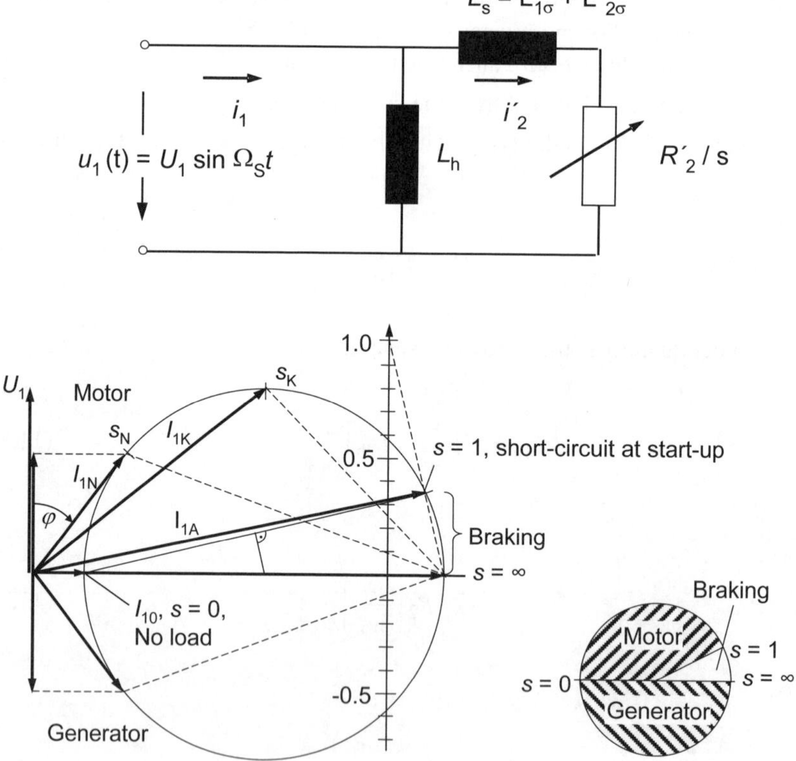

Fig. 11-23 Top: Simplified single phase equivalent circuit diagram of an induction machine; bottom: Heyland vector diagram; stator current vector dependent on slip s resp. phase angle φ; active current $I_{1S} = I_1 \cos \varphi$, reactive current $I_{1C} = I_1 \sin \varphi$ and slip scale s

In the *idling test* ($s = 0$) the rotor resistance approaches infinity because of the slip s in its denominator. The idling current I_{10} is purely inductive, has the phase angle $\varphi = 90°$ and is found on the horizontal axis. In the *short-circuit test* – very first moment of starting – the slip is $s = 1$. The inductances of stator and rotor together with the resistance R_2' determine the high starting current I_{1A} and its phase angle. This is shown in the circle diagram.

The case of *total braking*, $s = \infty$, would lead again to a current vector which is horizontal at $\varphi = 90°$ because the current path through the rotor does not contain a resistance any more. However, the practical realisation of this test is difficult. Nonetheless, the circle diagram can be determined using only the idling test and the short-circuit test [13]. An auxiliary line is drawn from the vector tip of the idling current I_{10} to the vector tip of the starting current I_{1A} . From the centre of the auxiliary line a perpendicular line is constructed towards of the horizontal axis.

The intersection of the perpendicular line with the horizontal axis gives the centre of the circle on which the current vector moves for motor, generator or braking mode:

    Motor mode: $0 < s < 1$
    Braking mode: $1 < s < \infty$
    Generator mode: $s < 0$

The slip is scaled linearly and its scale is obtained by a similar construction [13]. In Fig. 11-23 the slip scale shows for slip at pull-out torque a value of approx. 0.25.

The electrical power can be determined using the Heyland vector diagram:

$$\text{Reactive power::} \quad Q = 3\, U_1\, I_1\, \sin\varphi \tag{11.34a}$$

$$\text{Active power:} \quad P = 3\, U_1\, I_1\, \cos\varphi \tag{11.34b}$$

The factor 3 is due to the three phases of the stator coils.

If we ignore, somewhat generously, the copper and iron losses, the stator converts the active power into a torque which revolves with the synchronous frequency

$$P = 3\, U_1\, I_1\, \cos\varphi = M\Omega_S \tag{11.34c}$$

The rotor experiences this torque at its circumference but follows it, due to the slip, only with the angular frequency of $\Omega = \Omega_S\,(1 - s)$. Therefore, the motoric mechanical power is slightly smaller than the electrical power:

$$P_{\text{mech}} = \frac{3U_1\,I_1\,\cos\varphi}{\Omega_S}\,\Omega_S(1-s) = P(1-s) \tag{11.35}$$

In the generator mode ($s < 0$) it is larger than the electrical power. Slip represents loss. Precisely speaking: for both cases 3% slip equals 3% loss! Certainly, there are additional copper and iron losses. Nonetheless, large induction generators reach an efficiency of more than 95%.

Fig. 11-24 shows the torque-speed (resp. slip) characteristic of a large induction machine including current curve I_1 and efficiency curve. These curves may be measured directly or derived from the vector diagram (as approximation).

The starting and short-circuit currents are quite high, they reach multiples of the rated current. Therefore, the star connection is used for starting and then it is switched to the delta connection, Fig. 11-18.

The torque-speed curve which is important for wind turbine design can be described quite simply by the *Kloß'* formula referring to pull-out torque and pull-out slip

$$\frac{M}{M_\mathrm{p}} = \frac{2}{\dfrac{s}{s_\mathrm{p}} + \dfrac{s_\mathrm{p}}{s}} \ . \tag{11.36}$$

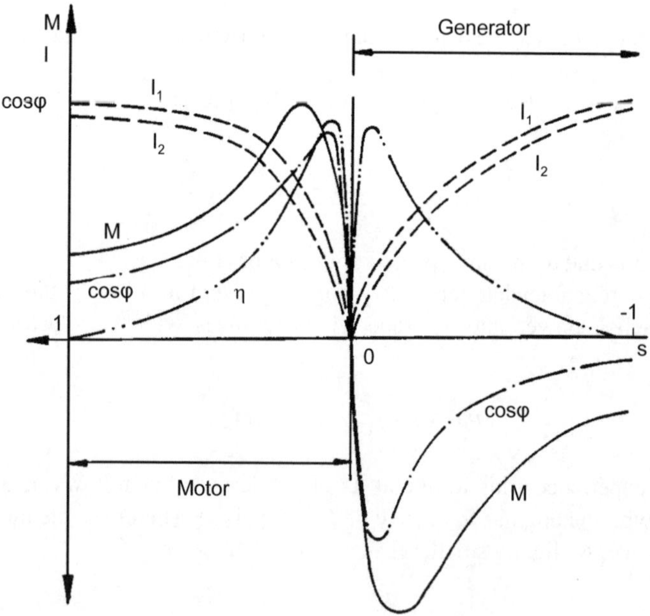

Fig. 11-24 Torque versus slip of an induction machine; stator current I_1, rotor current I_2, efficiency η and cos φ

During start-up ($s \gg s_\mathrm{p}$), the torque approximately follows a hyperbola, $M/M_\mathrm{p} = 2\, s_\mathrm{p}/s$. Near rated power, it is a straight line through the synchronous point. In the design phase, pull-out torque and pull-out slip can be calculated. For an existing induction machine, it will best be experimentally determined [10 to 15]. The big advantage of the induction machine, compared to the synchronous machine, is the self-starting behaviour and the easy transition from motor mode to generator mode. Synchronisation is not required. This and the ability for rotor speed switching (Fig. 11-26) made the induction machine very popular for the stall-controlled wind turbines of the Danish concept. The *reactive power consumption* $I_1 \sin\varphi$ is disadvantageous and occurs in both, motor and generator mode, as can be seen in Fig. 11-23.

Manipulation of the torque-speed characteristics

The position of the pull-out torque in the characteristics, Fig. 11-20 resp. 11-24, can be influenced by the resistance R'_2 as the equation for the pull-out slip shows:

$$s_\mathrm{p} = \frac{R'_2}{\Omega L_S} \quad \text{with} \quad L_S = L_{1\sigma} + L'_{2\sigma} \qquad (11.37)$$

It depends on the leakage inductances L_S of stator and rotor as well as the rotor resistance R'_2 [10, 14]. The pull-out slip also determines the slope of the torque characteristic close to the synchronous point ($s = 0$). Contrary to that, the magnitude of the pull-out torque M_p is not influenced by a change of R'_2, Fig. 11-25, since it is valid that [10, 14]

$$M_\mathrm{p} = \frac{3U_1^2}{2\Omega_\mathrm{s}^2 L_S} \,. \qquad (11.38)$$

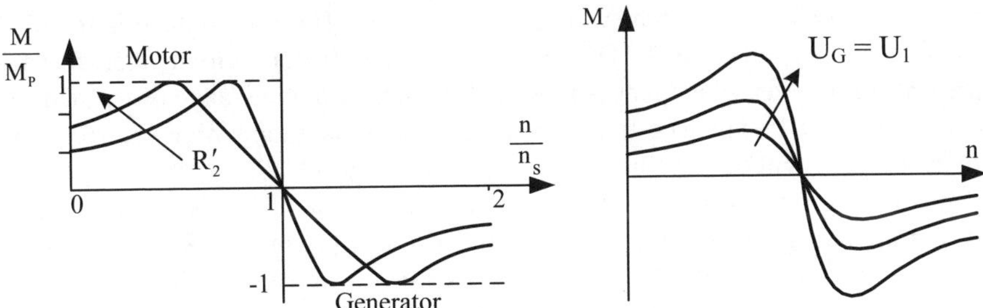

Fig. 11-25 Torque characteristics: influence of the rotor resistance and of the grid voltage U_G

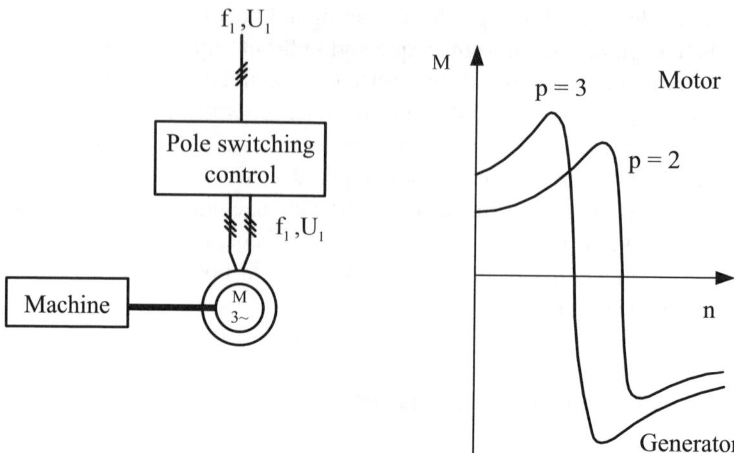

Fig. 11-26 Pole switchable induction machine, torque versus speed

If *additional resistance is introduced to the rotor circuit via slip rings* which increases R'_2, the characteristic becomes more flat. Consequently a change of the load produces a larger change of the speed, i.e. the slip increases.

This effect is used for the dynamic slip control, see Fig. 13-7 in section 13.11.2. A short-term increase of the rotor speed (and the slip) are allowed by increased R'_2, so that during strong wind gusts energy can be stored as kinetic energy in the rotor. This relieves the structure, but causes (short-term) higher losses. If the grid voltage U_G increases, the magnitude of the pull-out torque is changed, but not its position on the speed axis.

The possibility to change the synchronous speed of the induction machine by switching the number of pole pairs was already mentioned. Fig. 11-26 shows the practical consequences.

The speed of the induction (asynchronous) machine can be manipulated in a significantly larger range if the rotor frequency can be set independent of the grid frequency. Today, this is achieved using converters. The grid voltage is rectified and fed into an intermediate DC circuit. This intermediate circuit then feeds a three-phase inverter which makes it possible to supply a wide range of frequencies and voltages. If this is used to feed the *stator* of a machine with a squirrel-cage rotor, its speed will be variable over a wide range, Fig. 11-27.

In slip-ring rotor machines, the *rotor* can be fed so that the speed can be varied by manipulating the rotating field of the rotor [6], Fig. 11-28. This type of machine is called doubly-fed induction machine with variable-speed operation this is used in many wind turbines of the MW class, section 13.1.4. We will deal with this machine and the converter technology in the next section.

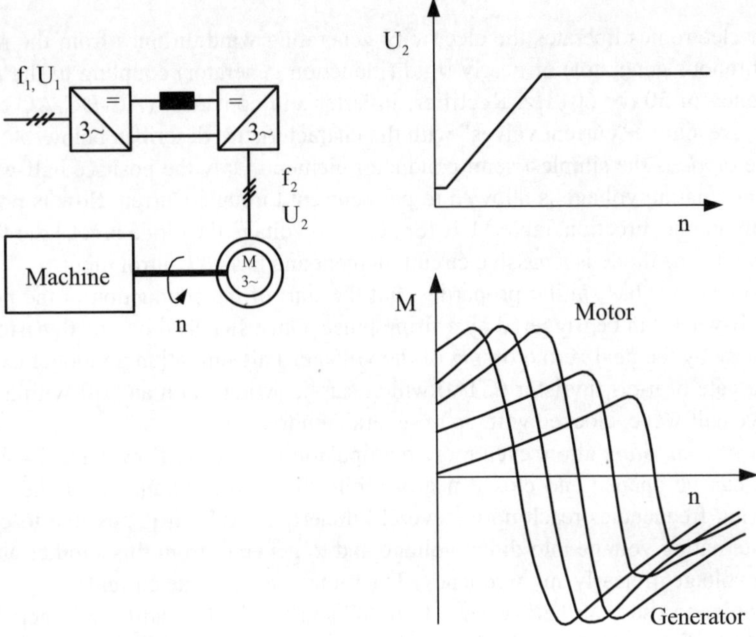

Fig. 11-27 Speed control of a squirrel-cage rotor induction machine by a frequency converter in the stator circuit; block diagram, voltage and torque

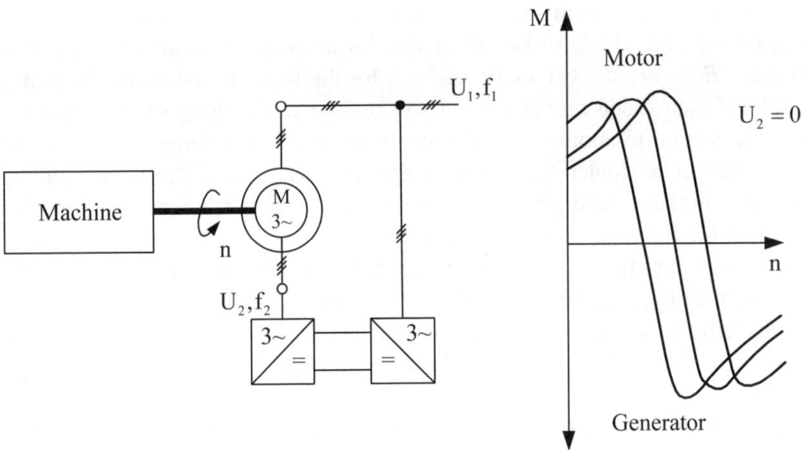

Fig. 11-28 Speed control of a slip ring rotor induction machine by a frequency converter in the rotor circuit, block diagram, torque

11.3 Power electronic components of wind turbines - converters

Power electronics liberates the electricity generating wind turbines from the rigid (synchronous generator) or nearly rigid (induction generator) coupling to the grid frequency of 50 (or 60) Hz. Rectifier, inverter and converter (AC-DC-AC converter) are simple "current valves" with the characteristics described below.

The *diode* is the simplest semi-conductor element. Only the positive half-wave of an alternating voltage is allowed to pass current through. Current flow is possible only in one direction, table 11.1. for negative voltage the diode blocks the flow of current. The diode is a passive circuit element and has no control input.

The *thyristor* has similar properties, but the start of the conduction in the positive half-wave can be triggered by a firing pulse. Once switched on, the thyristor is reset only by the next zero crossing of the voltage. This limitation no longer exists for the gate turn-off thyristor (GTO) which can be switched on and off within the positive half wave, clocked with up to several hundred Hertz.

Power transistors allow even more manipulation. They are flow-control valves which can be opened and closed nearly arbitrarily in the conducting range. The triggering frequencies reach up to several kilohertz (IGBT). It is possible to convert alternating voltage into direct voltage and to generate from this another alternating voltage of nearly any frequency. The same applies to the currents.

In order to use both half waves of an AC source, diodes can be connected to form a rectifier bridge, see Fig. 11-29. the bridge contains two diodes in series on each side and has the AC source feeding into the diagonal branch. Fig. 11-31 shows the extension of this configuration to three-phase inputs.

Rectifiers feed from an alternating or three-phase circuit of any frequency into a DC circuit. The most simple imaginable rectifier consists of four diodes to feed into a direct current circuit (bridge rectifier, Fig. 11-29). At any time the current flows through one diode of the upper and lower bridge section. For the positive half-wave these are the valves $V1$ and $V4$, for the negative half-wave $V2$ and $V3$. Therefore, each diode passes current for 180°. The changing of the conduction from one section to another is called commutation. The voltage is now rectified, but it is still quite rippled that for most applications it has to be smoothened by a capacitor. In that case the current flow time in the diodes is reduced, Fig. 11-30. The larger the capacitor and the smaller the load the smaller the ripple.

If the bridge rectifier is extended by another section the three-phase bridge is obtained in B6U connection, Fig. 11-31. The ripple of the voltage is even smaller with this circuit, compared to the simple bridge rectifier. The current flows as well through one diode of the upper and lower bridge section at a time, but now only for 120°.

Using controllable valves, the magnitude of the voltage in the intermediate circuit can be influenced actively through the delay angle. For a half-controlled bridge type B6H , e.g. the diodes $V1$, $V3$ and $V5$ are replaced by thyristors.

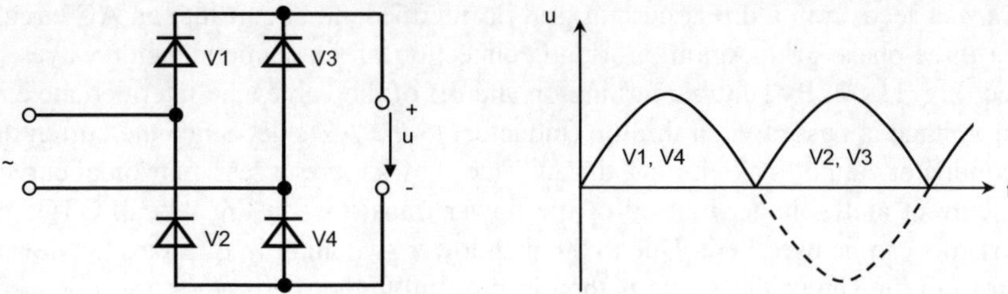

Fig. 11-29 Diode bridge rectifier, single phase

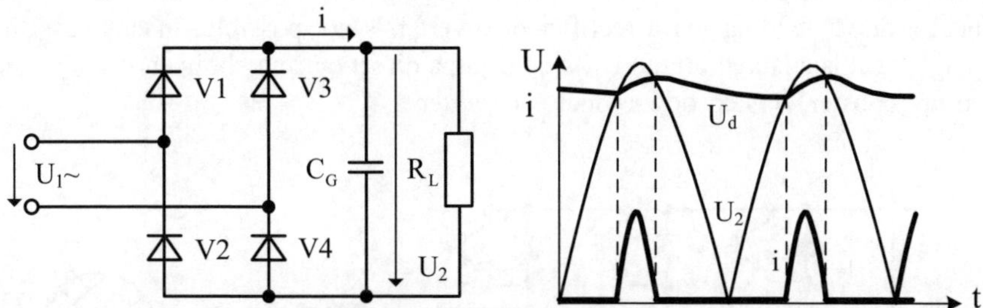

Fig. 11-30 Rectifier with smoothing capacitor C_G and load R_L; voltage u_2 without smoothing and voltage u_d and current i after smoothing [8, modified]

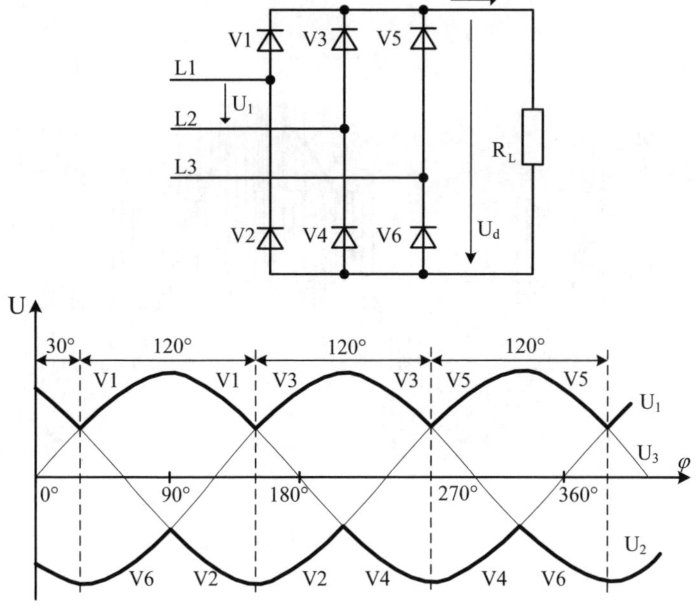

Fig. 11-31 Three-phase current rectifier (B6U arrangement); voltage after rectification [8]

Inverters feed from a direct current grid or intermediate circuit into an AC circuit or a three-phase grid. Again, a bridge connection of six semiconductor valves is used, Fig. 11-32. By suitable turning on and off of the valves the intermediate circuit voltage is passed via a throttle (inductor) to the AC side. Since the current in the inductor cannot make jumps, the AC side shows more or less sinusoidal curves for current and voltage. Instead of the power transistors in Fig. 11-32, GTOs or thyristors can be used here. Due to the then lower switching frequencies the deviation from the sinusoidal shape is then larger, and higher harmonics are produced, which are not wanted in the grid. Therefore, one will choose circuits which generate less higher harmonics.

Step-up or step-down converter: If the voltage levels of the grid differs too much, a direct feeding via a rectifier or inverter is not possible. In this case the voltage level is adapted on the DC side using a direct current chopper, often called step-up (booster) or step-down (buck) converter.

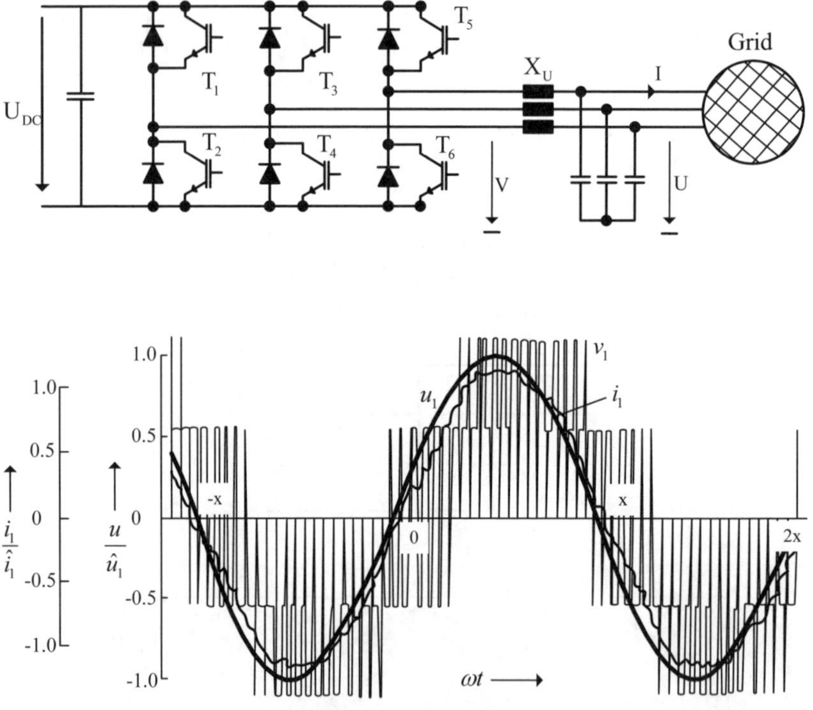

Fig. 11-32 Top: three-phase A.C. converter connected to the grid via six IGBTs with integrated free-wheeling diodes [9]; bottom: voltage and current and pulse width modulation [9]

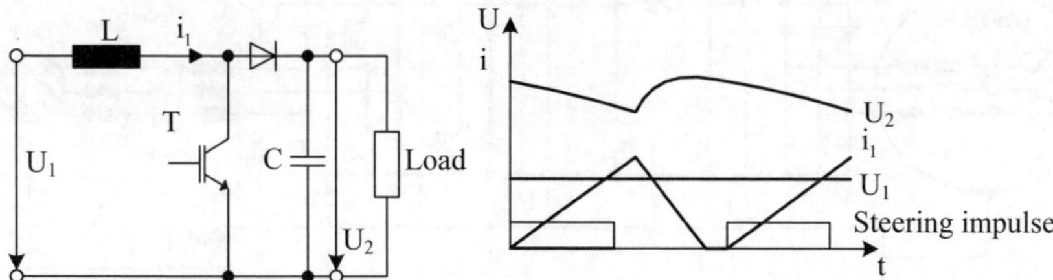

Fig. 11-33 Step-up converter (D.C. boosting unit)

The basic configuration of a step-up converter is shown in Fig. 11-33. The direct voltage source is applied to a series connection of inductor, diode and load where a power transistor is in parallel to diode and load. When the transistor is conducting current flows through inductor and transistor.

If the transistor is then shut off, so that it blocks the flow of current, the inductor, which cannot allow abrupt changes of current, will continue the current, which is now forced through the diode and the load. If a voltage larger than U_1 is needed to force the current through the load the inductor will provide the extra voltage. By a suitable choice of the switching moments the magnitude of the direct voltage at the load is adjusted.

Converter: Most of today's MW class wind turbines have in the normal wind range (approx. 4 to 12 m/s) a wind speed-dependent rotor speed and therefore operate always with λ_{opt} at the design point of the turbine's rotor. Then, the voltage supplied by the synchronous or induction generator has neither a frequency of 50 (or 60) Hz nor the common voltage level of 690 or 400 Volts. Therefore, the electrical power has to be conditioned by an AC-DC-AC converter, Fig. 11-34.

A converter consists of a rectifier and an inverter. If required, there is an additional step-up or step-down converter. A converter is able to transfer energy of any voltage and frequency from one grid to another grid of different voltage and frequency. In addition, any desired reactive power can be fed into both grids.

Fig. 11-35 shows the basic configuration of such a converter. Grid 1 consists of a synchronous machine. The frequency corresponds to the rotational speed of the synchronous machine. The voltage is determined by the rotational speed as well but it is variable within certain limits by setting the excitation of the synchronous machine. The energy supplied by the synchronous machine is fed into a D.C. intermediate circuit via a passive rectifier (RF) in a B6 bridge connection.

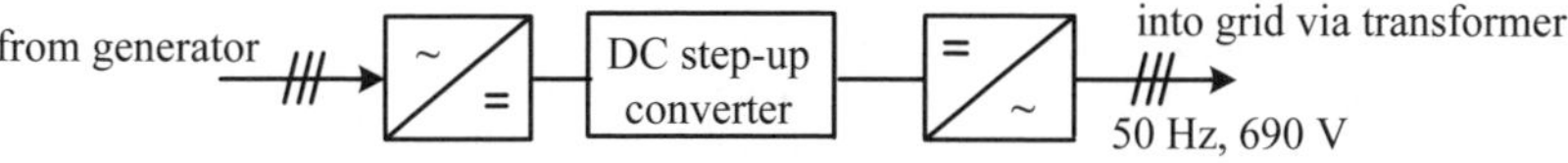

Fig. 11-34 AC-DC-AC conversion for a wind turbine

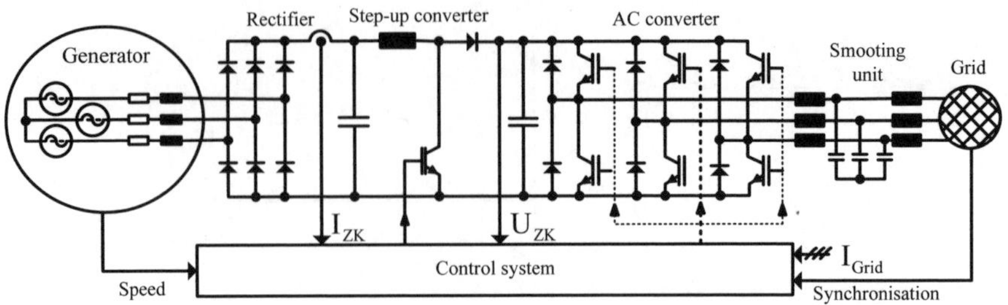

Fig. 11-35 Speed variable synchronous generator with AC-DC-AC conversion to the grid [9]

A step-up converter is integrated into the intermediate circuit. Thus energy provided at low rotational speeds low feed-in voltage levels can be used. Via the IGBT inverter and the power chokes produced energy is fed at constant frequency and voltage into the grid.

Multi-pole direct-drive annular synchronous generators are often switched for variable-speed operation to a six-phase mode. Then the rectifier in Fig. 11-35 needs twelve diodes. The advantage of this arrangement is obvious: the direct voltage has fewer harmonics, it is smoother than for a three-phase arrangement. A more detailed discussion of application of power electronics and converter technologies is found in chapter 13.

References

[1]   Nürnberg, W.: *Die Prüfung elektrischer Maschinen (Testing electrical machines)*, Springer Verlag, 1965

[2]   Taegen, F.: *Einführung in die Theorie der elektrischen Maschinen (Introduction to the theory of electrical machines)*; Vol. I and II, Vieweg + Sohn Braunschweig

[3]   Gerber/Hanitsch: *Elektrische Maschinen (Electrical machines)*, Verlag Berliner Union, Stuttgart 1980

[4]   Philippow, E. (Ed.): *Taschenbuch Elektrotechnik (Electrical Engineering Handbook)*, Vol. 1, 2 and 5, Carl Hanser Verlag München, Wien

[5]   Beitz, W., Küttner, K.-H. (Hrsg.): Dubbel, Taschenbuch für den Maschinenbau (Handbook of Mechanical Engineering), Springer-Verlag Berlin, Heidelberg, New York, 2002

[6]   Heumann, K.: *Grundlagen der Leistungselektronik (Basics on Power Electronics)*, Teubner Studienbücher Stuttgart, 6th revised and extended edition, 1996

[7]   *Handbuch der Physik (Handbook on Physics), Vol. XVII*, Verlag von Julius Springer, Berlin 1926

[8]   *Fachkunde Elektrotechnik (Knowledge on Electrotechnics)*, Europa-Lehrmittel (European teaching aids), Haan-Gruiten, 22. Auflage 2000

[9]   Heier, S.: *Windkraftanlagen (Wind energy Converters)*, B.G. Teubner GmbH, Stuttgart/ Lepzig/Wiesbaden, 3rd revised and enhanced edition, 2003

[10]   Fischer, R.: *Elektrische Maschinen (Electrical machines)*, Carl Hanser Verlag, München 1979

[11]   Weh, H.: *Elektrische Netzwerke und Maschinen in Matritzendarstellung (Electrical networks and machines representd by matrices)*, Hochschultaschenbücherverlag, Mannheim 1968

[12]   Pfaff, G.: *Regelung elektrischer Antriebe (Control of electrical drives)*, R. Oldenbourg Verlag, München, Wien

[13]   Fuest, K., Döring, P.: *Elektrische Maschinen und Antriebe (Electrical machines and drives)*, 6. Auflage, Vieweg-Verlag, Wiesbaden, 2004

[14]   Quaschning, V.: *Regenerative Energiesysteme(Renewable energy systems)*, 4th edition, Hanser-Verlag, München, 2006

[15]   Kremser, A.: *Elektrische Maschinen und Antriebe (Electrical machines and drives)*, 2nd edition, Teubner-Verlag, Wiesbaden, 2004

[16]   Quaschning, V.: *Understanding renewable energy systems*, 1st edition, Earthscan, London, 2005

[17]   Stiebler, M.: *Wind energy systems for electric power generation*, Springer-Verlag Berlin, 2008

12 Supervisory and control systems for wind turbines

The Western mill was the first wind turbine which operated completely automatic, "without a human supervisor". But the tasks of the control system and the supervisory system were strongly interwoven. The main vane which adjusts the rotor to the wind direction is in fact a simple "all-in-one" control system: it is sensor to register the deviation between the wind direction and the rotor axis, and at the same time it is the actuator producing the forces for the correction of the deviation from the demand value, Fig. 12-1. Together with the transverse vane it controls the turbine in the entire operating range from standstill to storm protection: For very strong winds the spring extends and the rotor is turned to increase the angle between rotor axis and wind direction. Rotational speed, power extraction and thrust are reduced, see also Fig. 12-7 (yaw angle β), and annex I of this chapter. In this annex additional examples of simple mechanical control systems for small wind turbines are presented.

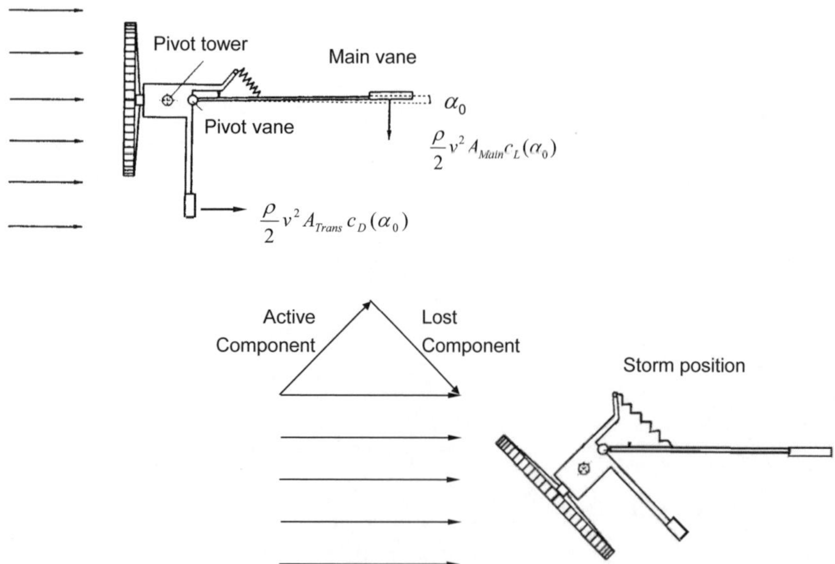

Fig. 12-1 Two-vane control of a western mill: Normal operation and power control by turning the rotor out of the wind for strong winds

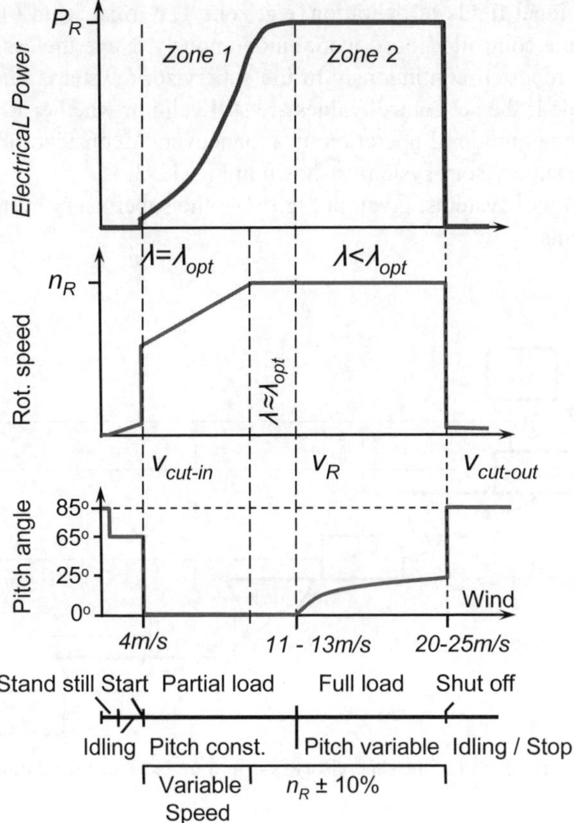

Fig. 12-2 Operating ranges and control of a variable-speed wind turbine with power control by blade pitching

For today's wind turbines of larger rotor diameters the range of functions of the machine supervision is far more complex, e.g.: Before a grid-connected wind turbine is started the following has to be checked:

- Is the oil pressure in the hydraulic systems ok?
- Are the actuators working all right?
- Is there enough wind and is it possible to supply power?
- Is there any voltage in the grid, is it possible to feed in?
- And so on.

All these checks are tasks of the supervisory system which also initiates the necessary changes and maneuvers. These are for example:

- The transition from low-speed idling to grid connection in normal operation, zone 1,
- The transition from normal operation, zone 1, to strong wind operation, zone 2,
- And so on, cf. Fig. 12-2.

Within the individual fields of operation, e.g. zone 1, normal wind (4 to 12 m/s), the control and the controller loops work autonomously. Nevertheless, the current status has to be reported continuously to the supervisory system, since only this system can decide if the set control values are still valid or whether it is necessary to shift to a different mode of operation by a maneuver. A considerably simplified flow chart of the supervisory system is shown in Fig. 12-3.

In the hierarchy of systems, given in Fig. 12-4, the supervisory system is above the control systems.

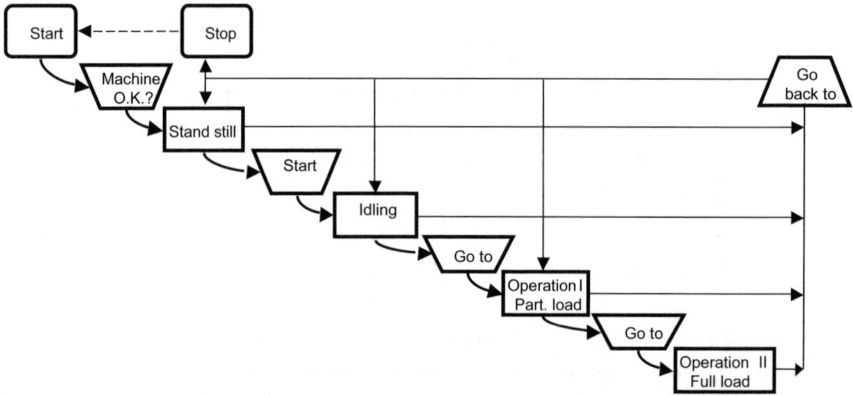

Fig. 12-3 Simplified flow chart of the supervisory system, variable-speed wind turbine

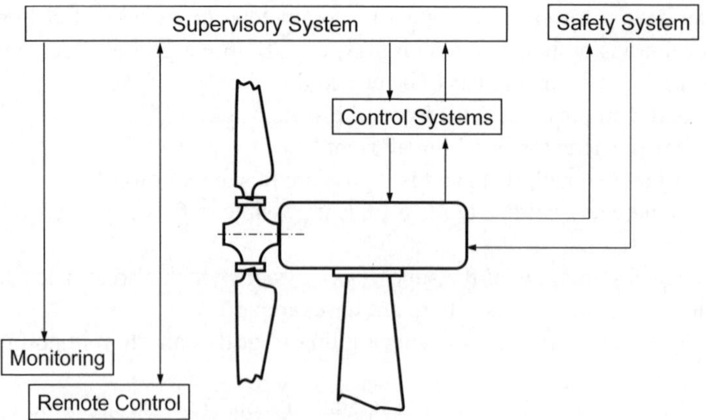

Fig. 12-4 Hierarchy of the systems

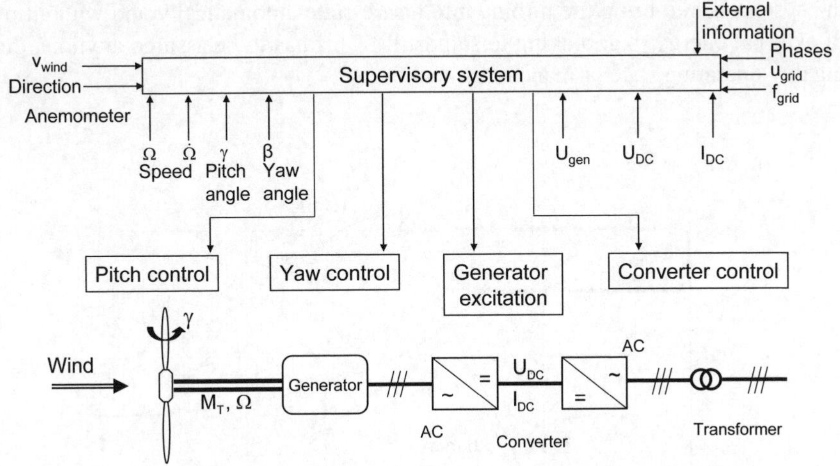

Fig. 12-5 Supervisory system - controllers - scheme of drive train with converter

Fig. 12-5 shows the most important controllers of a variable-speed wind turbine and the necessary sensor information.

The supervisory system governs also the *SCADA system* (SCADA = Supervisory Control and Data Acquisition) which consists of the monitoring and remote control modules. The *monitoring module* collects operational data (e.g. performance data and faults, etc.) and delivers required information "outwards", e.g. to the wind farm management office. The data transfer is either continuous or upon demand, e.g. via modem. The monitoring module cannot intervene in the turbine operation, therefore "it has no responsibility". The *remote control* serves e.g. to turn the wind turbine on and off from the wind farm management office or to reset the control after having checked and cleared a minor fault. For some machines, even control parameters or operating modes (e.g. noise reduced operation) can be set via the remote control.

The *safety system* has the responsibility for safe operation. It acts at once and directly if severe malfunctions occur, e.g.

- Excessive rotational speed (overspeed), Fig. 12-6 or
- Excessive power or torque,
- Differences among the pitch angles of the blades
- Excessive vibrations (unbalance, icing on the blades)
- Manual emergency shut-down, etc.

The safety system must assure that the wind turbine at no time damages or destroys itself. If a malfunction occurs it has to shut down the wind turbine fast and reliably and disengage it.

Therefore, this system has to be simple and robust (preferably hardware-based), and redundant in the most important components (e.g. two independent braking systems) and be of fail-safe design. In case of failures in the electronics or hydraulics the system has to bring the turbine into a safe state automatically and without use of external energy (e.g. braking to standstill). This has to be assured anytime, during any operating state or maneuver.

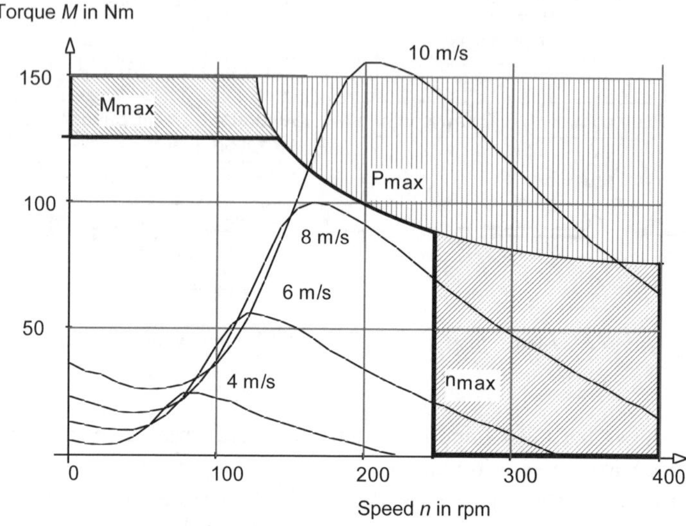

Fig. 12-6 Operating range limits for rotational speed, torque and power of a small wind turbine D = 4 m (see Figs. 6-8 and 6-9)

12.1 Methods to manipulate the drive drain

Basically, the drive train can be influenced either through aerodynamics at the rotor side, or by load manipulation, i.e. at the generator side (mechanical load side).

The wind turbine produces torque and thrust depending on the wind speed and the angle of attack of the blade sections, cf. chapters 5 and 6. The angle of attack itself depends on the rotational speed, the pitch angle and also on the yaw angle between rotor axis and wind. The driving torque feeds the mechanical load. If the driving torque is larger than the load torque the positive torque difference accelerates the drive train, Fig. 12-7.

For a quantitative consideration of the control processes the inertia of the drive train (and rotor) has to be considered, for larger wind turbines also the torsional-vibrations behavior. The build-up of aerodynamic forces at the blades (Wagner-Küssner' functions) is usually so rapid, that the blade performance characteristics, which were derived with the assumption of stationary behavior, can be used - see annex II of this chapter.

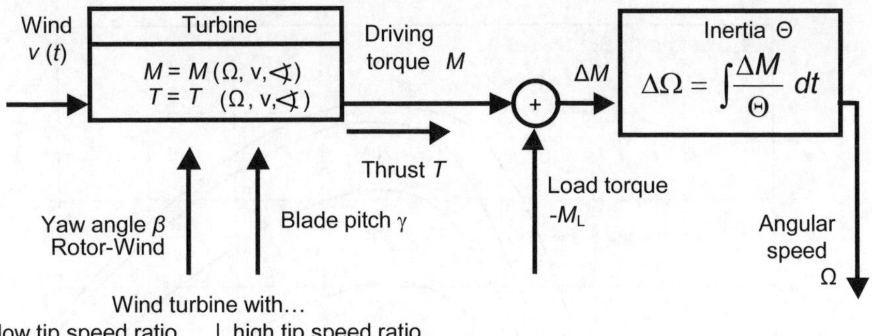

Fig. 12-7. Controlling the drive train of a wind turbine via aerodynamics and electrical load

12.1.1 Aerodynamic manipulation measures

The following are *simple methods* to reduce the power extraction of the rotor during strong winds:

- Turning the rotor out of the wind (Western mill),
- Holding the rotational speed constant in order to provoke flow separation at the blades during strong winds,
- Deploying spoilers or brake flaps.

A more elegant method is the *pitching of the blades* which will be discussed later.

Fig. 12-8 shows the effects of *turning the rotor out of the wind* (i.e. producing oblique flow) on the power characteristics of a small wind turbine with a high design tip speed ratio for strong winds. This is similar to the principle of the Western mill. With increasingly oblique flow, the maximum of the power curve decreases, but the corresponding rotational speed remains nearly constant.

If the *rotational speed is kept constant* for strong winds, the circumferential component in the velocity triangles does not change, Fig. 12-9. This can be used to limit aerodynamically the power extraction of the wind turbine, as already discovered in the 1950s by Johannes Juul, the ingenious inventor of the "Danish concept". The angle of attack α_A increases with stronger winds so much that the flow separates (stall effect). Thus, the drag grows strongly and rotates the resulting force R to point more downwind, Fig. 12-9. With a sophisticated blade design it is possible to keep its circumferential component in the strong wind zone nearly constant – and therefore the power as well since the rotational speed is already fixed.

The "Danish concept" was turned into practice by Juul using an induction (i.e. asynchronous) generator which has its rotational speed more or less rigidly tied to the three-phase grid frequency (50 or 60 Hz). This turns out to be a load control utilizing cleverly the effect of flow separation.

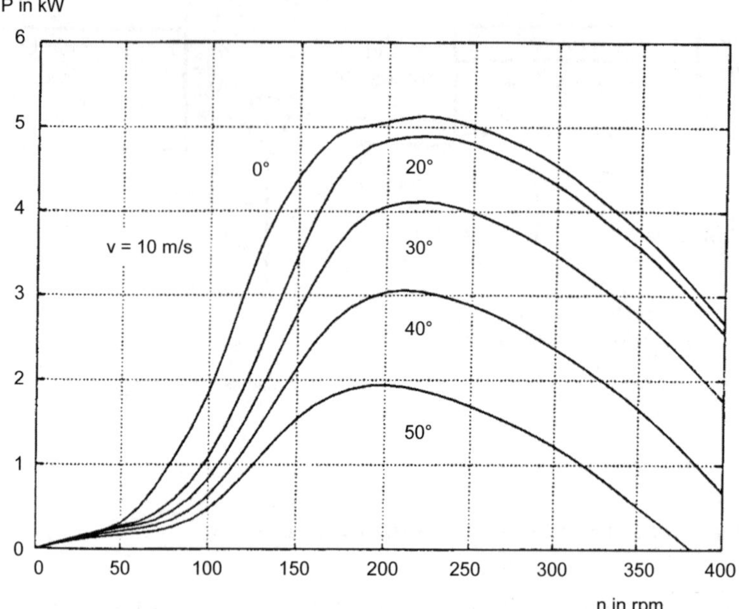

Fig. 12-8 Effect of turning the rotor out of the wind on the power characteristics of a turbine with a high tip speed ratio (angle between rotor and wind, see fig. 12-19 and 12-1)

Flaps and spoilers are aerodynamic brakes which protect against overspeed or serve as a simple means to limit and control power. They are activated by aerodynamic or centrifugal forces or are forced by a hydraulic control.

The braking torque M_{fl} from a flap areas A_{fl} can be easily estimated given the assumption that they are located at the outer radius R, and using the simplification that relative velocity equals circumferential speed ΩR. The braking torque M_{fl} of the flaps is then obtained by equating it to the driving torque M_R of the rotor, $M_R = M_{fl}$, since our goal is to assure that the rotor does not speed up to load-free idling

$$c_M (\lambda)\, A_R\, v^2\, R\, \rho/2 = c_{D.fl}\, A_{fl}\, (\Omega R)^2\, R\, \rho/2 \ , \tag{12.1}$$

where A_R is the rotor swept area resp. A_{fl} the flap area. The drag coefficient $c_{D.fl}$ is mostly equal to that of a rectangular plate and is in the range between 1.2 to 2.0 depending on the aspect ratio. $c_M (\lambda)$ is the torque moment coefficient of the rotor. Equation (12.1) may be transformed into the dimensionless equation

$$c_M (\lambda) = c_{M.fl}(\lambda) \ . \tag{12.2}$$

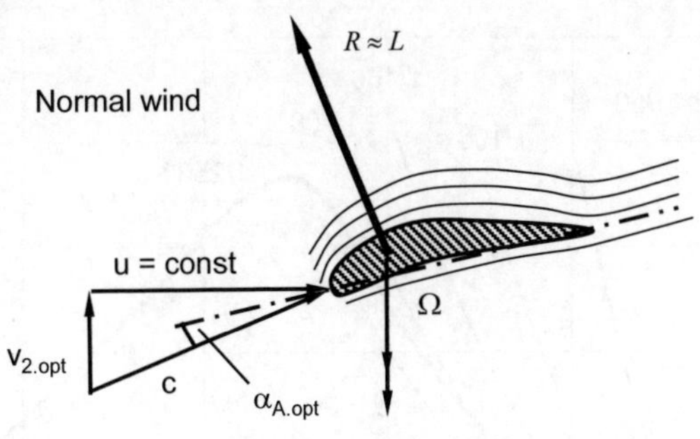

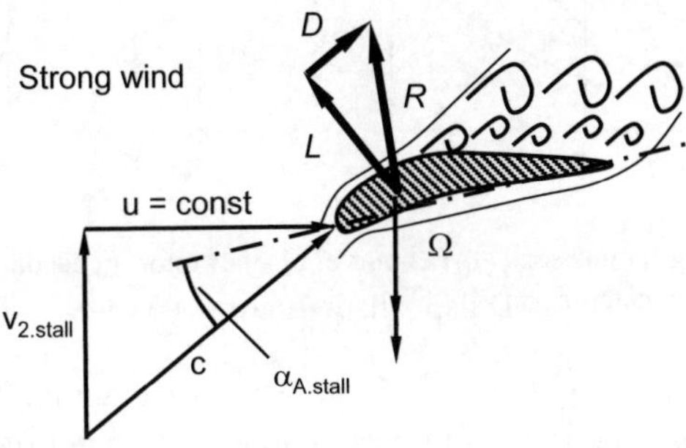

Fig. 12-9 Triangles of velocity at design point and under strong wind conditions: flow separation (stall effect) due to constant rotational speed and resulting reduction of lift and increase of drag

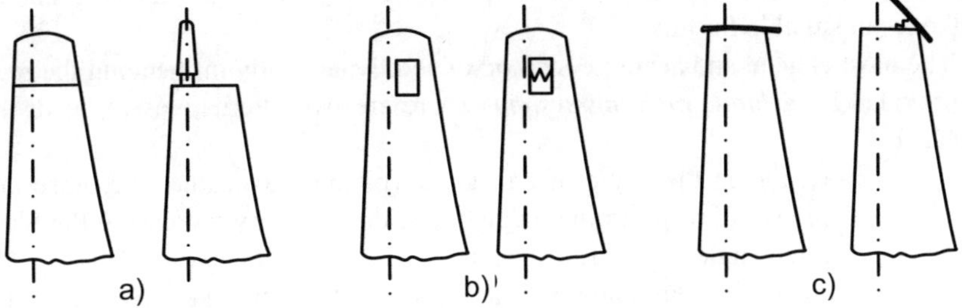

Fig. 12-10 Different types of braking flaps a) turnable blade tip, b) braking flap in the blade surface, c) fold-out end disk

The flap torque moment coefficient $c_{M.fl}(\lambda)$ is

$$c_{M.fl} = c_{D.fl}\,\lambda^2\,A_{fl}\,/\,A_R \equiv f \cdot \lambda^2 \ . \tag{12.3}$$

The factor f is determined by the ratio of flap area to rotor area.

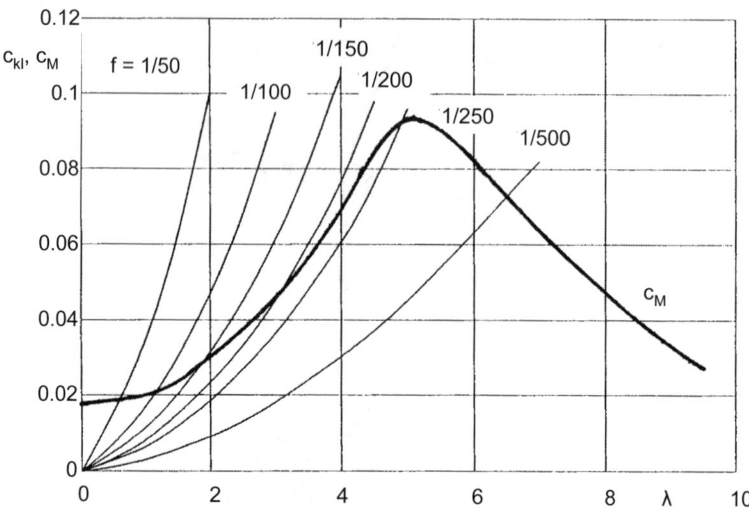

Fig. 12-11   Operating points on c_M (λ)-curve of the rotor depending on the flap size; parameter $f = c_{D.flap} \cdot A_{flap} / A_{rotor}$

After plotting the curve $c_{M.fl}(\lambda) = \lambda^2 f$ into the c_M-λ diagram, Fig. 12-11, the intersection of this curve and c_M (λ) gives the operating point of the system 'wind turbine with deployed flaps. Turbines with a high tip speed ratio require only small flap areas: for a wind turbine with a design tip speed ratio of $\lambda_D = 7$, the idling tip speed ratio would be reduced from $\lambda_{idle} = 13$ to approximately 6.5, provided that the area of the deployed flaps is 1/500 of the rotor area. In contrast to that, turbines with a low tip speed ratio require very large flap areas: a spoiler brake is not suitable for this.

The most elegant and accurate method of aerodynamically influencing the rotor is provided by a *blade pitching system*. There are two alternatives to be distinguished:

-   Reducing the angle of attack α_A (pitching to feather, i.e. nose into the wind, resp. trailing edge out of the wind) by increasing the blade pitch angle γ, Fig. 12-12, and
-   Increasing the angle of attack α_A (pitch to stall, i.e. nose out of the wind, resp. trailing edge into the wind) by reducing the blade pitch angle γ, Figs. 12-13.

If at constant wind speed, Fig. 12-12 top, the pitch angle is increased, the angle of attack at a blade section is reduced from the point of optimum flow conditions to smaller angles of attack. Therefore, the lift is reduced - and consequently the power output since the driving circumferential component of the lift, which wants to accelerate the rotor, gets smaller. In Fig. 12-12, bottom, the influence of the blade pitch angle on the power coefficient curve c_P (λ) is shown. The rotor characteristics of a wind turbine with a high design tip speed ratio for various pitch angles are given in the Figs. 6-15 to 6-17.

Pitch control systems which operate with a *reduction of the angle of attack* (pitch to feather) to smaller values have a good accuracy and produce a smoothly running rotor since for all occurring angles the flow remains attached to the blade. The disadvantage is that in the range od strong winds the pitch angle will require relatively large changes, see Fig. 12-14.

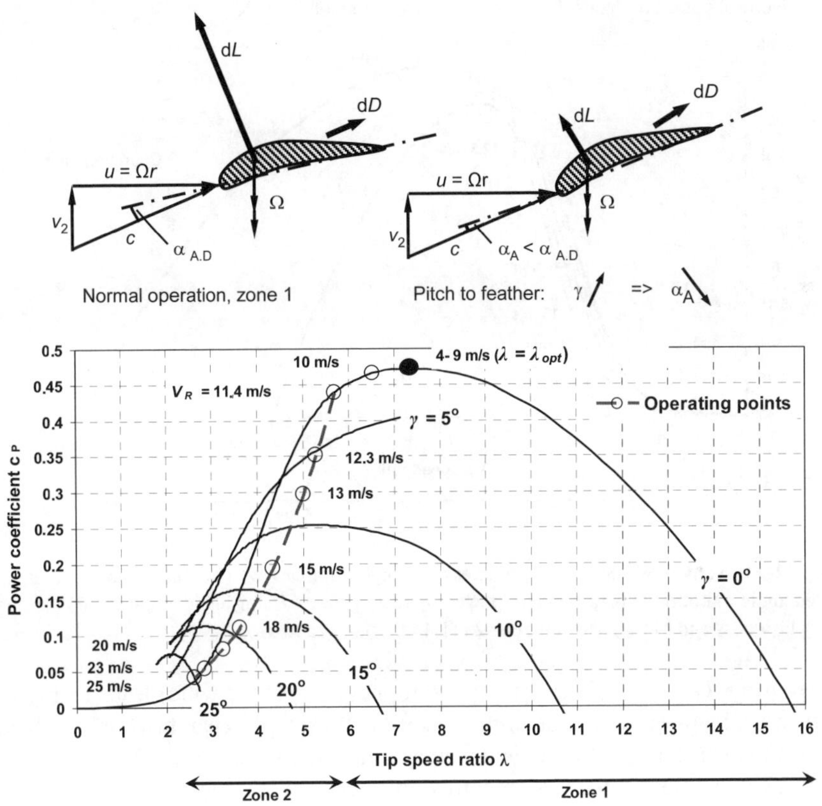

Fig. 12-12 Control by pitching to feather, top: reduction of angle of attack α_A by increasing the pitch angle γ, bottom: corresponding power characteristic $c_P(\lambda)$ for blade angle $\gamma = 0°$ in zone 1 ($v \leq 11{,}4$ m/s) and $0° < \gamma \leq 25°$ for zone 2 (11,4 m/s $< v <$ 25 m/s), see fig. 12-2

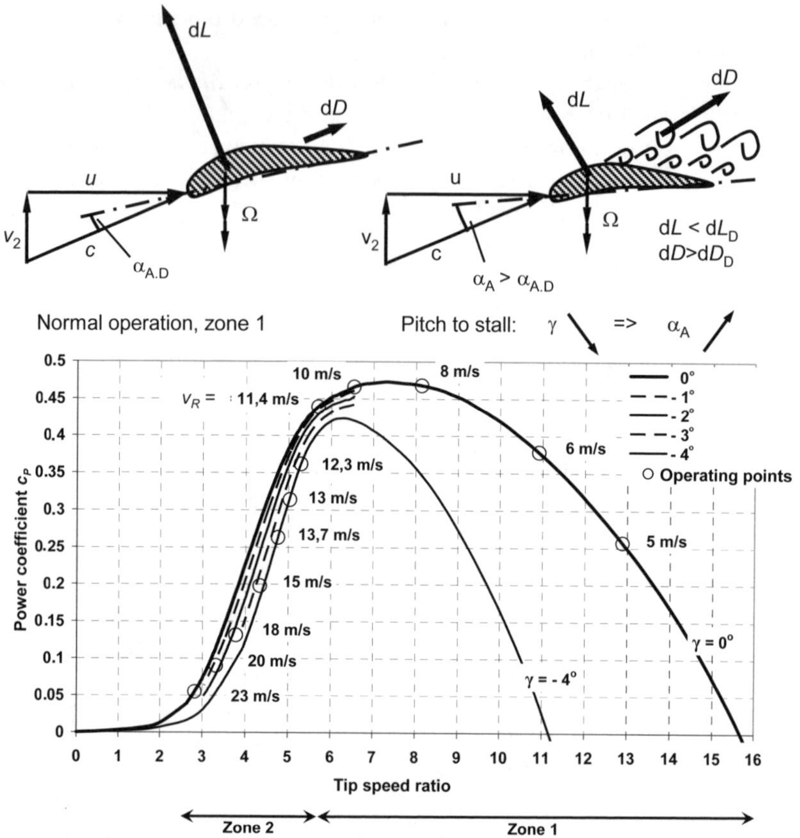

Fig. 12-13 Active-stall control, top: increase of angle of attack α_A (stall effect) by reducing the pitch angle γ, bottom: corresponding power characteristic $c_P(\lambda)$ for blade angle $\gamma = 0°$ in zone 1 ($v \leq 11{,}4$ m/s) and $-4° < \gamma \leq 0°$ for zone 2 ($11{,}4$ m/s $< v < 25$ m/s)

Increasing the angle of attack (active stall, nose out of the wind) leads to a reduction of the power output as well, since the flow will separate from the blade, which reduces the lift (a little) and increases the drag significantly, Fig. 12-13.

Provoking flow separation requires only small pitch angle changes and even achieving a controlled rotor standstill is easy. But the thrust remains quite large for this type of pitch control.

Fig. 12-14 compares the required pitch angle for constant power control in the range of strong winds (11.4 to 25 m/s). Obviously, the stall-controlled wind turbine requires only very small changes of the pitch angle to keep the power output, which was originally slightly rippled, really constant.

The summary of the different ways to influence the rotor by blade pitching in Table 12.1 gives an overview of the possibilities for wind turbine rotor operation. Practical examples exist for each field in this table.

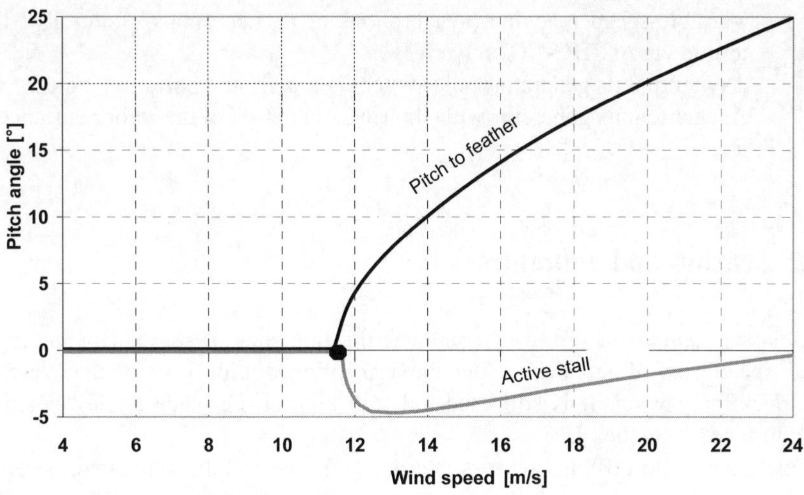

Fig. 12-14 Blade angle pitching to feather or to active stall in order to limit the power to rated power P_R for strong winds ($v > v_R =$ 11.4 m/s)

Table 12.1 Overview of possibilities for wind turbine rotor operation

<table>
<tr><td colspan="2" rowspan="2"></td><td colspan="3">Blade angle pitching</td></tr>
<tr><td>none</td><td>To feather</td><td>To stall</td></tr>
<tr><td rowspan="2">Rotor speed</td><td>constant</td><td>e.g. wind turbines of the classical "Danish concept "</td><td>e.g. early MW wind turbines</td><td>„active stall" wind turbines</td></tr>
<tr><td>variable</td><td>e.g. battery chargers</td><td>modern variable-speed wind turbines with A.C.-D.C.-A.C. converter</td><td>At present not a common wind turbine concept</td></tr>
</table>

12.1.2 Drive train manipulation using the load

In chapter 10 on wind pump systems we already discussed how the differences in the speed-torque characteristics of piston pumps and centrifugal pumps influence the behavior of the total system. Moreover, from chapter 11 we know several methods of drive train manipulation through the generator. They are briefly summarized here:

- Synchronous generator: variation of excitation,
- Variable-speed machine (synchronous or asynchronous): control of the torque via AC-DC-AC converter,
- Asynchronous generator: pole switching e.g. from 4 poles to 6 poles,
- Asynchronous generator with slip rings: variation of the rotor resistance,
- etc.

12.2 Sensors and actuators

The classic *centrifugal governor* combines the functions of sensor (for the rotational speed) and of actuator, as the above mentioned wind vane of the Western mill does for yawing. It is still used today to control the blade angles of small wind turbines, cf. annex I.

For larger wind turbines sensors and actuators are usually separated, because the signal of one sensor is often required in several different processing units. The most important sensors of a larger wind turbine are the following:

- Nacelle anemometer with wind direction indicator (e.g. wind vane),
- Rotational speed sensor,
- Electrical sensors for voltages, currents and phase angles,
- Vibration sensors,
- Sensors for oil temperature, pressure and oil level,
- Sensors for azimuth position of the nacelle and for the blade pitch angles,
- Limiting switches,
- etc.

The nacelle anemometer does not measure the real wind speed but only the wind behind the rotor, the so-called "nacelle wind" – whatever it may represent. Consequently, this signal is often only utilized to turn the generation of power on or off but not for the control of rotational speed and power. The better wind speed measuring device is the rotor itself, as we will learn in section 12.4.

The most important actuators for large wind turbines are:

- Hydraulic cylinders for yawing the nacelle and for pitching the blades, resp.
- Electrical servo-motors for these purposes,
- Torque manipulation via the generator,
- Actuators for the brakes
- etc.

12.3 Controller and control systems

The essential control systems for large wind turbines have already been presented in Fig. 12-5. They call for control of:

- AC-DC-AC-Converter,
- Excitation for the generator,
- Blade pitch and
- Yaw for the alignment of the rotor axis to the wind direction

The converter control has several objectives: e.g. maintaining the grid voltage level and the adaption of the torque demand of the generator to the (optimum) power output of the turbine rotor. At the time of entry into the grid it has to perform the synchronization and linking in the appropriate moment, etc.

For the design of machine controllers the necessary response speed of the controller system is of importance. Wind turbine controllers have approximately the following response speeds:

- Yaw system　　　　　　　　　　　360 degrees / 5 min
- Pitch system　　　　　　　　　　　2 to 8 degrees / sec
- Generator torque control　　　　　　fast
- Frequency control　　　　　　　　　very fast

Torque control by the generator is at least ten times faster than via blade pitch!

The controllers applied in wind turbines are mostly simple (gain scheduled) P-I-D controllers. More complex controllers like state-space controllers or observer based systems have not come into use up to now. Table 12.2 gives an overview of the most important properties of standard controllers.

A pure proportional controller loop is found already at the Western mill for the adjustment to the current wind direction with the wind vane and as well at the centrifugal governor: the adjustment y is proportional to the controller input $y = K_P x$. The P controller acts fast, but a small residual offset from the demand value x_P in the control path has to be tolerated.

A remedy for this is a (mostly small) integral portion in the feedback. It reduces the residual offset slowly to zero (PI controller).

If fast action is required to keep the control path under tight control, a differential portion in the feedback is useful, but the accompanying inevitable time delay T_D has to be taken into account.

In the past, PID controllers were analog devices built from amplifiers, resistors, capacitors and inductors, available at the electronics supplier. Today, the controllers are implemented in digital form in the control computer where the supervisory system is located as well.

Table 12.2 Overview of different controller types

Controller type	Time domain	Frequency domain	Numerical representation
P controller	$y = K_P \cdot x$	$\hat{y} = K_P \cdot \hat{x}$	
PI controller	$y = K_P \cdot x + K_I \int x\,dt$	$\hat{y} = \left(K_P + \dfrac{K_I}{s}\right)\cdot \hat{x}$	
PID controller	$y + T_D \dot{y} = K_P x + K_I \int x\,dt + K_D \dot{x}$	$\hat{y} = \left[K_P + \dfrac{K_I}{s} + K_D \dfrac{s}{1 + s \cdot T_D}\right]\cdot \hat{x}$	
			$y_0 = (a_0 x_0 + a_{-1} x_{-1} + a_{-2} x_{-2} \ldots)$ $+ (b_{-1} y_{-1} + b_{-2} y_{-2} + \ldots)$ **Attention:** Dead time Δt as well as missing or wrong data

Some manufacturers develop the supervisory and control system on their own. Others use standardized industrial controls like programmable logic controllers (PLC) and customize them for the special needs of their wind turbines.

The actual sensor information x_a is read in these digital systems with a kHz sampling rate and compared to the demand value x_D. Then the result, $x = x_a - x_D$, is weighted by the PID controller, cf. Table 12.2.

The sampling causes a small dead time. But it is worse that from time to time the data values of the sensor information x is distorted by noise from e.g. the power electronics or other noise sources. Therefore, it is not reasonable to base the controller action exclusively on the last two or three data values x_0, x_{-1}, x_{-2} . Commonly, the validity of the sensor information x is checked, e.g. by a least square technique, before further processing it in the controller. Values which are outside the confidence interval are identified, rejected and replaced by a reasonable estimate.

Recently, new control concepts known as "individual pitch" have been introduced. They manipulate the pitch of individual blades or vary the pitch angle in a cyclic fashion (vertical wind profile). They are already applied in some commercial wind turbines in order to reduce the blade loads. Of course, additional load sensors in the blades are then necessary.

12.4 Control strategy of a variable-speed wind turbine with a blade pitching system

What are the demand value settings for a wind turbine which operates in the zone 1 of normal winds always with the optimal power extraction ($\lambda = \lambda_{opt}$, $c_P = c_{P.opt}$, $\gamma = 0$) , see Fig. 12-12 (4 to 9 m/s) and Fig. 12-15 (approx. 4 to 12 m/s)? And what are the setting values in zone 2 of strong winds where the power extraction and the rotational speed are to be kept constant by blade pitching?

Since the wind signal of the nacelle anemometer is distorted by rotor and nacelle and therefore cannot be trusted, we have to rely on the signal for the rotational speed Ω. This makes sense since when operating at optimum tip speed ratio (zone 1, constant blade pitch angle 0°) it is proportional to the wind speed:

$$\lambda_{opt} = R\,\Omega/v = \text{const.} \tag{12.4}$$

Next we consider the torque-speed characteristics of the turbine given in Fig. 12-15. This figure contains additionally the curve of λ_{opt} (maximum power extraction). It defines the values of torque that we want to be absorbed by the generator for the normal winds, zone 1.

The corresponding control function for the torque demand values in its controller is

$$M_{gen} = [c_{Popt}\,\rho\pi R^5 / (2\,\lambda_{opt}^3)]\,\Omega^2 \tag{12.5}$$

$$M_{gen} = [\text{machine constant}] \cdot \Omega^2$$

This tracking function for the generator torque demand M_{gen} originates from the equation for the rotor power

$$P = c_P\,(\rho/2)\,v^3\,\pi\,R^2 \tag{12.6}$$

if we consider the torque equation $M = P/\Omega$ and insert equation (12.4). The wind speed no longer appears explicitly The machine constant is given by design data. The signal of the rotational speed replaces the very doubtful signal of the nacelle anemometer in the control strategy for normal winds. The rotor itself is the best wind measuring device!

For strong winds, zone 2, above 12 m/s the control strategy is different. In the simplest case we hold the torque constant by means of fast control actions of converter and generator. But at the same time the blade pitching system has to be activated in order to keep the rotational speed and consequently the power constant, cf. Fig. 6-18. Due to the control loop dynamics the rotational speed at rated power is slightly fluctuating. A certain speed "elasticity" during gusts is favorable for relieving the structure and smoothing the power output.

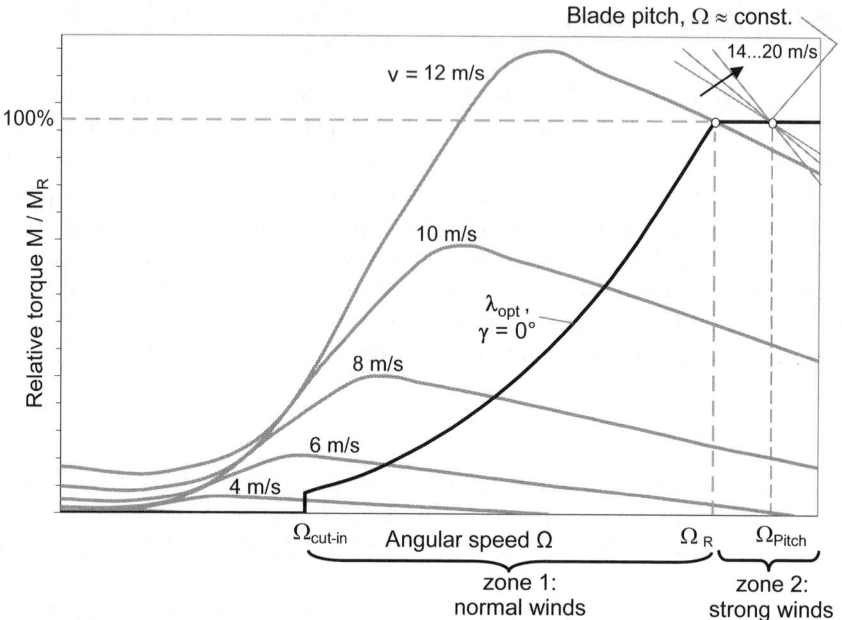

Fig. 12-15 Torque versus rotational speed and the curve λ_{opt} (best power coefficient) of a speed variable wind turbine with synchronous generator and AC-DC-AC converter. Variable rotational speed up to rated wind speed $v_R = 12$ m/s

In order to avoid switching repeatedly back and forth between the control regimes of normal and strong winds, the "constant" rotational speed of zone 2 is a little higher than the "transition corner". Morevoer, hysteresis is programmed into the transitions from the zones 1 and 2 and back.

A constant torque is not always the control target in the range of strong winds. It may be reasonable to choose the control target "constant power" in order to counteract brief power surges in the generator during wind gusts.

12.5 Remarks on controller design

At the start of a design it is often helpful to initially decompose the system into sub systems which are more or less independent of each other, e.g.
-     Yaw control of the nacelle,
-     Drive train control of torque and rotational speed,
-     Electric – electronic control of the torque via generator and converter system (fast) and moreover

- Electro-mechanical or hydraulic blade pitch control (slow)
- etc.

In the beginning, the application of classic analytical methods to the sub systems will be useful. Often a linearized approach is sufficient for the design of the PID controller. The controller settings for the coefficients K_P, K_I, K_D and T_D may be preliminarily selected according to classical rules (e.g. Ziegler-Nichols, etc.). Next more precise settings are obtained by digital simulation. If the solution seems to be quite good, the non-linear digital simulation is started. Amplification and amplitude limits and the non-linear behavior $P = P(v, \Omega, \gamma)$ is introduced, etc. At this point, the interdependences among the controllers will be taken into account. E.g., the drive train is influenced by a fast torque control from the electrical side and by the slower blade pitch control from the aerodynamic side. So torsional vibrations of the drive train may also have to be considered.

Software packages like SIMULINK are helpful here, since they provide a straight-forward interface between the model for vibrations of the drive train and the controller design work.

Structural dynamics also affect the control by way of the axial tower-nacelle vibrations, Fig. 12-16. Due to the tower vibration movements $u_T(t)$ the rotor no longer experiences the wind speed $v_{Wind}(t)$ but the difference between wind speed and vibration velocity of the tower. The blade pitching influences the thrust on the rotor and therefore the tower vibrations. An interaction occurs which may dampen the tower vibrations in the axial direction or - in an unfavorable situation - lead to controller induced self-excited vibrations.

Nowadays, the supervisory system is software based as well as the controllers. Often classical industrial controllers (PLCs) are applied which then have to be programmed accordingly. Only the safety system is hardware based.

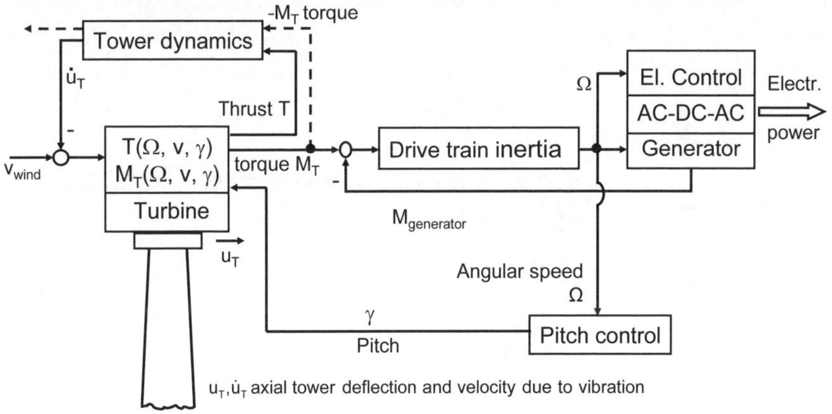

Fig. 12-16 Interaction of pitch-control, tower vibration and drive train control

Annex I

Examples for simple mechanical controllers

The simple controllers presented in this annex use the wind pressure (wind vane) or the rotational speed (centrifugal mechanisms) for power resp. rotational speed control. They have a good track record for wind turbines with up to 5 m rotor diameter.

Control by wind pressure of wind turbines with a low tip speed ratio

Fig. 12-1 shows the two-vane control of a Western mill. Fig. 12-17 presents the eclipse control where rotor thrust itself replaces wind pressure on the transverse vane. During normal operation, the aerodynamic moments of main and transverse vane of the two-vane control are in equilibrium,

$$l_{Trans} \frac{\rho}{2} v^2 A_{Trans} c_D(\alpha_0) = l_{Main} \frac{\rho}{2} v^2 A_{Main} c_L(\alpha_0) \qquad (12.7)$$

where l_{Trans} and l_{Main} are the corresponding lever arms.

The pre-stressed spring holds the main vane at the end stop. If a threshold of the wind speed is exceeded the spring lengthens. The start of control action is influenced by the geometry and the stiffness of the spring. If these are fixed, the control behavior may be calculated, but some empirical approaches are required to estimate the influence of the main vane [6].

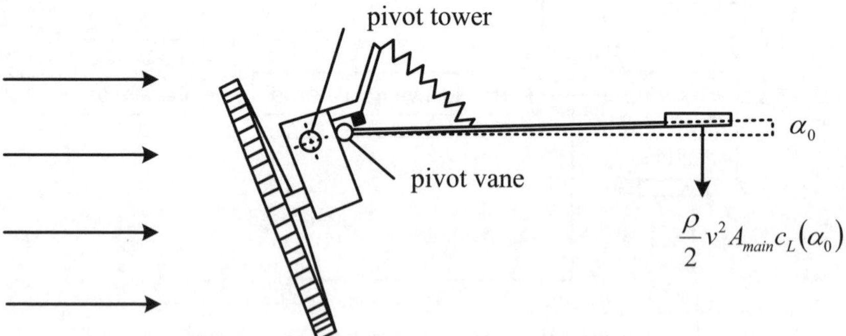

Fig. 12-17 Eclipse control

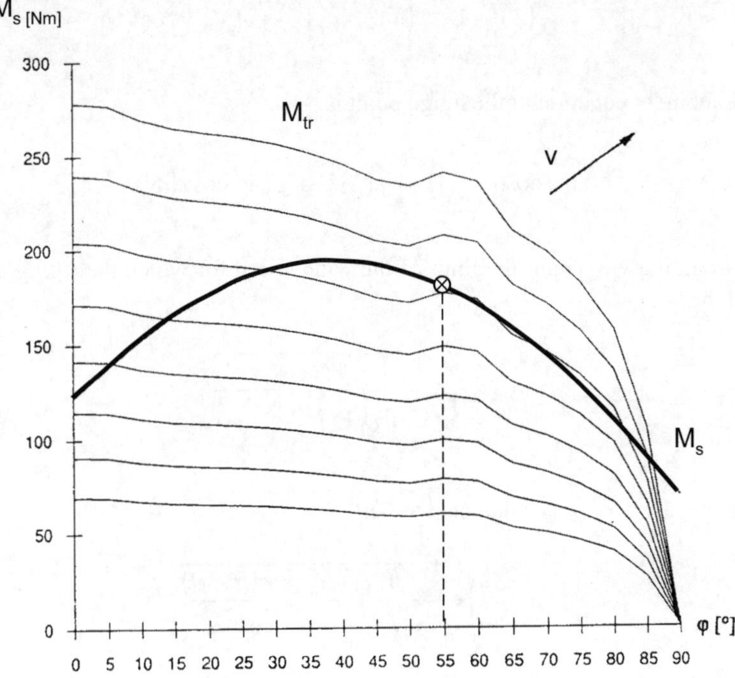

Fig. 12-18 Performance characteristics of a two vane control (see fig. 12-1); M_{tr} is the moment of the transverse wind vane, M_S is the spring moment

Fig. 12-18 describes the control behavior of a wind turbine with a diameter of 4 m. The intersections of the spring curve M_S and the wind vane moment curves represent the operating points where the moments are in equilibrium. For angles smaller than 55°, the control behavior is stable, above that angle, the rotor transitions abruptly into its storm position of $\varphi = 90°$.

For the eclipse control, Fig. 12-17, the thrust on the rotor replaces the force which in the above was produced by the transverse vane.

Control by wind pressure of wind turbines with a high tip speed ratio

Wind turbines with a high tip speed ratio can utilize the wind pressure for control, too. Fig. 12-19 shows such a turbine where rotor thrust and weight determine the position of equilibrium α (tilting angle).

The low location of the center of gravity can be created by means of a beam, which points downward and has a weight attached. The onset of the control can be approximated for a given geometry. To this end, the thrust of the tilted rotor has is estimated as:

$$T = c_T(\lambda)\frac{\rho}{2} A_{rotor}(v')^2 = c_T(\lambda)\frac{\rho}{2} A_{rotor} v^2 \cos^2 \alpha . \qquad (12.8)$$

The moment equation at the hinge point is:

$$W\, l \cos(\alpha + \beta) = c_T(\lambda)\frac{\rho}{2} A_{rotor} v^2 \cos^2(\alpha)\, e . \qquad (12.9)$$

Rearranging we obtain the limit of the wind speed for which the tilting angle α is still $0°$:

$$v_{begin} = \sqrt{\frac{W\, l \cos \beta}{c_T(\lambda)\frac{\rho}{2} A_{rotor}\, e}} \qquad (12.10)$$

The range $v > v_{begin}$ is calculated to be:

$$v = \sqrt{\frac{W\, l}{c_T(\lambda)\frac{\rho}{2} A_{rotor}\, e} \; \frac{\cos(\alpha + \beta)}{\cos^2 \alpha}} \qquad (12.11)$$

For turbines with a high tip speed ratio, the thrust coefficient is only slightly dependent on the load. According to Betz, we have for the design point $c_T(\lambda_{opt}) = 8/9$, and during idling, it increases to $c_T(\lambda_{idle}) = 1.0$ to 1.2, see section 6.3. Using the simplifying assumption of $c_T = 1$, the control behavior shown in Fig. 12-20 is obtained. If the center of gravity lies sufficiently low ($\beta < 0$), there will be a smooth transition to the tilted position. Usually, the results of this coarse calculation show a fairly good agreement with values observed in the field.

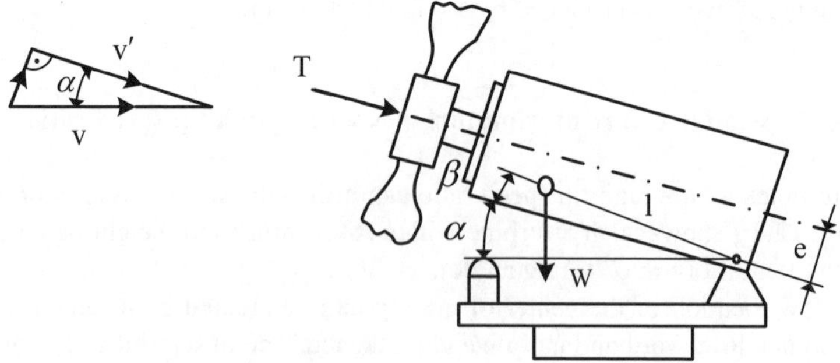

Fig. 12-19 Tilting rotor for a battery charger, active component v' and lost component

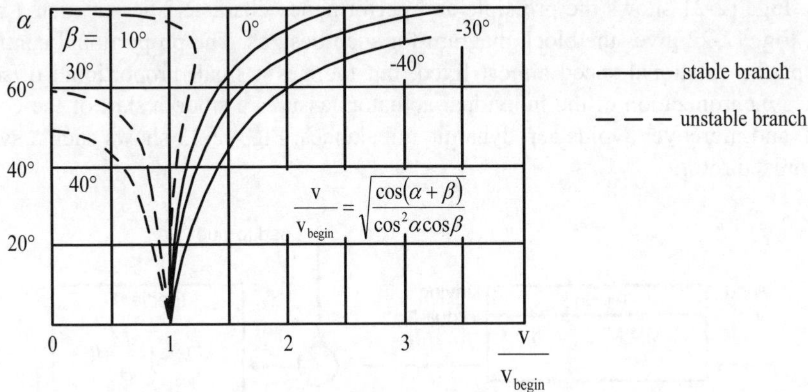

Fig. 12-20 Control behavior of a tilting rotor

Control of wind turbines with a high tip speed ratio using centrifugal mechanisms

Above a threshold of the rotational speed in the range of strong winds the centrifugal mechanism will adjust the blade angle or the braking flaps smoothly.

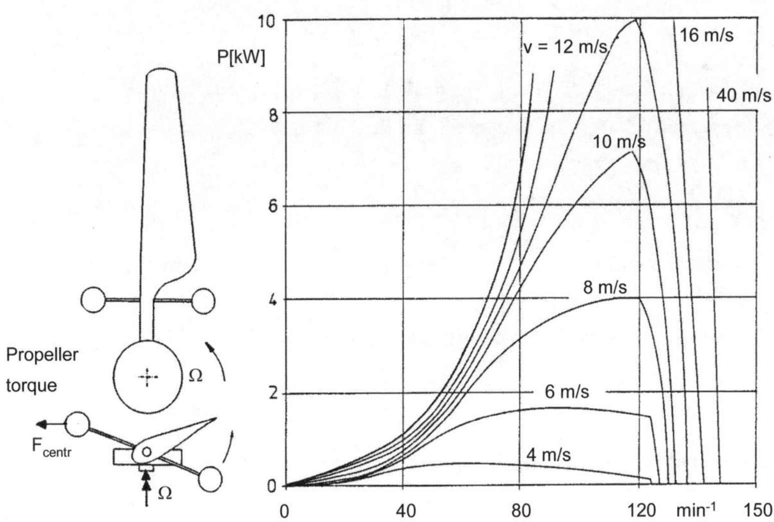

Fig. 12-21 Principle of a pitch controller using the propeller torque from centrifugal weights and the performance characteristics of such a rotor

Individual or coupled blade pitching by the propeller moment is part of this group, too. Fig. 12-21 shows the principle and performance characteristics of such a rotor. Fig. 12-22 gives the block diagram for such designs. The proportional control keeps the rotational speed almost fixed, but there is a small proportional offset. The synchronization of the individual actuators assures a smooth start of the control, and moreover avoids aerodynamic unbalances. Fig. 12-23 shows such a synchronized rotor.

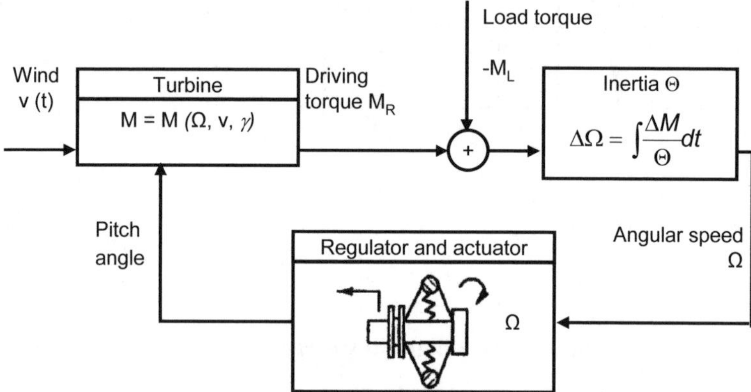

Fig. 12-22 Block diagram of the control system with a centrifugal governor

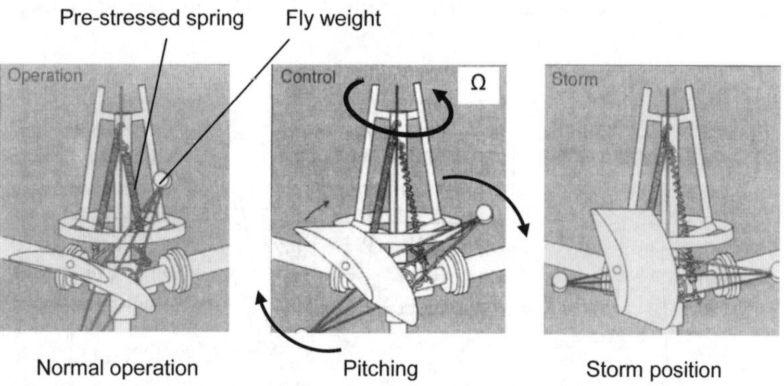

Fig. 12-23 Hub of a pitch rotor with centrifugal mechanism (see Fig. 12-21)

Passive control by aerodynamic forces

According to aerodynamics, the point of action of lift and drag forces is found at about 25% of blade chord length for a wide range of angles of attack. If a profile is not aligned on the radial line with its "$c/4$ point", a torque is generated. Depending on the chosen alignment of the chord, this torque tries to turn the profile into or out of the inflow. This torque is

$$M_{\text{reg}} = x F_{\text{T}} ,$$ (12.12)

where x is the distance between the radial pitch axis and the point of action of the thrust F_{T}. The thrust on one rotor blade is

$$F_{\text{T}} = \frac{\rho}{2} \frac{A}{3} v^2 c_{\text{T}} .$$ (12.13)

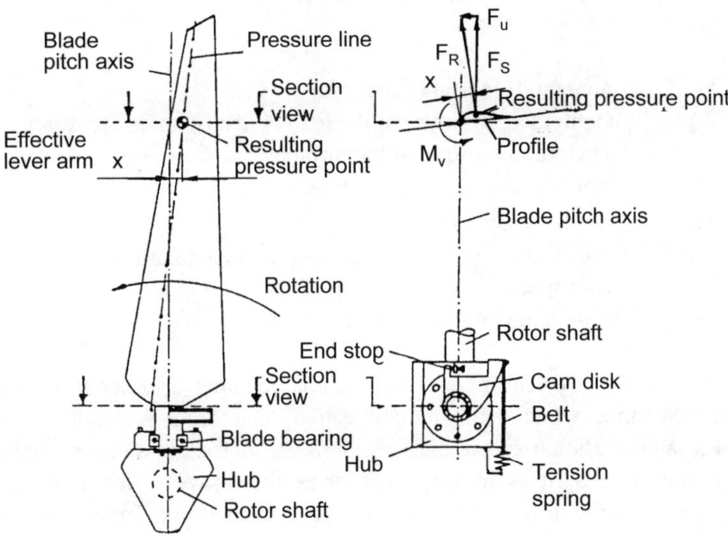

Fig. 12-24 Principle and design of a passive control by aerodynamic forces [7]

Similar to the aerodynamic forces, this torque is proportional to the square of the wind speed. It is, therefore, well suited as input of the control. For the common pitching towards feather, the point of action of the aerodynamic forces on the chord has to lie behind the radial line. Similar to the centrifugal pitch, a spring determines the start of the control action and the control characteristics as well. The blades should be synchronized in order to avoid aerodynamic unbalances. Combined with a synchronous generator, this mechanism provides a simple and easy system that operates independent of external energy.

Annex II

The differential equation for the control behavior of wind turbines and its linearization around the operating point

If pitch control is used, the control action has to be sufficiently fast to assure constant rotational speed for instance during gusts. In addition, it has to be stable and must not generate any unstable oscillations. Therefore, the differential equation describing this dynamic behavior has to be analyzed. Fig. 12-7 has already presented a simplified block diagram of a wind turbine. In the following, the corresponding differential equation of motion are presented:

$$\Theta \ \dot{\Omega} + M_L(\Omega, P_{el}) - M_T(\Omega, v, \alpha) = 0 \qquad (12.14)$$

with

Θ	moment of inertia of the rotor
M_L	counteracting torque of the load which opposes the rotation
M_T	accelerating torque of the turbine
Ω	angular velocity of the turbine
$\dot{\Omega}$	angular acceleration
P_{el}	electrical power output (in stand-alone operation)
v	wind speed
α	blade pitch angle (denoted γ in Fig. 12-7)

In a pitch control the pitch angle α will be manipulated in order to control the torque of the turbine, which is the main objective. This torque is also dependent on wind speed and rotational speed, cf. Figs. 6-15 and 6-16. For a closer analysis of the control dynamics, it is useful to linearize the above differential equation around a fixed operating point. M_T is therefore expanded in a Taylor's series. M_T can be calculated from the moment coefficient c_M which depends on the tip speed ratio and the blade pitch angle.

$$M_T = \frac{\rho}{2} \, A_R \, c_M(\lambda, \alpha) \, v^2 \, R \ = \ \frac{\rho}{2} \, A_R \, c_M\left(\frac{\lambda R}{v}, \alpha\right) v^2 \, R \qquad (12.15)$$

where

ρ	air density
R	rotor radius,
A_R	swept rotor area,
λ	tip speed ratio.

For the Taylor series, the partial differentials of M_T are required:

$$\frac{\partial M_T}{\partial v} = c_M(\lambda,\alpha)R\frac{\rho}{2}2v\,A_{\text{rotor}} + \frac{\partial c_M}{\partial \lambda}\frac{\partial \lambda}{\partial v}\frac{\rho}{2}v^2 A_R R$$

$$= \frac{\rho}{2}A_R R\left(2c_M(\lambda,\alpha)v - \frac{\partial c_M}{\partial \lambda}\Omega R\right) \tag{12.16}$$

$$\frac{\partial M_T}{\partial \alpha} = \frac{\rho}{2}A_R\frac{\partial c_M}{\partial \alpha}v^2 R \tag{12.17}$$

$$\frac{\partial M_T}{\partial \Omega} = \frac{\rho}{2}A_R\frac{\partial c_M}{\partial \lambda}\frac{\partial \lambda}{\partial \Omega}v^2 R$$

$$= \frac{\rho}{2}A_R\frac{\partial c_M}{\partial \lambda}R^2 v \tag{12.18}$$

The partial derivatives of M_T can be graphically interpreted. The common representation of the moment coefficients depending on the tip speed ratio with α as a parameter shows clearly $\partial c_M/\partial \lambda$ as the tangent of the corresponding $c_M(\lambda)$-curve at the operating point $\lambda = \lambda_{\text{op}}$. If the moment coefficients are plotted in the (λ,α)-plane, Fig. 12-25, it becomes clear that $\partial c_M/\partial \lambda$ corresponds to the slope of the tangent in direction of α at the operating point $\alpha = \alpha_{\text{op}}$ and $\lambda = \lambda_{\text{op}}$.

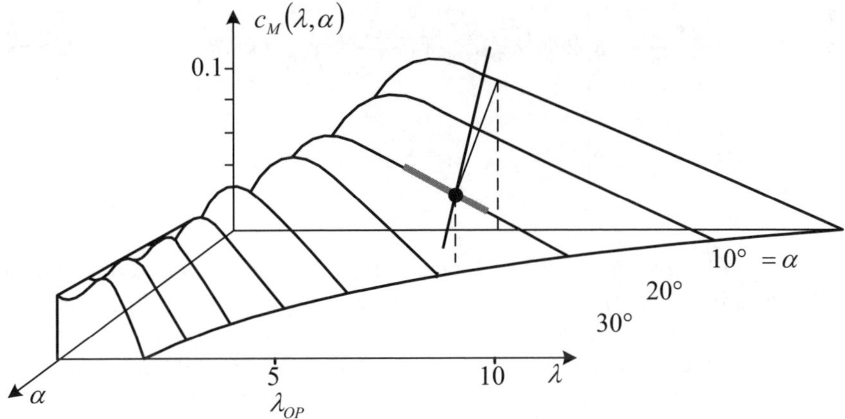

Fig. 12-25 Interpretation of the partial derivatives of c_M at the operating point as linear tangents, blade pitch angle α

From the Taylor expansion of the performance characteristics $c_M = c_M(\lambda, \alpha)$ we obtain a linearized representation of the curve of the turbine torque $M_T = M_T(\alpha, \Omega, v)$ in the neighborhood of an operating point which is given by λ_{op} and α_{op}:

$$\Delta M_T = \frac{\partial M_T}{\partial v}\Big|_{\substack{\alpha=\alpha_{op}\\\lambda=\lambda_{op}}} \Delta v + \frac{\partial M_T}{\partial \alpha}\Big|_{\substack{\alpha=\alpha_{op}\\\lambda=\lambda_{op}}} \Delta \alpha + \frac{\partial M_T}{\partial \Omega}\Big|_{\substack{\alpha=\alpha_{op}\\\lambda=\lambda_{op}}} \Delta \Omega$$

$$= \frac{\rho}{2} A_R R \left[2 c_M(\lambda_{op}, \alpha_{op}) v_0 - \frac{\partial c_M}{\partial \lambda}\Big|_{\substack{\alpha=\alpha_{op}\\\lambda=\lambda_{op}}} \Omega_0 R \right] \Delta v \qquad (12.19)$$

$$+ \frac{\rho}{2} A_R R \left[\frac{\partial c_M}{\partial \alpha}\Big|_{\substack{\alpha=\alpha_{op}\\\lambda=\lambda_{op}}} v_0^2 \right] \Delta \alpha + \frac{\rho}{2} A_R R^2 \left[\frac{\partial c_M}{\partial \lambda}\Big|_{\substack{\alpha=\alpha_{op}\\\lambda=\lambda_{op}}} v_0 \right] \Delta \Omega$$

If we assume that the consumer load also changes only in a linear way with the rotational speed, we obtain for the speed $\Omega = \Omega_0$ and for a wind speed v_0 the linearized differential equation:

$$\Theta \dot{\Omega} - \frac{\rho}{2} A_R \frac{\partial c_M}{\partial \lambda}\Big|_{\substack{u=u_{op}\\\lambda=\lambda_{op}}} R^2 v_0 \Delta \Omega + \frac{\partial M_L}{\partial \Omega} \Delta \Omega$$

$$= \frac{\rho}{2} A_R R \left[2 c_M(\lambda_{op}, \alpha_{op}) v_0 - \frac{\partial c_M}{\partial \lambda}\Big|_{\substack{\alpha=\alpha_{op}\\\lambda=\lambda_{op}}} \Omega_0 R \right] \Delta v \qquad (12.20)$$

$$+ \frac{\rho}{2} A_R R \left[\frac{\partial c_M}{\partial \alpha}\Big|_{\substack{\alpha=\alpha_{op}\\\lambda=\lambda_{op}}} v_0^2 \right] \Delta \alpha$$

The change of the torque is caused by the change of rotational speed (through the load and turbine curve), changes in wind speed, and changes in the blade pitch angle. Only the latter is manipulated by the control. The design must assure that the change in the rotational speed $\Delta\Omega$ is translated rapidly enough into a change of the angle of attack $\Delta\alpha$, without causing any oscillations in the control loop.

Fig. 12-26 shows how equation (12.20) may be represented in a block diagram using the common symbols of control engineering for proportional, integral and delay time terms.

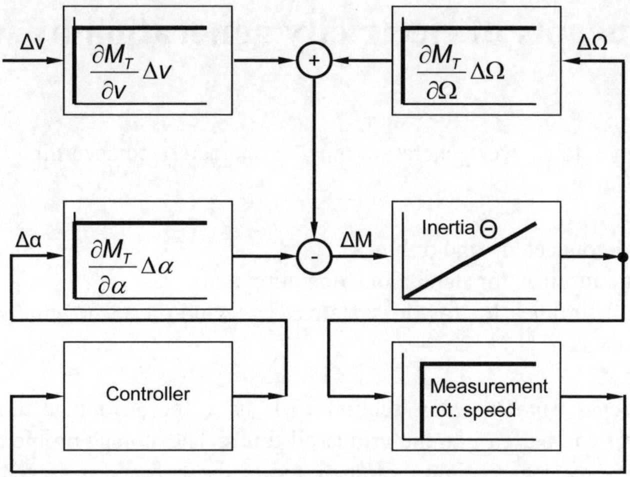

Fig. 12-26 Block diagram for the linearization of the turbine characteristics

References

[1]   Föllinger, O.: *Regelungstechnik (Control Engineering)*, 6th edition 1990, Hüthig Buch Verlag, Heidelberg

[2]   Manwell, J.F. et al.: *Wind Energy Explained*, J. Wiley & Sons, UK 2002

[3]   Burton, T. et al.: *Wind Energy Handbook*, J. Wiley & Sons, Chichester, UK 2001

[4]   Heier, S., *Windkraftanlagen (Wind Energy Converters):* 3rd edition, Teubner Verlag Stuttgart, 2003

[5]   Schörner, J. et al.: *Stand und Entwicklungsrichtung des Antriebsstranges von Windkraftanlagen (State of art and development tendencies of wind turbine drive trains)*, Windkraftjournal, Vol. 6/2001, p. 38-48

[6]   Franquesa, M.: *Kleine Windräder (Small wind turbines)*, Pfriemer Verlag, München, 1988

[7]   Frieden, P.: WENUS: *Innovative Konzepte zur Pitchregelung (WENUS: Innovative concepts for pitch control)*, DEWEK 1994, Tagungsband

[8]   Bossanyi, E. A.: *Individual Blade Pitch Control for Load Reduction*, Wind Energy, 6, 2, p. 119 – 128, 2002

[9]   Bossanyi E. A.: *Further load reductions with individual pitch control*, Wind Energy 8, 4, p. 481 – 485, 2005

[10]   Bianchi, F. D., De Battista H., Mantz, R. J.: *Wind Turbine Control Systems*, Springer – Verlag London, 2007

13 Concepts of electricity generation by wind turbines

Wind turbines for power generation may be characterized according to their type of application

- Grid-connected wind turbines
- Wind turbines for stand-alone operation and
- Wind turbines for hybrid systems, e.g. wind-diesel or wind-photovoltaic systems.

Grid-connected wind turbines, section 13.1, have the advantage that their produced power can be fed into the grid at all times. The storage problem for excess electrical energy, e.g. at night, is transferred to the grid and is solved there often by using hydro power plants with pump storage facilities. These have in Germany an installed capacity of approx. 6000 MW. Not only wind turbines produce sometimes excess electricity. Nuclear power plants which operate as base load plants also have times, e.g. at night, when they cannot deliver all the produced power directly to the consumer. Their excess "black" electricity is used to pump water "uphill" at night, and during the day to satisfy peak demands it flows "downhill" again, now delivering "green" electricity.

The problem of deficits in power generation from wind during periods of calm is passed on to the grid as well. The grid then depends on steam powered or gas powered generators which can be brought on-line within minutes. Thus, grid-connected wind turbines have it easy as far as mismatches between supply and demand for electricity are concerned. However after start-up, they have to be connected smoothly to a grid which operates with firm values of voltage, phase and frequency and tolerates only modest levels of harmonics. This demands substantial outlays for the control of the power electronics in the interface to the grid (see chapter 14).

Wind turbines in *stand-alone operation or small insular grids* can usually function with more relaxed rules, see section 13.2. For many devices it does not matter if they are run at 47 Hz or 52 Hz. The key issues are here: What to do with the electrical energy if no one needs it? Quite often, "dead loads" are turned on, which turn the excess power into heat. Even more problematic is a lack of power production: less important loads, e.g. the washing machine, are switched off by a priority control, and energy storage systems and accumulators are used where they are available. If the storage systems are empty and there is still no wind, one is on the safe side only with a *hybrid system*, e.g. a wind-diesel system, section 13.3.

13.1 Grid-connected wind turbines

The diameter of the rotor was around 10 m in 1980; today it exceeds 100 m for machines rated at 3 MW – an this is not the only significant change! The truly revolutionary development took place on the electrical side of the system. Initially, the rotational speed of the wind turbine was coupled rigidly to the grid frequency for both, the synchronous (SG) and the asynchronous generator (ASG, i.e. induction generator).

However, the discussion of aerodynamics in chapter 5 showed that optimal operation of a wind turbine requires that it must be running with the design tip speed ratio λ_{opt}. The turbine can deliver the largest possible amount of power only, if its rotational speed adapts to the varying velocities of the wind in such a way, that it remains at λ_{opt}. This is to say: double wind velocity requires double rotational speed.

Thanks to developments in power electronics during the 1980s and 1990s we now build variable-speed, "wind speed governed" wind turbines even for the MW class. Turbine and generator initially produce a "wild" three-phase current of variable frequency and voltage. This current is rectified and then converted back into three-phase current – but now of 50 or 60 Hz. Miraculously, for a whole decade the ratings of affordable AC-DC-AC converters increased in parallel with similar increases in the power ratings of the wind turbines.

The company Enercon pioneered the concept of variable-speed, "wind speed governed" operation. In 1993, this new wind turbine concept entered the market with their wind turbine E-40, which has no gears and uses an annular synchronous generator that feeds into the grid via a converter. This concept was and still is economically quite successful, cf. section 13.1.3.

In the beginning of the 1990s, the asynchronous machine had become quite flexible with regard to the rotational speed due to the Opti-Slip concept of Vestas: an adjustable resistance in the rotor circuit allows the machine brief increases in the rotational speed of up to 20% during wind gusts, cf. section 13.1.2.

The asynchronous machine eventually gained complete variability of speeds thanks to the introduction of a guided converter in the rotor circuit (doubly-feeding AS machine with a current converter cascade that may work above or below synchronous frequency), cf. section 13.1.4. The advantage of this concept, which entered the market approx. 1996 (Loher-SEG), is that not the entire generated power has to be converted as for the synchronous machine, but only that part of it which is required or produced in the rotor. Moreover, the asynchronous machine loses its disadvantage of reactive power consumption from the grid. Due to the converter in the rotor circuit it is able to provide controllable reactive power to the grid – like the synchronous machine with converter.

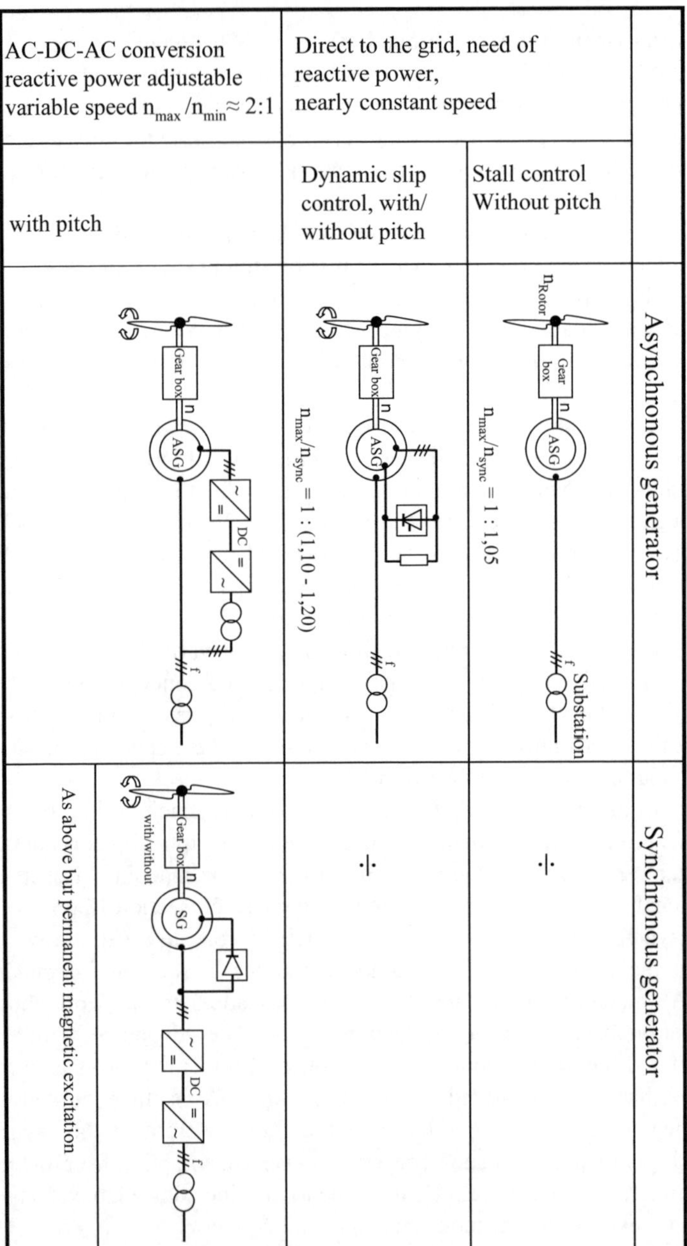

Fig. 13-1 Types of grid-connected wind turbines

Fig. 13-1 gives an overview of these four wind turbine types for grid-connected power generation:
- wind turbine with asynchronous generator feeding directly into the grid (Danish concept),
- its improvement with dynamic slip control,
- the variable-speed wind turbine with synchronous generator feeding the entire power via a converter into the grid and
- the variable-speed wind turbine with asynchronous generator and a converter in the rotor circuit feeding only a part of the power via a converter into the grid.

Each type will be discussed briefly below. The number of types of available machines was enlarged in 2005 by wind turbines which use permanent magnets to supply the excitation in a synchronous generator and provide AC-DC-AC conversion of the entire produced power [9].

13.1.1 The Danish concept: Directly grid-connected asynchronous generators

In the 1980s, the asynchronous generator (induction generator with a squirrel-cage rotor) that is directly connected to the grid dominated the market for wind turbines completely. This is known as the 'Danish concept', which was developed in the 1950s by the engineer Johannes Juul and tested at the Gedser wind turbine (1957-1967), cf. chapter 2.

The Danish wind turbines with a capacity of 30 to 450 kW (D = 12 to 35 m) were typically equipped with one small and one large asynchronous generator. In today's machines, only pole changing is used. If the wind speed is sufficiently high, the small asynchronous machine is connected to the grid. At first the turbine accelerates in the motor mode, but when it exceeds the synchronous speed it changes automatically to the generator mode. If the wind speed increases, the small generator is switched off and the large generator switched on. Its operating point lies at a higher rotational speed, to the right in the performance characteristics diagram, Fig. 13-2. Therefore, it follows the increasing supply of power by the turbine rotor and reaches once again the turbine optimum. This generator operates up to wind speeds of 25 m/s where the turbine cuts out because of severe storm conditions. The generator keeps the rotor close to the (gear transmitted) synchronous speed, if it was designed robust enough to prevent a 'pulling out' ($M > M_p$), cf. section 11.2.2. The power of the turbine is limited in a quite "natural" way, by flow separation at the rotor blades. Therefore, no pitching is necessary to control and limit speed and power. Only in case of a power failure in the grid, spoilers or tip brakes are deployed by centrifugal mechanisms. They serve to prevent over-speeding, cf. section 12.1.1.

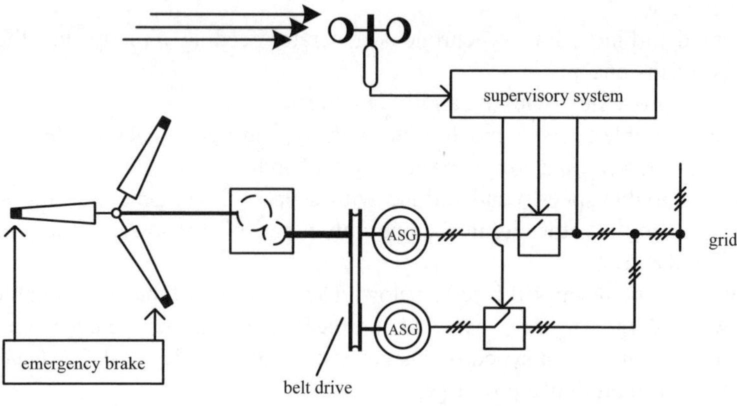

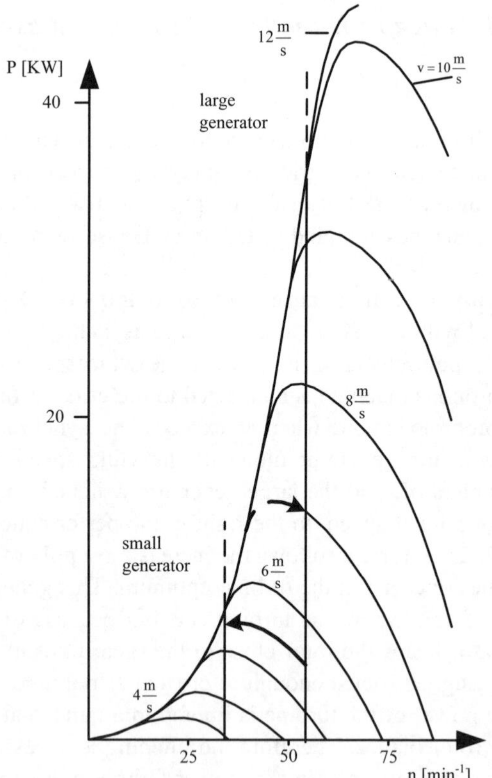

Fig. 13-2 Danish concept: Direct coupling of the asynchronous generator to the grid, nearly constant speed (e.g. Vestas 15/55); tip spoiler as emergency brake min[-1]

For large wind turbines ($P > 600$ kW) many companies have abandoned the tip spoiler in favour of pitching the entire blade. NEG Micon used tip spoilers, activated by centrifugal forces, up to a turbine size of $D = 64$ m, $P = 1500$ kW. Fig. 13-3 shows the simplified block diagram for wind turbines of the Danish concept which operate entirely in the passive mode.

Fig. 13-4 shows the aerodynamic forces of lift and drag as well as the circumferential force and the thrust in the section at half of the blade ($r = 0,5\,R$, uniform airfoil profile) of a small stall-controlled wind turbine designed for an optimum tip speed ratio of 5.6. The vectors are shown for the design wind speed of 7.5 m/s and for storm, i.e. a wind speed of 30 m/s.

From the direction of the relative velocities w_d resp. w_s relative to the profile chord it is apparent that the angle of attack increases for storm so much that the flow separates and the drag $F_\mathrm{D.s}$ becomes correspondingly high. Thus, there is only a moderate circumferential force $F_\mathrm{tang.s}$. The rotor "refuses" to exceed the maximum power P_max of the generator.

The stall effect can be used to hold the power more or less constant during strong winds ($v_\mathrm{Wind} > v_\mathrm{rated}$). The required blade design will then deviate slightly from the ideal configuration of Betz-Schmitz. Usually, it suffices to mount the ideal blade with a slightly "wrong" blade twist angle, differing by a few degrees, to achieve a sufficiently constant power until cut-out.

The first generation of commercial stall-controlled wind turbines had power ratings of less than 50 kW and a control system which was limited in its actions to starting up, switching over between the generators and shutting down the wind turbine as a function of the wind speed and power.

For larger wind turbines the control system has to operate more subtly since the starting (short-circuit) current of an asynchronous motor-generator at the very first moment before the squirrel-cage rotor starts to rotate reaches 6 to 8 times the rated current, cf. section 11.2.1. Therefore, the machine is started in star connection before switching to delta connection which uses the full power. This reduces the currents in the wires significantly (by 1/1.73). If operation is always started with the small generator, which has a rating of approx. 25% of the full rated power, the start-up currents exceed the rated current of the big generator only slightly, for which the electrical system was designed.

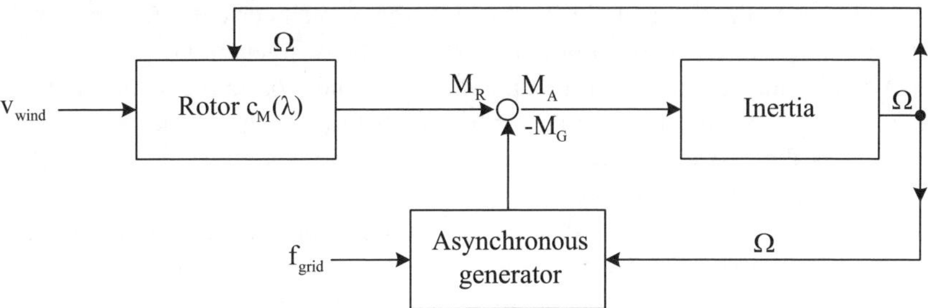

Fig. 13-3 Simplified block diagram of a wind turbine with an asynchronous generator at grid connected operation (Danish concept)

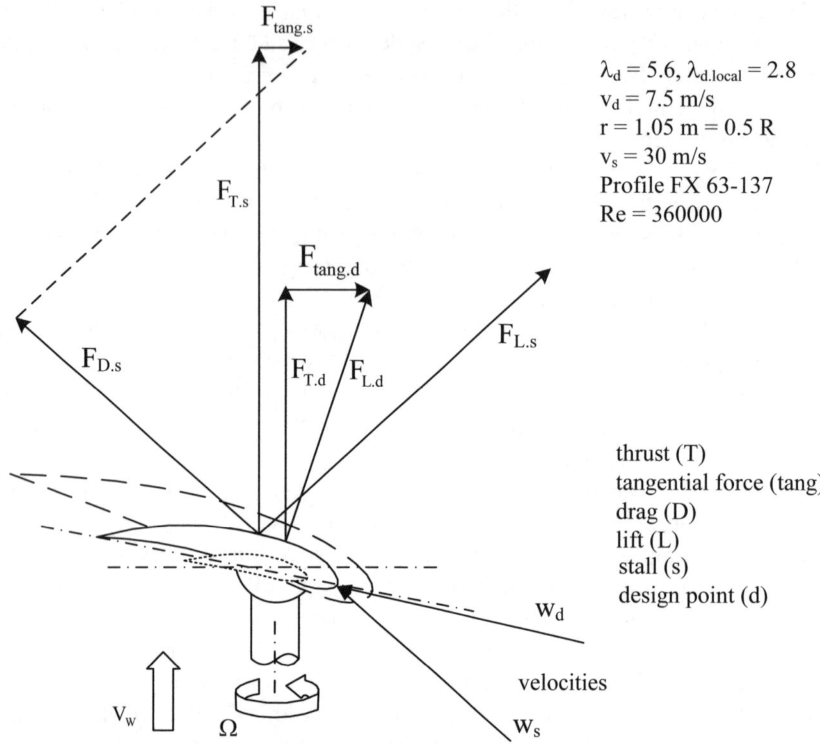

Fig. 13-4 Forces at the blade section 0.5 R at the design point (index d) and with control by stall effect during storm (index s)

Stall-controlled wind turbines of more than 250 kW use more complex supervisory and control systems. Not withstanding the passive limitations of power and rotational speed by the stall-effect itself, more complicated control procedures are required for switching under load.

In particular, the smooth transition with a large power demands control of the phase angle using thyristors. The transfer of power is built up gradually by manipulation of the firing angle. That way, large in-rush currents are avoided. The critical moments during switching over between the two generator speeds (see Fig. 13-2) can be controlled as well by operating the generator above its pull-out point under a given constant load torque with a small firing angle of the thyristor. Fig. 13-5 shows the tasks of a typical control and supervisory system of a medium-sized stall-controlled wind turbine (approx. 300 kW).

The turbines of Danish concept are very simple and robust, there are nonetheless three problems which led to further research and development.

To begin with, the grid operators dislike the reactive power consumption of the asynchronous machine, sometimes a penalty has to be paid for it. Therefore, for large stall-controlled wind turbines the reactive power compensation is done (in part) by capacitor banks. Certainly, it would be better to have controllable reactive power as is the case for synchronous machines.

Secondly, there are high structural loads and power fluctuations because the blades cannot be pitched in the strong wind range above the rated wind speed. This was shown already in Fig. 6-17 for the thrust coefficient c_T. Pitching the blade by 20° towards feather reduces the idling tip speed ratio from 13.5 to 4.5, and the corresponding thrust coefficient drops from 1.25 to approximately zero!

The third and most obvious disadvantage of stall-controlled wind turbines running at constant rotational speeds is shown in Fig. 13-2: the wind turbine works with optimum aerodynamics only for two rotational speeds in the normal wind range (at 5 and 8 m/s).

By contrast, an ideal variable-speed wind turbine would always run with the design tip speed ratio thanks to a wind speed dependent control until the rated generator power is reached, i.e. the rotational speed n has to be adapted to the wind speed $v(t)$.

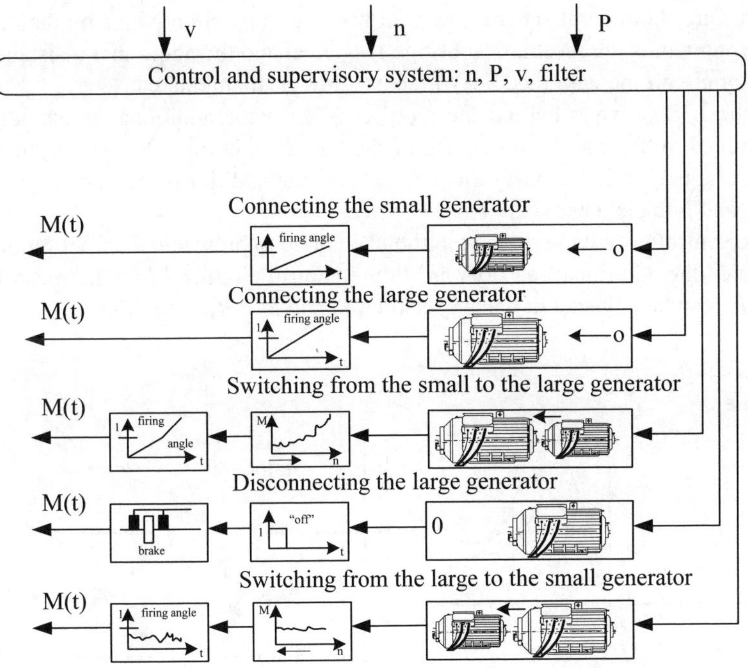

Fig. 13-5 Control and supervisory system of a wind turbine with stall control (approx. 300 kW)

13.1.2 Directly grid-connected asynchronous generator with dynamic slip control

The rigid coupling of the asynchronous generator (induction generator) to the grid frequency causes high structural load during gusts and high winds, especially for large asynchronous generators which operate with a very small slip s < 0.02 in order to reduce power losses. If during gusts a larger slip is allowed briefly, the structure is relieved and the power output is smoothened without permanent negative consequences for the efficiency.

This concept can be performed using a slip-ring asynchronous generator. It has a three-phase winding in the rotor, instead of the squirrel-cage, and, therefore, it is possible to connect variable resistances into the rotor circuit. The increased resistance in the rotor circuit leads to more slip, Fig. 13-6, as already discussed in chapter 11, Fig. 11-25.

The block diagram of such a dynamic slip control is given in Fig. 13-7. For normal winds the bridge at the slip rings is short-circuited mechanically. Thus, the resistors are shorted out. There is normal generator operation with a moderate slip. For strong winds the mechanical bridge is opened and the additional resistances in the rotor circuit increase the slip, which can now be manipulated.

If the IGBT switch behind the rectifier is open the additional resistances are fully added to the internal resistance of the windings in the rotor circuit, this produces a large slip. If the IGBT switch is closed the additional resistors are without effect, as for the mechanically closed bridge.

The controller will use a clock in the kHz range to prescribe the fraction of time for which the additional resistors are in the control circuits. Thus the mean resistance R_m can be adjusted to values between R_i and $R_\mathrm{i} + R_\mathrm{add}$, see Fig. 13-8.

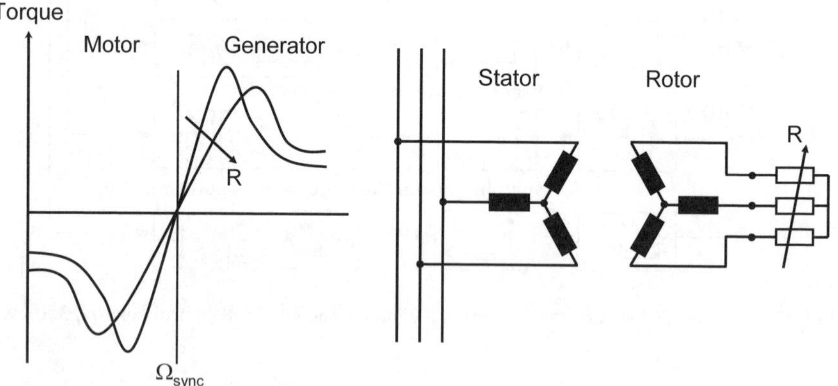

Fig. 13-6 Influencing the generator characteristic by additional variable resistances in the rotor circuit of an asynchronous machine

If there is a steep increase in the torque resp. the current due to a wind gust the controller will temporarily allow more slip. Afterwards it will return swiftly to the more rigid coupling to the grid frequency (only R_i).

This type of dynamic slip control has the disadvantage that during gusts power has to be passed through the slip ring. However, it is advantageous that the resistors which turn the power into heat may be located on top of the nacelle where the heat dissipation is not an issue. Other types of dynamic slip control are described in [5]. If now a blade pitching system is added to this type of control the work is shared as follows:

- The short-term response to gusts is provided by the slip control.
- The adaption to changes in the mean wind speed is performed by the blade pitching system (e.g. Vestas and other manufacturers).

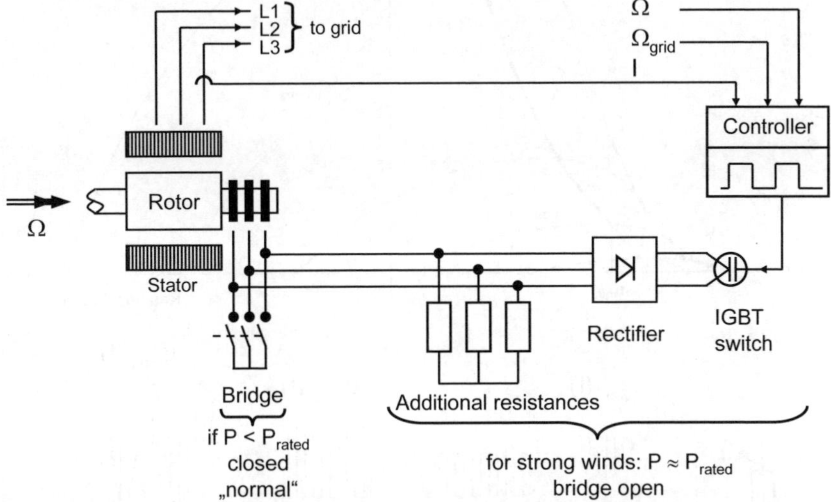

Fig. 13-7 Block diagram of the dynamic slip control by additional variable resistances in the rotor circuit of an asynchronous machine

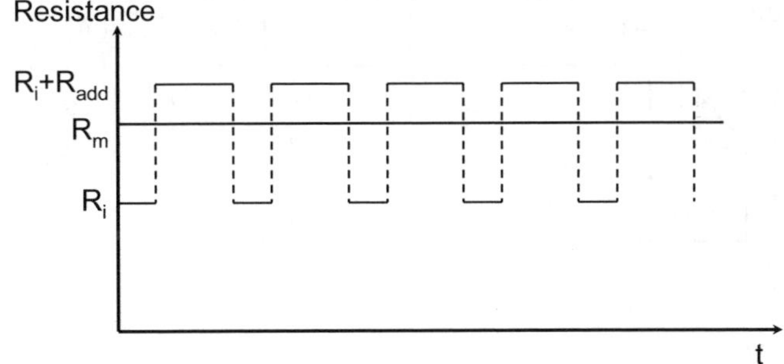

Fig. 13-8 Slip manipulation through clocked switching of the additional resistance R_{add}

13.1.3 Variable-speed wind turbine with converter and direct voltage intermediate circuit

The schematic view of this wind turbine concept is given in Fig. 13-9. No gearbox is required due to the direct-drive multi-pole synchronous generator of a large diameter.

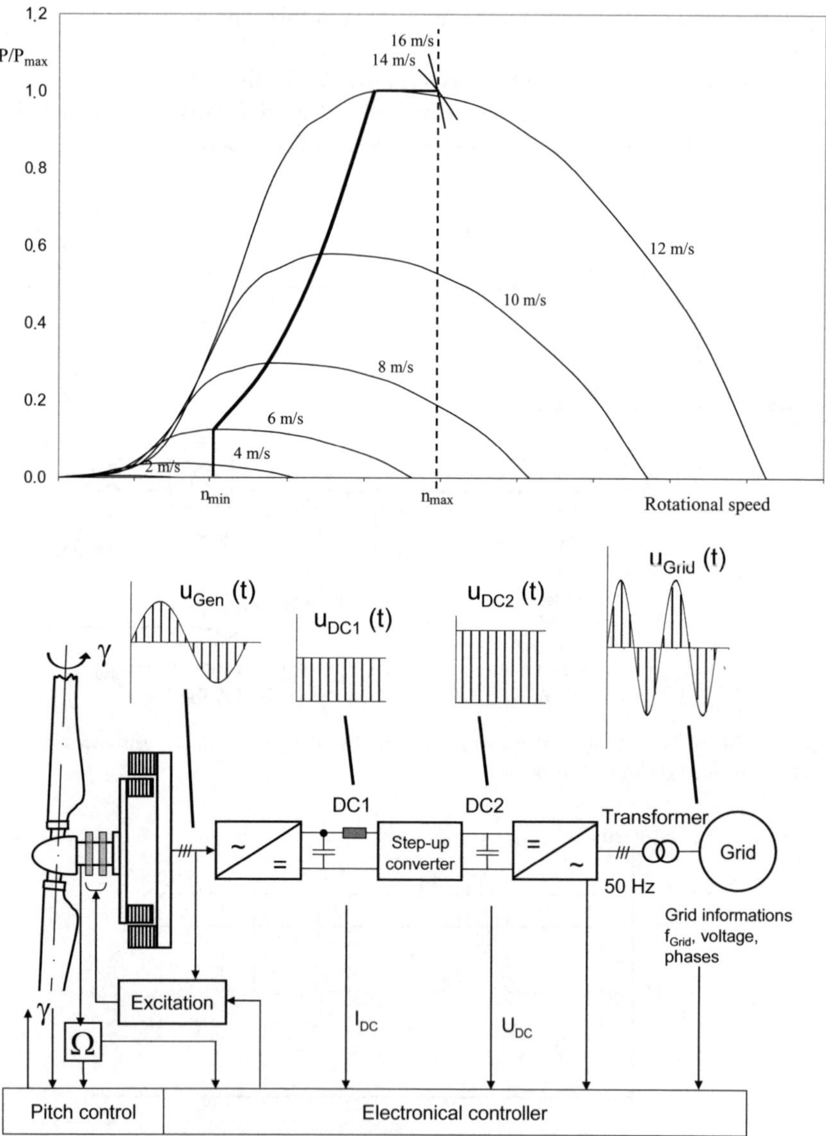

Fig. 13-9 Concept of a synchronous generator with AC-DC-AC converter

At point DC1 of the converter the three-phase current of variable frequency has been converted into direct current. For a low rotational speed the generator does not reach the output voltage of 400 V despite full excitation. Therefore, the step-up converter increases the voltage level (from DC1 to DC2 in Fig. 13-9). After that the direct current of 400 V is turned into three-phase current of 50 (or 60) Hz which is fed into the grid via the transformer.

Moreover, the DC intermediate circuit delivers information about the momentary power (current I_{DC} times voltage U_{DC}) which, together with the rotational speed, provides the control with the essentials of the operating state: the aim of the generator control in the normal wind range is to follow the optimum trajectory of "torque is proportional to Ω^2 ", cf. section 12.4.

The blade pitching system is inactive until the rated wind speed of 12 m/s is exceeded. For stronger winds it limits the rotational speed with a certain flexibility to the set value.

Since in the strong wind range the generator is operated with a (nearly) fixed torque, cf. section 12.4, the power output is approximately constant.

The mechanical implementation of this wind turbine concept (E-40, E-66, etc. by ENERCON) is impressively simple, cf. Fig. 3-30. The turbine is nearly "oil-free" because there are electrical drives for the blade pitching system as well as for the yaw system of the nacelle. The supply of reactive power can vary freely between inductive and capacitive: there is no problem of reactive power consumption as in the case of asynchronous machines.

Scaling the wind turbine up to rotor diameters of more than 100 m leads to an impressive weight of the nacelle (approx. 500 t for the E-112). Wind turbines using a gearbox have a significantly smaller weight, cf. chapter 3.

13.1.4 Variable-speed wind turbine with doubly-feeding asynchronous generator and converter in the rotor circuit

With dynamic slip control, section 13.1.2, the asynchronous machine is less rigidly coupled to the grid ($s = 0.02$ to 0.2). However, it requires the more complex slip ring rotor instead of the squirrel-cage rotor. Since one percent slip equals one percent power loss in the rotor, it is reasonable to allow high values of slip only for brief intervals of time.

If the power, that was exported from the rotor, is not converted into heat, but is instead fed into the grid via an AC-DC-AC converter the problem of heat dissipation is avoided. And the efficiency of the generator is improved. However, the latter is of little interest in the range of strong winds.

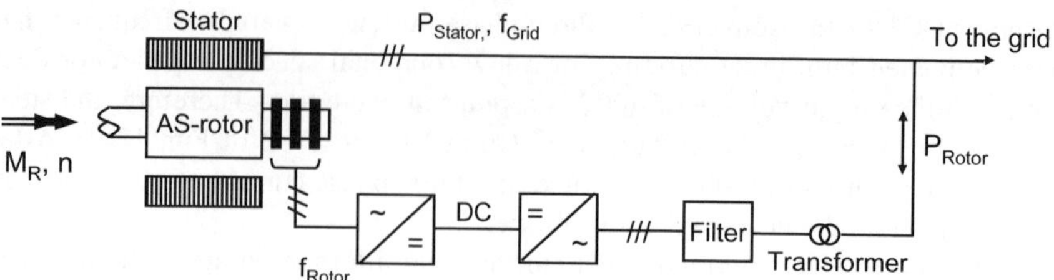

Fig. 13-10 Wind turbine with asynchronous generator and converter in the rotor circuit

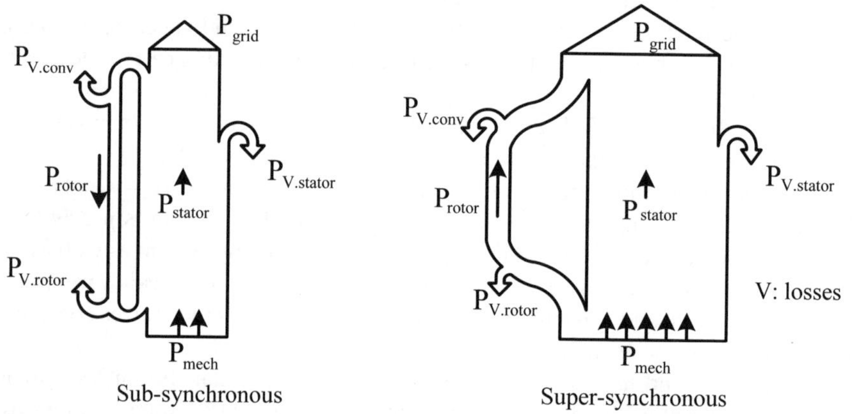

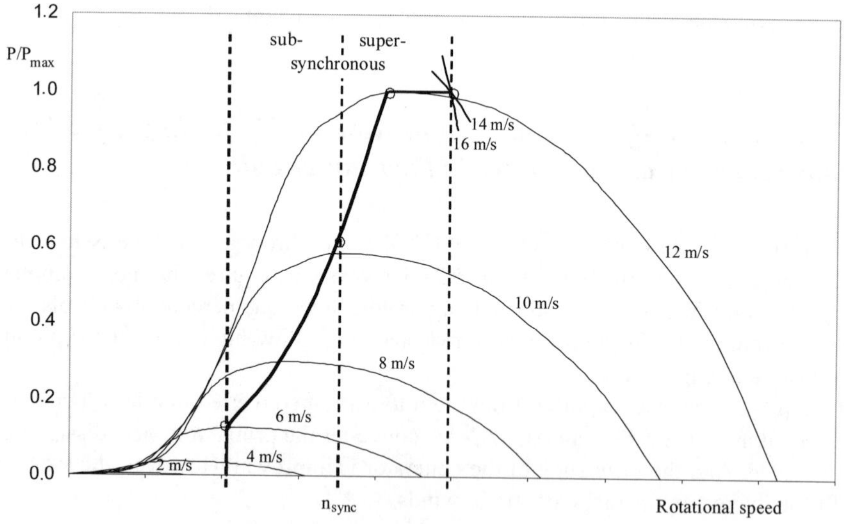

Fig. 13-11 Operation of a doubly-feeding asynchronous generator with converter in the rotor circuit below and above the synchronous speed (see Fig. 13-10),
Top: power flow; Bottom: relative power versus rotational speed

This concept is performed using the so-called super-synchronous (i.e. above-synchronous) converter cascade [1].

If the range of operation of the asynchronous generator below the synchronous speed is to be improved as well, a part of the stator power will have to be fed into the rotor with suitable values of frequency and voltage. This requires a certain control effort from the power electronics [2, 3, 5]. Fig. 13-11 shows the power flow in stator and rotor for operation below and above synchronous speed of the machine.

In this configuration, the asynchronous generator can have a speed range as large as that of the synchronous machine with a (full power) converter. However, the converter in the rotor circuit has to cover only approx. 20% of the rated power. Therefore, it is less expensive and also has smaller losses. The more complex control in the rotor circuit can also decide whether the asynchronous machine should consume or supply reactive power in order to meet the present needs of the grid.

13.1.5 *Power curves and power coefficients of three wind turbine concepts– a small comparison*

The Figs. 13-12 and 13-13 show for comparison the power curves $P(v)$ and the total power coefficients $c_{P.tot}$ (v) of three different wind turbine types. The total power coefficient includes rotor, gearbox, generator and converter efficiencies. The data were taken from type approval measurements, reported in market surveys [11, 15]. The power curves differ noticeably because the wind turbines have different rotor diameters:

- NORDEX N43, $D = 43$ m, $P = 600$ kW:
- Classical Danish concept with two fixed rotor speeds;
- asynchronous machine and stall control (section 13.1.1)
- ENERCON E-40/6.44, $D = 44$ m, $P = 600$ kW:
- Variable-speed synchronous machine with AC-DC-AC (full power)
- converter and pitch control (section 13.1.3)
- SÜDWIND S.46, $D = 46$ m, $P = 750$ kW:
- Variable-speed doubly-fed asynchronous machine with AC-DC-AC converter in the rotor circuit and pitch control (section 13.1.4)

From the power curves it is only apparent that the wind turbines with larger diameter deliver more power, except for the cut-in range where the curves intersect at 5.5 m/s.

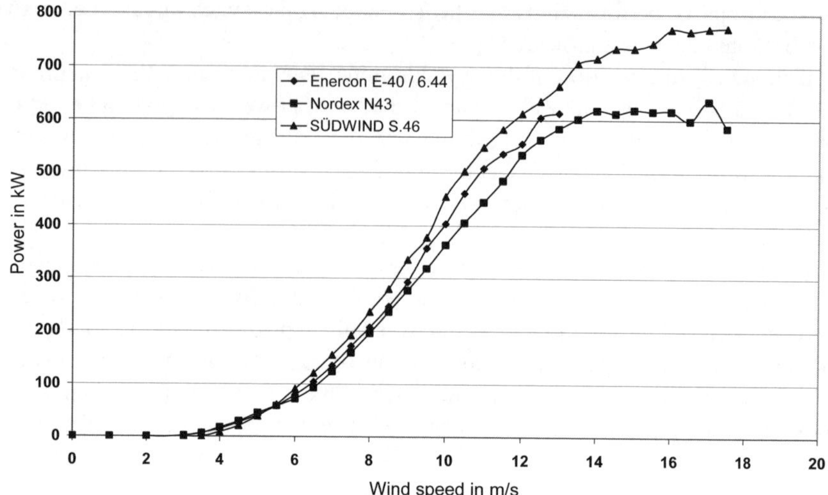

Fig. 13-12 Power curves P(v) of three different wind turbines

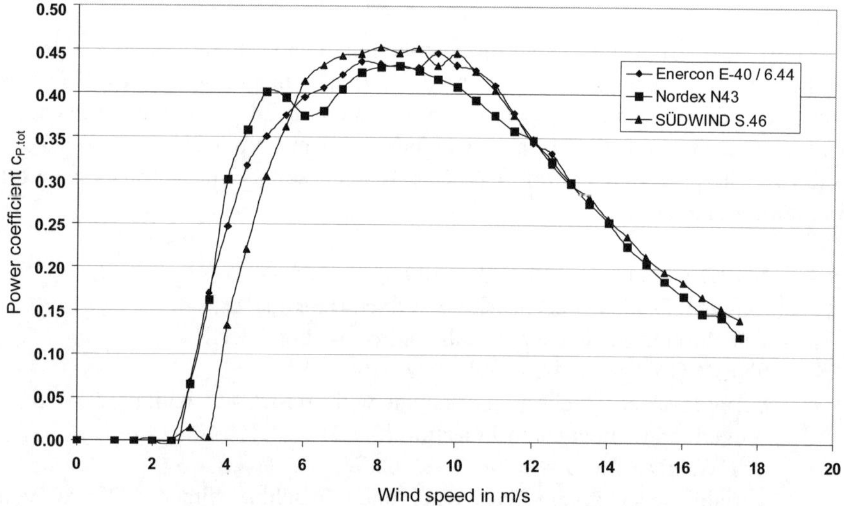

Fig. 13-13 Power coefficients $c_{P,tot}$ (v) of three different wind turbines

The dimensionless total power coefficients $c_{P,tot}$ reveal more. We compare first the wind turbine of the Danish concept by NORDEX (two fixed rotor speeds) with the ENERCON turbine which operates with variable rotor speed. For the range close to the cut-in wind speed (3 m/s) the curves of $c_{P,tot}$ are nearly identical. Then the NORDEX turbine shows two local optima which correspond to the design wind speed of the "small" resp. "large" asynchronous generator. The first maximum of

0.40 at a wind speed of approx. 5 m/s is significantly higher than the value of 0.35 for the variable-speed machine E-40. The full power conversion of the E-40 reduces its partial load efficiency significantly (at 10% of the rated power).

But for higher wind speeds the power coefficient of the E-40 which was designed for variable speed is always slightly better than for the N43. At their rated wind speed of 12 m/s the power is limited at both turbines. The N43 uses the stall effect, the E-40 a blade pitching system. The curves of $c_{P.tot}$ are nearly identical in this range. Since the control of the E-40 keeps the power output constant above 12 m/s there are no values shown for higher wind speeds.

The variable-speed wind turbine of SÜDWIND with a doubly-feeding asynchronous generator and a converter in the rotor circuit cuts in only at 3.5 m/s, which is barely visible in the power curve $P(v)$. Up to a wind speed of approx. 6 m/s the curve of $c_{P.tot}$ is significantly below the curves of the N43 and E-40. In the range between 6 and 9.5 m/s it is slightly above the values of the E-40. Both reach a maximum efficiency of $c_{P.tot} = 0.45$. Since the control limits the power in the strong wind range above the rated wind speed the differences are negligible.

In general, it can be stated that despite the differences in the three wind turbine concepts the $c_{P.tot}$ curves are astonishingly similar, except for differences in cut-in wind speeds. The maximum power coefficient of the "primitive" Danish wind turbine is only 2 percentage points below that of the wind turbines with a blade pitching system and variable rotational speed.

Thus, the relevant advantages of the modern variable-speed concepts are not found in their efficiency, but in

- The reduction of power fluctuations during gusts,
- Significant relief for the structure in the strong wind range,
- The flexibility in adjusting the reactive power,
- The adaptation of the blade tip speed to local conditions "by pressing of a button", etc.

In summary: we gain increased flexibility with regard to the actual on-site conditions.

13.2 Wind turbines for stand-alone operation

The most common stand-alone applications of wind turbines are the battery charger and the wind pump. Stand-alone wind turbines eke out a niche existence in Western Europe due to the omnipresent electrical grid. But they are of great practical importance in the rest of the world. In Mongolia, e.g., the nomads operate thousands of battery chargers providing electricity for light and television.

13.2.1 Battery chargers

Battery chargers have small power ratings. Typical values range from a few watts to approx. 1.5 kW. This implies small rotor diameters (0.5 to 3.0 m), and relatively high rotor speeds. The gearbox which is usually necessary between generator and rotor is not needed here. Directly driven synchronous generators with a medium or large number of poles (8 to 20 poles) are used.

For the selection of the aerodynamic profiles the low Reynolds numbers of the flow around the blades have to be taken into account ($Re = w\,c/v < 100\,000$). Suitable profiles for this range are found in catalogues of aerodynamic profiles for model airplanes (e.g. [4] in chapter 5 of this book).

In order to determine the stationary operating behaviour, the familiar equivalent circuit diagram of the AC synchronous machine is extended to include the load and the rectifier (RF). Fig. 13-14b shows the resulting load curves. Below the threshold speed, the generator voltage is lower than the sum of the voltage provided by battery charger and rectifier. Hence, there is no power output. Above the threshold speed, there is a steep increase in power transfer, similar to the power curve of the synchronous machine with a resistive load, see section 11.1, Fig. 11-9.

For a battery charger with a constant load, two types of operation can be distinguished. For low wind speeds, the battery provides the largest part of the required power. The wind turbine operates at its power optimum (A) if it is a well designed machine. For high wind speeds or small loads, the energy output of the wind turbine exceeds the demand of the load, and has to be limited (B), therefore.

For a generator with an excitation winding, this can be achieved by adjusting the excitation current so that the critical battery voltage is reached, but not exceeded. Fig. 13-14b shows that the power consumption is then limited, and that the wind turbine gradually transitions to idling for higher wind speeds. If the rotor is designed with a high tip speed ratio, limiting measures for the rotational speed are required, e.g. by tilting, see Fig.12-19 in chapter 12.

Small battery chargers which use *permanent magnets* for the excitation have no need for control of the charging current. In this case, the charging power is simply limited by the value of the generator inductance (Fig. 11-9) or by an additional inductance connected in series, Fig. 13-15. For these components, the impedance ($X_Z = \Omega\,L_{add}$) increases with the frequency. In this way the battery charging current is limited. This arrangement is shown in Fig. 13-15. If the battery storage is large in relation to the installed generator capacity, an overcharging of the battery is hardly possible.

In order to make better use of the advantages of excitation with permanent magnets (good efficiency and low wear), devices from power electronics have to be incorporated. Due to the constant excitation of the synchronous generator, the generated voltage is proportional to the rotational speed.

For high speeds, this voltage exceeds the rated charging voltage of the battery substantially. A step-down converter [4] is used to convert this high DC voltage to the lower voltage level needed for the battery. Generator and battery voltage are then decoupled, and the battery charging current can be adjusted to reasonable values in accordance with the available power and the needs of the battery

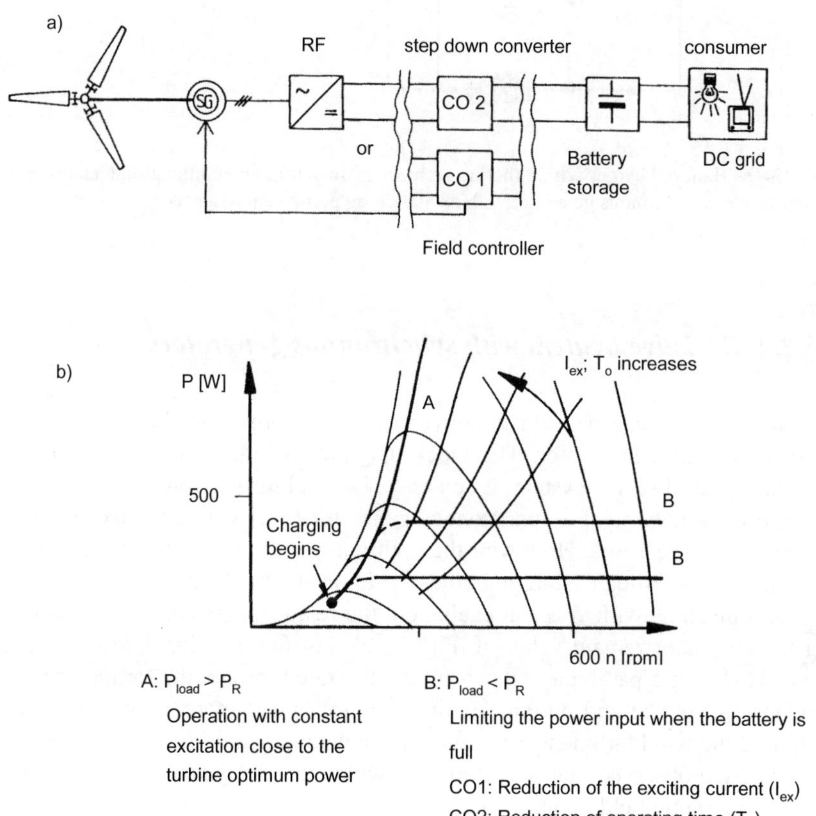

Fig. 13-14 Battery charger, a) Block diagram; CO 1 = Controller for a generator with a field winding excitation, CO 2 = Controller for a permanent-excited generator, b) Load curves of a battery charger

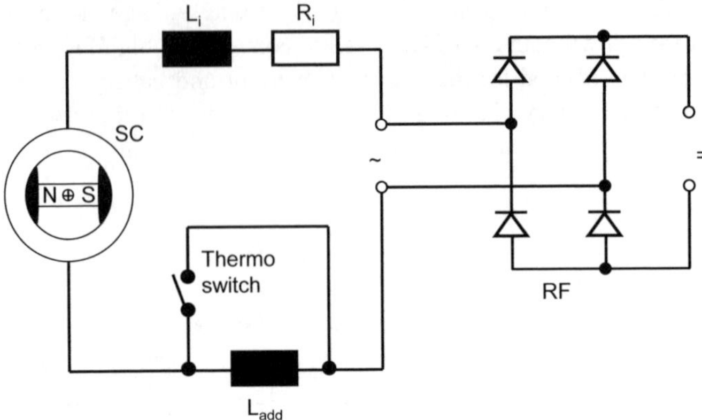

Fig. 13-15 Battery charger with a thermo element to switch in an additional inductance at high power levels; synchronous generator with excitation by permanent magnets

13.2.2 *Resistive heaters with synchronous generators*

Resistive heaters can be used to convert electric energy into thermal energy, for instance to make hot water. The block diagram in Fig. 13-16 shows the rather simple layout of such a system. It consists of a wind turbine rotor with only a simple speed limitation, of a synchronous generator (SG) with gearbox, and of the corresponding resistors. How well the optimum load curve of the wind turbine is matched can be influenced through the excitation and the load resistors.

For a machine with constant excitation, the relationships derived in chapter 11 for the alternator remain valid (cf. Fig. 11-9). For low rotational speeds the load curve is almost a parabola. After reaching the rated torque, the torque decreases, but power, voltage and current continue to rise, albeit more slowly. The power output of the wind turbine is well utilized in a wide range, Fig. 13-16, curve a.

It is often observed that wind turbines, which drive a generator, that has a permanent magnetic field, require wind speeds to be quite high, before they start turning. The culprit behind this is the "magnetic latching", i.e. the rotor appears to be locked into a rest position and needs torque to break loose. This possible short-coming can be dealt with during the design of the generator.

The demand for quickly increasing electrical power is another detriment to start-up at low wind speeds. This can be remedied by initially disconnecting the load.

Self-excited synchronous machines do not have this problem, since excitation only starts above a minimum rotational speed. Their load curve is steeper since the excitation current is caused by the stator voltage which increases with the rotational speed. The exact curve is somewhat influenced by the selected controller type. It will, however, always be quite similar to the curve b in Fig. 13-16.

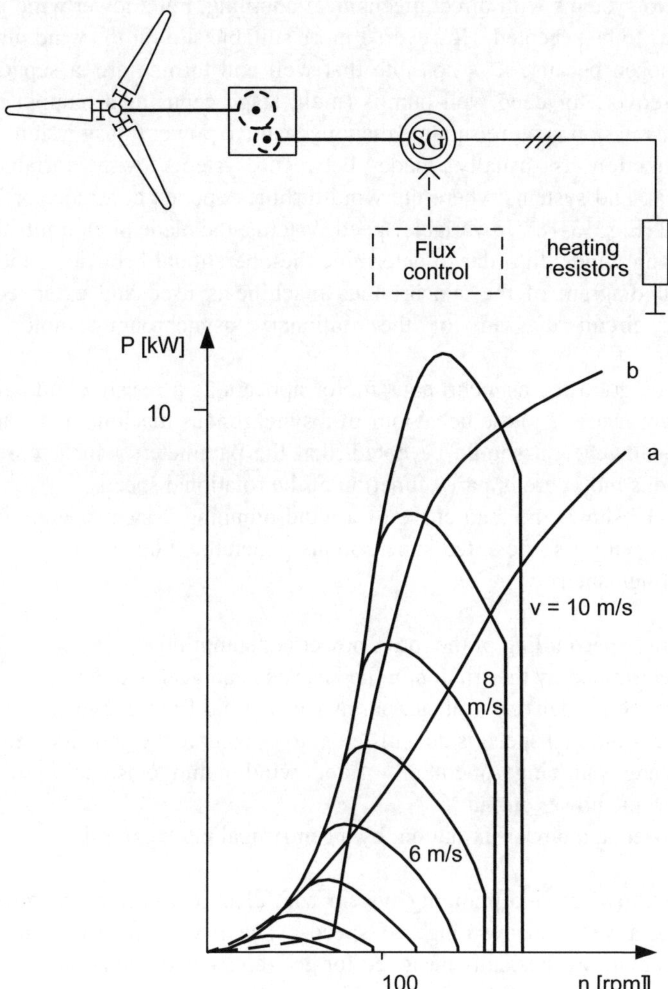

Fig. 13-16 Wind turbine for heating, speed control by centrifugal pitch, block diagram and load curves: a) generator with permanent magnetic field, b) self-excited generator

13.2.3 *Wind pump system with electrical power transmission*

Wind pump systems with electrical power transmission have some advantages compared to systems with direct mechanical coupling, but a lower wind pump efficiency has to be accepted. However, a more suitable site for the wind turbine can now be chosen because it is possible that well and turbine are at separate locations. Moreover, for deep well pumps (multi-stage centrifugal pumps) electrical power transmission is more suitable than mechanical power transmission.

A distinction is usually made between systems with variable speed (Fig. 13-17), and systems where the wind turbines operate at a more or less constant speed (Fig. 13-18). In variable speed systems, the blade pitch limits the speed during strong winds. In order to determine the operational behaviour, the equivalent circuit diagram of the synchronous machine is used and extended by the equivalent circuit diagram of the stationary asynchronous motor (ASM), Fig. 11-22.

In this diagram the asynchronous motor appears as a resistive-inductive load. The diagram leads us to the behaviour of a synchronous machine in a stand-alone operation. However, it should be noted, that the parameters which represent the asynchronous motor are in part a function of the rotational speed.

Fig. 13-17 shows the load curve of a wind pumping system operating at variable speeds with a self-excited synchronous generator. Four ranges of operation can be distinguished:

A   Low-speed idling of the rotor, power consumption
      determined by the friction in the bearings and gearbox, etc.
B   Generator delivers voltage: pump acts as fluid friction dynamometer since
      the rotational speed is not sufficient to overcome the geodetic head
C   Water pumping: operation of the wind pump close to optimal wind
      turbine power output
D   Speed and power limitation: by a centrifugal pitch control

Fig. 13-18 shows a wind pumping system with electrical power transmission that operates in a very narrow range of speeds. The pitch-controlled wind turbine MAN-Aeroman was basically designed for grid-connected operation. As the speed range is very narrow, the electrical machine and the centrifugal pump always work close to their rated operating points. The layout of the system is, therefore, comparatively easy, but the engineering and control effort is substantial.

When the wind speed changes, pumps are switched on or off in order to adapt to the current supply of energy. Above 10 m/s, the fast pitch control leads to a constant power output. As an alterative to pitch control, a stall control may be used to limit the power. In that case, the wind turbine's speed is limited by connecting a controllable heater, if the maximum allowed pump power would otherwise be exceeded.

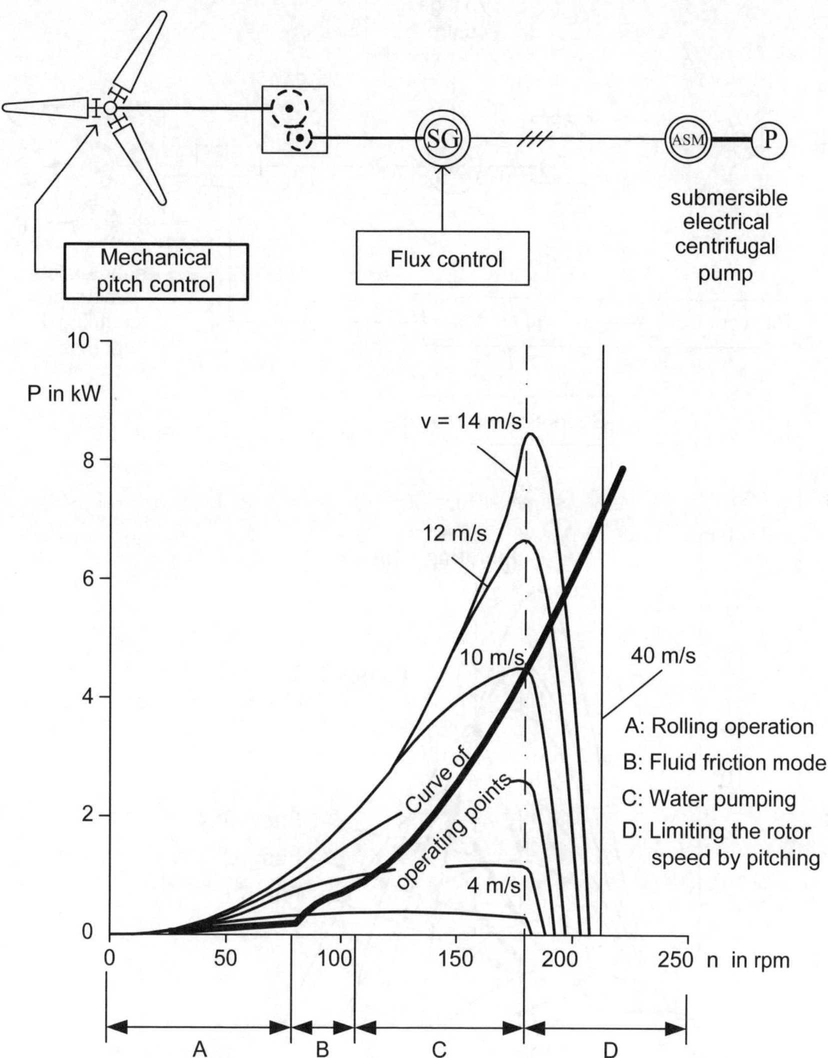

Fig. 13-17 Wind pumping system operating with variable rotor speeds, a) Block diagram, b) Load curve

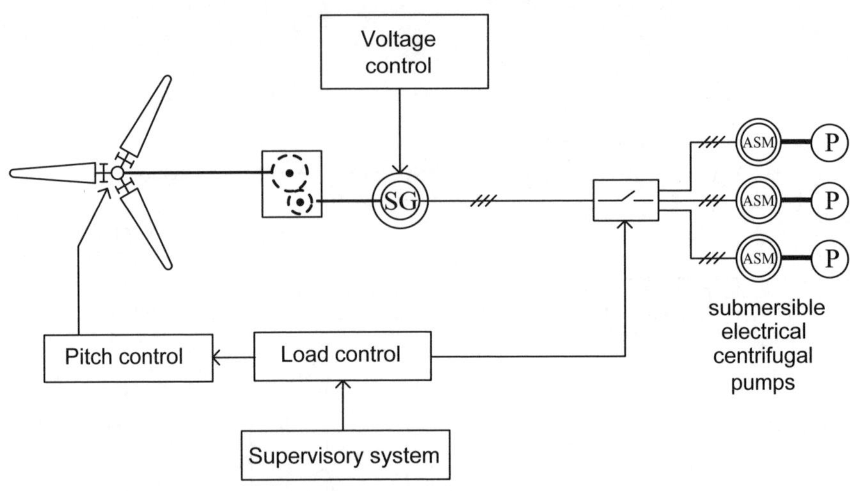

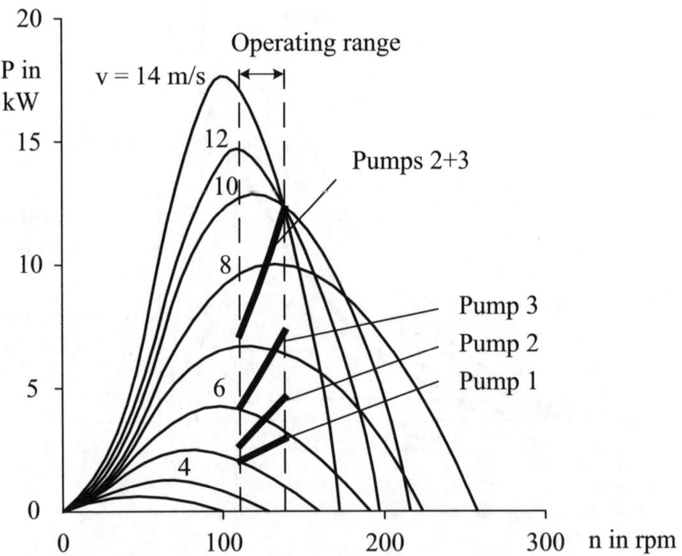

Fig. 13-18 Wind pumping system operating at nearly constant rotor speed by switching the pumps, a) Block diagram, b) Load curve

13.2.4 Stand-alone wind turbines for insular grids

When feeding insular or local grids, a wind turbine supplies a fairly constant frequency and voltage for isolated consumers, e.g. alpine huts, farms, or rural communities in the Third World. In this application, the supply of both active and reactive power is required, and voltage control is necessary. Therefore, synchronous generators (SG) are suitable. A fairly stable frequency is achieved by switching consumers on and off. A stall-controlled turbine combined with an adjustable heater or a fast pitch control is used.

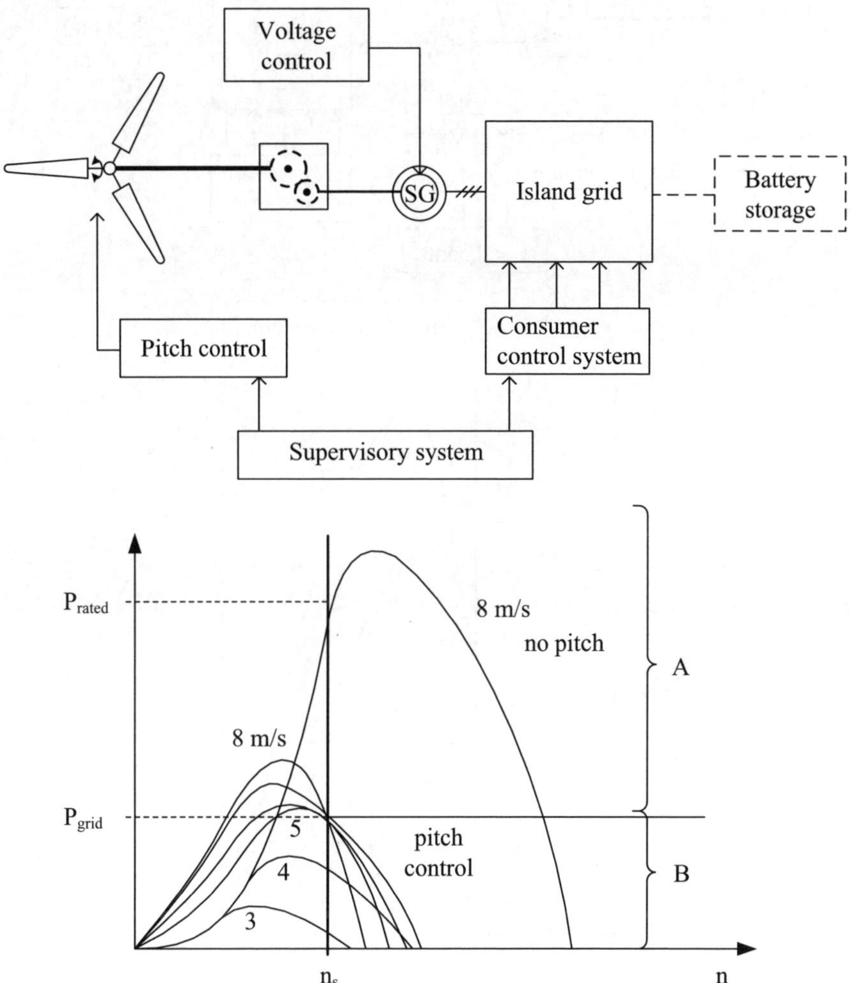

Fig. 13-19 Wind turbine with pitch control and consumer control system for stand-alone operation, a) Block diagram, b) Load curve

Fig. 13-19 shows the schematic design of a pitch-controlled turbine feeding an insular grid. The power demand P_{grid} can be met at a wind speed of 5 m/s if the optimal angle of attack is used. For higher wind speeds (A), the angle of attack reduces gradually thanks to the pitch control. This leads to a constant power output of the rotor at required synchronous speed.

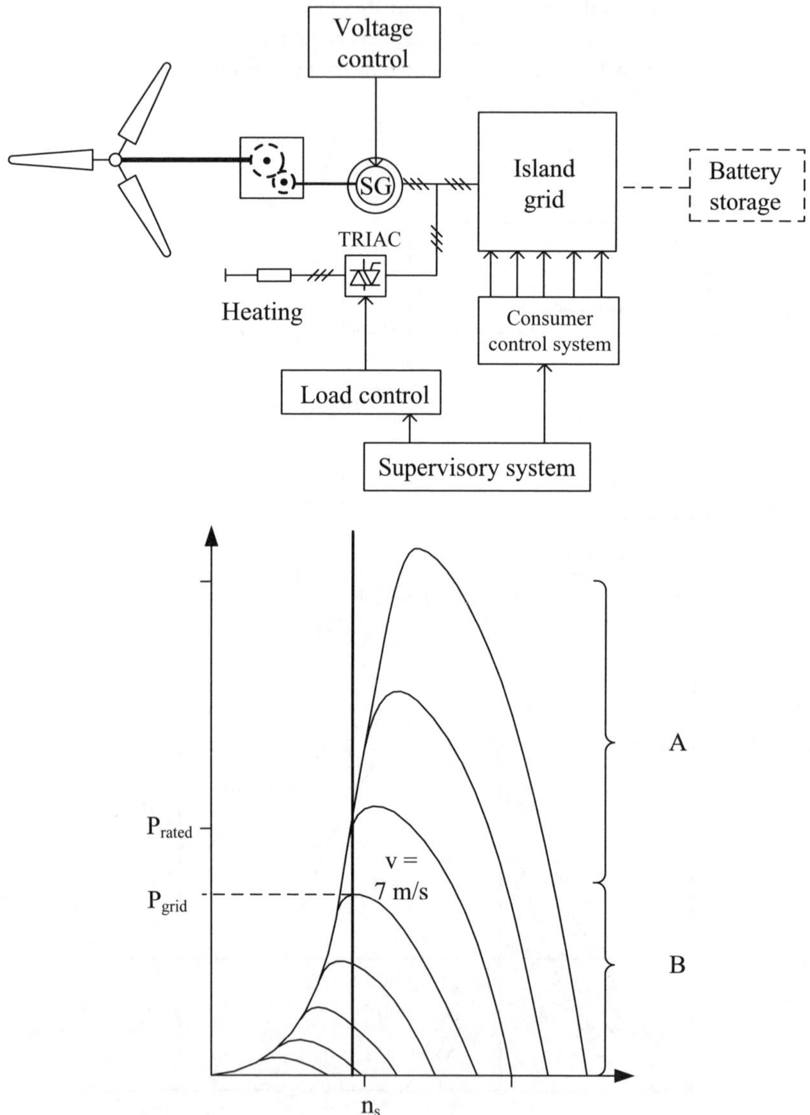

Fig. 13-20 Wind turbine with stall control and consumer control system for stand-alone operation, a) Block diagram, b) Load curve

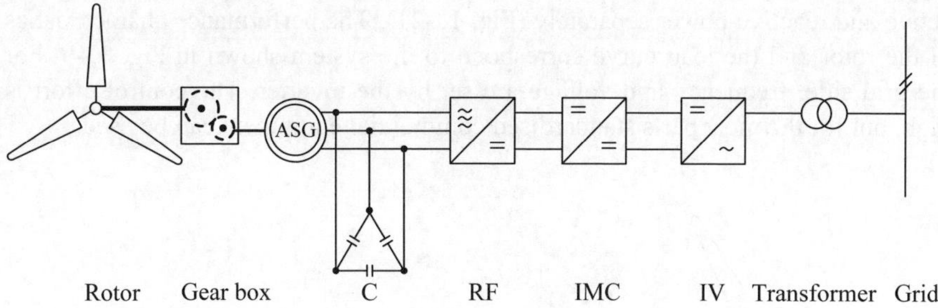

Fig. 13-21 Variable-speed asynchronous generator with an intermediate DC circuit

If the wind speed becomes so small that, even with the optimum angle of attack, the wind turbine cannot meet the demand (B), a control system has to decide which consumers are to be temporarily disconnected. This will often be loads, which serve a storage facility like pumps or refrigerators, etc.

Fig. 13-20 shows a stall-controlled turbine. It differs from the above system in that an excess energy output (A) is fed into resistors. The advantages are that the energy converted into heat by resistors can be utilized for heating, and that the rotor construction is more robust. If the wind speed decreases below the point where the rotor can supply all the power demanded by the grid, a consumer control system has to switch off lower priority loads (B), in this case too.

13.2.5 *Asynchronous generator operating in an insular grid*

If an asynchronous generator is connected to a grid, it draws reactive power required for its operation from the grid. In a local grid, this is not possible. The reactive power has to be supplied by capacitors or through fairly complex power electronic.

Capacitors provide reactive power that is dependent on frequency and on the square of the voltage. If one of these values changes the reactive power provided by the capacitors changes, too. Consequently, the usable range of rotational speeds is extremely limited. So, if an asynchronous generator is used in a wind turbine, the reactive power has to be provided by a number of switchable capacitor stages. A stall-controlled wind turbine is necessary (Fig. 13-22). Usually, a substantial amount of auxiliary equipment is needed. This is the reason, why this concept did not gain wider acceptance, in spite of inexpensive asynchronous machines.

Variable-speed wind turbines with an asynchronous machine often have an intermediate DC circuit. The converter on the generator side provides the reactive power for the asynchronous machine. If the start-up excitation is supplied by capacitors, a smaller frequency converter will suffice. It is possible to manipulate active and reactive power separately (Fig. 13-21). The performance characteristics of the rotor and the load curve correspond to the system shown in Fig. 13-9. For the grid side, frequency and voltage are set by the inverter. The control effort is high, but for the most parts standardised commercial equipment can be used.

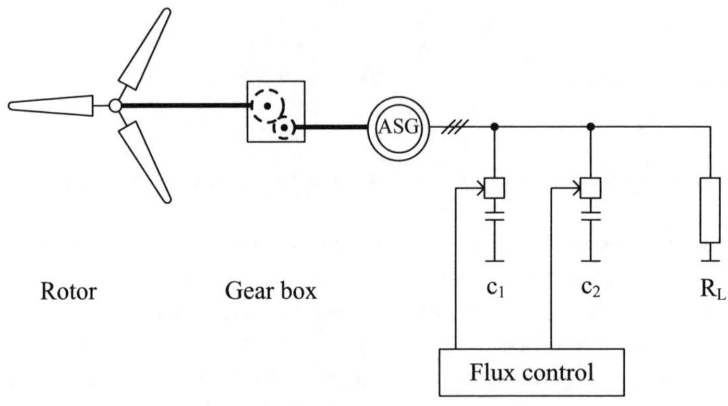

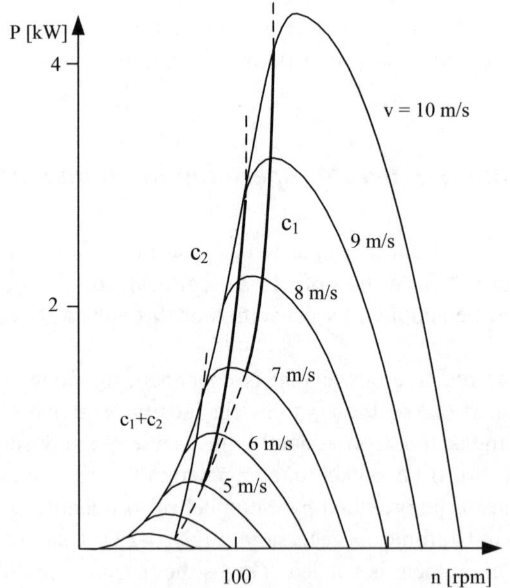

Fig. 13-22 Asynchronous generator in a local stand-alone grid for heating; reactive power is supplied by capacitor banks

13.3 Wind turbines in isolated grids

Outside of the urbanised areas all over the world, there are small towns, villages and single farms without a reliable electric power grid. The classical solution, a diesel generator, is becoming less and less affordable due to rising oil prices. The demand for wind-diesel systems and wind-photovoltaic systems is rising.

But due to the boom of grid-connected wind turbines in Western Europe since 1991 further development of the above systems was neglected. The concepts from the 1980s are no longer up-to-date: power electronics now offers a large variety of new possibilities.

Although there are several prototype plants in Greece and on Spanish and Portuguese islands in the Atlantic Ocean, there are few field-tested ones on the market:

Neither systems with the low power ratings desired for single farmsteads nor those of a few MW suitable for smaller towns are readily available.

In principle, variable-speed wind turbines with a synchronous generator can be used. However, the supervisory and control logic for a hybrid system is quite different from that of a grid-connected one.

Further, there is a lack of inexpensive storage systems which buffer the fluctuations of the wind energy. Flywheels for short-term storage are slowly appearing on the market (ENERCON [6], URENCO, etc.). Battery storage continues to be used for long-term storage even though they are good for only a limited number of charge / discharge cycles [4, 8, 14].

The diesel motor should not run in stand-by mode in parallel to the wind turbine for long periods of time. During idling the injection nozzles of the diesel engine collect soot and the fuel consumption is not negligible (5 to 10% of rated consumption, which is around 190 g /kWh for large motors).

To avoid cold starts of the diesel engine the coolant may be preheated electrically while the motor is turned off. This will reduce mechanical wear during start-up substantially and lead to a significantly longer service life of the motor. However, it does consume some energy.

A certain amount of voluntary load reduction can be achieved if LEDs are installed on the meter of the customer:

      green: cheap tariff, wind power,
      yellow: medium tariff, wind and diesel power,
      red: expensive tariff, diesel power.

This leaves it to the consumer to decide on his own, and at the same time he develops a certain feel for the system.

Fig. 13-23 shows the typical diurnal curve of the wind and diesel power production for a small grid on the Island of Brava (approx. 4000 inhabitants).

The wind turbine has a rated capacity of 120 kW. The "penetration" of the system by wind power is relatively high, at times up to 50%.

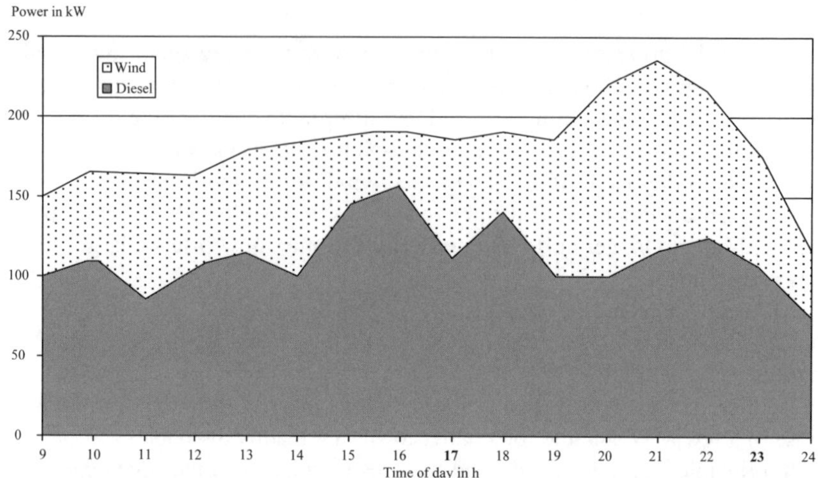

Fig. 13-23 Typical diurnal wind and diesel power production for a small local grid on the island of Brava, Cape Verde [12]

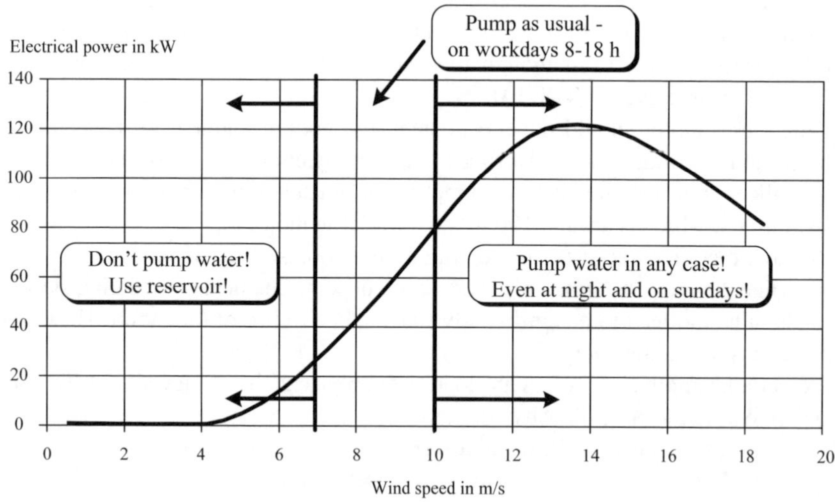

Fig. 13-24 Instructions for the manual control of the pump for the storage of drinking water of a small local grid according to the wind conditions [12]

However, manual load management is still used for these small systems. Fig. 13-24 shows the concept of the supervisory system. If the power demand of the inhabitants is sufficiently low, the storage reservoir (geodetic head of $H = 435$ m) of the drinking water system will be filled [12].

The diesel generator on Brava is the master machine of the grid despite the high wind power penetration since the wind turbine is of the classic Danish concept. A wind-diesel system, in which the diesel can be turned off automatically, requires substantially more control equipment.

13.3.1 Wind-diesel system with a flywheel storage

The wind turbine E-30 (250 kW) delivers power into a DC power line from which the power is then fed via an inverter into the three-phase grid, Fig. 13-25. As long as the supplied wind power is below the demand of the grid the diesel generator runs in parallel and governs the grid and matches the shortfall. Neither the fly wheel storage nor the master synchronous machine (MSM) is on-line.

If the available wind power exceeds the demand from the consumer, the fly wheel storage starts up (1,5 t rotating mass, rotational speeds between 1500 and 3300 rpm). It can store up to 5 kWh and can deliver on demand up to 200 kW for a short time. The grid is now governed by the master synchronous machine, and the diesel generator is switched off. Short deficits in the wind power supply are filled by the fly wheel storage and a battery bank if one exists. The power quality is excellent (i.e. voltage and frequency are constant), [13], but the costs are high.

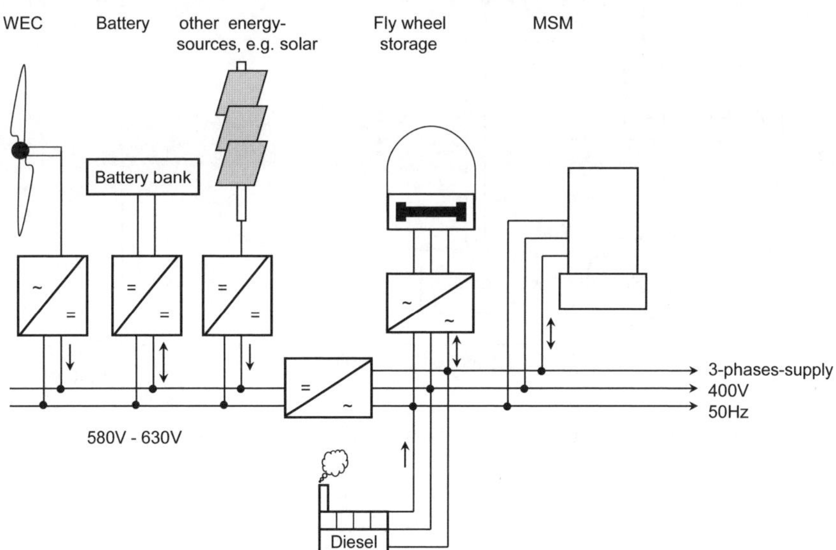

Fig. 13-25 Stand-alone system with a master synchronous machine (MSM), acc. [13]

13.3.2 Wind-diesel system with a common DC line

The system sketched in Fig. 13-26 was developed for an island in the Gulf of
Finland where there is a radar station and a homestead [10]. The synchronous generator is rated at 30 kW and uses permanent magnets for its field. It feeds – like
the diesel generator and the bank of batteries – into a common DC line. Finally, an
inverter converts from DC into 400 V three-phases AC for output.

The batteries are able to store approx. 100 kWh. The diesel generator may feed
the three-phase AC line directly via the bypass in case of long times without wind
or of a technical problem at the wind turbine or the batteries.

The rectifier will then be bypassed.

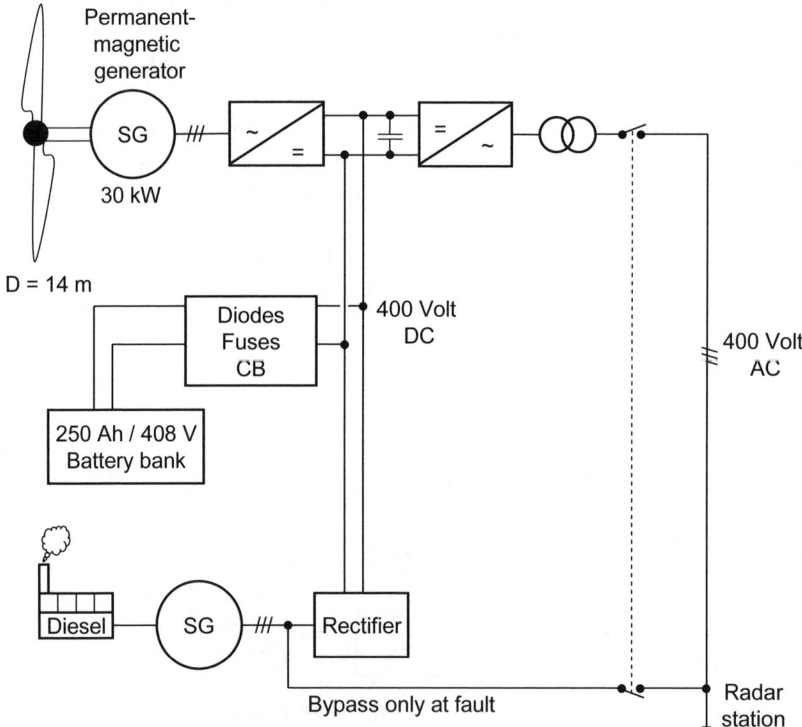

Fig. 13-26 Wind-diesel system for the island Osmussaare (radar station) [10]

13.3.3 *Wind-diesel-photovoltaic system (minimal grid)*

A control system offered by the company SMA is able to coordinate the power generation from various renewable energy sources. All these sources feed into a common three-phase line, as shown in Fig. 13-27.

13.3.4 *Final remark*

The variety of available hybrid systems with ratings between several kW and 1000 kW is at present quite small. However, their number will be expected to grow as the price of oil goes up and the hunger for electric power intensifies in countries that have many small towns without connection to a power grid. Hybrid systems will eventually reach a level of maturity comparable to that of today's grid-connected wind power plants.

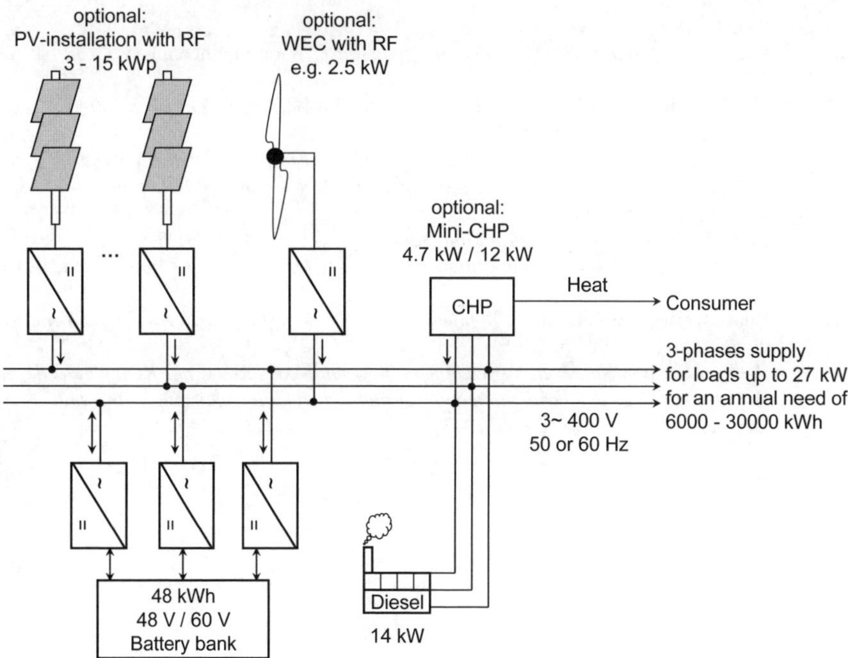

Fig. 13-27 Hybrid system with power generation from several energy sources in a small grid [7]

References

[1]   Dietrich, W.: *Drehzahlvariables Generatorsystem für Windkraftanlagen mittlerer Leistung (Variable-speed generator system for wind turbines of medium capacity)*, PhD thesis. Technical University of Berlin, 1990

[2]   Schörner, J. et al.: *Stand und Entwicklungsrichtung des Antriebsstranges von Windkraftanlagen (State of art and development tendencies of wind turbine drive trains)*, Windkraftjournal, Vol. 6/2001, S. 38-48

[3]   Alstom: *Technische Information (Technical information)*

[4]   Quaschning, V.: *Regenerative Energiesysteme(Renewable energy systems)*, 4th edition, Hanser-Verlag, München, 2006

[5]   Heier, S.: *Windkraftanlagen (Wind energy converters)*, 3rd edition, B.G. Teubner GmbH, Stuttgart/ Leipzig/ Wiesbaden, 2003

[6]   ENERCON: *Technische Information Inselsystem (Technical Information Stand-Alone System)*, 2001

[7]   SMA: *Technische Information Sunny Island System Kit (Technical information on the Sunny Island System Kit)*, 2003

[8]   Cruz-Cruz, I.: Vortrag: *Seminario Internacional de Energias Sustenables*, Puerto Natales, Chile, Okt. 2003

[9]   Akkmatov, V., Nielsen, A.H. et al.: *Variable-speed wind turbines with multi-pole synchronous permanent magnet generators*, Wind Engineering, Vol. 27, No. 6, 2003, S. 531-548

[10]   Ruin, S.: *New Wind-Diesel System on Osmussaare,* Wind Engineering, Vol. 27, No. 1, 2003, S. 53-58

[11]   Bundesverband Windenergie: *Windenergie 2002 - Marktübersicht BWE (Wind energy 2002 – Market survey by BWE), p.* 122-123

[12]   Jargstorf, B.: *"High Penetration" Windprojekte - Eine Alternative zu Wind/ Diesel-Sytemen (" High penetration" wind energy projects – an alternative to wind/diesel systems)*, DEWEK, 1996

[13]   ENERCON: *Technische Information Stand Alone System*, April 2003

[14]   ENERCON: *Technische Information Battery test for stand alone application*, April 2002

[15]   Bundesverband Windenergie: *Windenergie 2000 - Marktübersicht BWE (Wind energy 2000 – Market survey by BWE)*

[16]   Hacker, G.: *Wind ins Netz – Netzeinspeisung und Akkuladung mit Kleinwindrädern, (Wind into the grid – Grid-connection of and accumulator charging with small wind turbines)* ISBN 3-00-011545-5, 2003

14 Wind turbine operation at the interconnected grid

On the one hand, the grid-connected operation of wind turbines places demands on the wind turbines' operational behaviour and its technical equipment for grid connection. On the other hand, the operation of the interconnected grid is increasingly influenced by wind power production. Some years ago, the effects were negligible because the installed wind power was small compared to the grid capacity. But nowadays, at least in Germany, Denmark and Spain, the grid integration of wind energy has become a technical and economical challenge.

The following chapter first describes these challenges from the viewpoint of the grid. The interface between grid and wind turbine is then highlighted. Finally, the different wind turbine concepts, described in the previous chapters, are analyzed with respect to their grid compatibility. This chapter reflects the basics on the German interconnected grid in 2007, the year of publication of the 5th German edition of this book. The German electricity market is currently being transformed, among other because of regulations introduced by the EU; therefore, the keen interested reader should search additional sources for the up-to-date information on the electrical grid.

14.1 The interconnected electrical grid

14.1.1 Structure of the interconnected electrical grid

The *interconnected grid of Germany* is part of the synchronous European interconnected extra high voltage grid, Fig. 14-1, of the Union for the Coordination of Transmission of Electricity (UCTE). Therefore, it has to comply with the European UCTE guidelines. The structure and the technical standards of this interconnected grid are mainly based on requirements drawn, decades ago, from the structure of a *distribution grid* with only a few large-scale power generation plants supplying the distributed consumers.

Nowadays, decentralized wind power often feeds into this grid at its endings in weak and remote rural sub-grid sections, primarily in the coastal areas. Together with the liberalised international electricity market, this implies the need to transform the interconnected grid from a distribution grid to a transportation grid for large-scale transfers of electrical energy (not only at the highest voltage level).

As long as wind turbines provided only a relatively small share of the power generation capacity, they had very little impact on the grid operation. However, nowadays, wind power installation in Germany reached in 2006 a capacity of 20 GW (25 GW in 2009) which is a quite large proportion and therefore has a significant impact on grid operation. In Germany, the total capacity of conventional power generation was around 110 GW in 2006. The maximum grid load was approx. 75 GW, the minimum approx. 35 GW. Fig. 14-2 shows that during off-peak periods wind turbines were able to deliver nearly 30% of the total electricity generation already in 2003. From time to time in several regions of Germany this wind power share is even higher, e.g. in some grid sections in Northern Germany. On very windy days, the produced wind power exceeds the minimum off-peak load.

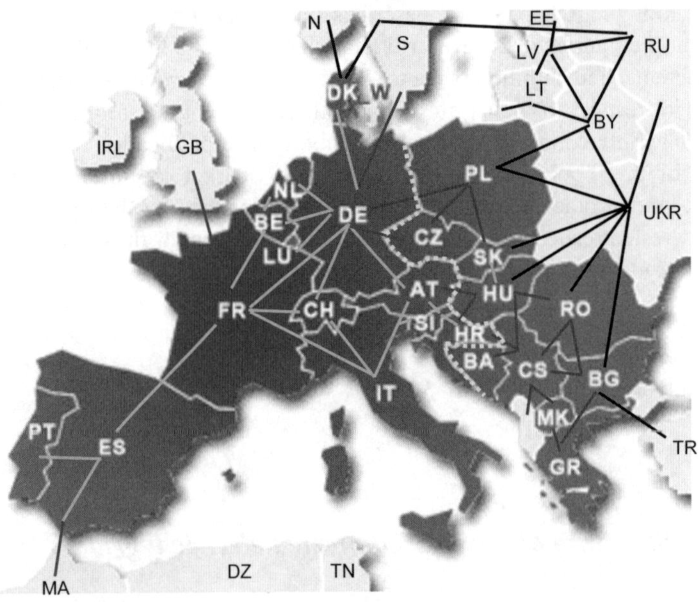

Fig. 14-1 European transmission grid UCTE (grey lines) with connections to the neighbouring countries (black lines), dark grey: UCTE members, acc. [1, 2]

Future wind power installation is planned in Germany preferably offshore, therefore new grid capacities have to be built. However, it is imperative that the existing grid also be revised and reinforced swiftly because the changing structures in the European electricity market require an *international transportation system*. In Northern Europe, the grid tends to reach its peak load generally in winter. In the Southern Europe this occurs partly in summer because of the increasing application of air condition [2].

Fig. 14-3 shows the *voltage levels of the interconnected grid* in Germany and the wind farm sizes suitable for connection. In general, the electricity from the transmission grid (380 kV and 220 kV) is delivered to the consumer via distribution grids. The larger the wind farm the more economic the connection to a higher voltage level. Grid operation is generally divided by voltage level and in Germany also by region. This does not apply to all European countries. In Germany, there are four large transmission grid operators and approx. 700 regional and local distribution grid operators (state of 2007).

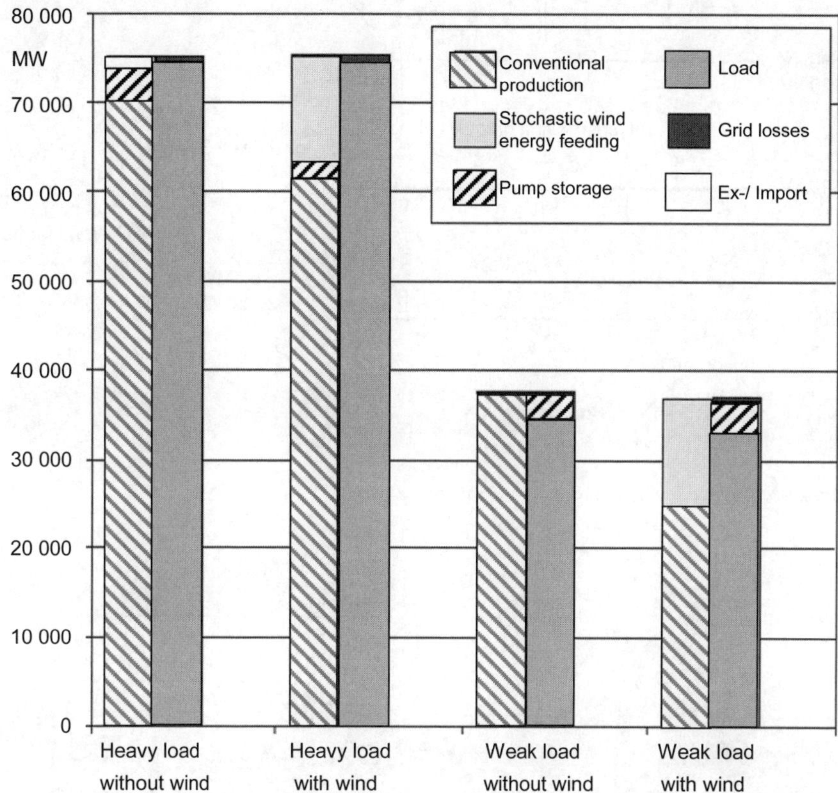

Fig. 14-2 Electricity generation (left column) and consumption (right column) in 2003, without and with wind power, acc. [3]

The planned wind farm capacity determines the required voltage level for the grid connection and has to be stated in the application to the corresponding grid operator, cf. chapter 15.

Fig. 14-4, left, shows the responsibility of the four transmission grid operators and their high voltage grids. In the control areas of E.ON and Vattenfall the ratio of wind power production to end user consumption is higher than e.g. in the region of EnBW. Therefore, wind power is often exported to this region. Due to the large-scale wind turbine installations in the Northern regions of Germany (Fig. 14-4 right, white areas) wind power production there is particularly high in comparison with local consumption.

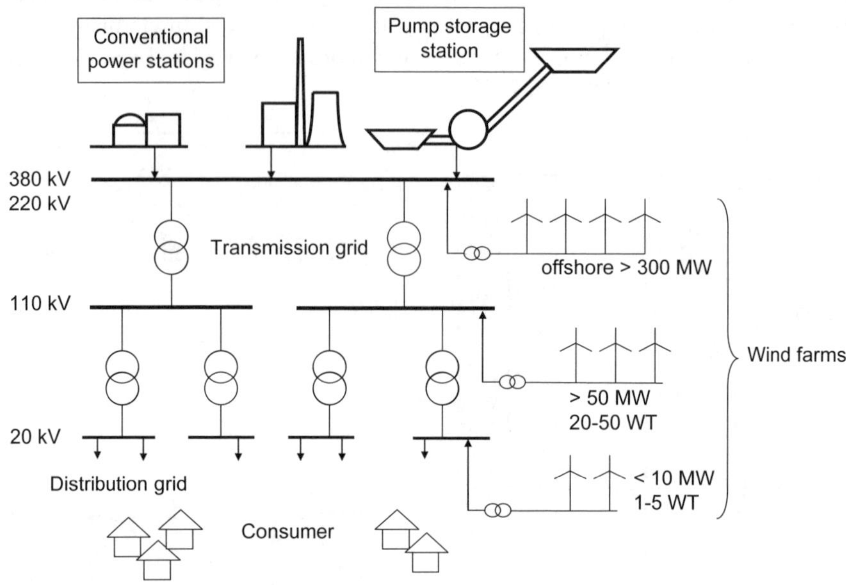

Fig. 14-3 Voltage levels of the grid in Germany

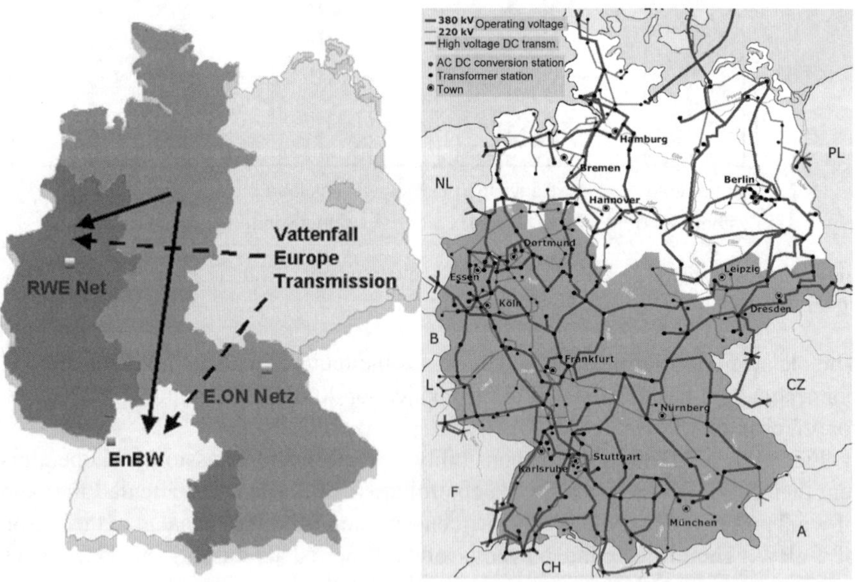

Fig. 14-4 Control areas of the four transmission grids and the high voltage grid in Germany [2, 4 left]

14.1.2 Operation of the interconnected grid

The operation of the electrical grid in the individual control areas from their control centres aims to *permanently adapt the generated power to the variable grid load*, i.e. the consumption. In Germany, the control areas correspond to the grid areas of the four transmission grid operators. At the distribution grid level, the schedules are produced based on contracts and experience values. These are then reported to the transmission grid operator concerned who makes his prediction for the next day based on the data received (i.e. day-ahead prediction).

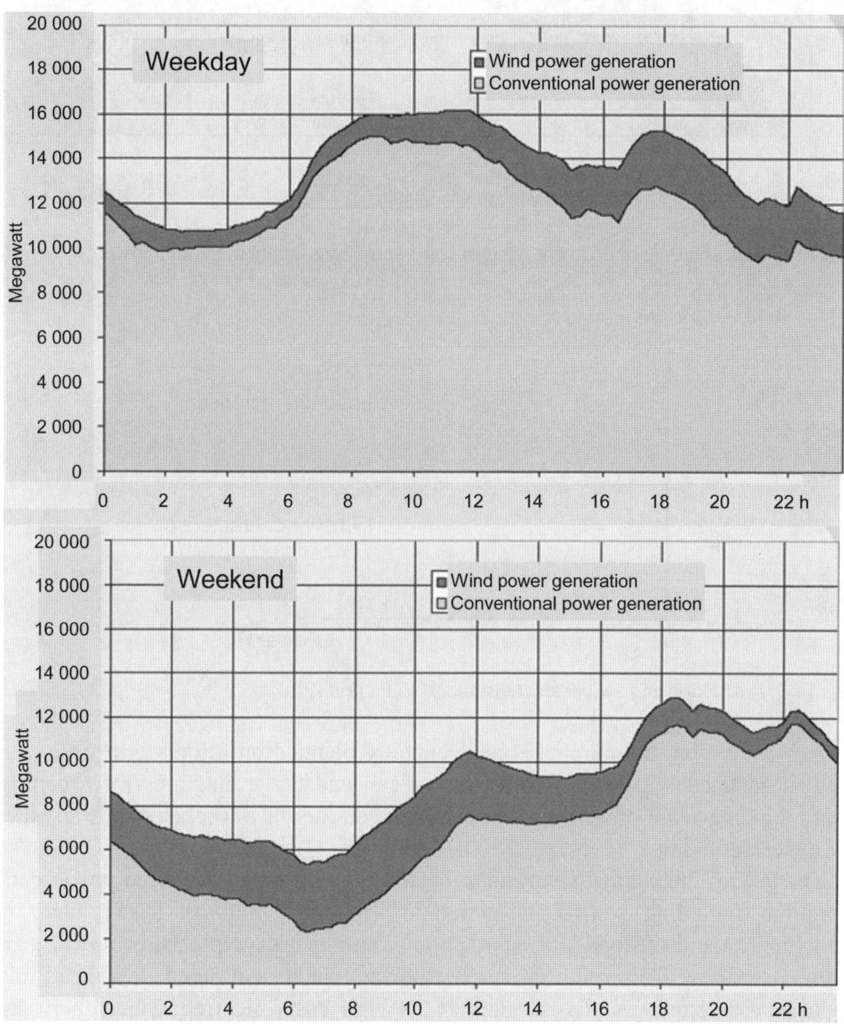

Fig. 14-5 Typical daily load profile, E.O.N grid, Germany [5]

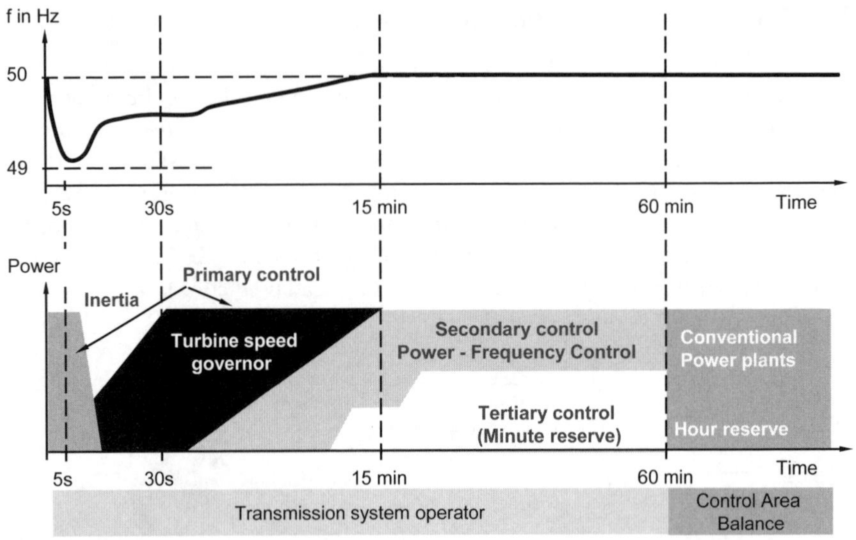

Fig. 14-6 Reaction time of different types of grid control and supply of control power [6]

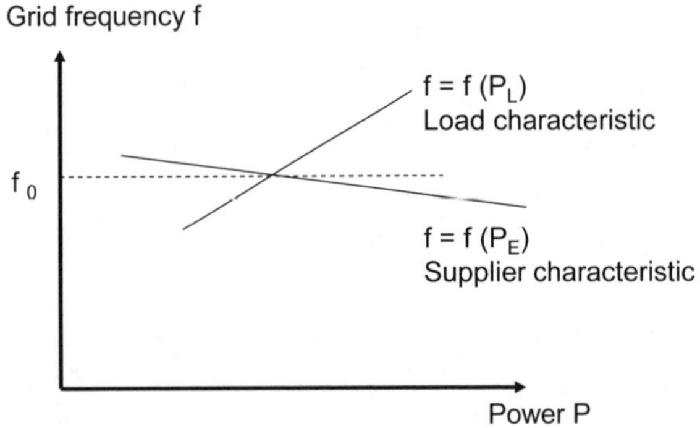

Fig. 14-7 Production and load characteristic curve of a grid [7]

Power generation and purchases are generally planned on a long-term basis and well in advance. Fig. 14-5 shows two typical load curve diagrams for a control area including wind power generation provided, one for a weekday and the other for a week-end day.

The task of the *control centre* is to balance any difference between power consumption (including export) and generation (including import). This is done by switching on and off loads, like pumps in pumping storage plants, or power generation capacity. This requires reserve power which is ordered day-ahead and should be as small as possible due to its high costs incurred. Since there are always differences between predicted and real load, control power is always required.

The required *control power* is divided by its reaction time into primary, secondary and tertiary control, Fig. 14-6. Primary control is immediately provided by the inertia of the rotating mass in the power generation systems: generators and turbines react with small changes in the rotational speed (i.e. conversion of kinetic energy) to counteract fluctuations of frequency and voltage due to load variation, cf. Fig. 14-7.

Secondary and tertiary control within the range of minutes, cf. Fig. 14-6, is provided by fast-reacting power plants such as gas turbines or water turbines (pumping storage plants). Nowadays, the share of the control power demand which may be estimated in advance is invited for tender in a bidding forum on the Internet.

The control areas contain so-called balance groups, consisting partially of several electricity agents, for which the responsible of the balance group has to balance also the 15-min average of generation and demand.

Nowadays, on the one hand, the attainable accuracy of the *prediction of the required reserve power* by the grid operator is reduced through the liberation of the electricity trade which causes unpredictably high electricity transmission capacities in the grids and, on the other hand, the fluctuation of wind power generation. Short-term fluctuations of wind energy are not a major technical issue for primary control since many effects in the time range of seconds and minutes are balanced by the decentralised large-scale installation of the wind turbines. Seasonal variations such as the curve of the full-load equivalence (capacity factor) have been well documented, Fig. 14-8.

If, however, in a certain region there is a large gradient in the wind power production due to rare weather conditions, Fig. 14-9, this may cause stability problems in the grid. Such rare occurrences may be a weather front, or a storm causing many wind turbines to shut down. Therefore, grid operators now work with established prediction methods in order to more accurately predict the curves of the wind power generation [8]. Fig. 14-10 shows the comparison between the predicted and the actual measured wind power generation. The difference between the day-ahead forecast (D+1) and the real wind power is around ± 10%. This uncertainty could be significantly reduced by short-term corrections, but these have not yet been applied because of the set day-ahead forecast (i.e. period of 24 hours) established in grid operation.

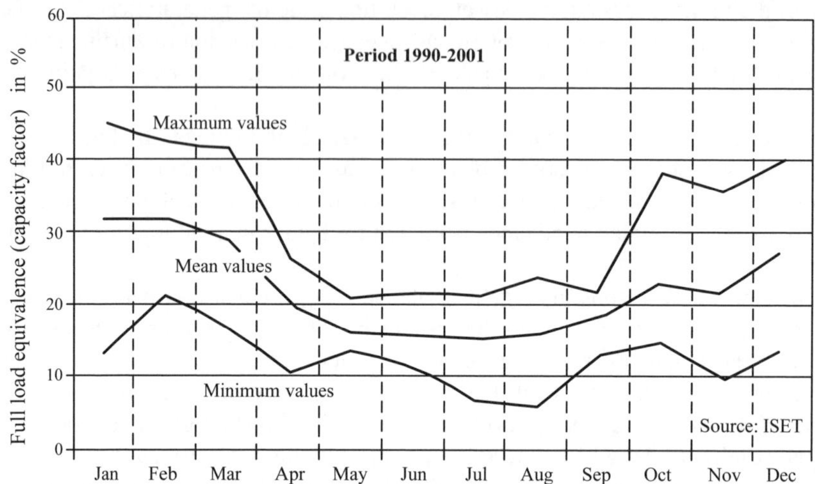

Fig. 14-8 Seasonal variation of wind power production [9]

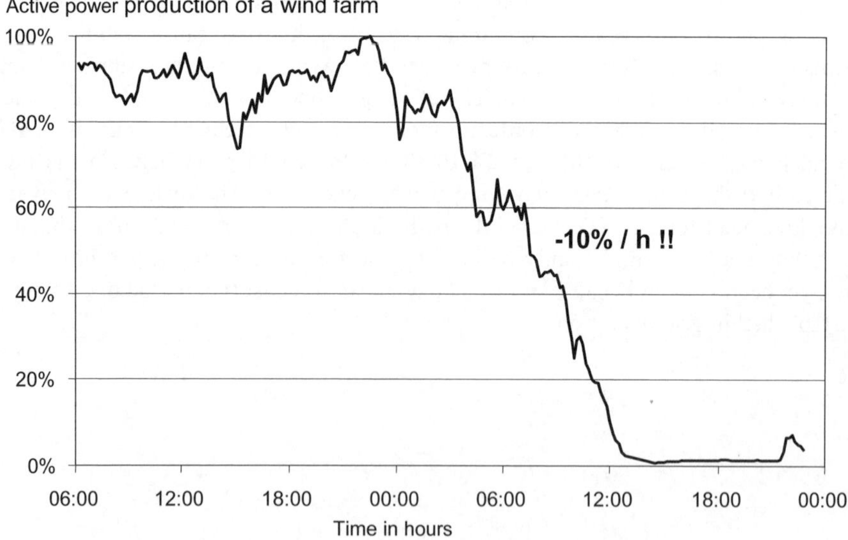

Fig. 14-9 Example of a critical drop in active power production of a wind farm [10]

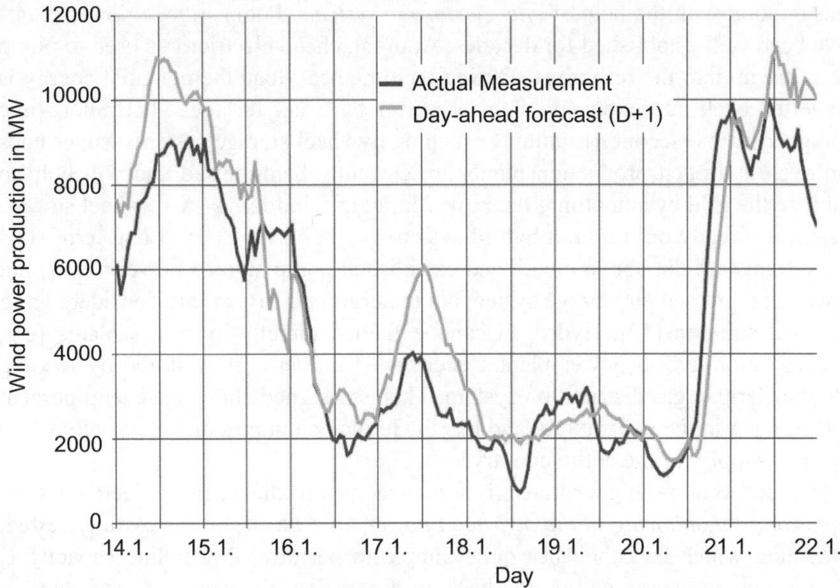

Fig. 14-10 Day-ahead (24 hours) forecast of wind power and real wind power production [8]

Another measure of more harmonic grid operation with high shares of wind energy input is the *power production management for wind energy* as required by the grid codes published in 2003 [11]. Requirements of wind turbine operation are derived from specific load situations in the grid. For example, if the grid load is too small, the wind turbine should be able to provide negative control power by reducing its power production.

If a relative high wind power production is predicted due to good winds, it is generally possible to operate the wind turbine with a reduced power output in order to be able to provide positive control power by a wind turbine cluster [8].

This is tested but is currently not being implemented (2007) since the fluctuations of the wind do not allow giving the required guarantee for the power production. Since July 2009, the German Renewable Energy Sources Act includes the Ordinance on System Services by Wind Energy Plants (System Service Ordinance – SDLWindV); therefore more and more wind turbine types are now being certified for actively support the grid.

Demand side management, i.e. control of the load, offers completely new possibilities for grid control. In order to avoid load peaks in the grid, large consumers, such as cold storage houses, whose operation is not time-critical could be temporarily switched off to provide positive control power. Smart metering is another method currently under development and test to provide decentralised intelligent devices for consumer load control. This requires new standardised communication to transfer and manage all the information.

A third method of optimising grid operation is balancing power generation and consumption with the help of *energy storage systems*. Pump storage power plants have been well-established for decades. At night, cheap electricity is used to pump the water up into the reservoirs. During daytime peak-load the potential energy is converted back into expensive control power by water turbines [12]. Short-time storage systems (seconds to minute) such as flywheel storage systems, super cups and even hydrogen production plants are currently being tested and will help to stabilize the grid by smoothing the power input of wind farms. A flywheel storage system is already offered in a hybrid system [13], cf. Fig. 13-25. Long-term storage systems are the above mentioned established pump storage power plants, but as well compressed air storage systems in underground caverns are considered as a technical solution [12]. Hydrogen can be fuelled directly in biogas plants (e.g. demonstration hybrid power plant Uckermark by Enertrag) or stored by feeding into the German gas distribution system. This is designed allowing several percent of Hydrogen in the gas mixture and has in Germany a huge storage capacity of 90 days gas supply for the entire country.

In order to allow higher transmission capacities in the existing electrical grid, *temperature monitoring of the overhead power lines* has been successfully tested, something which is technically quite simple to perform. Power line capacity is currently being determined on the basis of a worst-case scenario which occurs at special meteorological conditions without any wind. Wind farms would benefit from power line temperature monitoring because when the power production is high due to strong winds, the cooling of power lines by the wind is as well strong.

14.2 Wind turbines in the interconnected electrical grid

14.2.1 Technical requirements of the grid connection

Fig. 14-11 shows the electrical equivalent circuit diagram of a typical *wind turbine grid connection to the medium-voltage grid*. A distinction is drawn between the wind turbine connection point at the feed line with the transformer, which increases the voltage level, and the grid connection point. The latter is critical for determining the potential *grid connection capacity*, which is given by the short-circuit power at the grid connection point and the relevant properties of the feed line with the transformer given in Fig. 14-11 as well. The maximum admissible feed-in power P_{max} at the grid connection point in the German medium-voltage grid (20 kV) is 2% of the grid's short-circuit power S_k which is derived from the nominal voltage U_N and the short-circuit impedance Z_k between the source and the considered grid connection point:

$$P_{max} = 0.02 \cdot S_k$$
$$\text{where} \quad S_k = U_N^2 / Z_k$$
$$\text{and} \quad Z_k = (R^2 + X^2)^{0.5}$$

Z_k is given by the ohmic resistance R and the reactance $X = \omega L$ calculated from frequency $\omega = 2\pi f$ and inductance L. For larger wind farms which are connected to the high-voltage grid (110 or 380 kV) by individual sub stations, a maximum admissible power of 20% of the short-circuit power, or even more, may be allowed.

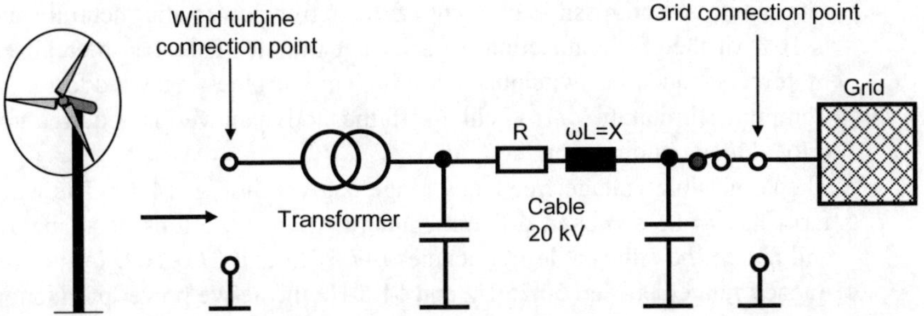

Fig. 14-11 Equivalent circuit diagram of the connection of a wind turbine to the medium-voltage power grid

During wind turbine operation, i.e. wind power input at the grid connection point, the short-circuit power changes and is influenced by the wind turbine, the transformer and the feed line properties. On the one hand, a larger section of the feed line and larger transformers allow higher admissible wind power input, but produce more installation costs, on the other hand. A more detailed calculation allows grid connection facilities to be optimised [7, 14].

The technical equipment of the grid connection comprises:
- Transformer for adaption of the voltage level
- Circuit isolator and switchgear
- Measuring and metering facilities
- Filter

Grid connection regulations (grid codes) which no longer allow a wind turbine is just shut down if grid faults occur have been in place since 2003. On the contrary, demands are now being made to support the grid at grid faults [11, 15, 16]. In this respect, only the behaviour of the entire wind farm at the grid connection point is considered, not the single wind turbine. Therefore, the requirements may be satisfied in different ways, and these grid codes mention following points:
- Switching-on procedure of the wind farm
- Reactive power production and consumption
- Active power production
- Behaviour at grid faults

For the *switching-on procedure* of the wind farm it is required that the switching-on current of the wind farm must not exceed 120% of the current corresponding to the grid connection capacity. Therefore, it may be required to connect the wind turbines to the grid one after the other.

With regard to *reactive power,* it is required that, by means of a set value signal sent by the grid operator, the power factor of the wind turbines at active power production can be adjusted between 0.975 (inductive) and 0.975 (capacitive) according to the situation. In this context the internal lines of the wind farm and the transformers must also be taken into consideration.

There are three requirements for *active power production*:

- The maximum admissible gradient of the active power after neutral state is 10% of the grid connection capacity per minute. This may be realised by forced sequential switching-on of the wind turbines in a wind farm.
- Temporal limitation or switching-off the active power is required for critical load conditions in the grid.
- The admissible voltage-frequency range is given in Fig. 14-12. The wind farm has to be switched off immediately when deviations occur below and above the admissible frequencies (47.5 Hz and 51.5 Hz). In the frequency range between 50.25 Hz and 51.5 Hz the active power production of the wind farm has to be limited at rising frequency.

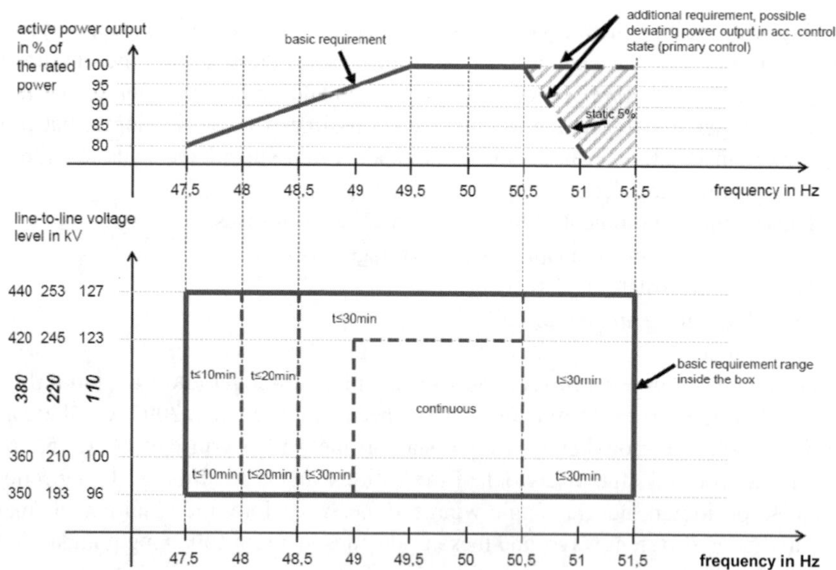

Fig. 14-12 Voltage-frequency-active power requirement (quasi-stationary, i.e. frequency gradient < 0.5% / min, voltage gradient 5% / min), from E.ON grid code April 2006 [11]

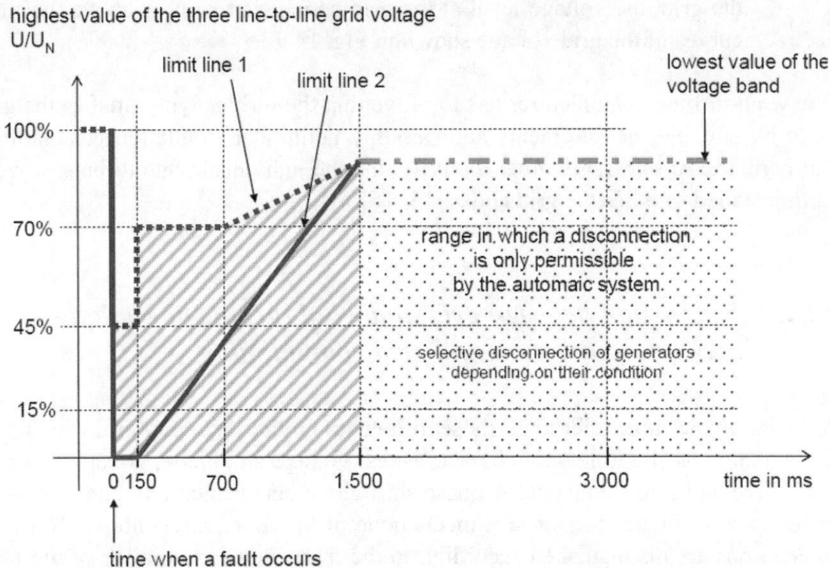

Fig. 14-13 Required behaviour of a wind turbine during a voltage drop due to grid fault, from [11]

As long as the wind power input into the grid was negligible, the turbines were simply shut down in case of *a grid fault due to short-circuit*. Today's rules are far more complicated since the proportion of electricity fed by the wind turbines may be significant, cf. Fig. 14-5. If the wind turbines are connected to the grid by converters they are still able to feed into the grid during a *grid short-circuit,* thereby stabilising the grid. By contrast, for stall-regulated wind turbines with a direct grid connection of the asynchronous generator, the excitation of the generator collapses during a short-circuit and, therefore, they are no longer able to provide power during a grid short-circuit.

Short-time short-circuits occur quite frequently in the grid at overhead lines. Snow, a wet branch or even a big bird, for example, may temporarily connect two of the lines and then fall down. Immediately, the automatic reclosure briefly disconnects the grid line and verifies one or two seconds later if normal operation can be relaunched upon rectification of the disturbance. If this is the case the wind turbines must also provide power immediately. Thus, the following *rules in case of grid faults* have been established:

- During a grid short-circuit (grid voltage above 60 kV) the highest possible reactive current has to be provided for up to 3 seconds in the 3rd quadrant. This is only required during grid faults and for a voltage range between 15% and 60% of the operating voltage U_0 at the grid connection point.

- There must be no disconnection from the grid if, in the event of a fault in the grid, the voltage level at the grid connection point is above the limit curves of the grid voltage shown in Fig. 14-13.

The wind turbine manufacturer has to prove that the turbine type satisfies the grid code by suitable measurements and possibly additional simulation calculations. Furthermore, rules exist in order to verify that the individual wind turbine or wind farm installed satisfies the grid codes [15, 16].

14.2.2 Interaction between grid and wind turbine operation - network interaction and grid compatibility

Network interactions refer to all those influences on the electricity grid which lead to deviations in the following characteristics: voltage amplitude, voltage and current curve and grid frequency. A phase shift alone between current and voltage is not a network interaction since it meets none of the four criteria above. Network interactions are distinguished according to the frequency and duration of the fault and the resulting effects in current and voltage, table 14.1.

Table 14.1 Network interactions, frequency range and causes [18]

Parameter	Peri-odic	Non-periodic	Frequency range in Hz	Cause
Harmonics	x		> 50	Non-linear consumers,
Intermediate harmonics	x		0 ... > 50	Switching
Sub harmonics	x	x	< 50	process
Voltage fluctuations	x	x	< 0.01	Power fluctua-
Flicker	x	x	0.005 ... 35	tions
Transient		x	> 50	Switching process
Voltage drop		x	< 50	Grid fault
Voltage collapse		x	< 50	
Asymmetric voltage	-	-	-	Asymmetric load

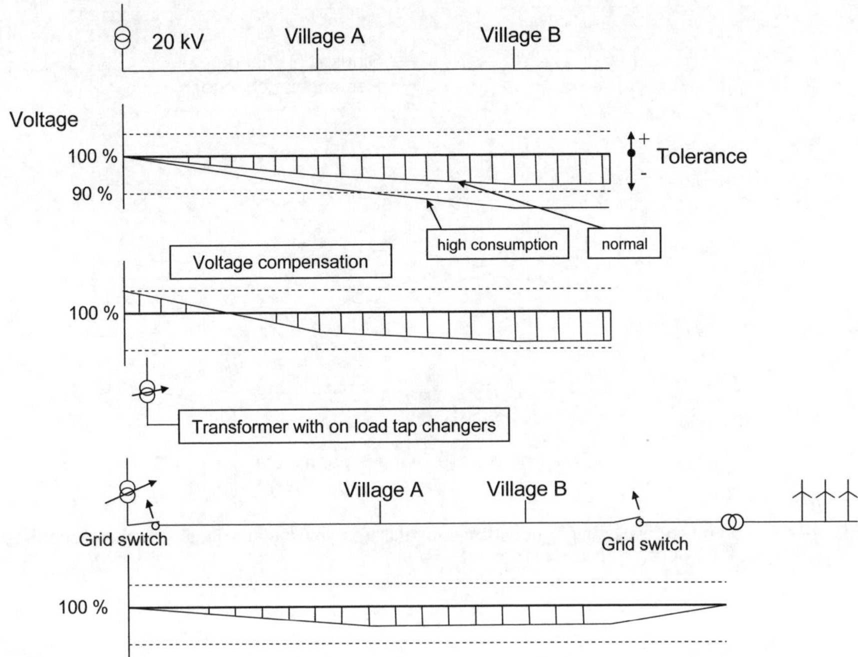

Fig. 14-14 Voltage compensation, top: by changing the transformer voltage (no wind turbines), bottom: by wind energy fed into the grid

Fig. 14-14, top, shows the voltage along a local overhead line at normal and at very high load. At high loads *voltage compensation* may be necessary using a transformer with on-load tap changer to increase the feeding voltage so that it is kept within the tolerated range between +5% and -10%, Fig. 14-14, middle. However, if there is additional power input by wind turbines at the line end, these measures may not be required any more, Fig. 14-14, below.

Another issue involved in grid-connected wind turbine operation are so-called *flicker*. Low-frequent voltage fluctuations may cause variations in the light density of light bulbs. Flicker frequencies between 8 and 9 Hz are quite disturbing for the human eye, Fig. 14-15. In course of the grid compatibility (i.e. power quality) measurements at a wind turbine, the flicker coefficient c is determined and has to be mentioned together with the other electrical characteristics in the wind turbine's specifications, see comparison of various wind turbines, e.g. in [19]. For most stall-regulated wind turbines, the flicker coefficient is higher than for pitch-controlled wind turbines. Moreover, the variation of the flicker coefficient with the phase angle has to be documented in the test reports.

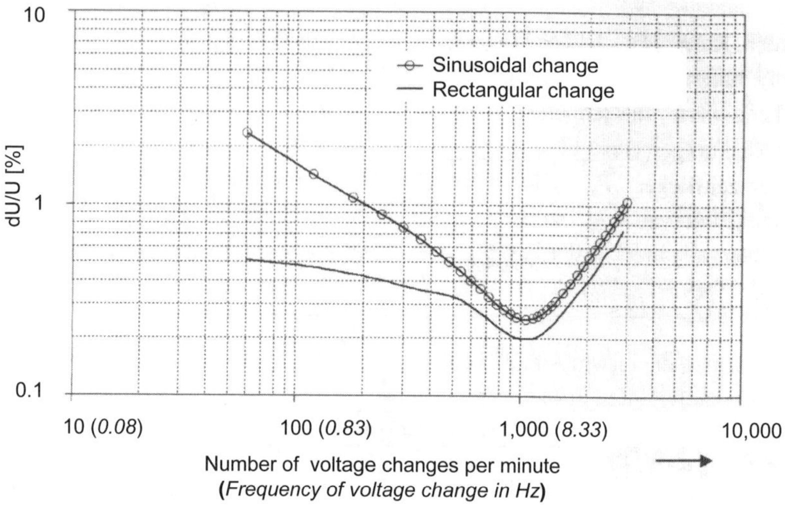

Fig. 14-15 Tolerance level for periodical rectangular and sinusoidal voltage changes according to EN 61000-3 3 and IEC 1000-3-3 [18]

14.2.3 Characteristics of wind turbine concepts for grid-connected operation

The wind turbine concepts presented in chapter 13 differ in terms of their grid compatibility. As already mentioned, one aspect of the effects of the wind turbine operation on the grid is its *demand for reactive power or the ability to provide it.* Basically, stall-regulated wind turbines with a directly grid-connected asynchronous generator require reactive power which has to be delivered by the grid. This demand may be reduced, but not eliminated, by capacitor banks for reactive power compensation (cf. chapter 13). For newly grid-connected wind turbines grid operators require that the power factor of the wind turbine (i.e. the ratio of reactive and active power) be adjusted depending on the load situation in the power line branch, Fig. 14-16. Modern wind turbines with converters fulfil this requirement. For older wind turbines with direct grid connection of the asynchronous generator, retrofit packages with power controllers exist for this purpose. Table 14.2 summarizes the key advantages and disadvantages of the different wind turbine concepts. For information on harmonic waves the interested reader should refer to the literature, e.g. in [7, 14, 17, 18, 20].

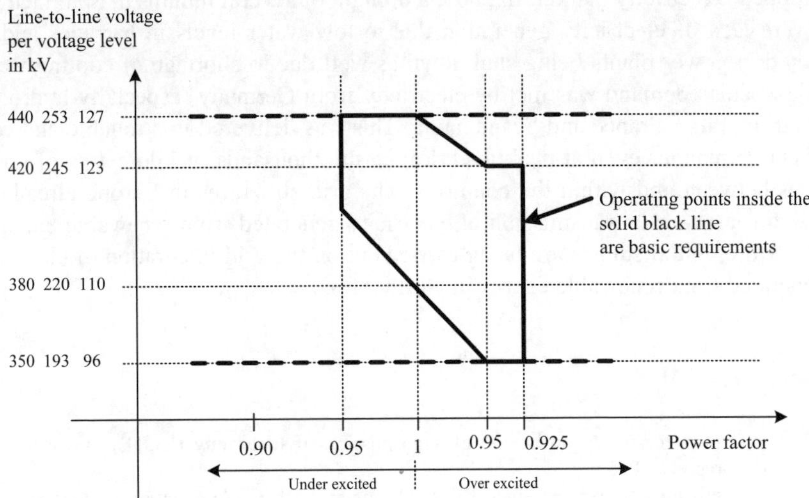

Fig. 14-16 Required range of the power factor during the turbine operation without limiting the effective power output [16]

Table 14.2 Assessment of wind turbine concepts concerning their grid compatibility [17, 19]

Generator type	Asynchronous	Asynchronous, Variable slip	Asynchronous, doubly-fed	Synchronous
Rotational speed	Rigid (nearly constant)	Soft	variable	variable
Grid connection	direct	direct	Via converter	Via converter
Reactive power	consumption	Consumption	adjustable	adjustable
Power at grid connection point $x*P_N$	Approx. 1.30 to1.50	Approx. 1.05 to 1.10	Approx. 1.05 to 1.10	Approx. 1.00 to 1.05
Flicker	High	Medium	Medium	Low
Harmonics	Not relevant	Existent	Partially existent	Not relevant

In conclusion, it may be emphasised that modern wind turbines fulfil the grid codes required by grid operators. Moreover, despite the unsteady wind being their energy source their electricity generation may nowadays be quite accurately predicted. This is a very important issue particularly in relation to the liberation of the

European electricity market. In 2005, a drought of several months in Spain led to a severe lack of electricity generation due to low water levels in barrages and resulted in power plants being shut down as well due to shortage of cooling water. The Spanish demand was met by electricity from Germany (especially hydro due to strong rain), France and Scandinavia. This was delivered in a reliable and cost-effective manner even at peak load despite the thousands of kilometres of power lines between and within the countries. The grid structures in Europe already allow for large-scale transmission of electricity generated from renewable energies. Therefore, initial studies are being carried out on the grid integration of electricity generated from renewable energy in North Africa.

References

[1]   Union for the Coordination of Transmission of Electricity (UCTE), www.ucte.org, Brussels, 2005

[2]   Verband der Netzbetreiber VDN e.V. beim VDEW (Association of German Grid Operators, www.vdn-berlin.de, Berlin, 2005

[3]   Deutsche Energieagentur (dena): *Energiewirtschaftliche Planung für die Netzintegration von Windenergie in Deutschland an Land und Offshore bis zum Jahr 2020 (On planning of the energy management of wind energy grid integration in Germany on- and offshore until the year 2020)*, Berlin 2005

[4]   Rohrig, K., et al.: *Online-Monitoring and prediction of wind power in German transmission system operator centres*, Proc. of EWEC 2003, Madrid 2003

[5]   Ensslin, C.: *Large-Scale Integration of Wind Power*, in "Grid-connected Wind Turbines", Post-Graduate Training Programme 2003, InWEnt gGmbH, Deutsche WindGuard Dynamics GmbH, ISET Kassel e.V.

[6]   Beck, H.-P., Clemens, M.: *Konditionierung elektrischer Energie in dezentralen Netzabschnitten (Conditioning electrical energy in decentral grid sections)*, etz, 5/2004

[7]   Heier, S.: *Windkraftanlagen – Systemauslegung, Integration, Regelung (Wind Energy Converters – System design, integration and control)*, Teubner Verlag, 3rd ed., 2003

[8]   Schlögl, F.: *Online-Erfassung und Prognose der Windenergie im praktischen Einsatz, (Practical applications of online acquisition and prediction of wind energy)* I/FS, Köln, Juli 2005, e.g. on www.iset.uni-kassel.de/prognose

[9]   *Windenergie Report Deutschland 2002, Jahresauswertung des "Wissenschaftlichen Mess- und Evaluierungsprogramms" (WMEP) zum Breitentest "250 MW Wind" (German Wind Energy Report 2002, Annual evaluation of the „scientific measuring and evaluation programme" of the large-scale testing programme „250 MW Wind")*, Institute for Solar Energy Supply Technology ISET e.V., Kassel 2002

[10]   Burges K., Twele J.: (Ecofys GmbH): *Wind farms participating in TSO's Power System Management,* Deutsche Windenergie Konferenz (DEWEK), Wilhelmshaven, 2004

[11]   Santjer, F., Klosse, R.: *Die neuen ergänzenden Netzanschlussregeln von E.ON Netz GmbH (The new additional grid codes by E.ON grid)*, DEWI-Magazine No. 22, February 2003

[12]   Franken, M.: *Sächsisches Wunderwerk (Miracle of Saxony)*, Janzing, B.: *Schnell aufladen, schnell einspeisen (Fast charging and feeding electricity)*, Lönker, O.: *Zukunftsspeicher (Future storage systems)*, in: Neue Energie (new energy) 04/2005

[13]   ENERCON AG: Brochure of the wind turbine E33, 2005

[14]   Ackermann, T. (Ed.): *Wind Power in Power Systems*, John Wiley & Sons, Ltd., Stockholm 2005

[15]   Verband der Netzbetreiber VDN e.V. beim VDEW: *EEG-Erzeugungsanlagen am Hoch- und Höchstspannungsnetz (Electricity generating plants powered by renewable energies and connected to the high and extra-high voltage grid)*, August 2004, www.vdn-berlin.de, Berlin

[16]   E. ON Netz GmbH: *Netzanschlussregeln (Grid Codes)*, Stand August 2003, Bayreuth

[17]   Schulz, D.: *Netzrückwirkungen – Theorie, Simulation, Messung und Bewertung (Network interconnections – theory, simulation, measurement and assessment)*, VDE series 115, 1st ed., VDE Verlag, Berlin 2004

[18]   Schulz, D: *Untersuchung der Netzrückwirkungen durch netzgekoppelte Photo-voltaik- und Windkraftanlagen (Investigations on network interactions of grid-connected photovoltaic and wind power plants)*, Ph.D thesis at the Technical University of Berlin, 2002, publisher: VDE-Verlag GmbH 2002

[19]   Bundesverband WindEnergie e.V., BWE Service GmbH (Ed.): *Windenergie 2005, Marktübersicht (Wind Energy 2005 – A market survey)*

[20]   Quaschning, V.: *Regenerative Energiesysteme (Renewable Energy Systems)*, publisher: Carl Hanser Verlag, 2003

15 Planning, operation and economics of wind farm projects

This chapter is structured according to the chronology of the individual working phases of a wind farm project: preliminary and project planning, erection and operation. Each phase may be differentiated according to the following aspects:

- Technical aspects
- Aspects of the permission regulation and
- Economic aspects.

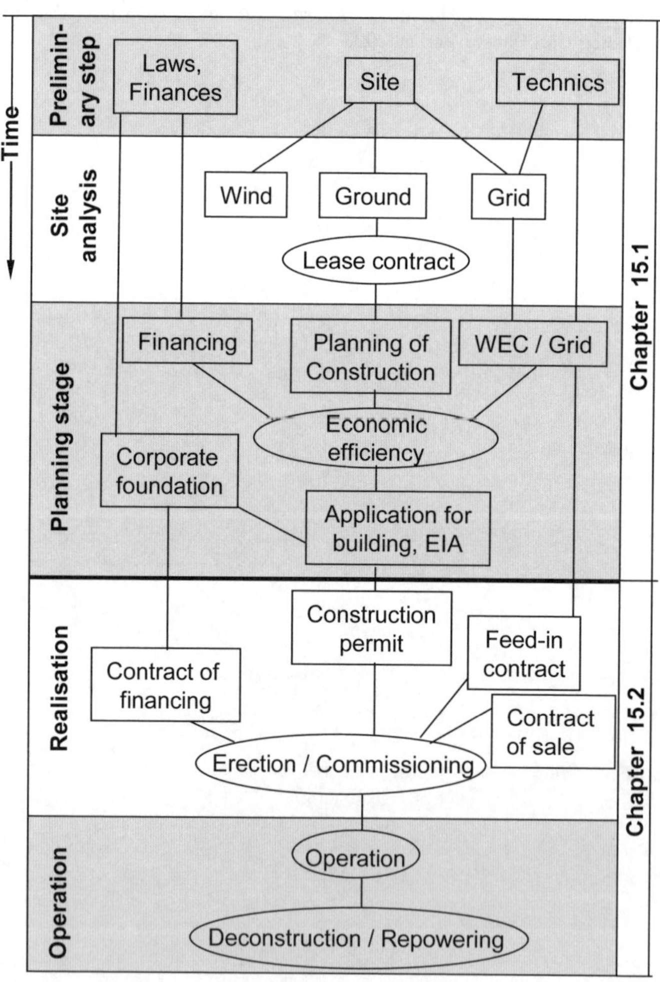

Fig. 15-1 Flow chart of a wind farm project

15.1 Wind farm project planning

Fig. 15-1 shows schematically the *chronology of the planning procedure* until project realisation and operation. In the preliminary investigation the project feasibility at the chosen site is analysed before formal and cost intensive steps are taken.

15.1.1 Technical planning aspects

Estimation of the wind regime

In course of the site selection the estimation and assessment of the wind regime is of prime importance since, as already mentioned several times, if the mean wind speed is 10% less than estimated the energy yield may be reduced by 30% and more. The first step includes the assessment of the determined wind speed values obtained e.g. by general meteorological data, and moreover the verification of the orography at the site, the terrain structure, the surface roughness (cf. chapter 4) and as well the type and size of the terrain boundaries. Moreover, single obstacles like rows of trees or other wind turbines have to be registered exactly. Even at this stage, it is necessary to consult an experienced planning expert who should then determine the further procedures and methodologies to be employed for the detailed determination of the wind potential. Different procedures for the site assessment were presented in chapter 4. Depending on the local conditions but also on the quality of possibly available regional wind data, e.g. from measuring stations, the applicable method is chosen and decided to which extent own wind measurements are required. The decision criteria have already been mentioned in section 4.5.

First estimation of the installed capacity and expected energy yield

Apart from the available area of the site, the available grid connection is a decisive factor in determining the number and rated power of turbines to be installed. Therefore, it is reasonable to send an enquiry to the local grid operator at an early stage in order to check the possible grid connection capacity, the distance to the next potential feed-in point and the voltage level of the grid connection (cf. chapter 14).

It may be reasonable or necessary to build a separate sub-station for larger capacities (> 20 MW). These two boundary conditions (available area of the site and grid capacity) may be used to estimate the number and rated power of the wind turbines which then forms the basis for the initial prediction of the expected energy yield.

In wind farm planning the yield is predicted for each wind direction sector individually (cf. Fig. 4-22) using its individual wind frequency distribution function and the power curve of the chosen wind turbines. This is necessary in order to find the optimum placing of the wind turbines in the mico-siting of the wind farm which gives the maximum wind farm energy yield and also reduces the inevitable interactions of the wind turbines. A resulting final wind farm layout is shown in Fig. 15-2.

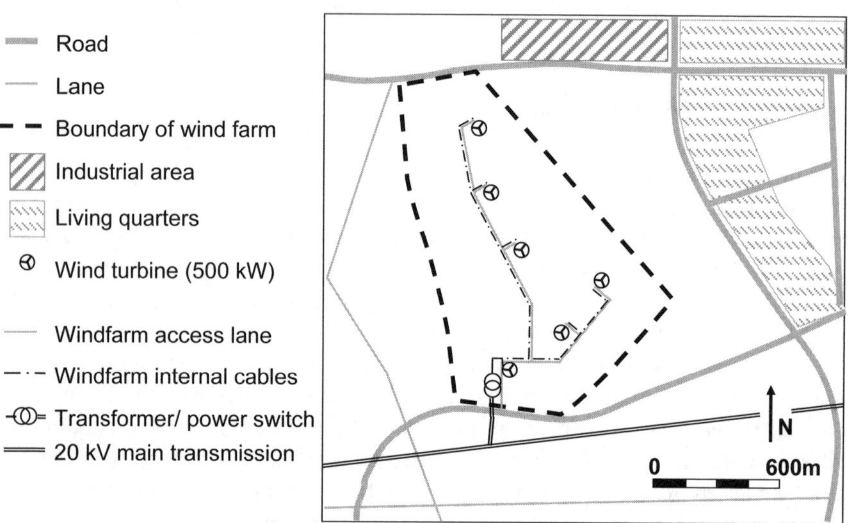

Fig. 15- 2 Wind farm layout and projected infrastructure

Wind farm micro-siting

The optimum arrangement of the wind turbines in the site area is obtained in the course of the micro-siting. The main criterion is the maximisation of the overall energy yield of the entire wind farm. However, the wind farm layout is as well influenced by the required infrastructure: the internal cabling from the wind turbines to the transformer and substations and, in addition, the access lanes for transport, erection, operation and maintenance costs (O&M) and service vehicles. Nowadays, established planning tools (e.g. WinPro, Windfarmer, a.s.o) exist, which help to determine the optimum wind farm layout fast end efficiently.

There might be further restrictions influencing the siting, e.g. minimum distances to living quarters and working areas, environmental protection, a.s.o. Further constraints, such as the allowed maximum total height, may be in place not only due to technical issues but as well due to laws and regulations stipulated by public authorities. It would be advisable to gather information on these issues in advance rather than having to change the planning later during the approval process due to imposed restraints.

Local conditions

Ownership of the relevant field parts has to be clarified and land lease contracts should be drawn up - or at least preliminary agreements.

Above all, investigation of the local conditions aims to ensure that the wind farm project is realised in practical terms. It is clarified whether a stable erection of the wind turbines is possible, the site can be accessed safely with all vehicles and machines required for transport, erection, service and O&M and whether for the grid connection a power line extension is required. These issues are discussed hereunder:

- Foundation:
  With regard to the wind turbine stability an individual geotechnical investigation report is required for each wind turbine to determine the load bearing capacity of the soil below the foundation. The type and shape of the foundation has to be chosen accordingly. Flat foundations are common, but if the soil is too soft, a pile foundation may be required.
- Accessibility and transportation to the site:
  Access to the site and the options for transport and wind turbine erection have to be checked and as well as the available suitable space for the crane. The transport of huge components such as tower sections and rotor blades to the building site may be limited by development, overhead clearances under bridges, overhead power lines, traffic signs, railways, antennas, waters, and many more.
- Investigation of the grid access:
  The location and type (e.g. voltage level) of grid connection have to be determined. The most simple case involves the connection of a short stub from the wind farm to the feed-in point. If required, the power line has to be extended directly to the next substation where either sufficient grid capacity exists or a separate cell has to be added. Large wind farms often have an individual substation which allows connection to a higher voltage level.
- Grid connection:
  The length and type of the cable route has to be checked having regard to technical and economical aspects. If the cable route is very long quite extensive planning with a separate environmental impact assessment is required, see below.

15.1.2 Legal aspects of the approval process

In Germany, the local community in whose area the site is located is responsible for the approval of the wind farm or wind turbine. Uniform German federal laws in conjunction with laws and regulations by individual federal states form the basis for the legal framework. An overview of the relevant federal laws and regulations of the federal states is presented in Fig. 15-3.

Federal laws

➢ Federal Building Code (Baugesetzbuch BauGB)
➢ Aviation Act (Luftverkehrsgesetz LuftVG)
➢ Road Traffic Regulation (Straßenrecht)
➢ Federal Nature Conservation Act (Bundesnaturschutzgesetz BNatSchG)
➢ Federal Act on Immission Protection (Bundesimmissionschutzgesetz BImSchG)
➢ Environmental Impact Assessment Act (Umweltverträglichkeits-Prüfungs-Gesetz UVPG)

Regulations by the federal states

➢ Federal Building Regulation (Landesbauordnung)
➢ Maximum height limitations and regulations on minimum distances
➢ Land use planning and regional development

Competence of the community
➢ Prescribed privileged areas
➢ Development of land use plans
➢ Verification of admissibility
➢ Issue of construction permit

Fig. 15-3 Overview of the general legal regulations for the approval of wind turbine installation in Germany

Since the amendment of the *German Federal Building Code* (BauGB) in 1998, the erection of wind turbines in the outskirts is a privileged project. Since then, most local communities designated suitable areas which were favourably examined for approvability in cooperation with the public agencies and regional planning authorities. If parcels are located in such a suitable area (or a privileged area for wind energy projects) then nothing stands in the way of building approval with respect to the building code. Approval is usually not granted to areas outside of these designated suitable areas.

The *German Aviation Act* (LuftVG) may impose constraints on the project. The approval of the aeronautical authority is a prerequisite for the permission to build in the protected areas around airports and in any place if the wind turbine would exceed a certain total height (hub height plus rotor radius). In general, a wind turbine with a total height of more than 100 m has to be equipped with obstacle warning lights, and the rotor blades have to be marked by red-white-red colouring. Red or white obstacle warning lights on the nacelle housing top are required during night or daytime. The latter is only required in the vicinity of airports.

The *Road Traffic Regulations* determine the minimum distances to freeways, highways, country roads. Wind turbine erection may be either restricted and need the approval of the local road construction authority or may be completely prohibited. However, in certain cases, the competent authority may make an exception.

The authorising agency decides on admissibility with regard to environmental protection based on the *Federal Nature Conservation Act* (BNatG) for nature conservation and rural conservation. Since wind turbines are privileged projects according to the Civil Building Code, contacting and informing the competent authorities should be sufficient. This means that a statement from the responsible nature conservation authority is requested in course of the approval process.

The erection of wind turbines constitutes interference with nature and landscape. This is considered to be permissible if it:

- Is in agreement with the aims of regional development and planning,
- Is inevitable in terms of significant and long term interference and
- Can be compensated by the adoption of appropriate measures.

According to the German Federal Nature Conservation Act, interference in nature and landscape has to be compensated. A survey report which determines the nature and extent of the compensation measures, called an accompanying landscape conservation plan, is therefore required.

The approval process for wind farm is a procedure under the German Federal Act on Immission Protection (German: Bundesimmissionsschutzgesetz, BImSchG). The procedure depends on the number of wind turbines in the wind farm, see Fig. 15-4.

If an environmental impact assessment (EIA) is not required, a simplified in camera approval process, i.e. without participation of the public, is carried out. If it is found that an EIA is required, a so-called formal approval process is carried out meaning that after an announcement in the newspapers the application documents have to be publicly displayed. Objections may be lodged up to 14 days after the end of public display. Thereafter, these objections are discussed in a meeting. The approval authority prepares this meeting by making enquiries to the concerned specialist authorities (e.g. nature conservation, waters or civil building authority). If all requirements for admission are fulfilled, approval of immission protection is granted which also includes the required building permission.

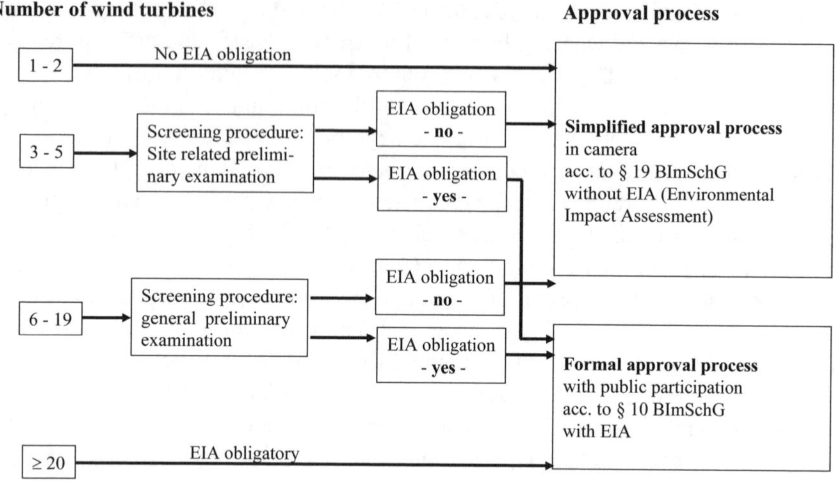

Fig. 15-4 Approval process for wind turbines with a total height of more than 50 m according to the German Federal Act on Immission Protection (Bundesimmissionsschutzgesetz, BImSchG)

Ecological effects of wind turbines

It is verified whether the disturbances caused by the erection and operation of wind turbines is acceptable with respect to nature conservation. If necessary, the protection of plants and animals is checked by additional expert reports. The general focus is on the protection of nesting and resting places of birds.

In Germany, the environmental impact assessment is not mandatory for wind farms of 3 to 19 wind turbines. On the contrary, the authority firstly assumes that the legal criteria are all met and checks for this case if there would still be significantly negative environmental effects. Only then, an EIA is requested. If there is an obligation for an EIA, biotopes, plants and animal species have to be mapped, e.g. bats, nesting, resting and migratory birds, and the landscape also has to be assessed.

In Germany, the following supporting documents on the effects of wind turbines on human and environment are integral parts of the admission documents:

- Mapping of the avifauna,
- Further aspects of animal and plant protection,
- Noise impact expertise,
- Shadow casting expertise and
- Aspects of protection of historical monuments.

Noise impact

The noise impact of wind turbines is investigated for the site based on the acoustic measurements results included in the wind turbine type approval. The acoustic measurements are usually performed by accredited institutes according to the international standard IEC 61400-11. They serve for the determination of the sound power level of the wind turbine type and penalty supplements for tonality (e.g. dominant single tones) and impulses (rhythmical pulses of low-frequency). According to [1], the noise impact L_P at a relevant *point of immission* at a distance d from the noise source is estimated by

$$L_\mathrm{P} = L_\mathrm{WT} - 10 \log_{10} (2 \cdot \pi \cdot d^2) - \alpha \cdot d + K \qquad (15.1)$$

where:

L_WT: Sound power level at the source (nacelle or hub of wind turbine) in dB(A)

d: distance between source and (measuring) point of imission in m

α: coefficient for absorption in dB(A)/m

K: penalty supplement for tonality and/or impulse in dB(A)

The sound power level L_WT is given on a logarithmic scale where an increase by 3.01 dB(A) is experienced as doubled volume. Depending on the soil properties the coefficient for absorption is in the range $\alpha = 0...1$.

The penalty supplements for tonality and impulses are given according to standards and are typically in the range $K = 1...2$ dB(A). The human ear is especially sensitive to tonality or impulses. Mostly, they originate from the gearbox, the generator or the converter and are also determined by the standardised acoustic measurements. The decay of the noise with increasing distance from the source can be calculated with equation (15.1). Fig. 15-5 shows the example of a 1.5 MW wind turbine with a typical total sound power level of 105 dB(A) at hub height (source point) including the penalty supplements.

The *noise emission* of the wind turbine varies strongly depending on the current wind speed and power production. A rough estimate is that for an increase of the wind speed by 1 m/s the sound power level rises by 1 dB(A). The sound power level relevant for the considerations on noise impact is given for a wind speed of 10 m/s, resp. at 95% of rated power. Furthermore, the noise emission varies with the spatial direction, i.e. upwind, sideways or downwind the rotor of the wind turbine. This must be taken into account in the acoustic measurements and also in the course of the noise impact study using wind farm planning software.

The maximum value of the sound power level at an immission point must not exceed the admissible limiting values provided in the legal regulations, table 15.1. During wind farm planning the isophone lines, i.e. lines of constant sound power level, are calculated and drawn on the site map. The superposition of the emissions from the different wind turbines gives the *total sound power level at the immission points* shown exemplarily in Fig. 15-6. If other noise emission sources already exist they also have to be considered when verifying the safety margin between noise impact and admissible values. In this way, critical areas are investigated and turbine locations in the micro-siting may possibly be changed to reduce the noise impact - or another more silent wind turbine type has to be chosen.

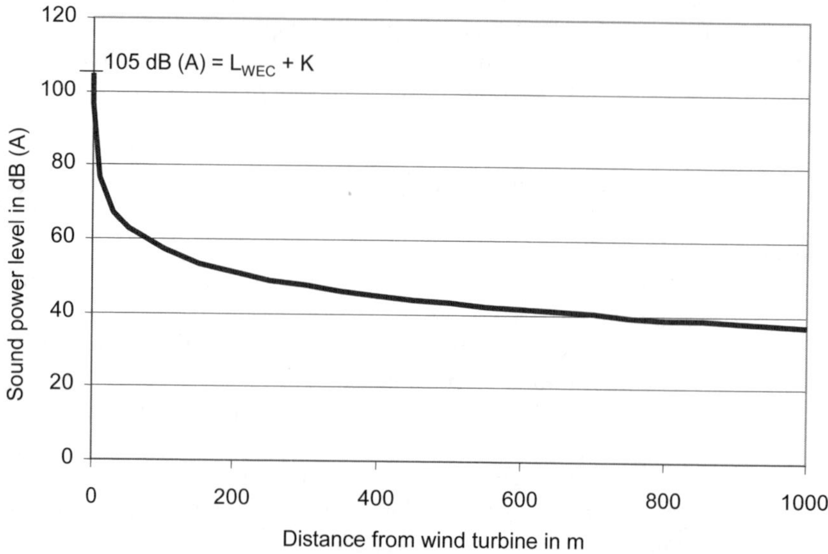

Fig. 15-5 Decay of the sound power level $L_P = L_{WEC} + K$ with the distance from the wind turbine

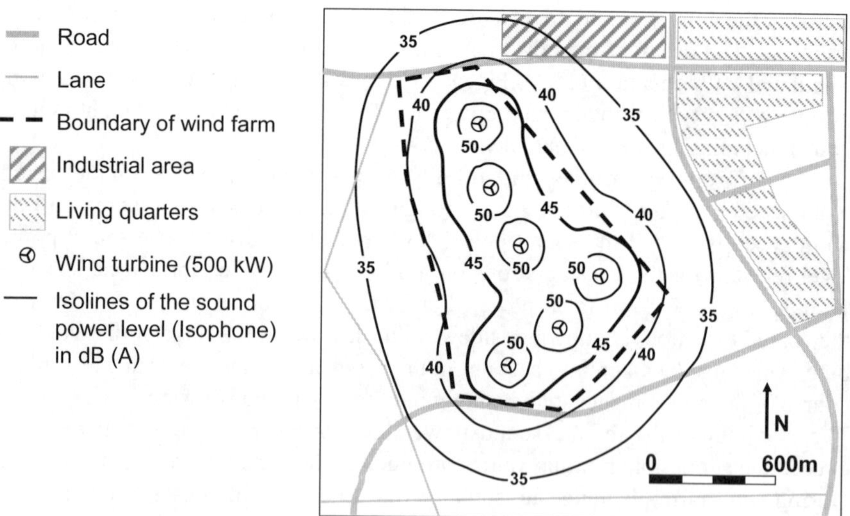

Fig. 15-6 Wind farm site map with isophone lines from the noise impact study

Table 15.1 Admissible sound power level according to the German Manual on Noise Impact Regulations (TA Lärm)

Area type	Day dB(A)	Night dB(A)
Industrial area	70	70
Majority commerce	65	50
Mixed (living/ commerce)	60	45
Majority living	55	40
Exclusive living	50	35
Hospitals, health resorts	45	35

If the actual sound power level of the operating wind turbines is questioned by inhabitants at a later stage, on-site measurements by the trade supervision authority are necessary. If for certain critical situations a problematic noise impact is detected, e.g. at night for certain wind directions, the operator is forced to take action. Variable-speed wind turbines with pitch control may then temporarily be switched to a noise-reduced operation with a smaller rotational speed (i.e. less circumferential speed). For stall-regulated wind turbines with a constant rotational speed this measure cannot be applied due to technical reasons. In the worst case there has to be a nightly shut-down which may cause significant yield loss. Therefore, the noise impact investigation should be carried out carefully and with a reasonable safety margin.

Shadow casting

Another aspect of the approval process involves investigation as to whether the turning rotor casts a disturbing periodical shadow on any problematic spot in the wind turbine's surroundings. This can be simulated on the basis of the site-specific sun's orbit, the hub height and the rotor diameter for the "worst-case scenario", i.e. without considering atmospheric opacity or clouds. Fig. 15-7 shows for two of the wind turbines of a wind farm the the isolines of annual shadow casting hours as the result of such a prediction.. In Germany, the maximum admissible duration is 30 hours per year in total and maximum 30 minutes per day for sensible inhabited locations. If this limiting value is exceeded, the wind turbine should be equipped with special sensors which shut it down automatically once the relevant critical operating parameters are present: clear sky, relevant position of the sun, relevant wind direction and a wind speed with the wind turbine in operation.

Further constraints

Apart from the uniform Federal German approval process, additional regulations
and writs have also been prescribed by the individual Federal States. The regulations of Schleswig-Holstein are mentioned by way of example in the following.
On November 25th, 2003 the Federal State Government of Schleswig-Holstein
amended the regulations on minimum distances for new wind turbines with a total
height h of more than 100 m (h = hub height + rotor radius). The given values are
minimum values and should be respected by regional planning authorities when
allocating the privileged areas for wind turbines. Table 15.2 lists some examples
of the distance regulations.

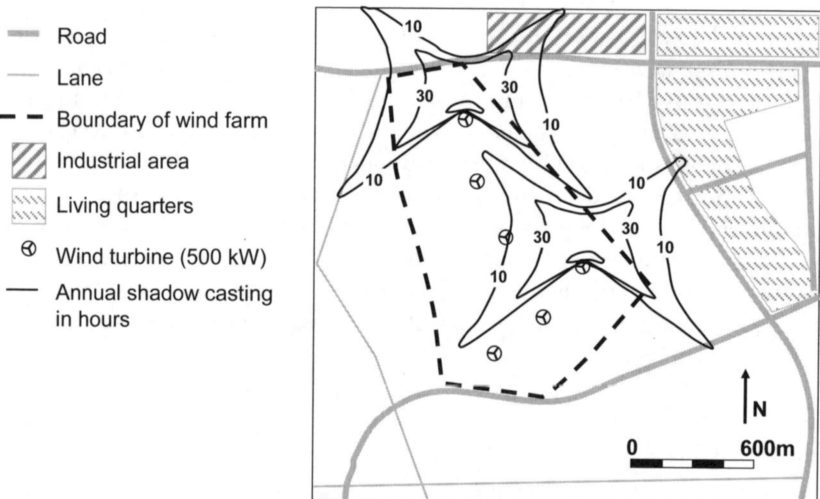

Fig. 15-7 Wind farm site map with isolines of the annual shadow casting in hours for two wind
turbines

Land-use planning is an instrument of regional development used to regulate the
expansion of wind energy utilisation which may also include the allocation of
privileged areas and precautionary land. Privileged areas for wind energy utilisation exclude other regionally significant land use if it is incompatible with wind
energy utilisation.

    At municipal level, a well organized *land use plan* offers the possibility of des-
ignating privileged areas for wind energy utilisation, thus keeping the remaining
planning area free of wind turbines as their erection at other sites would be
opposed in the interests of the general public.

Table 15.2 Extract of the distance regulations for wind turbines of the German Federal State Schleswig-Holstein

Type of land use	Minimum distance for Wind turbines with a total height of $h < 100$ m (Decree of 4th July 1995)	Minimum distance for Wind turbines with a total height of $h \geq 100$ m (Decree of 25th November 2003)
Single houses and scattered settlements	300 m	3.5 x h
Land settlements	500 m	5 x h
Urban settlements, areas with leisure residences and camping sites	1000 m	10 x h
Federal freeways, highways and country roads and rail roads	Approx. 50 m to 100 m	In general 1 x h
National parks, nature reserves, etc. and other woodland	min. 200 m, for individual cases up to 500 m	4 x h minus 200 m
Woodlands	200 m	In general 200 m
Primary waters and waters with a protected regeneration strip	min. 50 m	1 x h minus 50 m

15.1.3 *Estimation of economic efficiency*

A first prediction of the economic efficiency of the wind farm project should be performed already in the initial project phase (planning phase in Fig. 15-1). This is done in order to assess whether the project will be an economic success. If this initial assessment is negative, planning should be halted in order to avoid incurring any further costs. The expected *investment costs* and the *wind regime* are decisive parameters when undertaking this assessment. Literature may provide the basis for a rough estimation [2]. The composition of the investment costs varies slightly for each project. Fig. 15-8 shows the typical distribution of the investment cost shares according to a statistical evaluation of more than 500 onshore wind farm projects [2].

The wind turbine itself has the biggest share of the investment costs. Mostly, this share includes the costs for transport and on-site erection because these services are often offered as a bundle by the wind turbine manufacturer. Fig. 15-9 shows that with growing rated power, the specific investment costs per kW installed capacity decrease. The series effects and learning curves common in mechanical engineering are also observed for commercial grid-connected wind turbines. They lead to further cost reduction. The strongest reduction occurred in the first half of the 1990s during the production of wind turbines of the 500 kW class.

In the second half of the 1990s such a significant cost reduction was not achieved any more despite the increasing annual installation of wind turbines.. This was partly due to the strong market demand for larger wind turbines (cf. Fig. 1-1), which could only be satisfied by time reduction of the entire design process meaning less time for optimisation. Overall, the turbine costs fell by approx. 50% from 1990 to 2005.

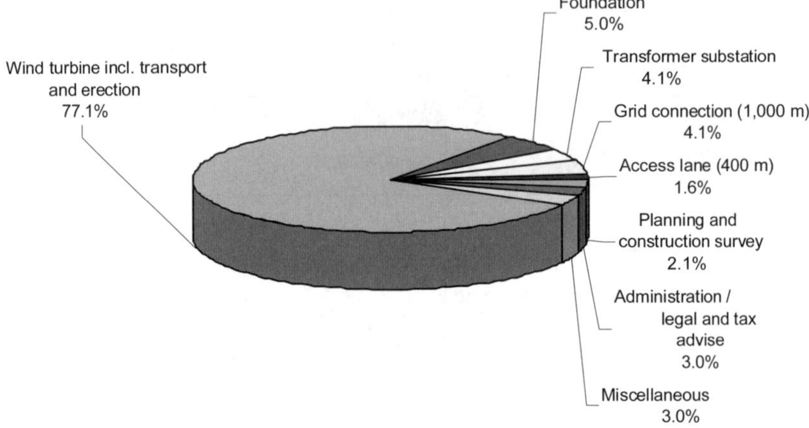

Fig. 15-8 Typical investment cost structure for a project with a single wind turbine in Germany (onshore)

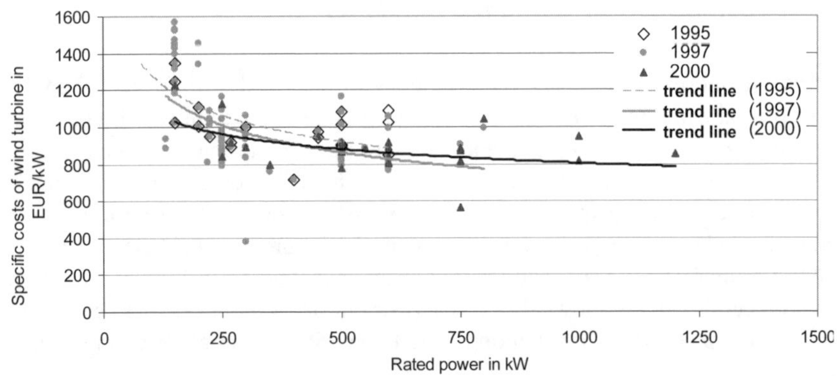

Fig. 15-9 Specific investment costs (turbine, transport and erection, see Fig. 15-8) versus rated power of wind turbines installed in Germany [3]

The *cost structure of the production costs* varies slightly depending on the wind turbine concept (stall or pitch control) and the wind turbine size. Table 15.3 shows the typical cost distribution. In general, the following tendencies can be stated:

-    With growing wind turbine size the portion of the costs of the rotor increases.

- For variable-speed wind turbines with pitch control the portion of the costs of generator and control increase.

In addition to the production costs, the *total wind turbine investment costs* include share the design and development costs, the profit and further costs attributable to warranty, transport and erection.

The other portions of the *total investment costs* mentioned in Fig. 15-8 are influenced to a large extent by local site conditions such as the wind farm project size (single wind turbine or wind farm), the type of soil and the distance to the grid-connection point. Therefore, when comparing economically efficient wind farm projects of the same size there will always a spreading of the actual project costs be observed.

The *grid connection costs* are often the second-biggest investment cost factor, Fig. 15-8. These vary strongly according to local conditions, i.e. the distance to the next grid-connection point, and the requirements of the local grid operator as well. The German Renewable Energy Sources Act (EEG) provides an unambiguous definition as to whom the costs are to be allocated. Basically, the grid connection costs are borne by the wind farm operator.

However, if the local grid is too weak for the planned wind farm capacity the *grid reinforcement costs* have to be borne by the grid operator who is legally obliged to permit wind power feeding. Therefore, the grid reinforcement costs do not form part of the investment cost estimation.

Table 15.3 Cost structure of wind turbine components and assembly , typical values according to [4, 5]

Rated power Component	600 kW €	Share in %	1,200 kW €	Share in %	Typical span of cost share in %
Gearbox and coupling	80,400	21.8	115,000	18.1	10 to 25
Rotor blades	63,800	17.3	150,000	23.6	15 to 25
Tower (incl. coating)	58,000	15.7	150,000	23.6	12 to 25
Generator and control	43,800	11.9	65,000	10.2	10 to 20
Nacelle frame and yaw bearing	31,600	8.6	35,000	5.5	5 to 10
Hub and main shaft	29,000	7.9	40,000	6.3	5 to 10
Housing	16,600	4.5	18,000	2.8	2 to 5
Cables and sensors	13,400	3.6	18,000	2.8	2 to 5
Hydraulics	11,600	3.1	15,000	2.4	2 to 5
Yaw system	8,700	2.4	10,000	1.6	1 to 3
Assembly	11.600	3.1	20,000	3.2	2 to 5
Total	**368,500**	**100.0**	**636,000**	**100.0**	
	614 €/W		**530 €/kW**		
Remark: The total wind turbine investment costs include additionally R&D costs, risk, profit, warranty, transport and erection					

For the purpose of estimating the *costs of the foundation,* typical values for the specific costs (per kW installed capacity) are available. For a flat foundation they vary between approx. 70 and 100 €/kW depending on the hub height of the wind turbine. If the geotechnical investigation report for the site reveals that the bearing capacity of the ground is not strong enough, additional measures are required (e.g. a piled foundation) which increase the costs significantly compared to those of a standard flat foundation.

The *costs incurred by the access lanes* and roads again vary strongly depending on the local site conditions and the optimum wind farm layout. The bearing capacity of the access lanes, and therefore the specific costs in € per metre, are determined primarily by the transport weight of the cranes. They also depend on the wind turbine size.

It must be differentiated between the access requirements which exist during erection and those emerging from wind farm operation. Nevertheless, access for cranes should be ensured in case larger repairs need to be carried out at a later stage.

The *costs for planning, permission and indemnification measures* and so-called „soft costs“ for the foundation and organisation of the operating company, as well as the financing costs, are in the order of 3 and 6% of the total investment costs for wind farm projects in Germany.

The first measure used to compare different wind farm projects is the power-specific investment costs index

$$SIK_L = \text{Total investment costs / installed capacity} \quad \text{in } €/\text{kW}. \qquad (15.2)$$

However, this index does not provide any information on the yield potential of the site, i.e. the annual energy yield or production. Therefore, the second measure yield-specific investment costs index:

$$SIK_E = \text{Total investment costs / annual energy yield in } €/\text{kWh}_a \qquad (15.3)$$

In Germany, typical values of the power-specific investment costs index are around $SIK_L = 1{,}100$ €/kW, and for the yield-specific investment costs index in the range of $SIK_E = 0{,}50...0{,}75$ €/kWh$_a$ [6].

The *type of financing plan* is of particular importance when assessing the economic efficiency of a wind farm project. A typical example of a financing plan for a wind turbine with a rated power of 1.5 MW is given in table 15.4.

The required share of equity is set by the financing bank which also submits an application to the KfW Bankengruppe (Kreditanstalt für Wiederaufbau) and Deutsche Ausgleichsbank (DtA, merged with KfW in 2003) for the loan with reduced interest under the umbrella of its ERP environmental and energy efficiency programme. The share of equity varies slightly depending on the site quality and the financial standing of the investor. A higher share may be fixed if there is a higher financial risk due to a lower site quality or reduced financial standing of the investor.

Table 15.4 Example of a financing plan for a wind farm project: wind turbine of 1.5MW rated power, conditions in July 2000

Financing	Condition	Value	Amount in 1000 €	Share in %
Equity			523	30
ERP loan by KfW bank	Interest rate in %	5.25	872	50
	Period in years	10		
	Cap in %	50		
	Years without re-demption	2		
Loan by DtA bank	Interest rate in %	4.25	349	20
	Period in years	10		
	Cap in %	25		
	Payout in %	96		
	Years without re-demption	2		
	Total		**1,744**	**100**

In order to calculate the balance of annual proceeds and expenditures, the *expected proceeds* from the electricity production are calculated using the predicted energy yield and the feed-in tariff stated in the German Renewable Energy Sources Act (German abbreviation: EEG). It is reasonable for safety reasons to assume a reduction in the expected proceeds by 10 to 15%. This should cover on one hand the uncertainty of the the wind regime prognosis for the site and on the other hand the statistical fluctuations of the annual wind regimes (cf. chapter 4, Fig. 4-17).

Apart from the return of capital on the basis of the financing conditions, *the annual operating costs* must also be estimated. The typical cost structure is shown in Fig. 15-10. In addition, some German Federal States require to include the *accrues for the decommission* already in the application for building approval.

Careful attention should be paid to the *portion of maintenance, repair and replacement costs*. The annual costs, stated in empirical studies [2, 6], are in the range of 2.5% of the wind turbine investment costs per year for the first decade of wind turbine operation, and approx. 4.0% for the second decade of operation . The higher costs in the second decade result from the replacement of worn components, Fig. 15-11. An increasing number of wind turbine manufacturers are offering full-service contracts (including a guaranteed availability). These contracts involve on the one hand increased running costs, on the other hand, they reduce the exposure of the operator to technical risks. This is taken into account by some insurance companies through an reduced insurance contribution [3].

According to an evaluation of the German scientific measuring and evaluation programme WMEP [7], the *power-specific value of the total annual operating costs* amounts to approx. 35€/kW for wind turbines of a rated power below 500 kW, and to approx. 15€/kW for wind turbines of the MW class.

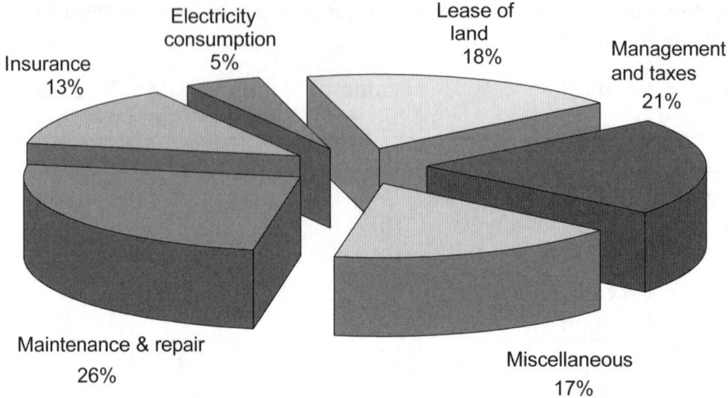

Fig. 15-10 Typical cost structure of the operating cost for a wind farm project in Germany; onshore and without capital service [2]

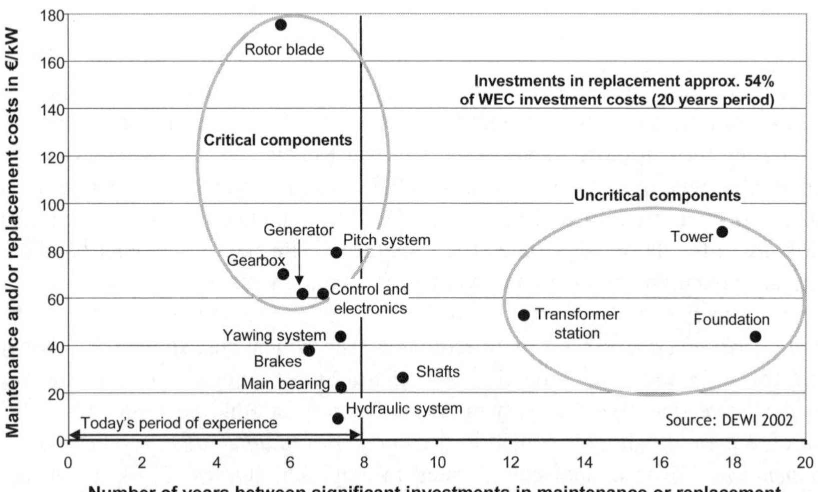

Fig. 15-11 Estimated maintenance, repair and replacement costs according to [2]

Table 15.5 provides an example of a rough calculation of a wind farm project with one 1.5 MW wind turbine erected in 2000 in the newly-formed Federal states at a site with an average annual wind speed of 6.4 m/s at hub height,. The imputed annual net income is given for capital costs calculated according to the annuity method for the first decade of operation and without considering taxes.

Only if the estimated economic efficiency satisfies the investor, the planned wind farm project will be continued and the next project phase, the implementation including erection and commissioning will be carried out.

Table 15.5 Exemplary calculation of a wind farm project with one 1.5 MW wind turbine at a site with an average annual wind speed of 6.4 m/s at hub height, erected in 2000 in the newly-formed German Federal states

Investment costs	Value	Unit	Share or comment
Wind turbine (incl. Crane and erection)	1,432,000	€	82.1%
Foundation	51,000	€	2.9%
Access lanes (400 m)	15,000	€	0.9%
Substation	77,000	€	4.4%
Grid connection (1000 m)	77,000	€	4.4%
Planning / Site supervision	31,000	€	1.8%
Site (land lease)	0	€	0.0%
Miscellaneous	31,000	€	1.8%
Business / tax and legal consultancy	31,000	€	1.8%
	1,744,000	**€**	100.0%
Financing			(see table 15.4)
Annual proceeds			
Wind speed at hub height	6.37	m / s	
Calculated annual electricity production	3,074	MWh / a	
Availability of wind turbine	98%		
Effective annual electricity production	3,013	MWh / a	
Reference yield (EEG-quality of site: 91.6%)	3,288	MWh / a	
Feed-in tariff (for 18 years until '09 2018)	0.091	€ / kWh	
Feed-in tariff (for remaining 2 years	0.062	€ / kWh	
Feed-in tariff (average for 20 years)	0.088	€ / kWh	
Annual feeding-in proceeds (average)	265,400	**€ / a**	
Annual expenses			
Capital costs (annuity)	158,000	€ / a	63,8%
O&M costs	7,700	€ / a	3,1%
Insurance (machine, 0.6% wind turbine cost)	9,200	€ / a	3,7%
Insurance (BMU, 0.9% of annual yield)	2,600	€ / a	1,0%
Power consumption / telecommunication	1,500	€ / a	0,6%
Land lease	7,700	€ / a	3,1%
Administration (1.0% of invest. costs)	17,400	€ / a	7,0%
Reserve / repair (2.0% of invest. costs)	34,800	€ / a	14,0%
Miscellaneous (0.5% of invest. costs)	8,700	€ / a	3,5%
	247,500	**€ / a**	100,0%
Imputed annual net income			
Annual feeding-in proceeds (average)	265,400	€ / a	
Annual expenses (calculatory)	247,500	€ / a	
Surplus	**17,900**	**€ / a**	

15.2 Erection and operation of wind turbines

The flow chart of a wind farm project, cf. Fig. 15-1, illustrates that we are now in the *phase of realisation*. All steps in the planning procedure are accomplished; all authorisations, at least preliminary ones, have been obtained. The erection starts with the construction of access lanes and the laying of cables. The foundation is then built. In accordance with the purchase contract, the wind turbine manufacturer will transport the wind turbine components to the construction site and erect the wind turbine. The final completing works in and around the wind turbine are followed by the commissioning procedure and testing phase. The wind turbine will then commence its permanent operation. The first two years of operation are usually the defect liability period in which the wind turbine is maintained and repaired by the manufacturer at his own expense.

It is only after this that the wind turbine is handed over to the operator who will assume full responsibility for it.

At the end of the time of operation designated in the building approval, the wind turbine has to be decommissioned, unless the operator applies on time for an extension of the building approval since he can give evidence of remaining useful life due to weaker wind conditions on site than assumed in the course of the wind turbine design. In 2009, GL Wind issued a guideline with general hints on how to verify whether a used wind turbine has still some potential (remaining useful life) for a continued operation beyond the 20 years of planned operating time [14].

An alternative, so-called "repowering", is the anticipated replacement of the wind turbine by a newer type of turbine with a higher rated power and correspondingly increased yield.

The technical aspects of a wind farm project are closely interwoven with its legal and economic aspects. Investors provided the equity, and the total funding was negotiated with the bank. The invested capital and the monetary return flows are organised in the operating company. Both, the energy production of the wind farm and the preservation of the wind turbine's value are important issues taken into consideration by banks and insurances companies when carrying out a risk assessment. Therefore, the main goal of wind farm operation is safe and reliable operation producing maximised energy yield and minimised loss in value.

15.2.1 Technical aspects of erection and operation of wind turbines

The type of wind turbines and their optimal arrangement in the wind farm (micro-siting) were already determined in the planning phase (cf. Fig. 15-2). This also provides requirements for the construction planning. The work of the technical wind farm operator begins already in the course of the preparatory construction works on the site and the erection of the wind turbines. Later, he will be responsible for any technical issues of wind farm operation. The following **four** steps are implemented during the phase of realisation and operation:

- Transport and erection,
- Commissioning and take-over
- Wind turbine operation
- Decommissioning or Repowering

Transport and erection of the wind turbine

In Germany, the typical duration of the *wind farm erection phase* is between three and six months. All services stipulated in the course of construction planning are now carried out.

Remote areas of the world pose far greater challenges for transport and erection. For these projects, e.g., the size of available cranes or the capacity or steepness of the roads may be the limiting factor. Substantial consideration must be given to this during already the technical and financial planning phase.

When the *road and foundation works* are finished and the areas for the cranes have been well prepared, the wind turbine components may be delivered. A big challenge for manufacturer and shipper is the timely delivery of the huge components. *Transportation* is usually carried out on the road by long vehicles and heavy trucks. Usually, tower sections, nacelle and rotor blades are delivered one after the other because they often come from different factories and suppliers and because for assembly they are needed in a certain order. More than ten flat-bed trailers, classified as as heavy transport, may be required, not counting the ones for transport of the crane. In Germany, they are only permitted to drive at night. When the route is planned, all the radii of curves, the overhead heights below bridges and other bottlenecks (e.g. inclinations) are checked to ensure the trailers can pass.

The designers try to reduce the weight of the *tower segments,* in particular for the lowest tower segment. This has the biggest diameter and the largest wall thickness. Moreover, the tower segment length should not exceed 22 m in order to allow the truck to be well manoeuvred and to avoid having to obtain special shipping permits with further requirements for escort vehicles and closing off the roads.

For the *transport of the rotor blades* of the MW class turbines, their length is a critical issue. The passage of villages or curvy roads in the low mountain ranges of Germany may pose a problem. Up to three rotor blades can be carried together on specially designed long vehicles for blades of a length in the range between 35 and 40 m (wind turbines up to 2.5 MW). Longer rotor blades are carried individually because their maximum chord length (blade depth) is the second limiting factor. If it exceeds 3.5 m the passage below most bridges (overhead height 4.0 to 4.2 m) is no longer possible with the blade in the vertical position. To solve this problem, special trucks are designed. E.g., the truck for the rotor blades of the ENERCON E112 (4.5 MW rated power, max. chord length 5.8 m) has a blade pitching mechanism which turns the blade into the horizontal position before passing below bridges.

Weight is mostly the critical factor when it comes to the *transportation of the nacelle.* Its dimensions depend strongly on the drive train concept, see section 3.2. The 2 MW wind turbine Nordex N-80, e.g., has a nacelle with the dimensions 10 x 3 x 3 m (length, width, height); those for the 5 MW class turbines may reach the dimensions of a single family house.

The entire process of wind turbine erection is strongly dependent on the *weather conditions on site.* While the wind should blow strongly during wind farm operation, low wind speeds are desired for the erection period in order to avoid the collision of components (e.g. rotor and tower) upon hoisting and the toppling of the crane.

Apart from the wind pressure on the components upon hoisting, there are further dangerous fluid dynamic problems. For example, as long as the nacelle is not mounted on the tower, periodic vortex shedding at the tower may occur, the so-called "van Kármán vortex street". Therefore, despite the requirement for accurate positioning in the range of millimetres during the assembly, unfavourable meteorological conditions may impose an intense time pressure on certain assembly steps. This is an even more critical issue for offshore wind turbine installation where additional risks emerge from waves and currents, cf. chapter 16.

Fig. 15-12 Transport of a tower segment [8]

The *mobile crane* hired to assemble the components and erect the wind turbine is very expensive, time is again a big issue. In addition to the main crane most erection procedures require a much smaller auxiliary mobile crane, cf. Figs. 15-13 and 3-55. For common wind turbines of 2 MW class the crane has to be able to lift approx. 70 tons to the hub height of 100 m, see Table 15.6. Due to the required reach of the crane its nominal load will then be 500 tons.

Table 15.6 Characteristic data of wind turbines and correspondingly required mobile cranes with guyed telescopic crane boom and jib [10]

Rated power of wind turbine in kW	300	600	1.000	1.500	2.000
Nacelle weight in t (without rotor)	15	20	40	50	70
Hub height in m	50	70	80	90	100
Maximum crane capacity load at reach	18 tons / 8 m	25 tons / 25 m	43 tons / 18 m	65 tons / 22 m	73 tons / 20 m
Maximum hook height in m (with jib)	56	76	84	94	104
Self weight of crane and ballast in t	60 50	72 87.5	84 100	96 165	96 160)*
Number of wheel axes	5	6	7	8	8

)* Mobile crane cannot drive individually, accompanying trucks for telescopic jib and ballast required.

The assembly and erection procedure, as described below, has its limitations. At present (2007), for wind turbines between 3 MW and 5 MW rated power the capacities of the trucks for transport and of the mobile cranes reach their bearing limits. The only solution will be a segmentation of the nacelle into components which are hoisted separately and then assembled on the wind turbine tower.

First, the *tower* is erected. Procedure and duration vary depending on the tower types. The erection of a conical steel tower with its 3 to 5 sections only takes up to 2 days. The assembly of a segmented concrete tower, with its approx. 100 pre-manufactured concrete segments, takes several days. For an in-situ concrete tower with jacking framework, the production even lasts several weeks, cf. Fig. 3-54.

The *nacelle* is then mounted on the tower. Depending on the manufacturer, either the entire nacelle, comprising nacelle frame, drive train and nacelle housing, is lifted or this is done component by component.

Finally, the complete *rotor* is lifted and fixed to the flange of the rotor shaft, on the premise that the rotor blades and the hub have already been assembled on the ground. Another way of final assembly is to fix the hub to the rotor shaft already at the ground and hoist the hub together with the nacelle. The rotor blades are then lifted one by one. Now the shell construction is finished and the wind turbine appears to an onlooker to have been finally erected. However, several days of work will have to be spent for completing the internal installations and the connections before initial operation.

Fig. 15-13 Hoisting the Rotor by means of main mobile crane and assisting mobile crane [9]

According to the German Renewable Energy Sources Act, the feed-in tariff valid for the individual wind turbine is determined by the date of the first electricity feed into the grid. Since the feed-in tariffs are usually reduced from year to year, it may be essential to set the wind turbine at least shortly into operation already in the end of December to get the higher tariff. This explains, among others, why the number of installed wind turbines is usually higher in the second half of the year than in the first half of the year despite the more unfavourable weather conditions for installation.

Commissioning and take-over

Several weeks of testing usually pass between the production of the first kilowatt-hour of electricity and the handing over of the wind turbine to the operator. The future operator/operational manager of the wind turbine checks himself or arranges for an inspection to take place in order to see whether the wind turbine was delivered and erected as agreed by contract and to ensure it proper and fault-less functioning.

The common procedure is:
- Test operation, surveyed by the manufacturer and adjustment of parameters to the site-specific conditions
- Inspection and elimination of deficiencies by the manufacturer
- Acceptance inspection and handing-over to the operator or operational manager which defines the begin of the defect liability period

After the elimination of "childhood diseases" (e.g. rotor unbalance) and the final adaption of control parameters to the site conditions, the wind turbine is officially commissioned and handed over tot the operator. An independent expert usually inspects the wind turbine for the operator to check and clarify whether the specifications of the contract have been all fulfilled. This inspection involves checking the technical condition of the wind turbine itself, as well as its auxiliary installations such as aviation safety markings or lights, a possibly modified power curve for the reduction of noise emission and so on. The requirements of the building permit are of prime importance in this context.

Wind turbine operation

Operation and inspection

The wind turbine shall now operate for a minimum of 20 years with the highest possible availability. The technical operator surveys all technical aspects of the wind farm project. After expiration of the defect liability period, any risks arising from the operation have to be borne by the operator, as well the costs for maintenance and repair if there is no full-service contract.

The safe operation of technical systems as a wind turbine has to be verified through *periodical inspections* of all mechanical and electrical components and testing their proper functioning. Moreover, the safety equipment for climbing the tower (ladder, lift, harness, rescue devices, etc.) is checked and all other aspects that are related to potential risks, such as fire hazard, environmentally hazardous materials, machine damage, etc.

In Germany, it is required that wind turbines are checked every two to four years by an *independent expert*. The interval depends on the wind turbine size and whether a service contract exists for the wind turbine. The aim is to minimise danger to the environment, humans and the machine itself, or to reduce these risks to an accountable degree at least.

High availability and high yield are irrelevant in the context of safety; however, these two issues are of upmost importance for the operator Hence, a "condition-oriented inspection" of the wind turbine should be carried out. All components are inspected to see whether faults, malfunction defects or damages announce or have already occurred. The focus is primarily on bearings and gearbox. Based on the recommendations of the inspection report, the operator decides whether maintenance, repair or even component exchange is necessary, e.g. of a bearing of the generator.

Full-service contracts offered by wind turbine manufacturer with the manufacturer continuing to be responsible for maintenance and repair of the wind turbine also provide an alternative. However, in this case as well, it is still the *duty of the technical operator* to care about a reliable operation with high availabilities and maximised energy yield. The following has to be surveyed by the technical operator:

- Operation during the defect liability period,
- Handing-over at the end of the defect liability period according to the purchase contract (usually 2 to 3 years),
- Operation after the expiration of the defect liability period,
- Maintenance and repair and
- Inspection and control of the operation.

Maintenance and repair

The necessary *service works* and intervals are documented in the technical specifications of the wind turbine. All bearings have to be lubricated periodically. All mechanical components, electrical devices and control components have to be maintained. Small repairs are carried out without delay. The technical operator will then be given a service protocol.

The operator is responsible for successful operation of the wind farm, but there is often a separate technical wind farm manager caring for technical tasks of O&M. Apart from a very good technical knowledge of wind turbines this person needs to be able to communicate in a clear and effective manner with the (more economically oriented) wind farm management team and all the other external partners concerned: wind turbine manufacturer, service companies, grid operator, inspectors and other experts, authorities, and so on. The duties are:

- Collection and management of all technical information (e.g. contracts, manuals, circuit diagrams, etc.),
- Remote monitoring of the wind turbine operation (e.g. energy production measured at the electricity meter, status messages and permanently recorded operating data),
- Regular visits to the wind turbines and wind farms for visual inspections
- Organisation and survey of service and repair works,
- Generation of periodical reports on successful operation (energy yield and availability) and technical condition (malfunction and repair) and
- Improvement and optimisation of wind turbine operation for increased availability and energy yield

Acquisition of operating data and condition monitoring

Wind turbines run usually in an automatic operation mode and are controlled by the internal supervisory and control computer. Typically, communication with the control room of the wind turbine manufacturer or the technical wind farm manager will be initiated only in the event of malfunction. Then, the wind turbine sends an email, fax or sms, and the service team has to take action.

Furthermore, the wind turbine stores its *operating data*, which includes current wind speed, wind direction, rotational speed, electrical power, several temperatures and further operational parameters. These can be read out with the System Control And Data Acquisition Software (SCADA) and are the data on which the technical wind farm manager bases his assessment of the wind turbine condition and any measures that may need to be carried out.

Fig. 15-14 shows the *yield data* versus the wind speed measured with the nacelle anemometer for a six-month period of operation. The 10-minute averages of the electrical power are sorted into wind speed class bins, forming the real energy histogram which shows the energy yield per wind class (cf. section 4.3). For comparison, the theoretical energy histogram is also displayed. This is calculated from the measured wind speeds (i.e. the measured wind histogram) and the power curve of the wind turbine manufacturer. The differences are clearly visible, not only between the histograms but also between the theoretical and real power curve which provides valuable information on site quality and wind turbine operation.

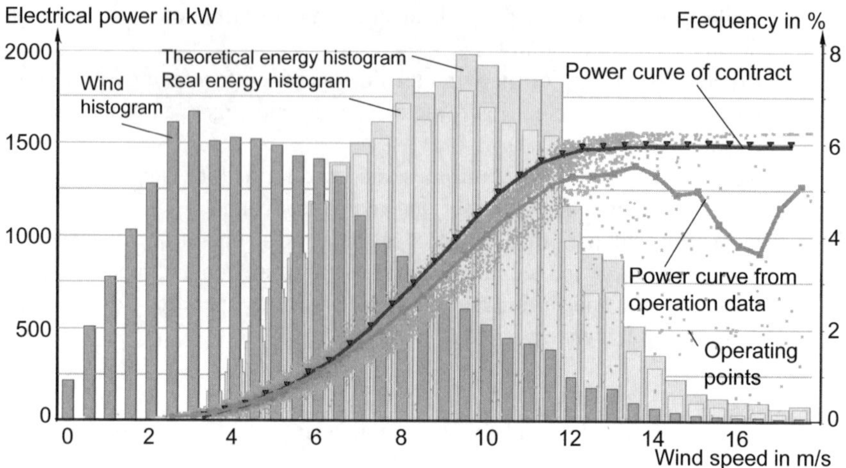

Fig. 15-14 Operating data of a wind turbine [11]

The *analysis of the operating data* reveals the efficiency of operation. Malfunction
statistics and trend analysis provide information on problems arising with certain
components. The *prevention of damages* can be improved by installing a condition
monitoring system (CMS). By vibration measurements, the condition of the main
drive train components bearings and gear box is permanently monitored and
assessed in order to prevent a machine breakdown with possible consequential
damages. Moreover, scheduling repair or replacement of the affected component
is facilitated, i.e. deciding whether an immediate component replacement is necessary or whether there is enough useful life remaining for waiting with the works
until the end of the windy season). The CMS offers the better alternative to periodical
short-time measurement and assessment by an expert but it requires some investment costs.

Insurances for wind turbines

The types of risks and who is to bear them are the most important questions for insurance companies. The manufacturer alone bears all risks before installation or
handing-over of the wind turbine to the operator. In all phases of the wind farm
project – from production and transport to the years-lasting operation – there is the
risk that proper operation may be disturbed or interrupted by unexpected events.
The wind turbines have to be insured against machine breaking and external dan-
gers (fire, lightning, etc.).
Additionally, it is advisable be insured against operation interruption to compensate for yield loss during downtimes. In some cases, the financing bank stipulates
such an insurance in order to make sure that the loan can be repaid. Practical
experience from wind turbine operation has revealed that prolonged downtimes

following an insured event of damage (e.g. lightning strike or generator damage) can result in significantly increased insurance costs. If the cause of a minor damage (e.g. a hot running bearing) was not detected a more serious consequential damage may result, e.g. a fire in the nacelle.

The third kind of required insurance is *liability insurance*. This covers events where somebody or something is damaged by the wind turbine which may even incluce damage due to a falling machine component or a piece of ice – or even fire in the field below the wind turbine caused by a fire in the nacelle.

At any rate, risks which cannot be insured include the following: wear due to operation, third party damages to the wind turbine, negligence by installation of already defective components as well as the effects of war and accidents in nuclear power plants.

Decommissioning and repowering of wind turbines

At the end of the operating period the wind turbine has to be decommissioned, cf. Fig. 15-1. Another economic alternative is so-called repowering which is the replacement of many small wind turbines by a few bigger ones. The local options for this should be checked carefully already in the wind farm planning because decommissioning or repowering should be considered in the contracts.

New wind turbines are not only bigger than smaller ones. Typically, they are more quiet and more efficient, showing higher availability and energy yield at the same site and as well as reduced service costs. Therefore, repowering brings technical progress and economic benefits. Moreover, a few big wind turbines have a reduced the visual impact on the landscape compared to the case of many small wind turbines. First and foremost, repowering is a decision based on economic considerations. Mostly, it is only economically viable after 20 years of operation. However, several successful repowering projects in Germany have already proven the increased economic benefit as well as the reduced visual impact.

15.2.2 Legal aspects

In order to implement the wind farm project and operate the wind turbines, it is necessary to establish a company, cf. Fig. 15-1, since operating wind turbines implies a commercial venture.

For a small wind farm with only a few participators a *partnership under the German Civil code* such as a BGB company is suitable. The formation of a BGB company does not require much effort as no formalities are involved except for registration with the tax office. There is no need for the minimum capital or the registration in the commercial register. However, the partners concerned are commercially liable, meaning that they together with their entire personal assets are liable for the company. The proceeds and losses are offset against other income in the income tax return.

In Germany, investment in wind farm projects has been primarily carried out by private individuals. Therefore, the legal form of a limited liability company should not be chosen for the operating company but rather a *limited partnership with a limited liability company* as a general partner. The limited liability company is responsible for the business and technical management. Private investors acquire the shares as limited partners in the limited partnership thus providing the required equity. This combines the advantages of both legal company forms. Limited partners get their share of proceeds and losses in the course of a separate determination of the annual accounts, declare them and are liable for the capital they have contributed.

If the wind farm project is started as a local or regional initiative, it is often quite easy to find investors without having to conduct a major marketing campaign. A nationwide search for investors would be appropriate if investment amounted to several hundred million Euros. A closed-end investment company is then launched and announced in a fonds prospectus. In this case the legal framework of the liability in respect of information contained in fonds prospectus has to be observed. The given information should follow the common standards (e.g. IDW 4, quality standard of the German Association of Certified Public Accountants) and have the attestation of a certified accountant.

All further legal implications (rights and duties of the limited partner, their right to say in a matter, etc.) have to be regulated in the company contracts of the operating company.

15.2.3 *Economic efficiency of operation*

If the financing has been secured and the wind farm project implemented, the revenue from the electricity sold are offset against the operation costs (cf. section 15.1.3). The legal framework of the German Renewable Energy Sources Act (German: EEG) stipulates the feed-in tariffs per kWh which form the basis of the 20-years feed-in contract [12]. In absence of such a set regulation regarding the feed-in tariffs, it would be difficult to finance a wind farm project as it incurs relatively high investment costs. The framework of the German Electricity Feed-in Act (German: Stromeinspeisegesetz, StrEG, 1991 to 2000) and the successor scheme of the EEG which followed in April 2000 is internationally regarded as a good example of a legal framework which helps to turn a wind farm project, for favourable site conditions, into a safe and profitable capital investment. The EEG differentiates according to site quality in order not to privilege coastal sites vis-à-vis inland sites which have disadvantageous wind conditions through lower mean wind speeds. Further details regarding the legal framework can be found in the references, e.g. [8, 12].

The usual *company accounts* with the balance sheet and the profit and loss statement are required for every year of operation. Apart from the economic development of the wind farm project, the development of liquidity (cash) plays a decisive
role as risk of insolvency lurks in the worst-case scenario. Illiquidity would mean
that the project has failed economically. Besides operating costs (see section
15.1.3 and Figures 15-16), the debt service (interest and repayment/amortisation)
for the external financing (approx. 75%) results in large cash flows during the financing period, usually the first ten years. In the first years, the dept service may
reach up to 2/3 of the proceeds.

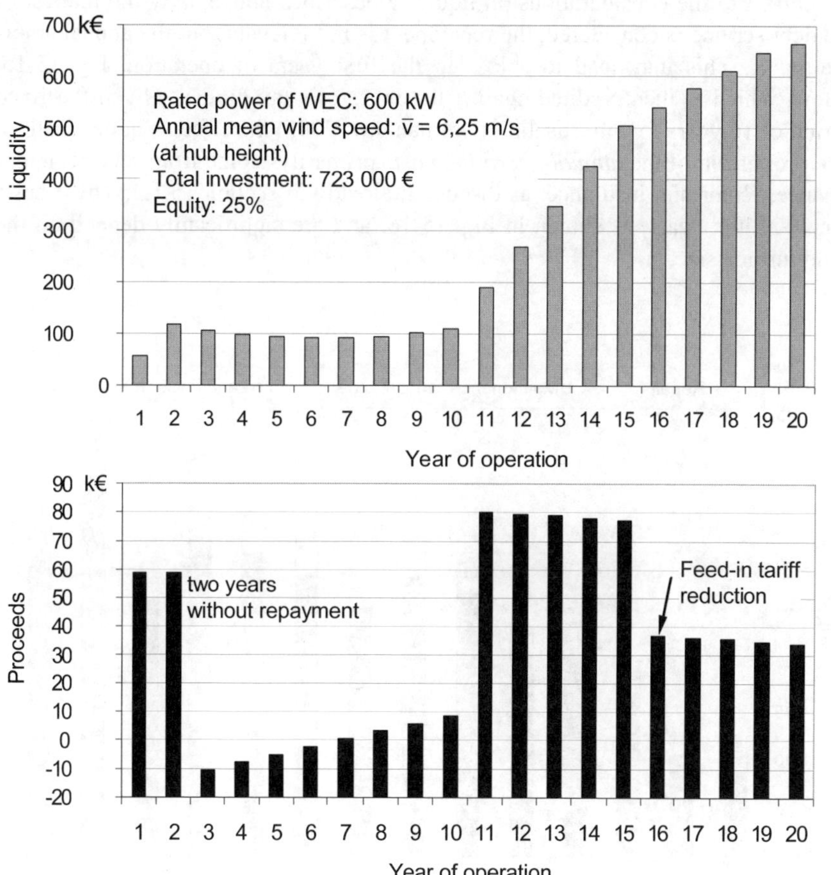

Fig. 15-15 Development of the liquidity and proceeds of a wind farm project with a 600 kW
wind turbine

Fig. 15-15, top, shows the development of the liquidity (available cash) of a sound wind farm project. In order to build up a cash reserve it is quite useful to have two years without repayment at the start of operation period. This is provided by the special financing programmes of KfW Bankengruppe for renewable energy projects. This can help to bridge years which may have a negative balance due to reduced energy yield in bad wind years. One should avoid promptly distributing company profits to investors so as to build up reserves for possible repairs and for the decommissioning at the end of the project. Nevertheless, a cumulative distribution of 200% to 300% at the end of the 20 years of operation is common for a sound wind farm project.

The *profit and loss statement* is decisive when evaluating the annual result. Here, in contrast to the considerations on liquidity described above, only the interest of the debt service is considered, the repayment is not relevant, but the annual amortisation is. This may lead to a loss in the first years of operation, Fig. 15-15, below, which is then credited against the investor's tax burden. Given the fixed period of 16 years for amortisation (German law, 2005) the effect is quite small.

The proportion of the *annual operating costs* primarilys come from service, maintenance, repair and insurance, as discussed already in section 15.1.3. These costs vary with the years, as shown in Fig. 15-16, and are significantly depend on the wind turbine size.

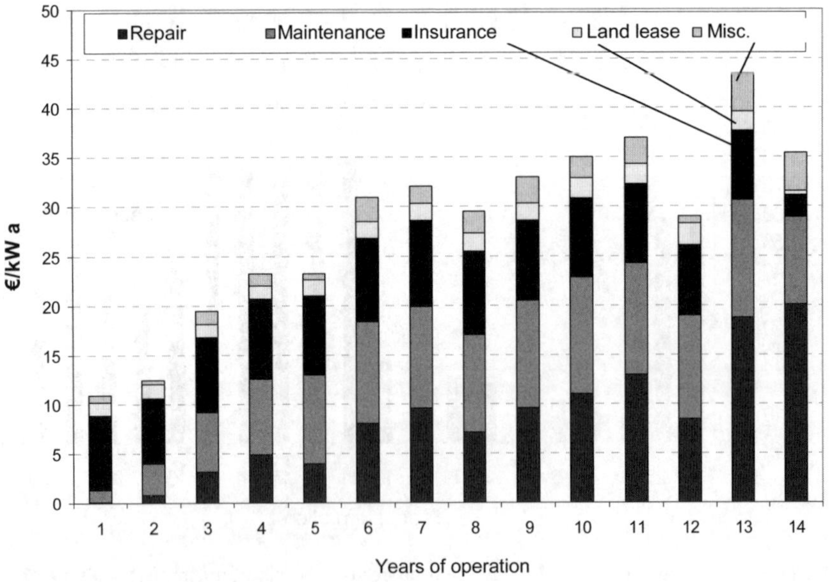

Fig. 15-16 Development of the operating costs for wind turbines with a rated power of less than 500 kW [7]

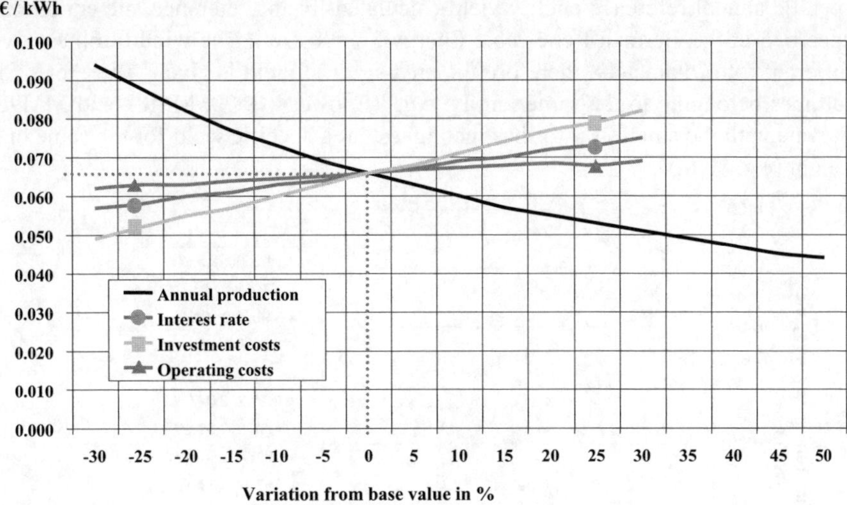

Fig. 15-17 Specific electricity generation costs, sensitivity to changes of the most important parameters [13]

Little experience has been gathered for the second decade of operation to date, hence a conservative estimate is useful [2]. A moderate increase in operating costs is expected due to inflation.

Fig. 15-17 summarizes the impact of the most important parameters on *electricity generation costs*. The effect of the operating costs is relatively low. Due to the high share of external financing (approx. 75%) the interest level and the level of the total investment costs are important factors. Nevertheless, there is no doubt that the wind itself is the biggest influencing factor. Annual production depends on the cube of the wind speed, as a result of which a 10% higher wind speed may increase annual production by up to 30%. With respect to the example shown in Fig. 15-17 this would result in a significant reduction in the specific electricity generation costs from 0.065 €/kWh to 0.050 €/kWh.

15.2.4 Influence of the hub height and wind turbine concept on the yield

The *suitable wind turbine* should be chosen according to the local wind conditions so as to achieve the highest possible annual yield. First and foremost, the *economically suitable hub height* has to be chosen according to the local wind profile (increase of the wind speed with the height above ground depending on the roughness length z_0). For some wind turbines with a rated power between 1.5 and 2.5 MW, Fig. 15-18 shows the significant influence of the hub height on the area-specific annual reference energy yield calculated for the reference site conditions specified in the German Renewable Energy Sources Act. The wind turbines have different rotor diameters; therefore the area-specific yield is given. Therefore, for turbines belonging to the same family, e.g. REpower MM70, MM82 and MM92, the one with the smallest rotor has the highest area-specific yield for the same hub height (e.g. 80 m).

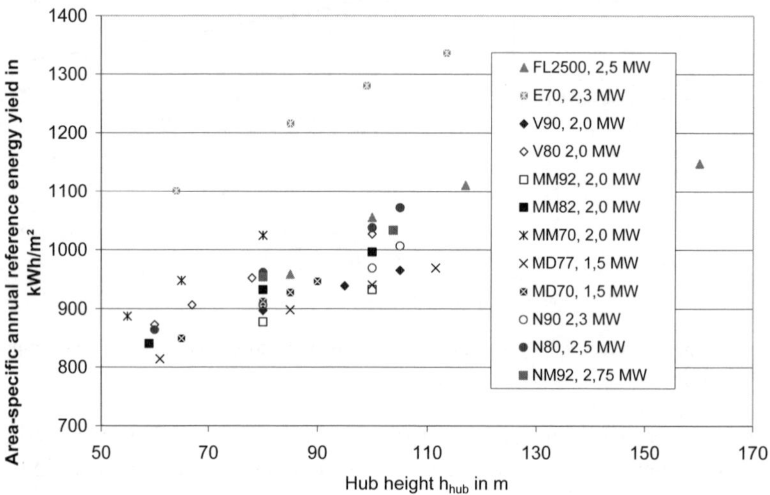

Fig. 15-18 Area-specific annual reference energy yield of wind turbines with a rated power between 1.5 and 2.5 MW (2006). Reference site conditions specified in the German Renewable Energy souces Act (EEG): Annual mean wind speed of 5.5 m/s at 30 m height, Rayleigh frequency distribution of the wind regime (k = 2) and roughness length $z_0 = 0.1$ m.

Table 15.7 Comparison of wind turbines of the 600 kW class [3]

WKA			Ecotecnia 44	Enercon E-40	REpower 48/600	Vestas V47/660
Rated power	P_N	kW	640	600	600	660
Power limitation		-	stall	pitch	stall	pitch
Rotor diameter	D	m	44.0	44.0	48.4	47.0
Hub height	H	m	65	65	65	65
Rotor swept area	A	m²	1,520.5	1,520.5	1,839.8	1,734.9
Area-specific power		W/m²	420.9	394.6	326.1	380.4
Rated wind speed v_N		m/s	14.0	12.5	13.0	15.0
Design wind speed $v_{opt} = v\,(c_{P.WKA.max})$		m/s	8.0	9.5	7.5	8.0
v_N/v_{opt}		-	1.75	1.32	1.73	1.88

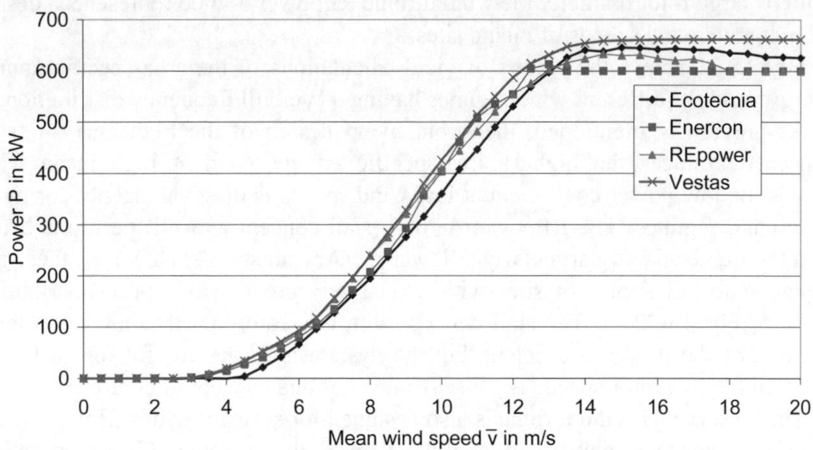

Fig. 15-19 Power curves of four wind turbines of the 600 kW class

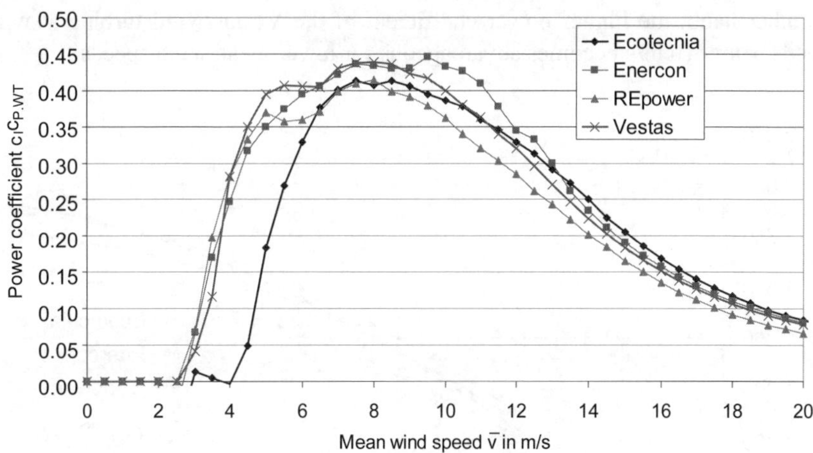

Fig. 15-20 Total power coefficient versus wind speed of four wind turbines of the 600 kW class

Moreover, the differences in the *power curves of different wind turbines* have a significant effect on the annual production. This will be discussed below for four wind turbines of the 600 kW class with the same hub height and the characteristic data given in table 15.7 and for variation in the wind conditions. Figs. 15-19 and 15-20 show the power curves and the total power coefficient characteristics (electrical power to wind power, i.e. including mechanical and electrical losses) of the four wind turbines. The wind turbine Ecotecnia 44 is designed for sites with stronger winds, therefore it shows the highest area-specific power of 421 W/m².

With its large rotor diameter the wind turbine REpower 48/600 represents a design for moderate wind speeds of inland sites.

Fig. 15-21 shows the results of yield calculations of the area-specific annual energy yield for different wind regimes having a Weibull frequency distribution.

As previously mentioned, the strong wind design of the Ecotecnia 44 (stall concept) produces the highest area-specific energy yield at high mean wind speeds. Its low power coefficient at low wind speeds is disadvantageous for moderate wind regimes. The REpower 48/600 (stall concept as well) performs better than the Ecotecnia 44 particularly at lower mean wind speeds. However, these advantages do not apply for sites with a relatively broad wind speed distribution ($k = 1.5$, Fig. 15-22 a). The stall concept with two rotor speeds shows two local maxima of the power coefficient, but the hysteresis in the control for switching between the two rotor speeds is a disadvantage of this concept of control.

The Vestas V47 wind turbine is also designed for stronger winds. However, despite its higher rated power and rated wind speed, the power coefficient curve lies below that of the Ecotecnia wind turbine, which is also designed for higher wind speeds. Therefore, the area-specific yield of the Vestas wind turbine is below that of the Ecotecnia wind turbine for higher mean wind speeds, on the one hand. On the other hand, the higher power coefficient of the Vestas wind turbine at wind speeds below 10 m/s becomes advantageous for lower mean wind speeds.

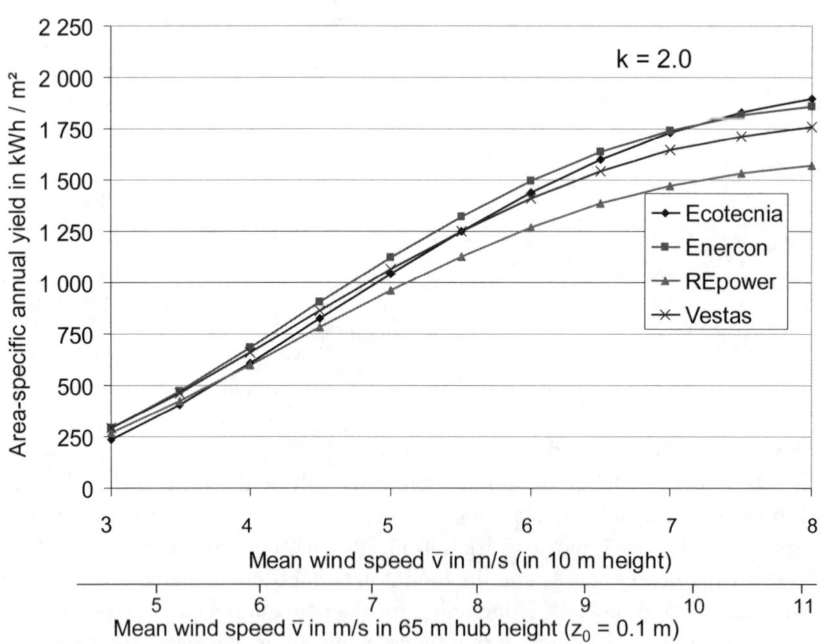

Fig. 15-21 Area-specific annual yield versus mean wind speed in 10 m height of four different wind turbines and shape factor k = 2.0 of the Weibull distribution

The ENERCON E-40 with its variable-speed operation and pitch control displays a relatively broad saddle around its maximum power coefficient. This results in an optimum yield for mean wind speeds between 4 and 7 m/s.

The shape factor k of the wind regime has a slight influence on the choice of the most suitable wind turbine, shown in Fig. 15-22 using the two wind turbines with stall control by way of example.

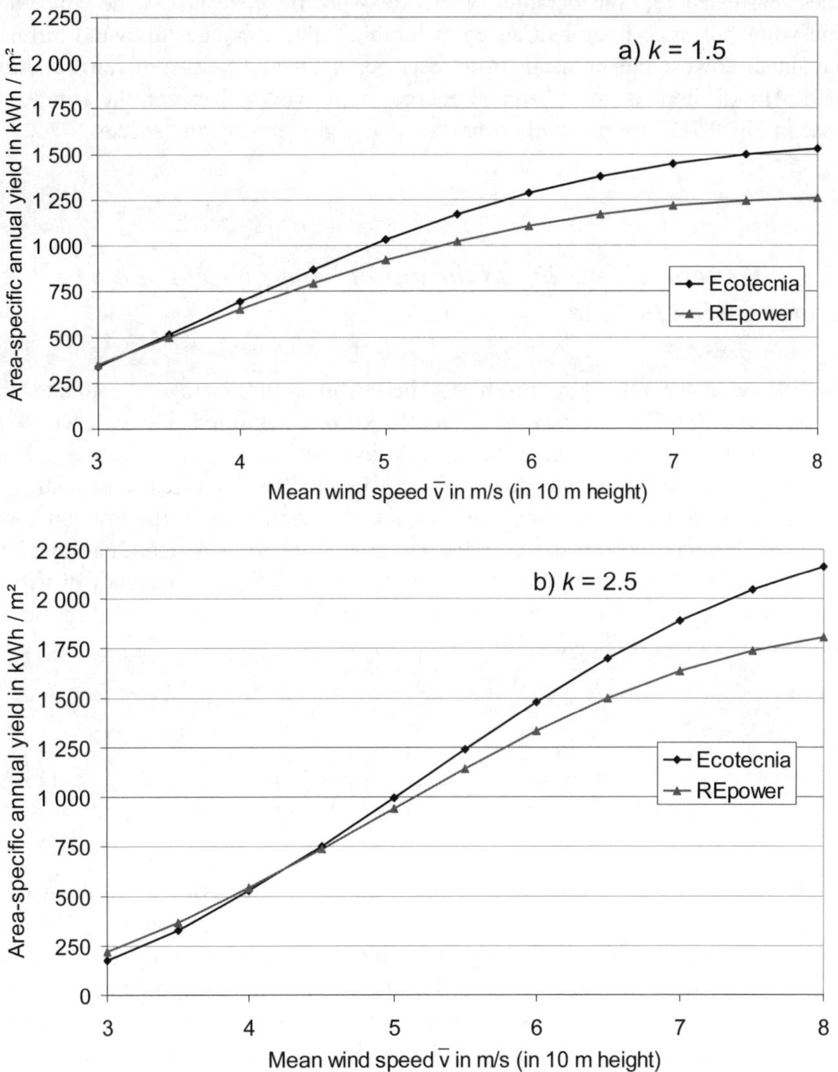

Fig. 15-22a, b Area-specific annual yield versus mean wind speed in 10 m height of two different wind turbines and a shape factor k=1.5 (a) resp. 2.5 (b) of the Weibull distribution

For a small shape factor (k = 1.5) the strong wind design of the Ecotecnia wind turbine gives a higher area-specific yield, for a more even wind frequency distribution (k = 2.5) the REpower wind turbine performs better for smaller mean wind speeds.

It has to be mentioned that the above discussion makes no reference as to which wind turbine would be the most economically efficient. This question can only be answered by taking into account the investment costs for the wind turbines. Furthermore, consideration of the area-specific yield places the wind turbine with the biggest rotor in an unfavourable light, although this wind turbine may incur lowest power generation costs per kWh and deliver highest energy yield. The differences in the annual energy yield in kWh between the wind turbines in Fig. 15-22 are relatively small, even at higher mean wind speeds.

15.2.1 General estimation of the annual energy yield of an idealised wind turbine

The following theoretical approach may be useful to provide a first estimate of the *annual electricity production* without a known measured power curve of a wind turbine. To simplify matters, the first assumption is that the histogram of wind conditions at the site can be described using the Rayleigh frequency distribution, cf. section 4.2.4. This suits well for the site conditions in the low lands of Northern Germany as well as e.g. in Patagonia in Southern America. The Rayleigh frequency distribution represents a special case of the Weibull frequency distribution for the factor $k = 2.0$, cf. equation (4.13) and Fig. 4-21a. The explicit formulation of the Rayleigh frequency distribution is given by

$$h(v) = \frac{\pi}{2} \frac{v}{\bar{v}^2} \exp\left(-\frac{\pi}{4}\left(\frac{v}{\bar{v}}\right)^2\right). \tag{15.4}$$

It only depends on the annual mean wind speed $\bar{v}$, and we assume that this value was already extrapolated to be representative of the hub height wind speed.

The wind turbine is modelled taking losses into account by a "real Betz-constant turbine", see small diagram sketch in Fig. 15-23. The meaning of this is as follows: the wind turbine operates between cut-in wind speed $v_{\text{cut-in}} = 0.25\ v_R$ and rated wind speed $v = v_R$ always at its best efficiency, λ_{opt}, due to a variable rotor speed (cf. chapter 13). In this wind speed range, the total power coefficient of the wind turbine is assumed to be $c_P = c_{P.\text{ges}} = 0.41$ to generally cover all losses due to the rotor, gearbox, generator and converter.

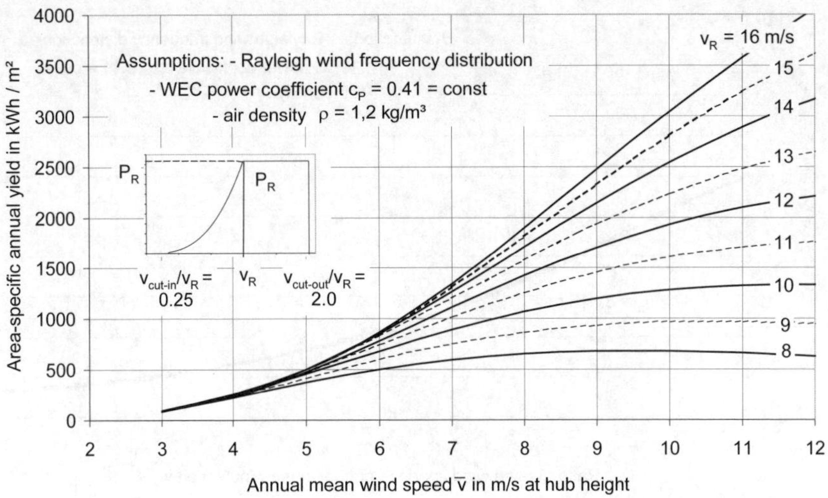

Fig. 15-23 Area-specific annual yield versus mean wind speed at hub height

At $v = v_R$ there is a sharp corner in the power curve $P = P(v)$. Between v_R and the cut-out wind speed $v_{cut-out} = 2\,v_R$ the wind turbine constantly operates at rated power P_R.

The integral of the energy yield, cf. equation (4.20), is now applied to the period of a year $T = 8{,}760$ h

$$E = T \int_0^\infty h(v,\bar{v})\cdot P(v)\,dv . \tag{15.5}$$

Using the area of one square metre as reference area for calculating the power gives the area-specific annual energy yield E_a of the wind turbine shown in Fig. 15-23 depending on its rated wind speed v_R (turbine) and the annual mean wind speed $\bar{v}$ (site). Here, it becomes apparent that reasonable designs will use a rated wind speed of $v_R = 2\,\bar{v}$, i.e. twice the annual mean wind speed (in praxi: $v_R = 11...15$ m/s). If the sharp corner in the power curve is moved to a higher rated wind speed v_R, i.e. if a bigger generator is installed, the annual energy yield is increased only slightly.

This is shown even more clearly in Fig. 15-24 where the annual energy yield - which contains a lot of partial load hours - is converted into the *full load equivalence* (i.e. capacity factor) which is the ratio of the annual energy yield E_a to the product of rated power and the time period of a year: $E_{a.v.R} = P_R \cdot 8{,}760$ h. The curves of full load equivalence versus the ratio of the rated wind speed to the annual mean wind speed, $v_R / \bar{v}$ in Fig. 15-24 reveal the following:

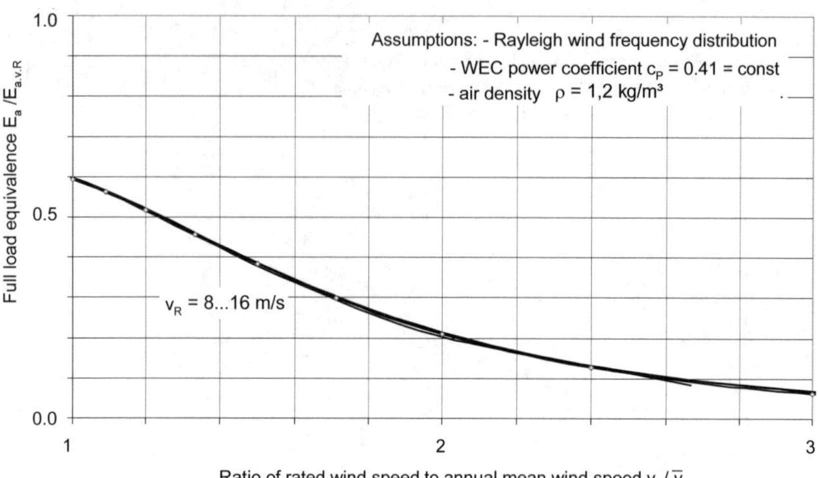

Fig. 15-24 Full load equivalence depending on the ratio of wind turbine rated wind speed to annual mean wind speed

A relatively small generator with a rated wind speed of $v_R = 1.5\ \overline{v}$ would show a full load equivalence of approx. 40%, but due to the quite small rated power, its power curve would not follow the cube power law $P \sim v^3$ for higher wind speeds, i.e. a significant portion of the wind power could not be harvested by this turbine. The more common wind turbine design with a design wind speed of approx. $v_R = 2\ \overline{v}$ has a yield which is only slightly smaller than the potential yield since the higher wind speeds only have a relatively low frequency. However, the full load equivalence is reduced to approx. 20%, the generator frequently operates at partial load.

References

[1]   Cremer C., Möser M.: *Technische Akustik (Technical acoustics)*, 5. Edition, Springer Verlag, 2003

[2]   Deutsches Windenergie-Institut (DEWI): *Studie zur aktuellen Kostensituation 2002 der Windenergienutzung in Deutschland (Study on the current cost situation of wind energy application in Germany in 2002)*, Wilhelmshaven 2002

[3]   Bundesverband Windenergie e.V.: *Windenergie 2004 und 2005, Marktübersicht (Wind energy market surveys of the German Federal Wind Energy Association, BWE)*

[4]   BTM Consult: *International Wind Energy Development*, World Market Update, BTM, March, 2002

[5]   Firmenangaben (company information)

[6]   Bundesverband WindEnergie e.V. (BWE): *Mit einer grünen Geldanlage schwarze Zahlen schreiben (How to be in the black with a green investment)*, Osnabrück, 2002

[7]   Hahn B., ISET: *Zuverlässigkeit, Wartung und Betriebskosten von Windkraftanlagen – Auswertungen des wissenschaftlichen Mess- und Evaluierungsprogramms (WMEP), (Reliability, maintenance and operating costs of wind turbines – Conclusions from the scientific measuring and evaluation programme (WMEP))*, Kassel, 2003

[8]   Bundesverband WindEnergie (*German Federal Wind Energy Association*, BWE), www.wind-energie.de, 2004

[9]   Hadel, J.: Assembly of the wind turbine Multibrid M5000, www.hadel.net, 2009

[10]   Liebherr: Firmenangaben (company information), www.liebherr.com, 2005

[11]   Deutsche WindGuard Dynamics GmbH: Firmenangaben, (company information), 2005

[12]   *Gesetz für den Vorrang Erneuerbarer Energien – EEG (German Renewable Energy Sources Act)*, BGBl. No. 13, Bonn, Germany, 31st March 2000

[13]   ISET (Institut für.solare Energieversorgungstechnik e. V.): *Fortbildungsunterlagen „Grid-connected Wind Turbines (Seminar documents)*, InWEnt gGmbH, 2004

[14]   Germanischer Lloyd Industrial Services GmbH: *Guideline for the Continued Operation of Wind Turbines)*, Hamburg 2009

Further literature: EWEA: Wind Energy - The Facts, Brussels, Belgium, 2004

16 Offshore wind farms

In the future, wind energy will be playing a dominant role in raising the share of renewables in electricity generation. This will significantly reduce the emission of carbon dioxide and the use of fossil fuels. Land use limitations in areas with high population density are hindering the installation of new large wind farms. Offshore, however, there is an enormous wind resource that has the advantages of both abundant space and dense winds. In 1995, a study found that the exploitable offshore wind resource theoretically exceeds the total consumption of electricity in Europe (Fig. 16-1).

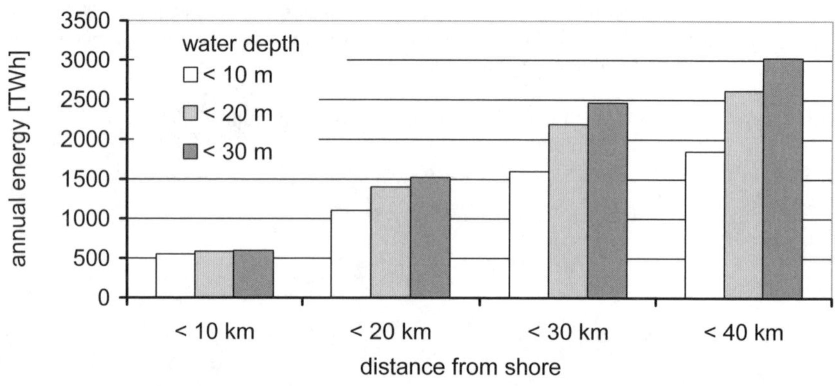

Fig. 16-1 Estimated offshore wind energy potential in the EU [1] (EUROSTAT estimated the EU 27 total electricity consumption for 2005 to be 2,756 TWh)

Though offshore projects were conceived as early as the 1970's, the first small-scale offshore wind farms were built between 1991 and 1997 in Sweden, Denmark and the Netherlands. Intended as demonstration projects, they were installed in relatively sheltered waters. The year 2000 marked the start of commercial offshore development with the erection of the first sea-based wind farms employing 1.5 to 2 MW wind turbines at water depths below 10 m. In 2002, the first large offshore wind farm, with 160 MW capacity, was commissioned in Denmark. 631 offshore turbines with a total capacity of 1,486 MW have been erected by the end of 2008. This corresponds to 1.2% of the total cumulative wind power capacity in the world. For the distant future, plans have been developed for Gigawatt-sized plants that employ multi-megawatt machines at sites as far as 100 km from shore and in water depths up to 30 m. The aggressive environment of the North Sea and Atlantic will present real challenges compared to the relatively benign Baltic. Eventually, the installed offshore capacity may amount to several times that installed on land.

The aforementioned demonstration projects proved the technological feasibility, but highlighted the economic challenges associated with sea-based development.

As a rule of thumb from the offshore oil and gas industry any work on sea is five to ten times more expensive than on land. In general, economically feasible offshore development depends upon viewing offshore wind farms as larger systems that are composed of integrated sub-systems (Fig. 16-2). This chapter is organised along similar lines and addresses the various sub-systems in order to construct an overview of offshore wind energy development. For further reading, refer to [2, 3, 4].

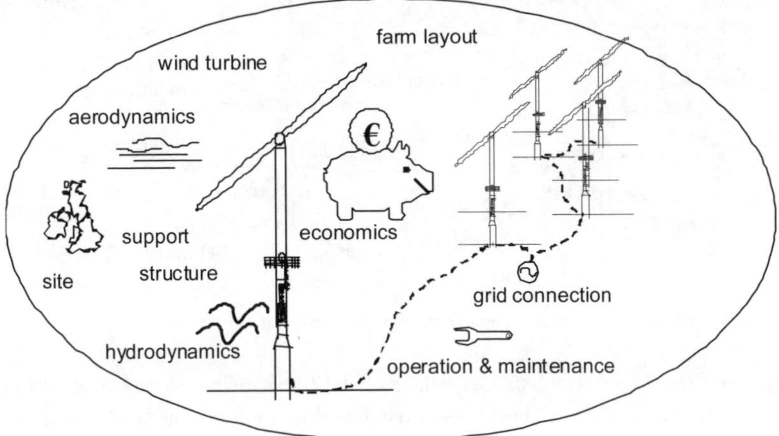

Fig. 16-2 Integrated design approach considering all sub-systems and techno-economic aspects of an offshore wind farm

16.1 Offshore environment

Various natural and man-made environmental conditions can affect an offshore wind farm (Fig. 16-3) and these conditions create both incentives and challenges. Offshore development is attractive because the annual average wind speed is considerably higher offshore than at most onshore sites (Chapter 4) and the lower surface roughness at sea causes a lower ambient turbulence. For large offshore wind farms, however, the latter advantage might be partially offset by the wake-induced turbulence and background turbulence of the wind farm itself. Though the favourable wind regime invites development, the offshore environment makes development technically complicated. Waves, currents and sea ice (if any) generate the most important hydrodynamic loads. The sea level varies according to tides, atmospheric conditions or long-term trends. Behaviour of offshore foundations is more complex for a variety of reasons. From both a practical and a theoretical viewpoint it is much more difficult to get reliable soil parameters for design purposes. Also, waves and currents may erode soil in the vicinity of the foundation (scour) or may destabilise the entire seabed. Finally, the high humidity and salinity increases corrosion of machinery and structure.

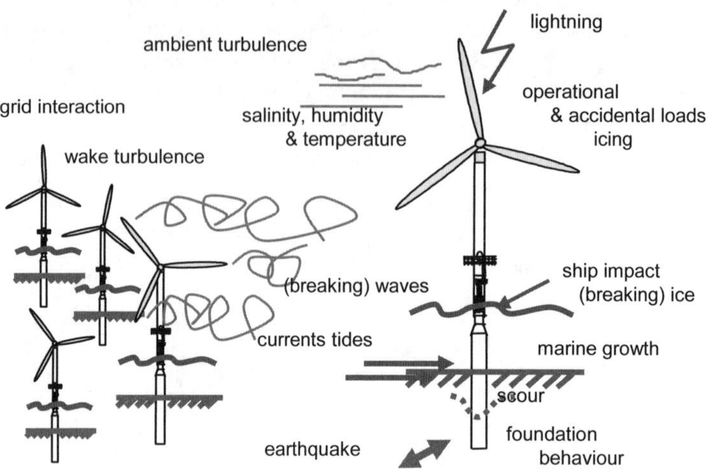

Fig. 16-3 Environmental impact on an offshore wind farm

Perhaps the most important consequence of the harsh offshore environment is the high design requirements. Turbines must be able to function unmanned over a long repayment period (up to 20 years) at a remote location that is difficult for maintenance personnel to access.

Using an example, we illustrate some fundamentals of wind-induced wave loading. Generally, one has to distinguish between sea conditions generated by local winds and swells originating from waves that have travelled a long distance. The topography of the North Sea and the Baltic Sea, for instance, leads to less significant swells.

The periods of energy-rich waves range between 2 s and 20 s. The energy of a sea state is proportional to the square of the significant wave height which itself is proportional to at least the second power of the zero-crossing period. Extreme waves have long periods, typically 7 to 13 s. Nonetheless, waves of moderate height are most important for *dynamic* wave loading and associated fatigue since their occurrence is high - 1 to 5 million waves per annum - and their periods are closer to the fundamental natural period of the structure.

The water surface profile of a wind-sea is random. For design calculations, such a sea state is decomposed into many elementary, regular waves. When considering extreme loading, one dominant, regular wave is often isolated. Assuming a wave height H, which is small compared to the water depth d, the linear, Airy-wave theory describes the water surface profile η as a sinusoidal function of the circular wave frequency ω and time t (Fig. 16-4).

$$\eta(t) = \frac{H}{2}\cos(-\omega t)$$
(16.1)

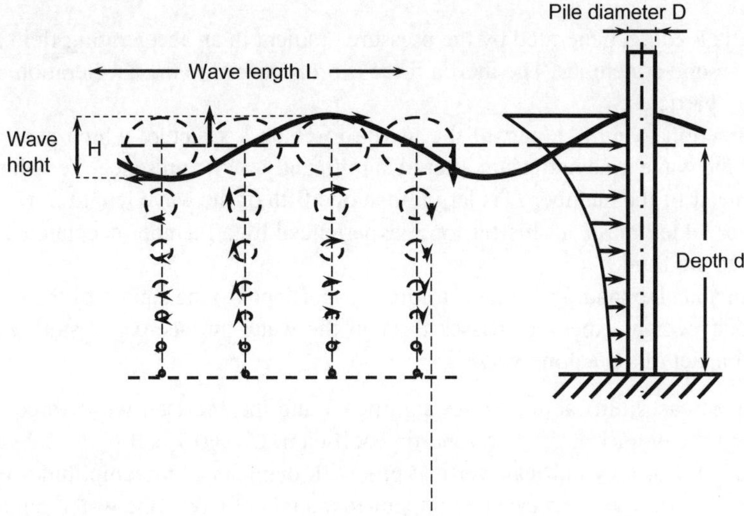

Fig. 16-4 Surface profile and water particle velocities of a small amplitude wave and wave forces on a slender vertical pile

It is more convenient to use the circular wave frequency ω instead of the wave period *T*. Analogously, the so-called "wave number" *k* is defined in respect to the wave length *L*.

$$\omega = 2\pi / T \qquad k = 2\pi / L \qquad (16.2)$$

The propagation of the wave profile results from the motion of the water particles in stationary, elliptical orbits. The extension of the paths, and the associated particle velocities, accelerations and forces, decay exponentially with depth (Fig. 16-4).

Waves approaching the shore become steeper as wave height increases and wave length decreases. The wave 'feels' the sea bottom and breaks when the height is equal to 78% of the water depth under ideal conditions. The transcendental dispersion equation expresses the relation between wave length or, more precisely, the wave number k and the wave frequency ω for a certain water depth *d*. To find the wave number, an iterative process may be started using the deep water case, $k = \omega^2 / g$, where *g* equals the gravity acceleration.

$$\omega^2 = gk \tan h(kd) \qquad (16.3)$$

Wave loading consists of:

- Drag loading, caused by vortices generated in the flow as it passes the structure. The associated forces are proportional to the square of the relative particle velocity both in steady currents and in waves.

- Inertia loading, generated by the pressure gradient in an accelerating fluid moving around a member. The inertia force is proportional to the acceleration of the water particles.
- Diffraction loading, a part of the inertia force on a structure which is so large that the wave kinematics are altered significantly by its presence, i.e. when the diameter of the member D is larger than one fifth of the wave length L.
- Water added mass, an inertia force experienced by a member accelerated relative to the fluid.
- Slam and slap loading, an inertia force proportional to the square of the relative velocity when a member passes through the water surface e.g. resulting from the impact of a breaking wave.

In most cases, diffraction is not significant and the Morison wave force equation with the empirical drag and inertia coefficient $C_D \approx 0.7$ and $C_M \approx 2.0$ can be applied [4]. For a cylindrical, vertical pile with diameter D the amplitudes of the drag and inertia wave force can be integrated analytically over the water depth.

$$\hat{F}_D = \frac{C_D \rho D H^2 \omega^2}{16k} \frac{\sinh(2kd)+2kd}{\cosh(2kd)-1} \tag{16.4}$$

$$\hat{F}_M = \frac{\pi C_M \rho D^2 H \omega^2}{8k} \tag{16.5}$$

Here ρ denotes the water density. The drag force varies with a $\cos^2$-function while the inertia force has a sinusoidal relation on time. The maximum of the total horizontal wave force can be found by numerically evaluating the quarter of the wave period in which both components reach their extremes.

$$\hat{F} = \max_{-T/4 \le t \le 0} \left\{ \hat{F}_D \left| \cos(-\omega t) \right| \cos(-\omega t) + \hat{F}_M \sin(-\omega t) \right\} \tag{16.6}$$

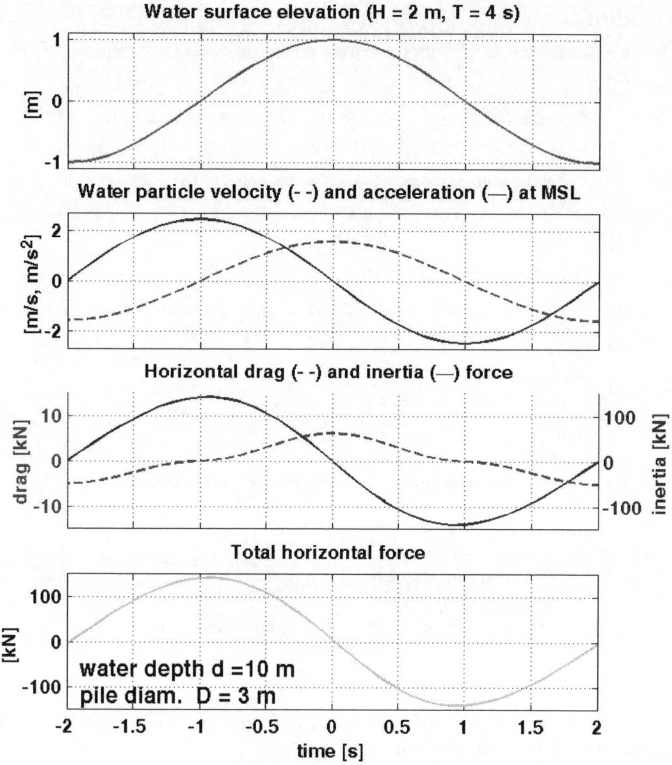

Fig. 16-5 Kinematics and forces of an inertia dominated fatigue wave calculated with linear wave theory

The particle kinematics and associated forces for two waves are shown in relation to a monopile with a 3 m diameter and in a water depth of 10 m. For a typical fatigue wave with a height of 2 m, the inertia force dominates and the behaviour is linear (Fig. 16-5).

In contrast, both the water kinematics and the loading become non-linear in extreme conditions. Fig. 16.6 shows the kinematics and loads associated with a wave of 6.4 m height computed with a non-linear wave theory. The wave crest of 4.95 m is more than three times higher than the trough. This relation yields high particle velocities in the crest and a drag force that is almost the same magnitude as the inertia force. The maximum horizontal wave force has a lever arm approx. equal to the water depth and the resulting overturning moment at the seabed contains several higher harmonics with significant excitation energy.

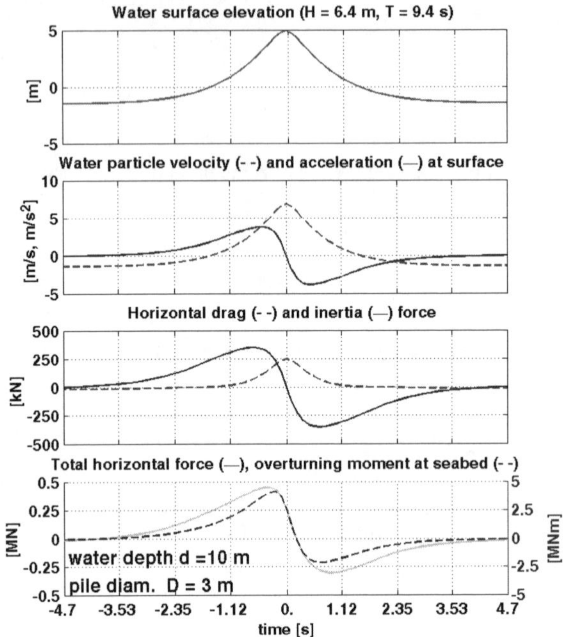

Fig. 16-6 Kinematics and forces of an extreme wave in shallow water calculated with the nonlinear, 10th order stream function wave theory

16.2 Offshore design requirements

What are the design drivers for offshore wind turbines? It is often argued that the economic viability of offshore projects will depend on the use of very large (i.e. 5 MW or larger) turbines. This certainly holds for exposed sites with water depths above 25 m and a large distance to shore which is both typical for German offshore sites. Easily, the reduction of the weight of the wind turbine, which increases disproportionately with $R^{2,5}$ to R^3 (see Table 7.1), has the most influence on the dimension of the structure. Therefore, the use of new control concepts, innovative drive train designs and innovative materials is often required. High infrastructure costs, e.g. for the high-voltage grid connection, result in a minimum profitable project size of about 100 MW and larger.

First of all, the high-quality requirements for offshore wind turbines have to be met before large projects with large wind turbines can become viable. Therefore, offshore-specific design requirements are often integrated into conventional wind turbine concepts. Such an approach would emphasise a reduction in operation and maintenance (O&M) costs, overall risk control, and the balance of investment costs over the entire wind farm.

Four qualitative rather than quantitative offshore design requirements can be formulated:

- Integrated design of offshore wind farms

  A more cost-efficient and reliable design can be achieved through an integrated design approach that considers the entire offshore wind farm as one system   (Fig. 16-2). The optimum offshore design will differ from land-based designs since the investment will be differently distributed among the sub-systems and since O&M costs will be higher. Dynamics and support structure design should be adjusted according to the particular site conditions and the turbine design.

- Design for RAMS (Reliability, Availability, Maintainability & Serviceability)

  Operation and maintenance aspects are a main design driver for offshore wind farms. Remote locations and poor weather conditions can postpone maintenance and result in longer downtime and greater losses of production. Therefore, the standard for offshore turbine reliability must be even higher than the already rigorous standards for land-based turbines. A commitment to reliability, in conjunction with dedicated machine design, comprehensive O&M strategies and specific maintenance hardware results in acceptable availability and reasonable operating costs.

- Smooth grid integration and controllability

  Future large-scale offshore wind farms will be operated as power plants, and will be a part of national, and possibly international, energy systems. Offshore wind farms must therefore be smoothly integrated into the grid and turbines must be controllable under normal operation and in case of grid disturbances. When combined with other considerations such as dynamic loading, these requirements rule out traditional stall control systems in favour of variable-speed, variable-pitch designs.

- Optimised logistics and offshore installation

  Sea-based wind farms afford the opportunity for considerable logistical optimisation despite high cost. Manufacturing, transport and installation can all be optimised when the number of heavy components to be lifted is reduced and pre-commissioning can be done in the harbour.

16.3 Wind turbines

It is clear that the offshore environment places demands on wind turbines that are not present in an onshore context. It is also evident, however, that offshore wind turbines, above the height of the wave impact, are not very different from onshore machines. Two technical approaches have been taken in developing offshore wind turbines: the marinisation of robust and proven onshore solutions and the integration of offshore-specific concepts into radical new designs.

In the marinisation process, modifications are made to onshore designs when they are transferred to offshore sites: maintenance aids that enable on-site repair of some components are added, climate and corrosion protection for the nacelle is improved, control capabilities are enhanced, and the specific rating, i.e. installed capacity per swept rotor area, is partially increased. An air-tight nacelle is sometimes proposed but this obliges to employ external heat exchangers for the gearbox, the generator and the power electronics. In addition, an air-tight nacelle requires an expensive and energy-consuming cooling system for the other smaller sources of radiated heat. For these reasons, most designs rely on a high standard of corrosion protection and the capsulation of individual components instead of an air-tight nacelle. The *REpower 5M*, which was first built in 2004, is an example for a marinized turbine using a conventional design concept in the 5 MW class.

Again two different trends have emerged from the integration of offshore requirements into the design process: a focus on reliability and serviceability and the development of advanced lightweight, high-tip-speed designs. Another example of new design concepts is the *Multibrid M5000* turbine which was first installed in 2004 as well. The weight of the nacelle is reduced by using a highly integrated drive train. Moreover, it features the avoidance of high-speed stages of the gearbox by matching a one-and-a-half-stage planetary gear with an intermediate-speed synchronous generator with permanent magnets (Fig. 3.13). So this drive train concept is situated between the traditional Direct-Drive design (e.g. Enercon) and the commonly used designs with a multi-stage gearbox with a high-speed generator. The nacelle of the *Multibrid M5000* is enclosed against the maritime environment and cooled by heat exchangers.

The ongoing rapid growth in turbine size and the need to base production on proven technology favours the horizontal axis, upwind-oriented turbine design that is used in land-based solutions. Though all offshore turbines are currently three-bladed designs, advanced two-bladed concepts might be applied at a later, matured stage. Two-bladed turbines have several advantages over three-bladed designs. First of all, two-bladed turbines are comparatively simple to install and maintain. Also, relaxed noise limitations offshore enable the use of a higher rotor speed, which reduces both the dimensioning torque for the drive train and the transmission ratio of the gearbox. Lower blade solidity is optimal for higher speeds and this can be achieved by omitting the third blade. Despite these advantages, however, the complicated dynamic behaviour and the alternation of high loads are major challenges for two-bladed machines. Furthermore, the offshore market must be large enough at the time of the marketability of the wind turbine in order to make the development of a product purely designed for offshore conditions profitable.

16.4 Support structures and marine installation

The turbine support structures, i.e. the integrity of tower and marine foundation, are the most obvious differences between offshore and land-based wind turbines. Ferguson classifies offshore bottom-mounted support structures according to three basic properties: structural configuration, foundation type and installation principle [5]. Any structural concept, including monotowers, braced towers and lattice towers, can be reasonably combined with either a piled or a gravity foundation. From a logistical point of view, however, it is beneficial to match piled designs with lifted installation and gravity foundations with floated installation (Fig. 16-7).

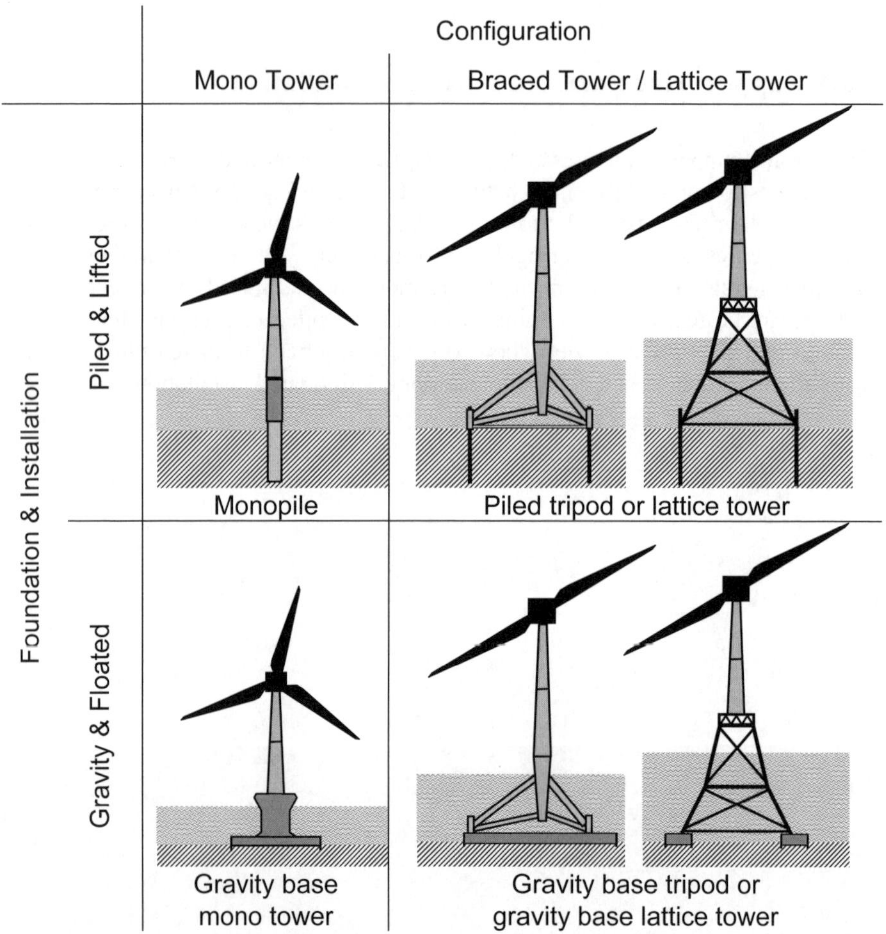

Fig. 16-7 Classification of bottom-mounted support structure types by structural configuration, foundation and installation principle

Fig. 16-8 Installation of an offshore wind turbine on a jacket support structure in 45 m water depth [REpower Systems AG]

Fig. 16-9 Rotor installation at the Utgrunden wind farm with two jack-up barges and two tug boats [Photo Uffe Kongsted]

Driven or drilled, steel monopiles are currently the preferred solution. They are economically attractive for use with the 2 to 3 MW class in water depths of approx. 20 m, and for use with 5 MW designs in depths up to 15 m. For larger depths, tripod, quadropod or lattice structures are needed. In 2006 the first jacket foundations for offshore turbines have been installed in 45 m water depth in Scotland (Fig. 16-8). Gravity foundations have been applied mainly at shallow Danish sites where high ice loads and particular soil conditions prevail. For other sites, such devices are more expensive. Dismantling and depositing might reduce the prospects for future use as observed recently in the oil and gas industry. Floating offshore wind turbines have been proposed for water depth of larger than approximately 50 m but are unlikely to materialise in the medium term because of the extraordinarily high costs for the moorings and the complex dynamic behaviour.

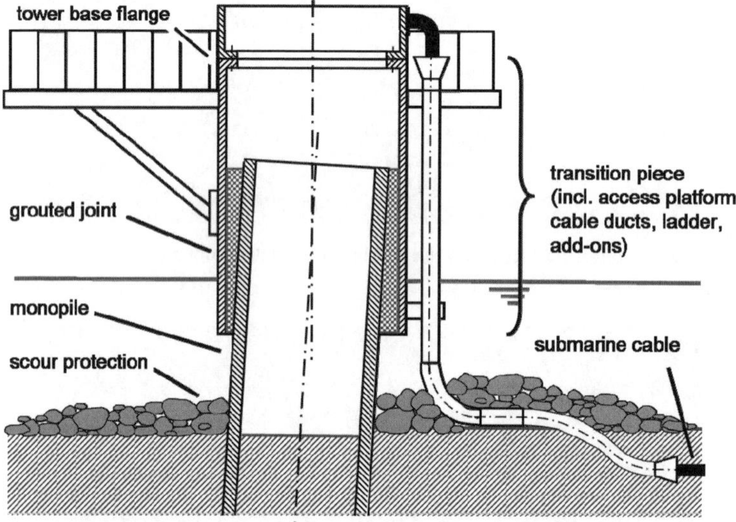

Fig. 16-10 Grouted joint (steel-grout-steel hybrid compound) of a monopile foundation between transition piece and pile as balancing connection

To reduce the high cost, the design must be optimised with respect to a particular turbine, local site conditions, batch size and the particularities of offshore wind energy. The dynamic characteristics of the support structure are of particular importance to offshore design. With this in mind, a compromise between the opposing requirements of the wind energy and offshore communities has to be found. Wind turbine towers in the megawatt class are slender and flexible structures with a first natural frequency below 0.45 Hz. Common offshore structures of the petroleum industry employ a significantly higher natural frequency since dynamic wave loading is strongly reduced for structural frequencies above approx. 0.4 Hz. Most designs for multi-megawatt turbines will be governed by aerodynamic fatigue or by a combination of aerodynamic and hydrodynamic fatigue. Sophisticated design approaches are required which integrate the experience from both the offshore oil and gas industry and the wind energy community [2].

Scarce equipment, high mobilisation costs, weather delays and safety requirements make any offshore operation very expensive and logistically complicated.

Custom-built installation vessels are able to transport and install pre-assembled modules in short time (Fig. 16-9). Another installation technique, illustrated in Fig. 16-10, involves a grouted joint between a driven monopile and a transition piece, with the tower base flange, service platform and other add-ons. With this construction, damage to the pile top and pile inclination from hammering can be quickly and easily compensated for.

16.5 Grid connection and wind farm layout

Integrating thousands of megawatts of (offshore) wind energy into the electricity system will present a great challenge to the transmission grid. In many countries, the present power system is dominated by centrally located, large-scale plants. Massive development of offshore wind power will lead to considerably greater periodic imbalances between production and consumption in the future. For reasons of operational security, a balance must be struck at any given time of day between production, exchange with countries abroad, and consumption. This means that other elements in the electricity system, e.g. biomass or pump storage power stations, will have to regulate the capacity variations. Wind farms will have to produce power on demand at variable feed-in prices. Reliable prediction of wind energy generation within a 24 hour and shorter window is therefore important and can result in a significant price bonus.

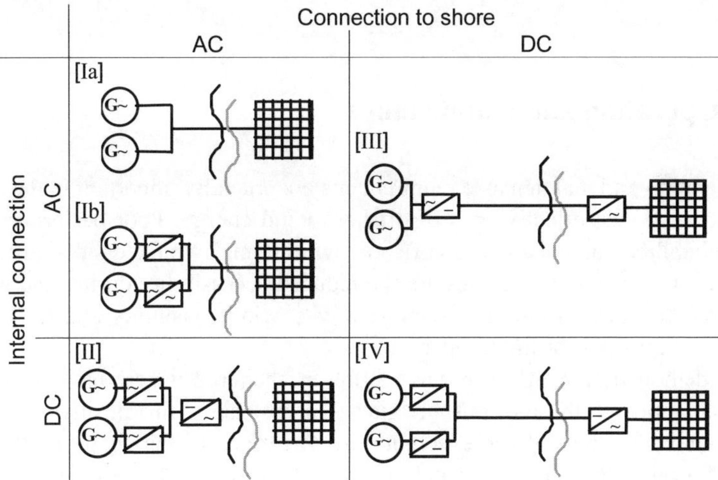

Fig. 16-11 Basic options for grid connection

The actual techniques for offshore power collection, transmission to shore, and grid connection are more straight-forward. Fig. 16-11 presents the basic options.

The main choice for shore connection is either AC or DC transmission. AC transmission involves high dielectric losses (the isolation material acts as a capacitor). These losses are proportional to the cable length and the voltage. On the other hand, DC transmission requires expensive converters. For short distances, AC transmission is most cost effective. The use of DC depends on the component costs, and usually becomes more economical at distances of 60 km. High-voltage DC transmission systems have increasingly been used in recent years to transport electricity from remote energy sources to the distribution grid.

With regard to power collection, the choice is once again between AC and DC. The first two options in Fig. 16-11, (Ia and Ib) are the ones commonly used for both onshore and offshore farms. Option II can be found rarely and only onshore. Option III, which combines an AC coupling of the wind turbines with a DC connection to shore, may cause stability problems in the 'AC island'. The pure DC technology system in Option IV is elegant, but practical only if reasonably priced equipment for direct DC voltage transformation becomes available.

Although the sea seems to offer infinite space, wind farm layout still needs to be carefully considered. Shipping routes, conservation areas, visual impact and local variations in water depth (e.g. sandbanks), restrict the area actually available for development. Theoretically, turbine spacing greater than 10 rotor diameters (D) apart is optimal when only cable lengths and wind farm losses are taken into consideration. In practice, however, spacing is only slightly wider than on land: 6 to 8 D in the prevailing wind direction and 4 to 6 D perpendicular to the prevailing direction.

16.6 Operation and maintenance

The operation and maintenance requirements are crucially important in the context of the overall cost effectiveness of offshore wind energy. Poor performance and low availability can jeopardise an offshore wind farm. Two factors that distinguish offshore sites from onshore sites are the reduced accessibility during bad weather conditions (resulting from unfavourable waves, winds, visibility and sea ice) and the costs of transport or lifting operations.

The demonstration offshore wind farms in sheltered waters have availability rates that approach those of onshore sites. Unfortunately, this positive data is not representative of real offshore conditions. Therefore in the scope of the Opti-OWECS project the operational and maintenance scenarios for large wind farms at remote offshore sites were analysed by Monte-Carlo simulations that took random failures and weather conditions into account [5, 6]. Although the analysis was made some years ago its results are qualitatively confirmed by the experience with the first large offshore wind farms.

As a qualitative illustration, the wind farm availability is shown as a function of the percentage of time that turbines are accessible on the horizontal axis and for three reliability levels in Fig. 16-12. In this case, reliability is defined as the mean time between failures. Reliability data for the state-of-the-art curve were derived from a large database of statistics from 500/600 kW-class turbines. The improved reliability level can be achieved with existing technology, but extra expenses will be incurred from higher specifications, redundant components and dedicated O&M strategies. Highly improved reliability can be achieved in several years through the development and testing of innovative technology.

A site without accessibility restrictions (i.e. onshore) reaches 97% availability for a state-of-the-art design, increasing to 99% for an extremely reliable design with an adequate maintenance strategy. This small difference in offshore availability requires a major effort to reduce the number of failures and caused downtime by a factor of four or higher. As weather conditions get worse, availability may fall anywhere from 50% to 80% depending upon the failure rate and the maintenance approach.

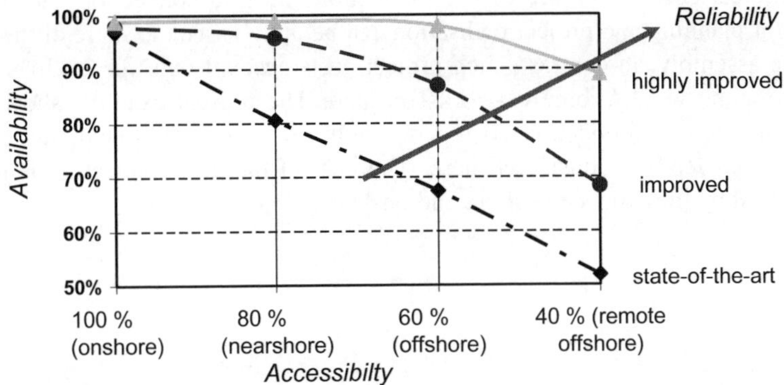

Fig. 16-12 Wind farm availability as function of the reliability of the wind turbine design for sites with different accessibility

16.7 Economics

Existing projects and various studies suggest a promising future for offshore wind energy developments. Nonetheless, the "gold rush" spirit prevalent at the beginning of this century has disappeared since it has become obvious that the generation of wind energy offshore is significantly more expensive than wind energy at good sites on land. However, the prospects of building large-scale offshore wind power plants and the more favourable wind conditions make such investments an interesting option for large players in the power business. Typical offshore capacity factors (full load equivalences), i.e. the ratio of average generated power and rated power, range from 30% to 40%, while a capacity factor of only approx. 25% is currently achieved on land.

Information about the actual costs of commercial offshore projects is hardly available. Table 16.1 compares the specific investment cost in €/kW and the specific cost of energy in €/kWh, i.e. the ratio between total investment and annual energy yield. On the one hand, a significant cost decrease is observed with respect to the first demonstration projects in the beginning of the 1990s. On the other hand, costs have increased again in recent years caused by the more exposed sites at larger water depths and larger distance to shore. Obviously, projects in the sheltered Baltic like Middelgrunden, Nysted and Lillgrund are less expensive than installations in the North Sea.

Typical relative cost contributions of an offshore wind farm with 5 MW machines are listed in Table 16.2. The absolute cost of the rotor-nacelle assembly and the tower are hardly affected by the site and project conditions. Obviously, the relative contribution of turbine cost fluctuates between a third and a half of the total project cost. The share of the foundation and grid connection and the soft costs for planning and project realisation can be up to twice the share of the rotor-nacelle assembly. In contrast, such project costs account only for 25-30% of the effort for the wind turbine ex-works on land. The present example shows only little variation in foundation costs since only one particular foundation type at large water depth is considered here. The cost of the grid connection is mainly influenced by the distance to shore and onshore grid connection.

Table 16.1 Cost comparison of actual build offshore wind farms [7]

Project, site & country	Year of installation	Power [MW]	Investment cost [M€]	Annual energy yield [GWh]	Spec. investment [€/kW]	Spec. Cost of energy [€/kWh/a]
Vindeby, Baltic, DK	1991	4.95	10	11.2	2,020	0.89
Tuno Knob, Baltic, DK	1995	5	10	15	2,000	0.67
Middelgrunden, Baltic, DK	2000	40	50	99	1,250	0.51
Horns Rev, DK	2002	160	270	600	1,688	0.45
Samsø, Ostsee, DK	2002	23	32	80	1,391	0.40
Nysted, Baltic, DK	2003	165.6	250	480	1,510	0.52
Arklow Bank, IRL	2003	25.2	50	NaN	1,984	NaN
North Hoyle, UK	2003	60	110	200	1,833	0.55
Kentish Flats, UK	2005	90	156	285	1,733	0.55
Burbo Bank, UK	2007	90	150	315	1,667	0.48
Lillgrund, SE	2007	110.4	167	300	1,513	0.56
Q7, NL	2007/2008	120	383	435	3,192	0.88

Table 16.2: Typical cost break-down of an offshore wind farm with 5MW wind turbines [8]

Component	Approx. cost share
Rotor-nacelle assembly incl. transformer, switch gear	33-50% *)
Tower	5%
Foundation (tripod)	15-18%
Foundation piles (site specific)	2-6%
Offshore installation (incl. weather risk)	5-7%
MV submarine cables (within cluster)	2%
HV submarine cables (shore connection)	2 - 20%
Offshore sub-station	4-10%
Onshore grid connection	4-10%
Planning, certification, due diligence	4-7%
Financing incl. interim financing	3-6%

*) actual contribution depending on amount of total project costs

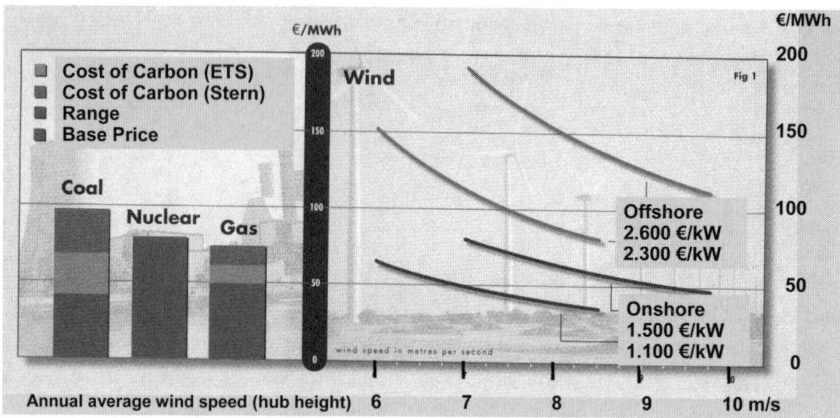

Fig. 16-13 Comparison of cost of electricity generation between newly installed conventional power plants and onshore and offshore wind farms [Windpower Monthly 1/2008]

The electricity generation cost for different power sources including financing and operation and maintenance costs are presented in Fig. 16-13. Increasing prices for raw materials like steel and copper as well as the high demand for wind turbines have caused an increase in wind farm costs of up to 30% during recent years. Even taking this into account, wind energy at good sites on land is cost competitive with newly installed conventional power plants. Offshore the specific investment costs in €/kW are about twice as high as onshore. Therefore, the higher energy yield of sea-based plants cannot compensate the larger investment and running costs. Typically, generation costs offshore are 50-100% higher than on land. Hence, further technological development and economy of scales are badly needed to improve economic viability.

Despite these disillusioning economic facts, the necessity to reduce carbon dioxide emissions and fossil fuel consumption results in political incentives to foster the future large-scale exploitation of offshore wind resources. The German Energy Feed-in Law (2009) grants a feed-in tariff which is twice as high offshore as for good onshore sites. Depending on the water depth and distance to shore the power purchase price range is 10-11 ct/kWh. In addition, the considerable cost for connecting offshore wind farms to the grid on land have to be covered by the transmission system operator (TSO) if the installation activities are started before the year 2012. This corresponds to an additional incentive of approx. 30% of the generation cost. Every European country operates a slightly different incentive system. Surprisingly, this is more driven by political considerations than by the actual natural resources and site conditions. For instance, in the United Kingdom the installation of new wind farms on land is difficult in the southern part of Britain due to lack of public acceptance, among other reasons. The exploitation of the enormous offshore wind resources is however booming because of both considerably higher feed-in tariffs and the much more favourable site conditions with lower water depths, shorter distances to shore and higher annual average wind speeds compared to other European countries.

References

[1]	Matthies, H., et al.Study on Offshore Wind Energy in the EC (JOUR 0072), Verlag Natürliche Energien, Brekendorf, 1995

[2]	Kühn, M., Dynamics and Design Optimisation of Offshore Wind Energy Conversion Systems, Doctoral Thesis, Delft Univ. of Technology, 2001

[3]	EWEA, Delivering Offshore Windpower in Europe, European Wind Energy Association (EWEA), Dec. 2007

[4]	Barltrop, N.D.P., Adams, A.J., Dynamics of Fixed Marine Structures, 3rd ed., Butterworh-Heinemann Ltd., Oxford, 1991

[5]	Ferguson, M.C. (ed.), Kühn, M., et al., Opti-OWECS Final Report Vol. 4: A Typical Design Solution for an Offshore Wind Energy Conversion System, Delft Univ. of Technology, 1998

[6]	van Bussel, G.J.W.; Development of an Expert System for the Determination of Availability and O&M Costs for Offshore Wind Farms, Proc. EWEC 1999, 402 - 405

[7]	http://www.offshorecenter.dk/offshorewindfarms.asp, accessed 20/11/2008

[8]	de Buhr, I., Darstellung der Kostenblöcke bei Offshore-Windenergieprojekten, Vortrag Husum Wind 2007.

Index

Printing: Ten Brink, Meppel, The Netherlands
Binding: Stürtz, Würzburg, Germany